Physiology of Soybean Plant

P Basuchaudhuri

Formerly Senior Scientist
Indian Council of Agricultural Research
New Delhi, India

CRC Press
Taylor & Francis Group
Boca Raton London New York

CRC Press is an imprint of the
Taylor & Francis Group, an **informa** business

A SCIENCE PUBLISHERS BOOK

T0174864

CRC Press
Taylor & Francis Group
6000 Broken Sound Parkway NW, Suite 300
Boca Raton, FL 33487-2742

© 2021 by Taylor & Francis Group, LLC
CRC Press is an imprint of Taylor & Francis Group, an Informa business

No claim to original U.S. Government works

Version Date: 20200520

International Standard Book Number-13: 978-0-367-54398-3 (Hardback)

This book contains information obtained from authentic and highly regarded sources. Reasonable efforts have been made to publish reliable data and information, but the author and publisher cannot assume responsibility for the validity of all materials or the consequences of their use. The authors and publishers have attempted to trace the copyright holders of all material reproduced in this publication and apologize to copyright holders if permission to publish in this form has not been obtained. If any copyright material has not been acknowledged please write and let us know so we may rectify in any future reprint.

Except as permitted under U.S. Copyright Law, no part of this book may be reprinted, reproduced, transmitted, or utilized in any form by any electronic, mechanical, or other means, now known or hereafter invented, including photocopying, microfilming, and recording, or in any information storage or retrieval system, without written permission from the publishers.

For permission to photocopy or use material electronically from this work, please access www.copyright.com (http://www.copyright.com/) or contact the Copyright Clearance Center, Inc. (CCC), 222 Rosewood Drive, Danvers, MA 01923, 978-750-8400. CCC is a not-for-profit organization that provides licenses and registration for a variety of users. For organizations that have been granted a photocopy license by the CCC, a separate system of payment has been arranged.

Trademark Notice: Product or corporate names may be trademarks or registered trademarks, and are used only for identification and explanation without intent to infringe.

Library of Congress Cataloging-in-Publication Data

Names: Basuchaudhuri, P., author.
Title: Physiology of soybean plant / P. Basuchaudhuri, Formerly Senior
    Scientist, Indian Council of Agricultural Research, New Delhi, India.
Description: Boca Raton : CRC Press, [2020] | Includes bibliographical
    references and index. | Summary: "The book provides in depth knowledge
    of understanding all aspects of physiology of soybean. It is written
    lucidly, systematically in depth, with recent information and findings
    explained with illustrations to express the ideas and concepts vividly
    to university students and researchers to further enrich the subject for
    improvement of the crop for enhancement of productivity to cope with the
    future demand. It describes the physiology of growth, development,
    flowering, Pod set and seed yield as well as C, O, N and Oil metabolisms
    -their hormonal regulations under normal and stress environmental
    conditions. Molecular approaches are also being mentioned"-- Provided by
    publisher.
Identifiers: LCCN 2020019438 | ISBN 9780367543983 (hardcover)
Subjects: LCSH: Soybean. | Soybean--Physiology.
Classification: LCC SB205.S7 B37 2020 | DDC 633.3/4--dc23
LC record available at https://lccn.loc.gov/2020019438

Visit the Taylor & Francis Web site at
http://www.taylorandfrancis.com

and the CRC Press Web site at
http://www.routledge.com

Dedication

To my elder son
Dr. Partha Basuchowdhuri MS, Ph.D.

Preface

Soybean is a miracle crop. The production of soybean is increasing gradually. Also the productivity of the crop is increasing. There is a great amount of genetic advancement in soybean. Soybean is important because it is rich in oil and protein content. Apart from oil numerous food products are produced by soybean. Thus it has a very good demand and market throughout the world.

Researches on soybean crop is long been performed to understand it vividly because of its unique characteristics. With the improvement of new varieties different from old and also genetically modified soybeans some characteristics have been changed. So, researches continued to improve the crop for better management and production. The productivity of the crop is low in many parts of the world. Thus, improvement of the productivity is another aspect needs much attention. Abiotic and biotic factors are also very much important for production stability. Specially, water scarcity of crop is a wide range problem.

Soybean is a leguminous crop. Nodulation is profuse in soils having nodule bacteria. Atmospheric nitrogen is fixed and supplied for growth of the plant. Thus, it is making mineral nitrogen economy. Hence, it is necessary to optimise the nodule growth with nutrients other than nitrogen, removing soil acidity and other abiotic stress factors.

Physiology of soybean is a fascinating subject. It is a short day C_3 plant with nonsynchronous flowering but monocarpic senescence nature. Pod growth is at different rate and duration. Numerous physiological aspects are coming forward to be known properly. Some are: Soybean seed storage and viability, flooding during germination, flowering characteristics, flower drop, pod setting, pod abortion, pod filling rate and duration, photosynthetic efficiency during pod filling, water stress during grain filling, nutritional aspects specially phosphorus and micronutrients such as iron, zinc and boron, nitrogen (mineral and atmospheric) utilization for grain protein, development of better quality oil in seeds through modification lipid biosynthetic pathway, water-temperature-salinity stress problems.

In the book I have written lucidly and elaborately almost all the important aspects of physiology of soybean crop based on the available findings in 13 chapters with many illustrations to clear the ideas. This book deals with only physiological aspects and will be suitable for MS and Doctoral students of universities and the researchers of the Institutes.

If you like the work my effort will be substantiated.

I am thankful to Publisher for shaping the manuscript to this book.

I am also thankful to my family members for their support during the preparation of manuscript.

P Basuchaudhuri
Kolkata

Contents

CHAPTER 1
Introduction

1.1 History and Origin

The first domestication of soybean has been traced to the eastern half of North China in the eleventh century B.C. or perhaps a bit earlier. Soybean has been one of the five main plant foods of China, along with rice, wheat, barley and millet. According to early authors, soybean production was localized in China until after the Chinese-Japanese war of 1894–95, when the Japanese began to import soybean oil cake for use as fertilizer. Shipments of soybeans were made to Europe around 1908, at which point the soybean attracted world-wide attention. Europeans had been aware of soybeans as early as 1712 through the writing of a German botanist. Some soybean seed may have been sent from China by missionaries as early as 1740 and planted in France.

The first use of the word "soybean" in U.S. literature was in 1804. However, it is thought that soybean was first introduced into the American Colonies in 1765 as "Chinese vetches". Early authors mentioned that soybeans appeared to be well adapted to Pennsylvania soil. An 1879 report from the Rutgers Agricultural College in New Jersey is the first reference that soybeans had been tested in a scientific agricultural school in the United States. For many years, most of the references to this crop were by people working in eastern and south eastern United States where it first became popular. Most of the early U.S. soybeans were used as a forage crop rather than being harvested for seed.

Before World War II, the U.S. imported more than 40% of its edible fats and oils. Disruption of trade routes during the war resulted in a rapid expansion of soybean acreage in the U.S. as the country looked for alternatives to these imports. Soybean was one of only two major new crops introduced into the U.S. in the twentieth century. The other major crop, Canola, was initially developed in Canada and is now grown on some U.S. acres. Soybean was successful as a new crop because there was an immediate need for soybean oil and meal, its culture was similar to corn, and it benefitted other crops in a rotation.

Following World War II, soybean production moved from the southern U.S. into the Corn Belt. The major soybean producing states of Iowa, Illinois, Minnesota, Indiana, Ohio, Missouri, and Nebraska produced 67 percent of the U.S. total in 2003; the southern and south eastern states of Arkansas, Mississippi, North Carolina, Kentucky, Tennessee, Louisiana, Alabama, and Georgia produced 14 percent. Other states with significant soybean acreage are South Dakota, Kansas, Michigan, Wisconsin, and North Dakota.

A record 2.9 million bushel soybean crop was produced in 2001 on 74.1 million acres with an average per acre yield of 39.6 bushels. The leading soybean states are Iowa and Illinois. In 2003, Iowa had 10.6 million acres of soybeans while Illinois had 10.3 million. The highest state yield ever achieved was 50.5 bushels per acre, produced by Iowa farmers in 1994.

The U.S. dominated world soybean production through the 1950s, 60s, and 70s, growing more than 75 percent of the world soybean crop. The U.S. was the major supplier of animal feed protein in the world during this period. A worldwide shortage of feed protein in the early 1970s led to the initiation of large-scale soybean production in several South American countries, most notably Argentina and Brazil. By 2003, the U.S. share of the world's soybean production had shrunk to 34 percent, while Argentina's and

Brazil's had increased to 18 and 28 percent, respectively. Most of the land suitable for soybean production in Argentina has been put into production. Brazil has an estimated additional 100 million acres of land that can still be put into soybean production.

1.2 Uses of Soybean

Early Uses. Soybeans were grown for centuries in Asia, mainly for their seeds. These were used in preparing a large variety of fresh, fermented and dried food products that were considered indispensable to oriental diets. Soybeans were not used to any great extent for forage in Asia.

Early use of soybeans in the United States was for forage and, to some extent, green manure. It was not until 1941 that the acreage of soybeans grown for grain first exceeded that grown for forage and other purposes in the United States.

Present Uses. Soybeans are the United States' second largest crop in cash sales and the number one export crop. In 2003, the export value of soybeans was more than 9.7 billion dollars, or about one-sixth of all agricultural exports. Normally, more than half of the total value of the U.S. soybean crop comes from exports as whole soybeans, soybean meal, and soybean oil. About 40 percent of the world's soybean trade originates from the U.S.

China has become the largest single country customer for U.S. soybeans with purchases totalling nearly $3 billion. Mexico, the European Union, and Japan are the second, third, and fourth largest international markets, respectively. Major export markets for soybean meal are the Philippines and Canada. Mexico and Korea are large customers of U.S. soybean oil.

The majority of the soybean crop is processed into oil and meal. Oil extracted from soybeans is made into shortening, margarine, cooking oil, and salad dressings. Soybeans account for 80 percent or more of the edible fats and oils consumed in the United States. Soy oil is also used in industrial paint, varnishes, caulking compounds, linoleum, printing inks, and other products. Development efforts in recent years have resulted in several soy oil-based lubricant and fuel products that replace non-renewable petroleum products.

Lecithin, a product extracted from soybean oil, is a natural emulsifier and lubricant used in many food, commercial, and industrial applications. As an emulsifier, it can make fats and water compatible with each other. For example, it helps keep the chocolate and cocoa butter in a candy bar from separating. It is also used in pharmaceuticals and protective coatings.

The high protein meal remaining after extraction can be processed into soybean flour for human food or incorporated into animal feed. Soybean protein helps balance the nutrient deficiencies of such grains as corn and wheat, which are low in the important amino acids, lysine and tryptophan. Use of vegetable proteins for human consumption continues to expand in the United States. They can be used as meat and dairy substitutes in various items. Most people are aware of the use of soy proteins in baby formula, weight-loss drinks, sport drinks, and as a low-fat substitute for hamburger meat. Soy flour and grits, made from grinding whole soybeans, are used in the commercial baking industry to aid in dough conditioning and bleaching. They have excellent moisture-holding qualities that help retard staling in bakery products.

A 60-pound bushel of soybeans yields about 11 pounds of oil and about 48 pounds of meal.

Soybeans (*Glycine max* L. Merr.) are part of the human diet and have been widely used in the food industry for centuries because they are an excellent source of protein, oil and other bioactive compounds, such as polyphenols and isoflavones. Soybean bioactive compounds act as antioxidants that play a role in human illness prevention by removing reactive oxygen species and thereby preventing oxidative damage in living tissue. However, soybeans and soy products contain antinutritional factors, such as trypsin inhibitors, oligosaccharides and phytic acid, which can negatively affect their nutritional value. For this reason, soybeans have to be processed prior to their consumption.

Germination has been identified as an inexpensive and effective technology for improving the nutritional quality of soybeans. During germination of seeds, the reserve materials of the seeds are degraded and used partly for respiration and partly for synthesis of new cell constituents of the developing embryo, this process causes important changes in the biochemical, nutritional and sensory characteristics of legumes. Numerous studies have reported the significant changes in biochemical and nutritional characteristics, for detail, germination increases nutritional (protein, acid amin, soluble sugars, various minerals, vitamins and dietary fibers) and antioxidant compounds (phenolics, vitamins E, vitamin C). Beside, this process reduces antinutritional factors in soybean seeds.

1.3 Food Security

World population is growing at an alarming rate and is anticipated to reach about nine billion by the end of the year 2050. On the other hand, agricultural productivity is not increasing at the required rate to keep up with the food demand. The reasons for this are water shortages, depleting soil fertility and various abiotic stresses. Therefore, minimizing these losses is a major area of concern for all nations when coping with the increasing food requirements.

In addition, the impact of global climate change on crop production has emerged as a major research priority during the past decade. Several forecasts for the coming decades project an increase in atmospheric CO_2 and temperature, changes in precipitation resulting in more frequent droughts and floods, and widespread runoff leading to leaching of soil nutrients and reduction in fresh-water availability. Each one of the abiotic stress conditions, alone or in combination with others, requires a specific acclimation response, tailored to the definite needs of the plant, and a combination of two or more different stresses might require a response that is also equally specific. Experimental evidence indicates that it is not adequate to study each of the individual stresses separately and that stress combinations should be regarded as a new state of abiotic stress in plants that requires a new defence or acclimation response.

The United Nations Sustainable Development Goals (SDGs) present an urgent and formidable challenge to scientists and society alike, highlighting the urgent requirement to transform agriculture and the food sector in order to achieve food and nutrition security, ecosystem sustainability, economic growth, and social equity over the coming decades.

Global food demand is predicted to grow by 70–85% as the population increases to over nine billion people by 2050 (FAO, 2017). A "next generation Green Revolution" is required in order to achieve future food security. Radical new concepts and approaches are needed in order to achieve a more sustainable development of agriculture. The next Green Revolution requires a much broader and systems-based approach, including environment, economy, and society, across all levels of organization. Transformative science across the agri-food sector is required if a major crisis in food production is to be avoided. Future agriculture requires tailored solutions that not only incorporate fundamental step changes in current knowledge and enabling technologies but also take into account the need to protect the earth and respect societal demands.

Climate change has far-reaching implications for global food security and has already substantially impacted agricultural production worldwide through effects on soil fertility and carbon sequestration, microbial activity and diversity, as well as on plant growth and productivity. Negative environmental impacts are exacerbated in current cropping systems by low diversity and the high intensity of inputs, climate associated yield instabilities being higher in grain legumes, such as soybean and broad leaved crops, than in autumn sown cereals. The predicted increased frequency of drought and intense precipitation events, elevated temperatures, as well as increased salt and heavy metals contamination of soils, will often be accompanied by increased infestation by pests, and pathogens are also expected to take a major toll on crop yields, leading to enhanced risks of famine. For example, the frequency and intensity of extreme temperature events in the tropics are increasing rapidly as a result of climate change. Tropical biomes are currently experiencing temperatures that may already exceed physiological thresholds. The ability of tropical species to withstand such "heat peaks" is poorly understood, particularly with regard to how plants prevent precocious senescence and retain photosynthesis in the leaves during these high temperature (HT) conditions. Such environmental stresses are among the main causes for declining crop productivity worldwide, leading to billions of dollars of annual losses. Throughout history, farmers have adopted new crop varieties and adjusted their practices in accordance with changes in the environment. However, with the global temperatures rising, the pace of environmental change will likely be unprecedented. Furthermore, with the expansion of crop cultivation to non optimal environments and non arable lands, development of climate resilient crops is becoming increasingly important for ensuring food security.

Sustainable innovation of the agricultural sector within SDG constraints is urgently required in order to improve the way that food and animal feed are produced. Current scientific advances offer considerable potential to meet the challenges of increasing agricultural production with conservation of the environment and the earth's ecosystems, compliant to the SDGs.

1.4 Soybean Production

The progenitor of soybean grows wild throughout eastern China, Korea, Japan and the far eastern portion of Russia. Domestication of soybean is believed to have occurred in the Yellow or the Yangtze River valleys of central or southern China somewhere between 3,000 and 5,000 years ago. There are numerous references to soybean in some of the earliest Chinese literature.

An important characteristic of soybean is that it is a legume and forms a symbiotic relationship with *Bradyrhizobium japonicum* (commonly referred to as rhizobia) bacteria that results in nodules forming on the roots. These nodules reduce atmospheric nitrogen gas to a form that the plant can utilize. A major advantage of soybean is that, because of nitrogen fixation, it does not require any nitrogen fertilizer.

A second important characteristic of soybean that has important implications for crop management is that it is a short-day plant. That is, soybean is triggered to flower as the day length decreases below some critical value. These critical values differ among maturity groups (MGs).

Due to the rapid rise in the commercial value of soybean in an international market, the total area under soybean cultivation has been increasing over the last three decades. Soybean is an important cash crop with a total production of over 313.05 million metric tons in 2015–2016 (USDA data). During this year, the USA has been the world's leading producer of soybean, representing 35% of the world production, followed by Brazil with 31%, Argentina with 17%, China with 4%, India with 3%, Paraguay with 3% and Canada with 2% (USDA data) (Fig. 1.1).

Fig. 1.1 Percentage of global soybean production by nation in 2010. Country data were taken from the USDA Foreign Agricultural Service's Production, Supply and Distribution (PSD) online database (http://www.fas.usda.gov/psdonline/psdHome.aspx). Tg, Teragram = 1 million metric tons.

It can be presented in detail in a nationwide data (Table 1.1)

This increase is associated with the improvement of soybean production; yield gain of the United States is 22.6 kg/ha/year from 1924 to 1997 and 12.1 kg/ha/year from 1950 to 1991 in China (Fig. 1.2). High soybean yield is also an important agricultural strategy, not only for the major producers but also the minor producers, such as India, Japan and other Asian countries. Many researches about soybean production were actively made to achieve high yielding. Improvements in weed control, planting and harvest machinery as well as the incorporation of disease and lodging resistance into elite soybean germplasm have contributed to the yield improvement in continental North America. In addition, many theories on the physiological traits associated with genetic yield improvement have been put forward; however, there are still some incongruencies among these theories. Therefore, it is important for soybean yield improvement to understand the relationship between yield and quantitative physiological traits.

Table 1.1 Country/region-wise area, yield and production of soybean.

Country/Region	Area (Million hectares)		Yield (Metric tons per hectares)		Production (Million metric tons)	
	2017/18	2018/19	2017/18	2018/19	2017/18	2018/19
World	124.59	125.14	2.74	2.87	341.62	358.77
United States	36.24	35.45	3.31	3.40	120.07	120.52
Brazil	35.15	35.90	3.47	3.26	122.00	117.00
Argentina	16.30	16.60	2.32	3.33	37.80	55.30
Paraguay	3.40	3.70	3.03	2.39	10.30	8.85
Bolivia	1.40	1.40	1.86	1.93	2.60	2.70
Uruguay	1.10	0.97	1.21	2.93	1.33	2.83
China	8.25	8.40	1.85	1.89	15.28	15.90
South Korea	0.05	0.05	1.87	1.75	0.09	0.09
North Korea	0.15	0.15	1.49	1.50	0.22	0.23
Japan	0.15	0.15	1.66	1.45	0.25	0.21
India	10.40	11.00	0.80	1.05	8.35	11.50
Canada	2.94	2.54	2.63	2.86	7.72	7.27
Russia	2.57	2.74	1.41	1.47	3.62	4.03
Ukraine	1.98	1.73	1.97	2.58	3.89	4.46
European Union	0.93	0.93	2.74	2.88	2.54	2.66
Indonesia	0.42	0.41	1.29	1.27	0.54	0.52
Vietnam	0.06	0.05	1.50	1.53	0.09	0.08
Thailand	0.04	0.04	1.57	1.57	0.06	0.06
Burma	0.15	0.15	1.07	1.07	0.16	0.16
Serbia	0.20	0.22	2.30	2.84	0.46	0.63
Mexico	0.26	0.19	1.65	1.75	0.43	0.34
South Africa	0.79	0.73	1.96	1.60	1.54	1.17
Nigeria	1.00	1.00	0.99	1.05	0.99	1.05
Zambia	0.23	0.19	1.56	1.58	0.35	0.30
Uganda	0.05	0.05	0.60	0.60	0.03	0.03
Iran	0.08	0.07	2.41	2.29	0.20	0.16
Turkey	0.02	0.03	3.75	3.80	0.09	0.10
Others	0.32	0.32	1.95	1.99	0.62	0.64

Office of Global Analysis, FAS, USDA, Circular Series WAP 10–19, October 2019.

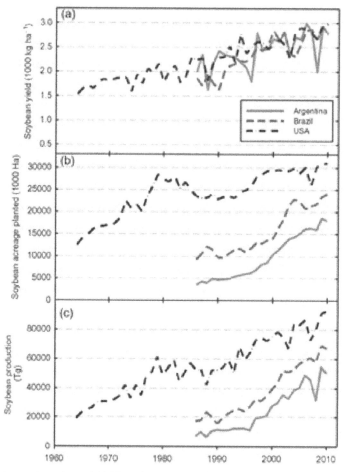

Fig. 1.2 Historical changes in soybean yield (a), soybean acreage (b) and soybean production (c) in the United States, Brazil and Argentina. Country production data are from the USDA Foreign Agricultural Service's Production, Supply and Distribution (PSD) online database (http://www.fas.usda.gov/psdonline/psdHome.aspx).

1.5 Yield Criteria

Production genetic can increase yield from soybean cultivar development has been 10 to 30 kg ha^{-1} yr^{-1}, factors responsible for this increase have not been clearly defined. Selection for yield during this process has largely been done through empirical yield trials across a range of different environments. Desirable lines are selected as future cultivars based on high and stable yields across years and locations. As indicated by coefficient of parentage (0.25) and molecular marker studies, genetic diversity within the southern U.S. cultivar germplasm pool is not great. Approximately 37% of parentage for modern southern cultivars comes from a single F, plant derived from a cross of ancestral cultivars CNS and 5–100. Development of cultivars resistant to soybean cyst nematode has further diminished genetic diversity.

In an effort to identify indirect yield criteria for streamlining cultivar development, scientists have endeavoured to determine the pertinent factors related to genetically-induced yield enhancement in the soybean cultivar development process. Yield, whether affected by genetic and/or environmental factors, is controlled by an interplay between growth dynamic and yield component parameters. Growth dynamic parameters are rates and levels of dry matter, leaf area, and light interception that characterize soybean seasonal growing patterns. Yield components are morphological characteristics whose formation is critical to yield production.

For soybean, examples are seed m^{-2}, seed size, seed per pod, pod m^{-2}, pod per reproductive node (reproductive node contains at least one pod having at least one seed), reproductive node m^{-2}, percent reproductive nodes, and node m^{-2}. Yield components in soybean can be organized into a sequential series (Fig. 1.3) of causative relationships where: Yield is controlled by primary yield components seed size and seed m^{-2}; seed m^{-2} is controlled by secondary yield components seed per pod and pod m^{-2}; pod m^{-2} is controlled by tertiary yield components pod per reproductive node and reproductive node m^{-2} is controlled by quaternary yield components node m^{-2} and percent reproductive nodes.

Yield is basically a function of intercepted light (fraction of the sunlight intercepted by the crop), the dry matter produced from this light [which is controlled by radiation use efficiency (g of dry matter/units of light energy intercepted)], and the percentage of this dry matter transferred to the seed harvest index (g of seed yield/g of total dry matter). Yield components are the vehicle through which dry matter increases affect yield. Based on studies conducted on the environmental level, dry matter accumulation was shown to affect yield through control of seed m^{-2}, pod m^{-2}, reproductive node m^{-2}, and node m^{-2}. In contrast, seed size, seed per pod, pod per reproductive node, and percent reproductive nodes appeared unrelated to the yield formation process (Fig. 1.3).

It is hypothesized that a similar mechanism may be operating on the genetic level to explain yield increases due to cultivar development.

However, previous studies involving new and old cultivars have provided ambivalent results and usually dealt only with the primary yield formation level (seed size and seed m^{-2} affecting yield). At the primary level, yield increases were attributed to greater seed number by some, while others have reported greater importance for seed size. On the secondary level, it reported that seed per pod played a role in explaining greater seed m^{-2} in new vs. old midwestern U.S. cultivars.

It was found that pod per plant explained little of the yield improvement shown by new vs. old mid western U.S. cultivars. A similar result was reported with southern U.S. cultivars. In contrast, some reported that pods per plant did play a role for greater yield during cultivar improvement for Indian and Chinese cultivars, respectively. Because of divergent results from different soybean germplasm pools, lack of data representing southern U.S. cultivars, and little yield component information beyond the primary level the contention is unclear.

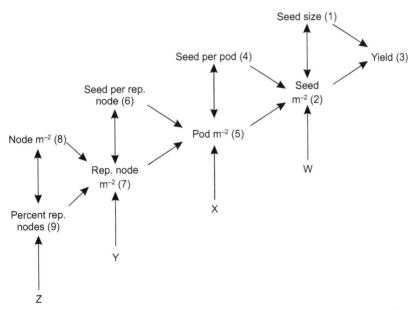

Fig. l.3 Path diagram showing interrelationships among primary level traits (traits 1,2–+3), secondary level traits (traits 4,5–+2), tertiary level traits (5'7–+5), and quaternary traits (traits 8,9–+7). W, X, Y, and Z represent residual effects, in the primary, secondary, tertiary, and quaternary levels, respectively.

The period of reproductive growth in soybean is very long (70 to 90 days after the start of flowering) when compared with cereals. The pod shells in soybean develop after flowering, whereas the husks in rice or wheat, those function is the same as that of pod shell in soybean, are formed before flowering. Therefore, the seed filling begins about four weeks after flowering in soybean, but about one week after flowering in rice or wheat. In some legumes, such as mung bean, cowpea and common bean, the period from flowering to maturation is also shorter than that in soybean.

Several studies documented that the pod formation is very slow after flowering in soybean. It was reported that the time of pod appearance (the ovary developed more than 10 mm in length) varied from 6 to 18 days after the start of flowering among 71 late maturing cultivars. It was also reported that there is difference in time of pod growth between genotypes. The delay of pod growth seems to lengthen the reproductive growth period. On the other hand, there are indications that the delay of pod growth is much longer in the low order racemes (early opening flowers) and is reduced in high order racemes (late opening flowers). It is suggested that the timing of pod formation plays a role in the regulation of synchronous pod maturation in soybean.

United States average soybean [*Glycine max* L. (Merr.)] grain yields have increased from the earliest record of 739 kg ha^{-1} in 1924 to a high of 2956 kg ha^{-1} in 2009 (USDA-NASS, 2013). While this increase in soybean yield over time is substantial, both researchers and growers have documented yields much greater than the reported nationwide averages (Table 1.1). In converted rice paddies in Japan, there have been reported yields of 6490 kg ha^{-1}. In New Jersey, a soybean yield of 7923 kg ha^{-1} was recorded in 1983 and a 5-yr average irrigated yield of 6921 kg ha^{-1}. In 1982, a researcher was able to achieve yields of 6817 kg ha^{-1} in research in Ohio. In Queensland, Australia, yields of up to 8604 kg ha^{-1} were reported. Researchers in China recorded yields of up to 9200 kg ha^{-1}, but yields were based on only a small number of individual plants (14 to 28). Without further details and without a more representative yield sample, this report remains questionable. Physiological experiments in Argentina extended the photoperiod by 2 h in field experiments from 1 to 35 days after R3 and reported yields of up to 8957 kg ha^{-1}.

A few soybean growers have also achieved exceptional soybean yields. In 1968, the winner of the United States National Soybean Yield Contest did so with 7310 kg ha^{-1}. In 1997, a grower achieved yields near 6719 kg ha^{-1} in the Nebraska irrigated contest category. In 2008, the winner of the Missouri Soybean Association's non-irrigated contest category had a yield of 7324 kg ha^{-1} in southeast Missouri. And finally, the highest soybean yields reported from yield contests were from the Missouri Soybean Association Yield Contest, with yields of 9339 kg ha^{-1} (2006), 10,388 kg ha^{-1} (2007), and 10,791 kg ha^{-1} (2010). Yields from these same fields in southwest Missouri were reported in 2011 to 2013, and each year individual cultivars had yields between 6979 and 7953 kg ha^{-1}.

The high yields reported in yield contests have created some controversy and skepticism because of the lack of supportive, quantitative data that would provide a mechanistic explanation. Additional concerns are associated with the uncertainty in what constitutes the potential yield of soybean. Potential yield is defined as the "yield of a crop cultivar when grown with water and nutrients non-limiting and biotic stress effectively controlled". This becomes an issue when attempting to estimate future prospects for yield increases via yield-gap analyses using current farmer average yields and the potential yield. High potential yield estimates will often result in large yield gaps; however, attaining such potential yields are likely not economically or sustainably possible. Potential yield also varies geographically due to changes in climate. Furthermore, non-irrigated production systems may be consistently constrained by water supply and a water-limited potential yield may be more realistic in those production systems. However, the focus Soybean (*Glycine max* (L.) Merr.), which contains about 200 g oil and 400 g protein/kg seed dry matter, is the major oilseed and protein crop in many regions worldwide, providing approximately 60% of the world supply of vegetable protein. In Europe, more than 1.2106 ha of soybean was grown in 1998, mainly in Italy, France, Russia, Romania, Yugoslavia, Croatia and Austria. In Central European countries, such as Austria, the interest in soybean is due to its high seed protein content, whereas vegetable oil is preferably produced by cultivating oilseed rape, sunflower or other species. Consequently, the soybean crop processed in Austria is mainly applied in livestock feeding, e.g., as full-fat soybean for pig fattening, whereas a smaller proportion is used in the food industry or in tofu making.

For both feed and food utilisation, a high and stable seed protein content is desirable. However, in northern regions of soybean production, protein content is reduced owing to climatic conditions, such as low temperatures and high amounts of precipitation. In the U.S., seed protein content in the Western and Eastern Corn Belt was clearly lower than in southern regions of production over a number of seasons. Similarly, protein content was reported to be low in the northern locations of north-east China and in northern sites of Europe, where large seasonal variations were observed in protein content. Apart from other findings, it was recently demonstrated that low root zone temperatures reduce nitrogen fixation, which might explain the low protein contents commonly found in northern areas of soybean cultivation.

Nevertheless, several soybean cultivars of early maturity groups have been developed with considerably improved seed protein content. Moreover, agronomic research has been initiated in many soybean production areas focusing on the management of seed quality characters.

Recent information on the seed protein content of Central European-grown soybean, which might be of interest to both agronomists and processors, is not available. For this reason, the magnitude of environmental and genetic sources of variation in seed protein content is being studied in a number of performance trials, nitrogen supply experiments and protein screenings within a soybean breeding programme carried out in the soybean growing area of Austria.

1.6 Maturity Group

Soybean maturity selection is an important management decision. Maturity group zones represent regions where a cultivar is best adapted without implying that MG-specific cultivars cannot be grown elsewhere. Hypothetical MG zones were first developed by Scott and Aldrich (1970), followed by the work of Zhang et al. (2007) who redefined the optimum MG zones using yield variety trial data from 1998 to 2003. Most recently, Mourtzinis and Conley (2017) re-delineated MG zones across the United States using 2005 to 2015 yield variety trial data. In their study, although the zones were generated using a vast amount of information, the results are restricted to the PD range of the variety trials.

Since the 1970s, the length of the growing season has increased, most notably in the northern Corn Belt where producers are planting 1 to 3 wk earlier. Soybean yield increases have been documented with earlier planting, where early May PDs consistently result in the greatest yield. In Wisconsin, researchers saw a 21.2 kg ha^{-1} d^{-1} yield decline when planting was delayed past the first week in May. Since MG selection does not increase input costs, with earlier PDs, many producers question the optimum MG for planting soybean in their region.

While early planting is a prudent management practice to increase soybean yield, logistical, equipment, environment, and labor challenges can delay planting. However, when early planting is possible, soybean is exposed to a greater risk of a spring killing frost, early season insects and seedling diseases, and damaging rainfall events that may result in suboptimal stand. In such years, replanting may be necessary. Furthermore, the climate variability that is affecting state and regional soybean yields may also cause more frequent replanting situations. Proper replanting methods and optimal final plant stands (> 247,000 plants ha^{-1}) have been determined, and yet, the proper MG to use in replant or late planting scenarios are unclear.

A large-scale, widespread soybean seed industry has developed since the early 1950s in response to the rapidly increasing demand. Primary emphasis during this period of rapid growth has necessarily been given to expansion of capacity. Only slight improvements have been made to the quality of soybean seed offered in the market. Seed quality in soybeans is only moderately good, and periodically poor. Moderately good quality seed, however, are not now satisfying the expectations of farmers, who are becoming increasingly aware of the importance of high quality seed for efficient, maximal production. The most chronic seed quality problems in soybeans relate to germinability and vigor. Soybean seed are inherently short lived and structurally weak as compared to other kinds of seed. Substantial losses in germinability and vigor are caused by hot, dry weather during seed maturation, weathering from rainfall and warm temperatures during the harvest period, and mechanical abuse during harvesting and handling operations.

Production of high quality soybean seed requires timely harvest followed by aeration and/or drying as necessary to reduce seed moisture content to 12% or less, and careful combining and handling to minimize mechanical damage.

Germination percentage is not a reliable index of the stand and crop-producing potential of soybean seed. Seed with lots of good germination but low in vigor can and do perform poorly in the field even under rather favorable conditions.

1.7 Biological Nitrogen Fixation

For soybean, BNF requires 6–7 g C g^{-1} N in comparison to 4 g C g^{-1} N for assimilation of mineral N; integrated over the growing season the difference in cost is substantial, with potential implications for seed yield and seed protein or oil concentrations. The cost of BNF can be partially compensated by increase in photosynthesis of plants associated with rhizobia or shifts in allocation of biomass. For instance, nodulated roots accumulated less biomass compared with plants growing with high soil N supply and lower biomass partitioning to seeds associated with increasing BNF. Thus, the crop can accommodate the cost of BNF by five non-mutually exclusive mechanisms, whereby N fixation: (a) reduces shoot growth and seed yield, or maintains shoot growth and seed yield by (b) enhanced photosynthesis, or (c) reduced root:shoot ratio, or maintains shoot growth but reduces seed yield by (d) reducing seed oil and protein concentration in seed, or (e) the fraction of shoot biomass allocated to seed (i.e., harvest index; HI).

Furthermore, there is an agronomic interest in the role of mineral N to support high seed yield and avoid protein dilution. A recent review of Mourtzinis et al. concluded that N fertilization has a small and inconsistent effect on soybean seed yield. This conclusion is, however, largely based on generic trials where coarse fertilization regimes were established in order to shift the contribution of mineral N and BNF. In contrast, a full-N treatment devised with a careful experimental protocol to ensure an ample N supply during the entire crop season increased soybean seed yield by 11% in relation to unfertilized controls, with a range from no effect for stressful environments (ca. 2500 kg ha^{-1}) but increases of 900 kg ha^{-1} in high potential environments (ca. 6000 kg ha^{-1}).

Legumes rely on soil mineral nitrogen (N) and biological N fixation (BNF). The interplay between these two sources is biologically interesting and agronomically relevant as the crop can accommodate the cost of BNF by five non-mutually exclusive mechanisms, whereby BNF: Reduces shoot growth and seed yield, or maintains shoot growth and seed yield by enhanced photosynthesis, or reduced root:shoot ratio, or maintains shoot growth but reduces seed yield by reducing the fraction of shoot biomass allocated to seed (harvest index), or reducing concentration of oil and protein in seed. It was explored that the impact of N application on the seasonal dynamics of BNF, and its consequences for seed yield, with emphasis on growth and shoot allocation mechanisms. Trials were established in 23 locations across the US Midwest under four N conditions. Fertilizer reduced the peak of BNF up to 16% in applications at the full flowering stage. Seed yield declined 13 kg ha^{-1} per % increase in RAUR. Harvest index accounted for the decline in seed yield with increasing BNF. This indicates that the cost of BNF was met by a relative change in dry matter allocation against the energetically rich seed, and in favor of energetically cheaper vegetative tissue.

Many soybean fertility recommendations are derived from research conducted during the 1930s to 1970s, and may not be adequate in supporting the nutritional needs of the greater biomass accumulation and seed yield associated with current soybean germplasm and production systems. Furthermore, no recent data that document the cumulative effects of improved soybean varieties, fertilizer source and placement technologies, and plant health/plant protection advancements on the rate and duration of nutrient accumulation in soybean exist. A more comprehensive understanding of soybean's nutritional requirements may be realized through this evaluation of the season-long nutrient uptake, partitioning and remobilization patterns in soybean.

Table 1.2 Nutrient accumulation associated with producing, on average, 60 bu/A of soybean grain.

Parameter	Maximum total uptake (lbs/A)	Removal with grain (lbs/A)	Harvest index (%)	Nutritional removal coefficient (lbs/bu)
Macronutrients				
N	245	179	73	2.98
P	19	15	81	0.25
P_2O_5	43	35	81	0.58
K	141	57	41	0.95
K_2O	170	70	41	1.17
S	17	10	59	0.17
Mg	45	8	18	0.13
Ca	101	9	9	0.15
Micronutrients				
Zn	4.78	2.00	42	0.033
B	4.64	1.58	34	0.026
Mn	5.30	1.31	25	0.022
Cu	0.90	0.56	62	0.0093

† Multiply grain yield by nutrient removal coefficient to obtain the quantity of nutrient removal. Maximum total nutrient uptake, removal with grain, and harvest index (percentage of total nutrient uptake present in the grain) of macro- and micronutrients were averaged over treatments at DeKalb (2012 and 2013) and Champaign (2013).

1.8 Varietal Improvement

Soybean (*Glycine max* Merr.) is the world's most widely grown leguminous crop and an important source of protein and oil for food and feed (Liu et al., 2008). World soybean production increased from 117 Mt in 1992 to 316 Mt in 2015. Countries producing the highest soybean yields are the USA, Brazil, Argentina and China (data from FAO, www.fao.org/faostat/en/ #compare). China's soybean yield increased from 0.61 Mt in 1949 to 1.89 Mt in 2002 and then levelled off. Soybean area under cultivation had decreased earlier but increased in recent time (Fig. 1.4) despite an increase in the domestic consumption of soybean foods as the standard of living in China increased. In 2015, China imported 81.69 Mt of soybeans, which accounted for ~ 70% of the world's soybean trade (http://finance.people.com.cn/n1/2016/0407/c1004-28257572.html).

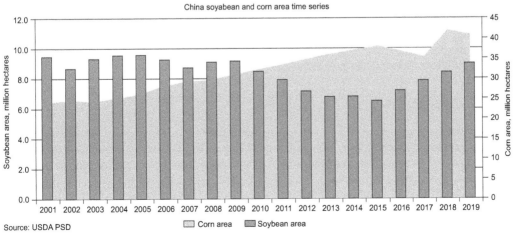

Fig. 1.4 Soybean and corn area in China with time.

Breeding and variety improvement are important for increasing crop yield. Soybean yield has increased substantially in the last century in the USA, Canada, India and China. In Canada, yields of short-season soybean cultivars have increased by ~ 0.5% year^{-1}. In southern USA, yields have increased linearly by 16.8 kg ha^{-1} year^{-1} for soybean cultivars released from 1928 to 2008, and throughout the USA by 26.5 kg ha^{-1} year^{-1} for cultivars released from 1923 to 2007.

Soybean breeding in China commenced in the Northeast Jilin Province as early as 1913. The first cultivar, Huang-Bao-Zhu, was released in 1923. From 1923 to 2013, > 1400 soybean cultivars were released in China. A positive correlation between seed yield and year of cultivar release has been reported, with an average annual increase of 0.58% from 1950 to 2006 in Northeast China. In the Yellow-Huai-Hai (YHH) summer soybean region, yields increased by 9.97 kg ha^{-1} year^{-1} for soybean cultivars released from 1929 until 2004.

Plant breeders in China have improved grain yields by reducing plant height, enhancing lodging resistance, and decreasing levels of disease and pest-infested seed. Seed number per plant was the most significant contributor to yield gain in Northeast China. Most studies in China have been based on limited field experiments, often growing representative historical soybean cultivars for 1 or 2 years to measure yield progress and to determine the cause of yield improvement. However, yield increases have not come only from breeding efforts but also from changes in cultivation technologies and environmental factors. Changes in cultivation technologies can result in older cultivars responding differently under modern agronomic practices. Breeders may over- or under-estimate the rate of change in traits through breeding by growing historical soybean cultivars in the same growing environment. Use of a large number of historically released varieties offers an important reference value for breeding with regard to the actual trends of past characters that will supplement existing experimental research.

Furthermore, ecological regions vary in their environmental conditions (including geography, soil, climate) and farming culture. Ecological regionalisation of soybean in China has been divided into three main areas: North spring soybean region, YHH summer soybean region and South soybean region. The North spring soybean region produces ~ 45% of China's total soybean yield, the YHH summer soybean region ~ 30% and the South soybean region ~ 25%. At present, China has many soybean varieties and complex cropping systems. It is important to analyse the trends of relevant yield characters in different ecological areas by using as many released varieties as possible.

Soybean production has increased sharply in the past few decades. The yield increase per unit area is the main goal of cultural practices and genetic improvement of soybean cultivars. It was pointed out that the comparisons of cultivars in test environments were the most direct estimate of genetic progress. There were many researches showing that soybean yield increased linearly with the progress of release year. At the same time, agronomic characteristics, growth features and physiological traits were assayed to examine the reason why the yield of different soybean cultivars differed despite having the same cultural and management environment.

Leaf photosynthesis is a basis of yield formation in crops. A number of researchers have demonstrated that there are great differences in photosynthesis level of cultivars. 14 soybean cultivars were grown, representing the genetic improvement of 58 years (1934–1992), in order to investigate the physiological changes associated with yield. They found that the increase of yield with year of release was significantly correlated with the harvest index, photosynthesis and stomatal conductance.

1.9 Physiological Aspects

Field studies were conducted in two years using 638 F2 and 1185 F3 lines of selected 16 F1 and 15 F2 parent lines ($\geq$ 80 pods plant^{-1}) to evaluate pod number and CO2 exchange rate (CER) as selection criteria. Pod and seed number, and seed weight of individual lines were observed during harvesting time, and CER of randomly selected 32 F2 and 30 F3 lines was measured at initial seed filling stage. The selection of F2 lines based on pod number to generate F3 lines increased the average of seed yield by 39%, and pod number by 77% in F3 lines compared with F2 lines. A close relationships was found between seed weight and pod or seed number per plant. Net CER responded sensitively to a reduction of light in a short-term and showed 78% of F2 lines and all F3 lines with maximum CER (Pmax) ≥ 20 $\mu molCO_2.m^{-2}.s^{-1}$. The ratios of pod number per plant and Pmax varied between lines and were used to group lines resulting in close relationships between Pmax and pod number. It is concluded that the use of pod number and CER (Pmax) as selection criteria offers an alternative approach in soybean breeding for high yield.

Researchers suggested that soybean yield increase was linear post several decenniums. This linear increase demonstrated that a yield plateau has not been reached. However, the yield of new soybean cultivars did not show linear increase in recent years in China. This may be due to the lack of effective physiological indices for the selection of high yield soybean cultivars.

The results of two years study showed that number of days to maturity decreased from 140.3 d to 125.6 d depends on the year of release, but soybean seed yield increased significantly from 1282.3 Kg to 2310.7 Kg during the 82 years of breeding. The net photosynthetic rate (Pn), stomatal conductance (Gs), apparent mesophyll conductance (Pn/Ci), transpiration (Tr), chlorophyll content and SLW all increased in bred cultivars during the improvement, however, the leaf area and ratio of Ci/Ca (intercellular CO_2 concentration, Ci; ambient CO_2 concentration, Ca) decreased in the same time. Yield was positively correlated with Pn, Tr, Gs, Pn/Ci, chlorophyll content and SLW, whereas the leaf area, Ci/Ca and WUE were negatively correlated. Results revealed that Pn and Pn/Ci are the effective selection indexes for seed yield and can be used in future soybean breeding programs. It was also found that the increase of Tr is higher than Pn. As a result, WUE is decreased as the increase of yield with year of release. Therefore, the main cost of high yield in new cultivars is water expense.

Plants are constantly confronted with both abiotic and biotic stresses that seriously reduce their productivity. Plant responses to these stresses are complex and involve numerous physiological, molecular, and cellular adaptations. Recent evidence shows that a combination of abiotic and biotic stress can have a positive effect on plant performance by reducing the susceptibility to biotic stress. Such an interaction between both types of stress points to a crosstalk between their respective signalling pathways. This crosstalk may be synergistic and/or antagonistic and include, among others, the involvement of phytohormones, transcription factors, kinase cascades, and reactive oxygen species (ROS).

In certain cases, such crosstalk can lead to a cross-tolerance and enhancement of a plant's resistance against pathogens.

Soybean (*Glycine max* Merr.) is the world's most widely grown legume and provides an important source of protein and oil. Global soybean production and yield per hectare increased steadily over the past century, with improved agronomy and development of cultivars suited to a wide range of latitudes. In order to meet the needs of a growing world population without unsustainable expansion of the land area devoted to this crop, yield must increase at a faster rate than at present. Here, the historical basis for the yield gains realized in the past 90 years are examined together with potential metabolic targets for achieving further improvements in yield potential. These targets include improving photosynthetic efficiency, optimizing delivery and utilization of carbon, more efficient nitrogen fixation and altering flower initiation and abortion. Optimization of investment in photosynthetic enzymes, bypassing photorespiratory metabolism, engineering the electron transport chain and engineering a faster recovery from the photoprotected state are different strategies that can be used to improve photosynthesis in soybean. These potential improvements in photosynthetic carbon gain will need to be matched by increased carbon and nitrogen transport to developing soybean pods and seeds in order to maximize the benefit.

Better understanding of control of carbon and nitrogen transport along with improved knowledge of the regulation of flower initiation and abortion will be needed in order to optimize sink capacity in soybean. Although few single targets are likely to deliver a quantum leap in yields, biotechnological advances in molecular breeding techniques that allow for alteration of the soybean genome and transcriptome promise significant yield gains.

Previous greenhouse studies have demonstrated that photosynthesis in some cultivars of first-(GR1) and second-generation (GR2) glyphosate-resistant soybean was reduced by glyphosate. The reduction in photosynthesis that resulted from glyphosate might affect nutrient uptake and lead to lower plant biomass production and ultimately reduced grain yield. Therefore, a field study was conducted to determine if glyphosate-induced damage to soybean (*Glycine max* L. Merr. cv. Asgrow AG3539) plants observed under controlled greenhouse conditions might occur in the field environment. The study evaluated photosynthetic rate, nutrient accumulation, nodulation, and biomass production of GR2 soybean receiving different rates of glyphosate (0, 800, 1200, 2400 g a.e. ha^{-1}) applied at V2, V4, and V6 growth stages. In general, plant damage observed in the field study was similar to that in previous greenhouse studies. Increasing glyphosate rates and applications at later growth stages decreased nutrient accumulation, nodulation, leaf area, and shoot biomass production. Thus, to reduce potential undesirable effects of glyphosate on plant growth, application of the lowest glyphosate rate for weed-control efficacy at early growth stages (V2 to V4) is suggested as an advantageous practice within current weed control in GR soybean for optimal crop productivity.

Despite the impressive gains made in soybean yields over the past century, the current pace of crop improvement is projected to fall short of the 2050 target of doubling crop yield. The physiological basis for seed yield improvements in historical soybean cultivars was examined. What has this retrospective analysis taught us about the prospects of enhancing future yield? It demonstrated that the efficiencies of light use (εi) and utilization (εc) along with harvest index (εp) were important in achieving the gains in yield made by traditional breeding. Although it appears that neither εi or εp are beginning to plateau, modern cultivars are nearing the maximum values that have been predicted for these parameters, therefore, there may be little room for further improvement of these efficiencies. However, the amount of solar energy intercepted by a soybean canopy could be increased with earlier planting date, more rapid canopy closure, and lengthening of the growing season. While εc has been increased through traditional breeding, it is still below the theoretical maximum, suggesting that it is an important target for future crop improvement. However, greater absolute values of εc do not always result in greater seed yields in soybean, and do not consistently correlate with yield across different years of study. Thus, more research is needed in order to better understand how εc is influenced by the environment and how changes in εc impact seed yield to guide future breeding strategies.

Experiments also provided evidence that the improvements in εc were driven by gains in photosynthetic daily carbon gain, and that the increases in photosynthesis were sustained through greater stomatal conductance. Because greater carbon acquisition came at the expense of increased water use, improvements in photosynthesis were only apparent during periods of ample water availability. This is important as a majority of soybeans are grown under water-limiting conditions and irrigation of this acreage is unsustainable. The data support the need to improve water-use efficiency (WUE) in soybean in order to maintain production in the future. Strategies for improving WUE in crops include breeding for high yielding cultivars that maintain a low transpiration rate and identifying cultivars that achieve constant transpiration rates at high vapor pressure deficits. While some of these strategies have been successfully incorporated into wheat, improvements in WUE do not always lead to greater yields and are often associated with decreases in photosynthesis. Because of the intimate connection between photosynthesis and water use, it is imperative that strategies to improve εc are in the context of a warmer future.

The correlation between transcript abundance of putative yield enhancement genes (YEG) and yield within historical germplasm was investigated. While several of these YEG had correlations with yield, none of these proposed YEG fell in genic regions identified to be QTL for seed yield (http://soybase.org). This discrepancy perhaps illustrates that seed yield is a complex, multi-genic trait, and research is only beginning to understand the underlying genetic mechanisms governing yield formation. However, the continually growing genetic resources in soybean, including the development of a Nested Association Mapping population in soybean, should improve the ability to understand the genetic architecture of complex quantitative traits like seed yield.

Some public breeding programs have placed an emphasis on increasing the genetic diversity by developing germplasm with increased productivity that is at least partially derived from exotic sources. It has identified Chinese cultivars which yield 80 to 88% as much as elite U.S. cultivars. High yielding cultivars and germplasm have been developed using these and other introductions from the U.S.D.A.

germplasm collection. Standard plant breeding practices have been used for this, e.g., hybridization between adapted high yielding cultivars, followed by inbreeding and selection, sometimes with backcrossing.

A major achievement for soybean improvement is the development of consensus maps for the 20 linkage groups of the soybean genome and DNA sequencing of the genome. Four consensus maps have been published, beginning with the first in 1995. The most recent map, published by Hyten et al. (2010), has between 175 and 418 markers on each linkage group. There are 3792 SNP markers, 1006 SSR markers, 664 RFLP markers, and 38 other markers for a total of 5500 markers. In addition, they developed a subset of 1536 biallelic SNP markers called the Universal Soy Linkage panel (USLP).

1.10 Herbicide Tolerance

Herbicide tolerance has been a mixed blessing for soybean improvement, but it has had a major impact on soybean production. The first herbicide tolerant soybeans were resistant to ALS (acetolactate synthase) inhibiting herbicides, e.g., sulfanyl-urea. This was a naturally occurring mutant discovered by Dupont. STS (sulfanyl-urea tolerant soybeans) cultivars were introduced in 1994. RoundUp Ready (glyphosate resistant) cultivars were released in 1996 by Monsanto. Glyphosate resistance is provided by a transgene of microbial origin, which inhibits a phosphate synthase enzyme (EPSPS). It was rapidly adopted by farmers throughout soybean producing regions in the U.S. and Argentina, and some years later, in Brazil. It provided some production advantages:

1. Glyphosate is a low-cost herbicide with low environmental impact.
2. It enables the use of minimal tillage. Soybeans can be drilled and glyphosate applied over the plant canopy, eliminating cultivation with lower fuel costs.

But there are also disadvantages:

1. Seed costs are high. Technology is proprietary and farmers are unable to save seeds for replanting.
2. Yields, at least initially, were lower, because the transgene was backcrossed into existing cultivars.
3. Continual use has selected glyphosate resistant weeds in some fields—particularly Amaranthus species—making the herbicide essentially useless in some regions.
4. Round Up resistant volunteer plants from a previous crop have to be dealt with.

Because glyphosate herbicides were so popular, research on other herbicides was discontinued. Now, older herbicides are being used, in addition to glyphosate with, some concern that weeds which are resistant to those herbicides will develop. Other types of herbicide resistant soybeans are being developed. Currently there are two other herbicide resistant soybeans being used, GAT soybeans, which detoxify glyphosate, and glyphosinate resistant (Liberty Link) soybeans.

Varieties with resistance to two different herbicides have been developed. There are other herbicide transgenes which might be used in the future, including resistance to 2-4-D and Dicamba. These could probably be deployed in soybean fields if the need arises, but research and the approval process would be costly.

Grain-type soybeans are processed into two commodities: Oil and protein meal. Both are economically important. In the U.S., there has been a continual increase in the use of both. There have been breeding efforts to increase the concentrations of both oil and protein in seeds. There is considerable genetic variation in both traits. In the USDA germplasm collection protein varies from 35.1% to 56.8% on a dry weight basis, and the oil concentration range is 8.3% to 27.9%. The two tend to be negatively correlated. In recent research with two populations that were randomly intermated for 26 generations, thereby dissipating most of the linkage disequilibrium, that negative correlation between the two traits was maintained at about $r = -0.6$. Nevertheless, some progress has been made in changing the two traits separately. A major protein QTL, found in *Glycine soja*, has been mapped to linkage group I and increases protein by two percentage points. A high oil germplasm was developed by recurrent selection, N98-480, and was used as a parent or the high oil cultivar, NC-Raleigh. QTL for both protein and oil content have been published and a compilation of references has been made. Although efforts to improve the two traits simultaneously have been challenging, a recent study provides some insights on how to address this issue. They identified a QTL that has a positive effect on both oil and protein concentration and also several QTL that increase oil but do not decrease protein or yield.

CHAPTER 2
Seed Germination

Germination begins with the water uptake of dry seed, and ends with the emergence of the radicle (Bewley, 1997). Generally, this can be divided into three phases, based on the style of water uptake. Phase I is a rapid water uptake phase, in which DNA damage repairing (Macovei et al., 2011) and resuming of glycolytic and oxidative pentose phosphate pathways occur (Howell et al., 2006). Phase II is a plateau phase, in which mitochondria synthesis (Howell et al., 2006) and translation of storage mRNA occur (Dinkova et al., 2011). Phase II is also regarded as a metabolism active phase, during which reserves mobilization is initiated. Phase III is the post-germination stage, in which the radicle begins to grow. Mobilization of reserves is one of the most critical events in germination, which could provide not only precursors but also energy for the biosynthetic processes. Although mobilization of the reserves may not be necessary for germination (Pinfield-wells, 2005), it is crucial for germination efficiency and post-germinative seedling establishment (Eastmond et al., 2000).

Seed germination is a complex physiological process that is regulated by different external and internal factors, such as temperature (Penfield et al., 2005), light (Piskurewicz et al., 2009), soil salinity (Kim et al., 2008; Park et al., 2011), gibberellic acid (GA) (Ogawa, 2003) and abscisic acid (ABA) (Gubler et al., 2005). During germination, the environmental and hormone signals integrate together to play regulatory roles (Penfield et al., 2005; Chen et al., 2008). The environmental factors could affect the germination through regulating the biosynthesis and catabolism of phytohormones, such as GA and ABA (Finch-Savage and Leubner-Metzger, 2006).

Phase II of soybean seeds germination was defined as 12–24 h after imbibition (Fig. 2.1A). At the end of this period, germination was almost complete (Fig. 2.1B).

Compared with rice seed, it took less time to germinate (Fig. 2.1). It is known that rice seed has tiny embryo and belongs to morphological dormancy category. Its embryo has to grow inside the seed before protrusion (Linkies et al., 2010), which is also supported by the observation that the embryo expands its size before radical emergency (Fig. 2.1B).

As a dicot, the soybean embryo has two cotyledons, which are the first leaves that develop within the seed. Temperature, moisture, oxygen, and soil conditions within the seed zone can affect soybean germination and emergence. The radicle root is the first part of the embryo to penetrate the seed coat, followed by elongation of the hypocotyl, which pulls the cotyledons and epicotyl to the soil surface.

Soybean seed are physiologically mature at the time maximum dry weight is reached (40%–50% moisture content). At this stage, germinability and vigor are highest even though the seed first become capable of germination when about one-third of the dry weight has been accumulated. The mature soybean seed is generally spherical in shape and has a relatively thin seed coat. The hilum, point of attachment of the seed in the pod, is linear to elliptic in shape and located on the ventral face of the seed coat. It may be variously pigmented. The endosperm is represented only by a thin layer of cells immediately beneath the seed coat. The remainder of the interior of the seed is occupied by the embryo, which consists of a short radicle-hypocotyl axis, two fleshy cotyledons (lateral organs) and a well-developed plumule growing point with two leaves, which is terminal on the radicle-hypocotyl axis and between the cotyledons.

The short radicle-hypocotyl axis is curved so that it lies against the basal margins of the cotyledons with its tip pointed in same direction as the apices of the cotyledons. The position of the radicle-hypocotyl

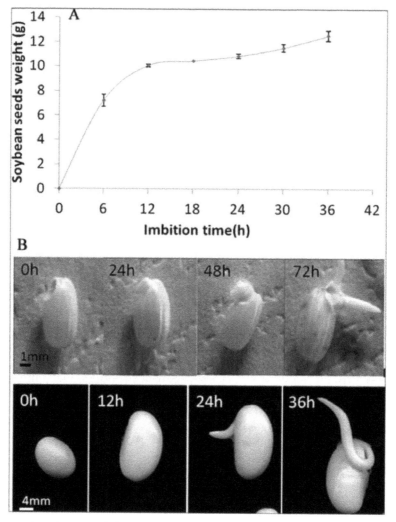

Fig. 2.1 (A) water absorption curve during soybean seed germination; (B) images of rice and soybean seeds during germination.

axis and the delicacy of the seed covering, which is its only protection, makes the seed especially vulnerable to injury by mechanical abuse from any source-harvesting, conveying, processing, etc. Since the radicle-hypocotyl axis is essential for normal germination, any substantial damage to it can be disastrous to seed quality.

Seeds are living organisms that respire at a very low rate. They are in a state of quiescence, or rest, where they stay dormant until desirable conditions that trigger germination occur. Seeds can stay viable for a year under cool, dry conditions.

Soybean seed can begin to germinate when soil temperatures are approximately 10°C (50°F); however, germination is likely to be slow until soil temperatures warm to near 25°C (77°F). Upon being placed into the soil, the seed begins to absorb or imbibe water and, as a result, starts to swell. When enough water (approximately 50% of the seed's weight) is taken in, and with favourable temperatures, the radicle breaks through the seed coat and rapidly develops into the primary seedling root, which has the ability to force the cotyledons toward the surface. Lateral roots quickly emerge from the radicle as it elongates and root hairs grow from the radicle and lateral roots. Root hairs are barely visible and should not be confused with later developing and easily seen branch roots. The root hairs become the main absorbing structures. Soon after the radicle appears, the hypocotyl starts elongating and forms a hook that pushes toward the surface.

The cotyledons are attached to the hypocotyl and progress upward with the growth of the hypocotyl. The epicotyl contains small leaves, buds, and the growing point.

Moisture. Planting into a moist seedbed with good seed-to-soil contact is necessary as moisture needs to move into the seed for germination to occur. Planting into dry soil with rainfall occurring too soon after can result in crusting and poor soybean emergence.

Soil Conditions. Soil crusting can delay or prevent seedling emergence and cause soybean hypocotyls to be swollen or broken when trying to push through the crust. Fields with fine-textured soils, low organic matter, and little surface residue can be vulnerable to crusting, especially where excessive tillage has taken place.

Temperature. Cold soil temperatures can cause seeds to remain dormant, causing them to become increasingly vulnerable to feeding by wildlife that dig up the seeds, insects, and seed/seedling diseases. Once emerged, soybeans can tolerate a temperature dip down to −2.8°C (27°F) for a short period of time. However, the growing point of a soybean plant is near the top and newly emerged plants could suffer permanent frost damage with a late spring frost.

Oxygen. Saturated, flooded, and compacted soils can reduce germination and emergence due to the lack of oxygen. Soil pore spaces filled with water reduces the amount of oxygen available for seed respiration. Compacted soil reduces the availability of water and oxygen required for germination, root and plant growth, and nutrient uptake.

The effects of grain damage on soybean increase if the damaged seed is stored, as most often be done in practice. During storage, harvesting and threshing, the mechanically damaged seed loses its germination capacity more rapidly than the undamaged seed and, as a result, the seedling percentage decreases. Therefore, a study introduced the effects of impact velocities (IV), number of impact loadings (NL), and time (T) on percentage of grain damage and percentage of loss in germination to soybean seeds. The result showed significant influenced percentage of grain damage and percentage of loss in germination. As the extent of grain damage increased, capability of germination decreased.

Environmental stress during seed development may influence soybean [*Glycine max* (L.) Merrill] seed quality. Two greenhouse experiments were conducted in order to study the effect of drought stress on soybean (cv. 'McCall') seed germination and vigor. Two stress treatments (moderate and severe) were imposed at growth stages R5 and R6 and compared with well-watered plants. Drought stress significantly reduced weight per seed and yield per plant and increased stomatal resistance. It had little effect on seed shape; however, and few shrunken or wrinkled seed were produced. Drought stress had no effect on seed germination and little effect on seed vigor, as measured by the accelerated aging, conductivity and cold tests. In one experiment, drought stress increased the proportion of hard seed (especially in the smaller seed size fraction) which lowered 3-day germination, but did not affect final germination or vigor. Although the most severe drought stress treatment reduced seed weight and yield substantially (34% and 38%, respectively) it had little or no effect on seed quality.

2.1 Temperature

Low positive temperature has an inhibiting effect on the growth, development and other physiological processes of cold-sensitive plants, including soybean. Huang and Yang (1995) report that soybean seeds generally germinate at a temperature between 10 and 30°C. However, the rate of germination increases with increasing temperature and reaches its maximum at 30°C. Other authors also report that the optimal temperature for germination and hypocotyl elongation in soybean is around 30°C (Liao Fang Lei et al., 2011). It was also found that earlier hydration of seeds at higher temperatures accelerated the rate of germination, while at lower temperatures (10°C), it slowed the rate of germination. An experiment in Petri dishes investigated the effect of temperature: 28/28°C (control), 10/28°C, 28/10°C, and 10/10°C (imbibition/germination), on germination of seeds of eight soybean cultivars. The obtained results showed that a temperature of 10°C used during germination (28/10°C), and even to a larger extent during imbibition and germination (10/10°C), clearly reduced the speed of germination, percentage of germinated seeds, and radical length relative to the control, but it increased catalase activity in sprouts (Table 2.1).

Table 2.1 Effect of chilling temperature (10°C) during the period of imbibition or germination as well as in both studied periods on the speed of seed germination in eight soybean cultivars (plant day[-1]).

Temperature °C	Cultivars								
	Aldana	Jutro	Progres	Mazowia	Nawiko	Augusta	OAC Vision	Dorothea	Mean
28/28	11.1	13.4	12.2	12.1	13.9	12.5	11.5	11.3	12.2
10/28	10.6	11.6	10.9	11.2	11.5	11.3	11.3	10.7	11.1
28/10	8.9	9.2	6.9	6.5	10.9	7.5	6.4	5.1	7.7
10/10	5.0	6.8	6.8	6.0	6.5	6.1	6.0	4.2	5.9
Mean	8.9	10.2	9.2	9.2	10.7	9.3	8.8	7.8	
LSD0.05: for temperature – 0.50; for cultivars – 0.84; for interaction – 1.98.									

The influence of germination temperature and time on the antioxidant activity of soybean seeds was investigated. The MTĐ 760 soybean cultivar was germinated in dark condition at 22, 25, 28°C and ambient temperature for 0, 12, 23, 36, 48, 60 and 72 hours. The total phenolic content (TPC), total flavonoid content (TFC), vitamin C and α-tocopherol contents as well as antioxidant activity (AA) assayed by DPPH radical-scavenging activity in terms of IC50 of germinated soybean were determined. These values increased with the increase of germination time for all applied germination temperatures and they tend to reach the maximum values after a period of 60 hours. Germination temperature also influenced these parameters and, generally, they reached the highest values at 25°C. The TPC, TFC, vitamin C and IC50 values of soybean seed germinated at 25°C for 60 hours were, respectively, 8.11 ± 0.03 mg GAE/g (DW); 6.45 ± 0.09 mg QE/g (DW); 12.80 ± 0.02 mg/g (DW) and 5.49 ± 0.03 mg/ml. The maximum content of α-tocopherol after 72 hours of germination was 0.247 ± 0.004 mg/g (DW). Correlation studies indicated significant negative correlation between the values of IC50 with TPC ($r = -0.968$) (Lien et al., 2016).

2.2 Flood Stress

Seed germination is a critical developmental phase in plant life cycle and reproductive success (Donohue et al., 2010). In general, seed germination capacity is determined by genetic factors and environmental cues, such as light, water, temperature, drought, and oxygen (Bewley et al., 2013). Under flooding, soybean seeds exhibit poor survival and germination in the field due to quick loss of viability in hypoxic environment because oxygen supply is required for germination activation (Parolin, 2001). On the other hand, the presence of soil-borne diseases, caused by Phomopsis, Pythium, Phytophthora, Rhizoctonia, and Fusarium, significantly impacted on soybean seed germination and seedling emergence (Heatherly, 2015). Using seeds covered by an appropriate fungicide increases seed germination by about 10%, resulting in a large plant emergence in the field (Schulz and Thelen, 2008). Apron Maxx RTA (Syngenta Crop Protection Inc.) is a broad-spectrum fungicide widely used in the United States for seed treatment and it can control or suppress pathogens Phomopsis, Pythium, Phytophthora, Rhizoctonia, and Fusarium.

The study of flooding influence on soybean seed germination after planting in the field showed that seed germination rate (SGR) of each genotype, without flood stress, was significantly different and ranged between 64.7% to 84.0% and 69.0% to 90.7% while using untreated and fungicide-treated seed (P < 0.0001), respectively.

Results indicated that fungicide treatment improved soybean seed survival and germination in the field. The average of SGR of high-yielding soybean group was significantly higher than those of non-high-yielding soybean (P < 0.0001). The results indicated that a high-yielding trait of each genotype was correlated with seed germination and survival. Under flood stress in the field, SGR means of untreated and fungicide-treated seed significantly decreased over eight flooding treatment times (P < 0.0001). Flooding effect on germination between untreated and fungicide-treated seed was not significantly different (P = 0.1559). Furthermore, comparing the high-yielding and flood tolerant soybean groups showed no difference in their SGR means over eight flooding treatment times (P = 0.7687 and P = 0.8490), indicating that soybean seed germination did not depend on genotype, yield, and flood tolerance trait, and seed treated by fungicide did not increase its germination in the field under the flood stress (Table 2.2 and Fig. 2.2).

Table 2.2 Seed germination rate (SGR) of twenty soybean genotypes in untreated and fungicide-treated seed without or with flood stress tests.

Variety	SGR% (US/N)*	SGR% (TS/N)#	SGR% (US/F)†	SGR% (TS/F)‡
UA5615C	84.0	90.7	25.0	25.1
UA5612	83.3	87.3	24.3	24.7
Osage	81.7	86.7	23.7	25.4
UA5414RR	78.7	84.0	25.1	24.0
UA5715GT	78.7	84.0	24.5	24.6
UA5014C	78.7	82.7	25.8	23.3
R11-6870	78.3	86.7	24.2	24.9
UA5213C	78.3	88.0	25.2	23.9
UA5115C	76.3	83.3	25.4	23.7
R07-6669	75.0	76.3	24.3	24.8
Walters	73.0	74.3	23.6	25.5
R10-4892	71.0	74.3	24.7	24.5
R04-342	70.3	74.7	23.2	25.9
R10-2379	68.7	72.3	24.8	24.4
R13-12552	68.3	69.0	23.8	25.8
RM-22590	68.3	72.3	25.3	23.8
R09-4095	67.3	70.3	24.9	24.3
R01-2731F	66.0	72.7	25.0	24.1
R06-4433	65.3	78.7	24.2	25.0
R99-1613F	64.7	72.7	25.1	24.0

* SGR, seed germination rate; US/N, untreated seed without flood stress; # SGR, seed germination rate; TS/N, fungicide-treated seed without flood stress;

† SGR, seed germination rate; US/F, untreated seed with flood stress; ‡ SGR, seed germination rate; TS/F, fungicide-treated seed with flood stress.

Water is necessary for germination. However, too much water, which enters through cracks in the seed coating, can lead to over-absorption. This becomes a bigger problem if conditions are not right for rapid germination. As seeds soak up water, water absorbed too quickly can cause cell walls to rupture. That can cause cell death. An intact seed coat slows water absorption. High water content in the soil can be associated with colder temperatures but also lower oxygen levels in the soil. Water in soil pores excludes oxygen needed for seedling growth. Initial water absorption by seeds is not dependent on oxygen. Even dead seeds can absorb water.

Once water content exceeds 50 percent, continued absorption depends on energy released by the seed respiration rate. When growth starts, oxygen demand increases rapidly. That oxygen must come from air in soil pores.

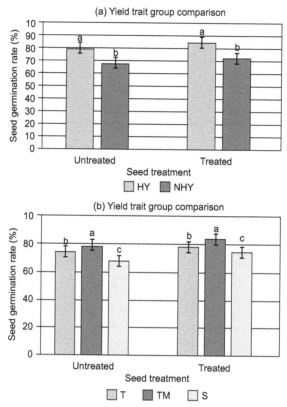

Fig. 2.2 Seed germination rate (SGR) means of untreated and fungicide-treated seed tests without flood stress: (a) Yield trait groups with high-yielding (HY) and non-high-yielding (NHY) traits; (b) Flood trait groups with flood-tolerant (T), flood-moderately-tolerant (MT), and flood-sensitive (S) traits.

2.3 Salt Stress

Salinity in soil or water is one of the major stresses and can severely limit crop production, especially in arid and semi-arid regions. Salt stress negatively affects seed germination; either osmotically through reduced water absorption or ionically through the accumulation of Na^+ and Cl^-, causing an imbalance in nutrient uptake and toxicity effect.

The results showed that salinity stress caused by NaCl and Na_2SO_4 reduced both germination and seedling growth of both varieties of soybean. JS-335 appeared more tolerant under different NaCl concentrations and more sensitive under different Na_2SO_4 concentrations than BSS-2 (Kumar, 2017) (Tables 2.3 and 2.4).

Increasing salinity delayed the beginning and ending of germination and reduced the final germination percentage, inhibiting germination completely above 0.3M salinity. Salinity stress caused by NaCl and Na_2SO_4 reduced both germination and seedling growth in both the soybean varieties. JS-335 appeared more tolerant under different NaCl concentrations and more sensitive under different Na_2SO_4 concentrations than BSS-2. Obviously, acceptable growth of plants in arid and semi-arid lands which are under exposure of salinity stress is related to ability of seeds for best germination under unfavourable conditions, so necessity of evaluation of salt resistance soybean plant species are important at primary growth stage.

Salinity stress has adverse effects on soybean development periods, especially on seed germination and post germinative growth (Basuchaudhuri, 1990). Improving seed germination and emergence will have positive effects under salt stress conditions on agricultural production. It was reported that NaCl

Table 2.3 Effect of different NaCl concentration on the germination and early seedling growth of soybean varieties (BSS-2 and JS-335).

Treatment	Germination (%) BSS-2	Germination (%) JS-335
Control	90	85
0.05M	65	70
0.10M	70	90
0.15M	70	70
0.20M	45	60
0.25M	35	45
0.30M	30	20
0.40M	--	--
0.50M	--	--

Table 2.4 Effect of different Na_2SO_4 concentration on the germination and early seedling growth of soybean varieties (BSS-2 and JS-335).

Treatment	Germination (%) BSS-2	Germination (%) JS-335
Control	90	90
0.05M	50	65
0.10M	65	75
0.15M	70	45
0.20M	50	30
0.25M	35	10
0.30M	15	10
0.40M	--	--
0.50M	--	--

delays soybean seed germination by negatively regulating gibberellin (GA) while positively mediating abscisic acid (ABA) biogenesis, which leads to a decrease in the GA/ABA ratio. This study suggests that fluridone (FLUN), an ABA biogenesis inhibitor, might be a potential plant growth regulator that can promote soybean seed germination under saline stress. Different soybean cultivars, which possessed distinct genetic backgrounds, showed a similar repressed phenotype during seed germination under exogenous NaCl application. Biochemical analysis revealed that NaCl treatment led to high MDA (malondialdehyde) levels during germination and the post-germinative growth stages. Furthermore, catalase, superoxide dismutase, and peroxidase activities also changed after NaCl treatment. Subsequent quantitative Real-Time Polymerase Chain Reaction analysis showed that the transcription levels of ABA and GA biogenesis and signalling genes were altered after NaCl treatment. In line with this, phytohormone measurement also revealed that NaCl considerably down-regulated active GA_1, GA_3, and GA_4 levels, whereas the ABA content was up-regulated; and therefore ratios, such as GA_1/ABA, GA_3/ABA, and GA_4/ABA, are decreased. Consistent with the hormonal quantification, FLUN partially rescued the delayed-germination phenotype caused by NaCl-treatment. Altogether, these results demonstrate that NaCl stress inhibits soybean seed germination by decreasing the GA/ABA ratio, and that FLUN might be a potential plant growth regulator that could promote soybean seed germination under salinity stress (Shu et al., 2017).

2.4 Herbicides

The effects of different concentrations of glyphosate acid and one of its formulations (Roundup) on seed germination of two glyphosate-resistant (GR) and one non-GR variety of soybean were investigated.

As expected, the herbicide affected the shikimate pathway in non-GR seeds but not in GR seeds. It was observed that glyphosate can disturb the mitochondrial electron transport chain, leading to H_2O_2 accumulation in soybean seeds, which was, in turn, related to lower seed germination. In addition, GR seeds showed increased activity of antioxidant systems when compared to non-GR seeds, making them less vulnerable to oxidative stress induced by glyphosate. The differences in the responses of GR varieties to glyphosate exposure corresponded to their differences in enzymatic activity related to H_2O_2 scavenging and mitochondrial complex III (the proposed site of ROS induction by glyphosate). Results showed that glyphosate ought to be used carefully as a pre-emergence herbicide in soybean field crop systems because this practice may reduce seed germination.

2.5 Seed Size

The most popular soyfoods, like soymilk, tofu, and vegetable soybean (edamame), are all produced mainly from large-seeded (> 20 g/100 seeds) soybeans (Gandhi, 2009). Large-seeded soybeans are reported to increase quantity and strength of tofu and had higher total content of monounsaturated oils (Bhardwaj et al., 1999; Bhardwaj et al., 2003). Soymilk sales reached 210.5 million in 2013, while tofu as a traditional soyfood had sales of approximately $274 million in 2013. Differences in seedling emergence and vigor and the subsequent crop stand are reportedly attributed to prevailing seedbed conditions (Johnson and Wax, 1978). The effects of seed size on germination and subsequent yield also vary, with some showing increased yield with seed size (Smith and Camper, 1975) and others showing no effect (Hoy and Gamble, 1987). It has been reported that seed size correlates with seed vigor and that large seeds tend to produce more vigorous seedlings and better stand (Roy et al., 1996; Cookson et al., 2001) and are more likely to emerge from greater depth than those from small seeds (Mandal et al., 2008). Plants from larger seeds have been reported to produce bigger seed yield than those from small kernels (Stobbe et al., 2008). However, other results indicate either better germination and vigor in small and medium sized seeds (Peksen et al., 2004) or no relationship between seed size and seedling emergence and final yield (Johnson and Luedders, 1974). Also, addition of fertilizer to hydro-primed seeds could affect plant growth and yield, as shown for sulphur, which resulted in an increase in yield in both primed and non-primed seeds (Bejandi et al., 2009). Soil factors play a role in germination and subsequent yield of a crop. Under ideal soil moisture conditions, large soybean seeds showed better physiological quality than small seeds; with more vigorous seedling but at moisture content of −0.2 MPa, small seed performed better (Pereira et al., 2013). Other edaphic factors, like salinity, affect germination by creating osmotic potentials, preventing seed water uptake, as well as direct toxic effects of Na^+ and Cl^- (Harris et al., 2001). At moisture content ranging from 20 to 30%, all soybean seed sizes showed no germination at 20% moisture content, but small and medium sized seeds showed better germination and greater root developments than large sized seeds at all higher moisture content (Edwards and Hartwig, 1971).

Seed size class	Variety	Hours after start of experiment	
		48	72
Small	MFS-561	88.0a	99.7a
Small	V08-4773	79.1a	93.1a
Medium	V03-4705	75.7a	92.1a
Medium	Glenn	70.5a	91.3a
Large	V07-1897	29.4b	45.5b
Large	MFL-159	28.9b	44.9b
Numbers in a column followed by a similar letter are not statistically significant.			

2.6 Effects of Chemicals

Studies were carried out on the effects of coumarin, ferulic acid and naringenin on soybean seed germination and on the growth of seed-borne fungi at concentrations of 50 and 100 mg.L^{-1}. The compounds showed good inhibition of seed germination, especially at 50 mg.L^{-1}, but little fungistatic activity. In an experiment, primary roots reached a maximum of 1–2 cm in ferulic acid and coumarin treatments and about 0.5 cm in naringenin treatments. In the first treatments, stunted roots with necrotic tips were observed (control showed normal seedling development). This morphology suggested a possible common mechanism of action for ferulic acid and coumarin (Colpas et al., 2003).

Table 2.5 Analysis of variance of germination percentage of phenolic-treated BRS-155 seeds, under constant white light and 25°C temperature.

Treatment	% germinated seed[1]
Ferulic acid 50 ppm	3.5 c
Ferulic acid 100 ppm	29 b
Coumarin 50 ppm	1 c
Coumarin 100 ppm	30 b
Naringenin 50 ppm	0 c
Naringenin 100 ppm	2 c
Control	94.5 a
F test	243.5*
C.V. (%)	13.47

[1] Means, followed by same letter not statistically differ by Tukey's test at 5%.

These compounds seemed to act like mitosis disrupter herbicides on microtubules. Microtubules were responsible for setting the plane for cell division and cellulose deposition (Vaughn and Lehnen, 1991). Wall deposition accounted for a cylindrical cell shape (Green, 1962), which is responsible for elongation.

Auxin is an important phytohormone which mediates diverse development processes in plants. Published research has demonstrated that auxin induces seed dormancy. However, the precise mechanisms underlying the effect of auxin on seed germination require further investigation, especially the relationship between auxins and both abscisic acid (ABA) and gibberellins (GAs), the latter two phytohormones being the key regulators of seed germination. It was reported that exogenous auxin treatment represses soybean seed germination by enhancing ABA biosynthesis, while impairing GA biogenesis, and finally decreasing GA$_1$/ABA and GA$_4$/ABA ratios. Microscope observation showed that auxin treatment delayed rupture of the soybean seed coat and radicle protrusion. qPCR assay revealed that transcription of the genes involved in ABA biosynthetic pathway was up-regulated by application of auxin, while expression of genes involved in GA biosynthetic pathway was downregulated. Accordingly, further phytohormone quantification shows that auxin significantly increased ABA content, whereas the active GA$_1$ and GA$_4$ levels were decreased, resulting in significant decreases in the ratios GA$_1$/ABA and GA$_4$/ABA. Consistent with this, ABA biosynthesis inhibitor fluridone reversed the delayed-germination phenotype associated with auxin treatment, while paclobutrazol, a GA biosynthesis inhibitor, inhibited soybean seed germination. Altogether, exogenous auxin represses soybean seed germination by mediating ABA and GA biosynthesis (Shuai et al., 2017) (Fig. 2.3).

The study revealed that elevated dose of lead concentrations reduces the growth parameter as compared to control. Lead concentrations of 1000 mg/kg significantly decreased the percentage of germination and root length. However, at low levels of zinc (250 and 500 mg/kg) showed increased germination percentage and also increase root and shoot length. However, at high levels (750–1250 mg/kg), a detrimental effect on the growth parameter and germination was observed.

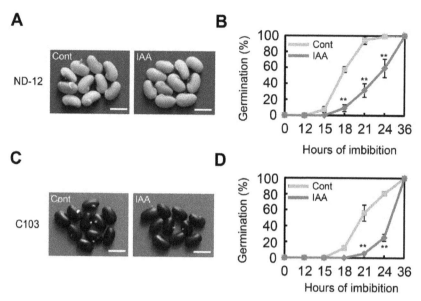

Fig. 2.3 Exogenous IAA treatment represses soybean seed germination under dark conditions. Healthy and elite soybean seeds (cultivars ND-12 and C-103) were incubated on two layers of filter paper in Petri dishes. The concentration of IAA used was 1 μM, and the equivalent amount of ultrapure water was added as control (Cont). The germination rates under dark conditions were recorded using a safe green light. Quantitative analysis of germination rates is shown in the right panels. The representative images (21 hours after sowing) are shown (left panels). (A,B) for cultivar ND-12; (C,D) for cultivar C-103. Bar in panel A and C = 10 mm. The average percentages of four repeats ± standard error are shown. ** Difference is significant at the 0.01 level.

2.7 Metabolism

Macromolecules accumulated in seeds are used as an energy source for early seedling development and seed germination. Germination begins with water uptake by the seed (imbibition) and the emergence of embryonic axis, usually the radicle, through the structures surrounding it (Bewley et al., 2013). When seed germination begins, starch and proteins are converted to sugars and amino acids within the starch granules and protein storage vacuoles, by diastase and protease enzymes, respectively (Wilson, 2006). TAGs are hydrolysed by lipases, enzymes catalyzing the hydrolytic cleavage of the fatty acid ester bonds, to yield glycerol and free fatty acids (Theimer and Rosnitschek, 1978). Free fatty acids enter the glyoxysome for conversion to oxaloacetic acid (OAA), which passes into the mitochondrion, and ultimately into the cytosol for conversion to sucrose, which is then transported as an energy source from cotyledons to the growing axis of the seedling (Graham, 2008).

Starch is the main carbohydrate of plant storage organs. Starch has been extensively characterized in many cereal, root and tuber crops, as well as many legumes, but has not been studied in leguminous soybeans because of its low content at maturity. The only research on soybean starch focused on starch content (Wilson et al., 1978) or granule morphology (Nakamura, 1974) (Tables 2.6, 2.7 and 2.8).

The HPLC elution profiles of soluble sugars extracted from germinating soybean axes and cotyledons were distinctively different. For instance, in d 2 of seed germination, axes contained much smaller amounts of stachyose and sucrose, and a much larger amount of glucose than cotyledons. Fructose and an unknown compound, labelled peak No. 7, were present in axes but absent in cotyledons (Fig. 2.4). The putative sorbitol was not detectable in the mature, dry seeds of 'Williams 82' (data not shown), but a substantial amount accumulated during germination. To determine the identity of the compound, material was isolated from 2-d germinating soybean axes by repetitive injections on HPLC and 11.5 mg of the unknown was obtained from 1.5 g of dry axis. A SugarPak I column was used because this column was able to separate sorbitol from a mixture containing some naturally occurring six-carbon polyols, such as

Table 2.6 Seed weight as a proportion of total soybean pod weight, and water, starch, protein and oil content of soybean seeds collected 20 d prior to harvest.

Soybean variety	Seed weight (% of total pod)	Water content (% dry weight)	Protein content[1] (% dry weight)	Oil content[1] (% dry weight)	Starch content[1] (% dry weight)
High-protein	49.3	64.0	39.5	18.0	11.7
Lipoxygenase-free	52.0	64.4	40.8	17.3	11.5
Low-linolenic acid	52.6	63.7	39.4	17.0	10.9
	P = 0.082	P = 0.53	P = 0.54	P = 0.21	P = 0.24

[1]   Starch, protein and oil contents were averaged over two analyses from each of three replicates.
P represents the probability of F-statistic exceeding expected for each comparison between soybean varieties in the respective column.

Table 2.7 Soybean varietal characteristics of crude fat, crude protein and total sugar.

Soybean	Crude fat (%)	Crude protein (%)	Total sugar (%)
JS-335	20.33	41.33	7.60
DS-228	18.50	42.05	8.20
MAUS-71	20.50	39.60	7.10
MAUS-81	20.53	36.53	6.20
Mean	19.96	39.88	7.28
CD at 5%	2.56	2.99	0.73

Table 2.8 Water, protein, oil, fiber and carbohydrate content of soybean seeds collected at commercial harvest maturity. Values after 6 represent standard deviation.

Soybean variety	Water [%]	Protein [%]	Oil [%]	Fiber [%]	Carbohydrate [%]
High-protein	12.3	60.1	39.6	6.1	17.6
Lipoxygenase-free	11.7	38.6	17.8	4.7	20.8
Low-linolenic acid	10.8	37.1	18.9	4.8	21.2

galactitol, L-iditol, myoinositol and D-mannitol (Table 2.9). In addition, the germinating soybean axes did not contain detectable D-mannitol, galactitol and L-iditol, three polyols that elute near sorbitol under these HPLC conditions. Hence, a highly purified preparation of the soybean axis compound was obtained by repetitive HPLC runs without further purification (Kuo et al., 1990).

Enzymes of sucrose metabolism were assayed from the same samples used for the analysis of soluble carbohydrates. In cotyledons, invertase activity was present in low but detectable levels throughout the examined period of germination (Tables 2.10 and 2.11). Sucrose synthase activity was not detectable at d 0 and 1 but increased to a relatively low level thereafter, whereas sucrose phosphate synthase activity was present at a relatively constant level. Activities of these enzymes in germinating soybean axes, however, exhibited different patterns from those observed in cotyledons. In axes, invertase activity increased rapidly from d 0 to d 2 and remained at fairly high levels thereafter. Sucrose synthase activity was present at fairly high levels during the first 2 d of germination and declined at d 3 and 4, whereas the activity of sucrose phosphate synthase, which had the greatest variation among the examined activities, was low or undetectable in the first 2 d and showed some increase in d 3 and 4.

The interrelationship among sugars, sorbitol, and enzyme activities in germinating soybean cotyledons and axes during incubation is shown in Table 2.9. Germination time was highly correlated with the decrease in sucrose content and weakly with the increase in sucrose phosphate synthase activity. The decrease in sucrose content was weakly correlated with the increase in invertase activity and glucose, fructose, and sorbitol content. Sorbitol levels were highly correlated ($P < 0.001$) with the levels of glucose, fructose, and invertase activity, whereas all these compounds and invertase activity had no correlation with the activities of sucrose synthase and sucrose phosphate synthase.

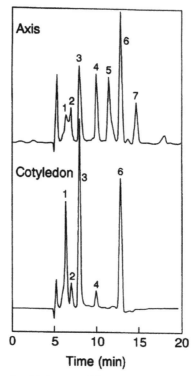

Fig. 2.4 HPLC separation of sugars and sorbitol extracted from axes (top) and cotyledons (bottom) of 2-d germinating soybeans. The elution peaks were identified by the same retention time of each authentic standard. Mannitol was added as an internal standard prior to extraction. 1, Stachyose; 2, raffinose; 3, sucrose; 4, glucose; 5, fructose; 6, mannitol; 7, sorbitol.

Table 2.9 Sugar and sorbitol content in cotyledons and axes of germinating soybean seeds values represent mean of four determinations (± SE) from two separate experiments.

Tissue	Days of germination	Sucrose (mg.gdw^{-1})	Glucose (mg.gdw^{-1})	Fructose (mg.gdw^{-1})	Sorbitol (mg.gdw^{-1})
Cotyledon	0	55.5 ± 1.2	5.8 ± 1.2	ND	ND
	1	53.6 ± 2.4	5.3 ± 0.4	ND	ND
	2	50.5 ± 2.7	7.8 ± 1.3	ND	ND
	3	34.7 ± 1.6	5.1 ± 0.4	ND	ND
	4	23.3 ± 7.6	5.1 ± 0.7	ND	ND
Axis	0	42.7 ± 4.1	5.0 ± 0.9	ND	ND
	1	55.8 ± 3.0	14.6 ± 1.6	11.0 ± 2.6	6.5 ± 0.8
	2	26.0 ± 3.6	38.9 ± 3.0	51.4 ± 8.8	17.3 ± 3.1
	3	23.6 ± 2.2	34.5 ± 3.5	57.2 ± 9.0	17.7 ± 4.7
	4	19.0 ± 0.5	33.9 ± 1.1	46.4 ± 10.2	16.3 ± 6.1

Table 2.10 Changes in a-amylase (µg maltose min^{-1}g^{-1}) during germination of soybean.

Soybean	1st day	6th day	12th day
JS-335	0.012	0.045	0.027
DS-228	0.027	0.051	0.029
MAUS-71	0.020	0.039	0.022
MAUS-81	0.018	0.029	0.017
Mean	0.019	0.041	0.024
CD at 5%	0.0017	0.004	0.004

Table 2.11 Activities of enzymes (μmol.min⁻¹gdw⁻¹) of sucrose metabolism in germinating soybean cotyledons and axes. Values are the mean (± SE) of four extractions, two from each separate experiment.

Tissue	Days of germination	Sucrose synthase	Invertase	Phosphate synthase
Cotyledon	0	0.21 ± 0.04	ND	0.98 ± 0.05
	1	0.23 ± 0.03	ND	0.82 ± 0.05
	2	0.14 ± 0.04	2.12 ± 0.91	1.35 ± 0.58
	3	0.11 ± 0.02	0.44 ± 0.12	0.77 ± 0.10
	4	0.14 ± 0.04	0.63 ± 0.26	1.78 ± 0.26
Axis	0	0.08 ± 0.02	1.81 ± 0.13	ND
	1	1.17 ± 0.03	1.58 ± 0.19	0.75 ± 0.24
	2	5.74 ± 1.40	1.81 ± 0.13	ND
	3	5.81 ± 1.19	0.87 ± 0.21	1.31 ± 0.09
	4	6.72 ± 0.99	1.49 ± 0.49	1.02 ± 0.11

Most of the investigators found heat to be injurious; however, in the case of legume protein, heat was found to be beneficial. However, the results suggest that protein concentration decreases on boiling but this also inhibits the antinutritional compounds and certain proteases. In addition to this, the protein profile through SDS-PAGE shows that 6 main proteins are present in soybean seed extract. Protein profile shows little difference with germination period. It also got the similar profile and identified these bands as subunits of 7S and 11 S globulins.

The decrease in sugars might be due to their utilization in respiration. The starch content in 6 day germinated seeds was found to be maximum and its content decreased in non-germinated seeds. The presence of starch and the corresponding enzymes for its hydrolysis have already been well documented in soybean. The increase in starch content of seeds after germination as compared to mature seeds support the earlier observations of a number of workers that soybean produce starch during imbibition and germination which is a transient reserve material for germinating soybean cotyledons (Adams et al., 1980). Lastly, activity of alpha amylase also increases with germination time. This suggests that the starch content should decrease but the newly formed starch is probably produced by gluconeogenesis using precursor from oil reserves (Table 2.12).

Raffinose family oligosaccharides (RFOs), which include raffinose and stachyose, are thought to be an important source of energy during seed germination. In contrast to their potential for promoting germination, RFOs represent anti-nutritional units for monogastric animals when consumed as a component of feed. The study was to compare the germination potential for soybean seeds with either wild-type (WT) or low RFO levels and to examine the role of RFO breakdown in germination of soybean seeds. There was no significant difference in germination between normal and low RFO soybean seeds when imbibed/germinated in water. Similar to the situation in pea, soybean seeds of wild-type carbohydrate composition experienced a delay in germination when treated with a chemical inhibitor of a-galactosidase activity (1-deoxygalactonojirimycinor) (DGJ) during imbibition. However, low RFO soybean seed germination was not significantly delayed or reduced when treated with DGJ. In contrast to the situation in pea, the inhibitor-induced germination delay in wild-type soybean seeds was not partially overcome by the addition of galactose or sucrose. It was concluded that RFOs are not an essential source of energy during soybean seed germination (Dierking and Bilyeu, 2009) (Table 2.13).

Table 2.12 Activities of enzymes of hexose metabolism in 2 day germinating soybean cotyledons and axes.

Enzyme	Cotyledon	Axis
NADH-Ketose reductase	ND	0.76 ± 0.14
NADPH-Ketose reductase	ND	0.78 ± 0.08
NADH-aldose reductase	ND	0.43 ± 0.06
NADPH-aldose reductase	ND	0.53 ± 0.06
Fructokinase	0.39 ± 0.09	0.79 ± 0.21
Glucokinase	0.63 ± 0.12	0.23 ± 0.09

Table 2.13 Oligosaccharide content of wild-type and low RFO dry seeds.

Line/Genotype	Sucrose (% DW)	STD	Raffinose (% DW)	STD	Stachyose (% DW)	STD
1WT	4.85	1.76	1.13	0.27	5.99	0.69
2WT	4.86	0.89	0.99	0.25	5.79	0.43
3WT	5.05	1.12	1.13	0.13	4.77	0.64
4WT	3.79	0.54	1.07	0.09	5.96	0.57
5WT	3.51	0.71	0.85	0.14	4.65	0.81
Mean	4.41	1.00	1.03	0.18	5.43	0.14
6LowRFO	6.06	1.08	0.19	0.04	2.26	0.34
7LowRFO	6.19	1.36	0.20	0.03	1.47	0.44
8LowRFO	9.35	1.03	0.18	0.03	1.78	0.24
9LowRFO	7.74	0.61	0.18	0.07	1.87	0.60
10LowRFO	7.00	1.40	0.18	0.05	1.58	0.41
Mean	7.27	1.09	0.19	0.04	1.79	0.41

DW, dryweight; STD, standard deviation; WT, wild type; RFO, raffinose family oligosaccharide.

Seed storage proteins are proteins that accumulate significantly in the developing seed and whose main function is to act as a storage reserve for nitrogen, carbon, and sulphur. These proteins are rapidly mobilized during seed germination and serve as the major source of reduced nitrogen for the growing seedlings. In general, seed storage proteins do not carry out any enzymatic functions. Seed storage proteins are generally not found in non-seed organs. These proteins accumulate within membrane-bound organelles called protein bodies (Fig. 2.5). The sequestration of storage proteins within protein bodies ensures that these proteins are separated from the metabolic compartments of the cell. The expression of

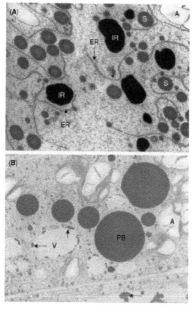

Fig. 2.5 (A) Transmission electron micrograph of rice endosperm showing spherical (S) and irregular-shaped (IR) protein bodies. The spherical protein bodies store prolamins and the irregular-shaped protein bodies accumulate glutelins. Note the direct connection between the rough endoplasmic reticulum (ER) and spherical protein bodies. (B) Protein bodies (PB) in the cotyledons of developing soybean seed. Note the occurrence of protein deposits (arrows) within the vacuoles (V). Large amyloplasts (A) are also seen.

storage proteins is also regulated by nutrition. For example, the synthesis of sulfur-rich proteins may be restricted when the plants are grown in soils having low sulphur.

Seed germination is a complex physiological process that is regulated by different external and internal factors, such as temperature (Penfield et al., 2005), light (Piskurewicz et al., 2009), soil salinity (Kim et al., 2008; Park et al., 2011), gibberellic acid (GA) (Ogawa, 2003) and abscisic acid (ABA) (Gubler et al., 2005). During germination, the environmental and hormone signals integrate together to play regulatory roles (Penfield et al., 2005; Chen et al., 2008). The environmental factors could affect the germination by regulating the biosynthesis and catabolism of phytohormones, such as GA and ABA (Finch-Savage and Leubner-Metzger, 2006). Some genes, such as embryonic identity genes LEAFY COTYLEDON1/ LEAFY COTYLEDON2/FUSCA3 (LEC1/LEC2/FUS3) and maternal gene DNAbinding with one zinc finger AFFECTING GERMINATION (DAG1/DAG2), are involved in the signalling of environmental factors or phytohormones and regulate the seed germination (Finkelstein et al., 2008) which are actively synthesized in germinating seed. Of the enzymes from soybean axes that were tested, acid phosphatase activity increased the most during germination. In the experiments with detached axes, the elevated activity of acid phosphatase was not the result of transfer of the enzyme molecules from cotyledons. *De novo* synthesis of this enzyme occurred before most of the protein synthesis in the embryonic axes, which suggested that this enzyme was particularly important in germination. One possible role of acid phosphatase in germinating embryonic axes is to cause turnover of the phosphate esters in the seed. L-Malate dehydrogenase activity also increased greatly at an early stage of germination, but the increase was not inhibited by cycloheximide or chloramphenicol. These results indicated that the increase in L-Malate dehydrogenase activity was due to the activation of the enzyme protein by imbibition, not the synthesis of the enzyme protein. Therefore, acid phosphatase is a suitable enzyme for the study of gene expression during germination of soybean seeds. L-Malate dehydrogenase was activated within 3 hr of the start of imbibition even when both cycloheximide and chloramphenicol were present. This is consistent with observations that cytochrome oxidase (EC 1.9.3.1) and L-Malate dehydrogenase levels rise. Preformed proteins, both structural and enzymic, seem to be transferred into preformed immature mitochondria following imbibition, resulting in active, efficient mitochondria with a complete membrane and respiratory system.

Changes in thiamine pyrophosphotransferase activity and thiamine pyrophosphate content were followed for six days in soybean (Menf.) seedlings. Maximum enzyme activity occurred 48 to 96 hours from imbibition. Thiamine pyrophosphate content peaked sharply at 36 hours and was preceded by increased thiamine pyrophosphotransferase activity. Addition of pyrithiamine, an inhibitor of *in vitro* thiamine pyrophosphotransferase activity, to the imbibition medium at various times inhibited subsequent fresh weight gains of soybean seedlings. These results indicated that, although not among the earliest phosphorylation events after initiation of water imbibition by soybean seeds, a substantial increase in thiamine pyrophosphate content did precede the onset of rapid seedling growth and development (Molin et al., 1980).

Proteome profiling was conducted through one-dimensional gel electrophoresis, followed by liquid chromatography and tandem mass spectrometry strategy in the germinating seeds of soybean (*Glycine max*). Comprehensive comparisons were also carried out between rice and soybean germinating seeds. 764 proteins belonging to 14 functional groups were identified and metabolism-related proteins were the largest group. Deep analyses of the proteins and pathways showed that lipids were degraded through lipoxygenase dependent pathway and proteins were degraded through both protease and 26S proteasome system, and the lipoxygenase could also help to remove the reactive oxygen species during the rapid mobilization of reserves of soybean germinating seeds (Table 2.14).

The kernel is comprised of two parts, the embryo and the endosperm. Lipase activity is investigated during seed germination where it is maximum value (Paques et al., 2006). Triacylglycerols are stored in oleosomes and comprise in range from 20 to 50% of dry. As germination proceeds, triacylglycerols are hydrolyzed to produce energy required for the synthesis of sugars, amino acids (mainly asparagine, aspartate, glutamine and glutamate) and carbon chains required for embryonic growth (Quetter and Eastmond, 2009) (Table 2.15).

Table 2.14 Sub-categorization of the metabolism related proteins in germinating seeds of soybean and rice.

Functional class	Soybean	Rice
Photosynthesis	12	13
Major carbohydrate	10	23
Glycolysis	21	22
TCA cycle	10	15
Fermentation	10	06
Gluconeogenesis/glyoxylated cycle	02	05
Mitochondrial electron transport/ATP synthesis	15	08
Cell wall metabolism	10	08
Lipid metabolism	48	16
Nitrogen metabolism	03	01
Amino acid metabolism	34	47
Secondary metabolism	14	07
Cofactor and vitamin	03	02
Nucleotide	13	15
C1 metabolism	06	03
Minor carbohydrates	06	06
Phosphate pentose pathway	00	06
OPP cycle	01	00
S-assimilation	00	01
Tetrapyrrole synthesis	01	02
Poly-amine synthesis	00	02

Table 2.15 Lipase activity (meq ffa $min^{-1}g^{-1}$) during germination of soybean.

Soybean	1st day	6th day	12th day
JS-335	2.85	4.49	2.12
DS-228	2.91	6.04	2.58
MAUS-71	2.03	3.63	1.91
MAUS-81	2.64	3.14	1.78
Mean	2.61	4.33	2.10
CD at 5%	0.17	0.20	0.21

Fifty nine lipids metabolic proteins were detected, much more than what was detected in rice seeds. Most of the enzymes were lipoxygenase (LOX), which indicated that the oils in soybean seeds might be degraded through a LOX-dependent pathway. LOXs are non-heme iron-containing dioxygenases which catalyze the oxidation of polyunsaturated fatty acids by adding molecular oxygen at C9 and C13 of the acyl chain in linolenic or linoleic acid (Brash, 1999). It was reported that there were four major branches of LOX pathway: (a) The peroxygenase (POX) or hydroperoxide isomerase pathway, (b) The hydroperoxide dehydratase (AOS) pathway, (c) The hydroperoxide isomerase (HPL) pathway, (d) The divinyl ether (DES) pathway. The detection of some P450 monooxygenases implied that the AOS branch might be the major pathway for lipids degradation during soybean germination.

Storage lipid mobilization in germinating seeds begins with hydrolysis of triacylglycerols in oleosomes by lipases into free fatty acids and glycerol. Fatty acids then undergo β-oxidation in peroxisomes. Next, glyoxylate cycle will proceed partially in the peroxisome and partially in the cytoplasm. Three of the five enzymes of the glyoxylate cycle (citrate synthase, isocitrate lyase and malate synthase) are located in peroxisomes, while two other enzymes (aconitase and malate dehydrogenase) operate in the cytoplasm (Pracharoenwattana and Smith, 2008). Succinate is transported from peroxisome to mitochondria and there is converted to malate via the Krebs cycle. Malate, in turn, after transport to the cytoplasm, is converted to oxaloacetate. Finally, gluconeogenesis and the synthesis of sugars proceed, these are the processes which are a form of carbon transport especially in germinating seeds (Quettler and Eastmond, 2009; Borek and Ratajczak, 2010).

The greatest storage form of total phosphorus (about 50–80%) is phytic acid ($C_6H_{18}O_{24}P_6$), also known as inositol hexophosphate (IP6) in legumes and cereals seeds (Jacela et al., 2010). Phytic acid is regarded as antinutrient because it has the ability to form complexes with proteins and bind with cations (especially Fe, Ca, K, Mn, Mg, Zn) via ionic association to form a mixed salt called phytin or phytate with the reduction of their digestive availability (Lott et al., 1995). On the other hand, phytate may play an important role as an antioxidant by forming iron complex that cause a decrease in free radical generation and the peroxidation of membranes, and may also act as an anticarcinogen, providing protection against colon cancer (Thompson and Zhang, 1991). Since it is regarded as antioxidant, anticarcinogen or vitamin-like substance, it is essential to measure and manipulate phytate content in food grains, such as beans (Okazaki and Katayama, 2005).

Phytin in germinating seeds is hydrolyzed by an acid phosphatase enzyme called phytase (Hubel and Beck, 1996), with releasing of phosphate, cations, and inositol which are utilized by the seedlings. Little changes were found in extractible P_i in hazel seeds during chilling accompanied with IP6 mobilization, which might suggest the rapid conversion of P_i into organic form (Mukherjee et al., 1971). These results were discussed as evidence of active metabolism in germinating seed (Silva and Trugo, 1996). In agreement, phytase is strongly and competitively inhibited by P_i, while the decrease in phytase activity coincided with maximal IP6 turnover (Andriotis and Ross, 2003). It was found that about 87% of IP6 is digested during the first 6 days of germination (Azarkovich et al., 1999). In this respect, Ogawa et al. (1979) postulated that the early axiferous IP6 digestion is essential for metabolic activity of the resting tissue via supplying P_i and minerals for physiological and metabolic requirements, for example, enzymes of starch metabolism. In addition, IP6 related compounds, such as pyrophosphate-containing inositol phosphates (PP-IP), play a potential role in providing P_i for ATP synthesis during the early stages of germination before complete dependence on aerobic mitochondrial respiration, the main source of ATP production (Raboy, 2003).

During seed maturation in higher plants, phytic acid (*myo*-inositol-6-phosphate) is synthesized by the condensation of *myo*-inositol-1-phosphate synthesized from glucose 6-phosphate and five phosphate molecules provided by adenosine triphosphate (ATP) hydrolysis (Loewus and Louewus, 1983). Phytic acid then accumulates in the protein body. Phytic acid is found in cereal grains, legume seeds, and oilseeds and accounts for 1%–8% of the dry weights of these materials (Lott et al., 2000). About 90% of phosphorus in seeds is bound to phytic acid, and this source of phosphate is essential for nucleotide and phospholipid synthesis as immature roots of younger individuals are not able to take up sufficient inorganic phosphate from the soil. Moreover, phytic acid strongly chelates various metal ions and is, therefore, involved in the preservation and detoxification of metals. During germination, phytic acid is broken down by phytase, releasing phosphorus, metal ions, and *myo*-inositol for use by the growing individual (Lazali et al., 2014) (Table 2.16).

Seed germination is an important process in plant development which is complicated by several factors. Recently, as one of such factors, the relationship between seed germination and reactive oxygen species (ROS) in species such as *Arabidopsis thaliana* (Liu et al., 2010; Leymarie et al., 2012), sunflower (Oracz et al., 2007), wheat (Ishibashi et al., 2008), cress (Müller et al., 2009a) and barley (Ishibashi et al., 2010a; Bahin et al., 2011) has been reported.

Table 2.16 Phytic acid content and ratio of phytic acid to protein in soybean seed germination.

Day of germination	Phytic acid (% dw)	Phytic acid (% fw)	Phytic acid/protein
0	1.70 ± 0.200	0.731 ± 0.0557	0.0314 ± 0.00453
2	1.85 ± 0.121	0.643 ± 0.0246	0.310 ± 0.00300
4	0.499 ± 0.310	0.119 ± 0.0722	0.00973 ± 0.00616

In general, ROS, such as O_2, hydrogen peroxide (H_2O_2) and OH^-, cause oxidative damage to lipids, proteins and nucleic acids. Indeed, seed deterioration is due, in part, to peroxidation of membrane lipids by ROS and the resulting leakiness of the membranes (Sung and Jeng, 1994; Bailly et al., 1998). Seed longevity is enhanced through elimination of ROS by over accumulated ROS scavengers in transgenic seeds (Lee et al., 2010; Zhou et al., 2012). However, they also play various important roles in cellular signalling in plants, notably acting as regulators of growth and development, programmed cell death, hormone signalling, and responses to biotic and abiotic stresses (Mittler et al., 2004). In seed physiology, several studies have reported that exogenous H_2O_2 promotes seed germination in many plants (Chien and Lin, 1994; Fontaine et al., 1994). On this basis, ROS produced after imbibitions appear to regulate seed germination. Indeed, in barley seeds, NADPH oxidase, which is one of the major sources of ROS, acts as a key enzyme in germination and subsequent seedling growth (Ishibashi et al., 2010a).

Plant hormones, which are one of such factors, are important in the regulation of seed dormancy and germination (Koornneef et al., 2002; Finkelstein, 2004). The interactions among abscisic acid (ABA), gibberellins, ethylene, brassinosteroids, auxins and cytokinins in regulating the interconnected molecular processes that control dormancy release and germination have been reported (Kucera et al., 2005). There are many reports on the interaction of ROS with plant hormones in plant.

In guard cells, ROS are considered second messengers in the ABA transduction pathway (Wang and Song, 2008), and exogenous ABA leads to an increase in H_2O_2 in guard cells regulating ion channels leading to stomatal closure (Schroeder et al., 2001). In addition, ethylene receptor ETR1 plays an important role in guard cell ROS signalling and stomatal closure (Desikan et al., 2005). In seed physiology, exogenous H_2O_2 increased ABA catabolism by enhancing the expression of CYCP707A genes, played a major role in ABA catabolism and enhanced gibberellic acid (GA) biosynthesis genes in *Arabidopsis* dormant seeds (Liu et al., 2010). ROS regulated the expression of ethylene response factor ERF1, a component of the ethylene signalling pathway in sunflower seed germination (Oracz et al., 2009). In barley dormant seed, H_2O_2 enhanced GA synthesis genes, such as $GA_{20}ox1$, rather than repression of ABA signalling in embryo (Bahin et al., 2011). Recently, it was also shown that ROS regulate the induction of α-amylase through gibberellin–ABA signalling in barley aleurone cells (Ishibashi et al., 2012).

In soybean seeds, ROS are produced in the embryonic axis during germination (Puntarulo et al., 1988; Puntarulo et al., 1991), and the production and scavenging of ROS during ageing were related to vigour and cell death, respectively, during accelerated ageing of the embryonic axis (Tian et al., 2008). In addition, low temperatures led to oxidative stress and lipid peroxidation caused by ROS in the embryonic axis (Posmyk et al., 2001). Although the negative role of ROS in soybean seed is now well documented, there have been fewer studies of a positive role of ROS in soybean seed (Table 2.17).

H_2O_2 promoted germination, which N-acetylcysteine suppressed, suggesting that ROS are involved in the regulation of soybean germination. H_2O_2 was produced in the embryonic axis after imbibition. N-Acetylcysteine suppressed the expression of genes related to ethylene biosynthesis and the production of endogenous ethylene. Interestingly, ethephon, which is converted to ethylene, and H_2O_2 reversed the suppression of seed germination by N-acetylcysteine. Furthermore, morphological analysis revealed that N-acetylcysteine suppressed cell-induced cell hypertrophy, but not hyperplasia. Results suggest that the ethylene produced in response to ROS regulates the length of the embryonic axis by increasing the size of root tip cells without increasing their number, and thereby regulated soybean seed germination.

Table 2.17 Changes in peroxidase activity (units $min^{-1}g^{-1}$) during soybean germination.

Soybean	1st day	6th day	12th day
JS-335	9.70	36.50	15.09
DS-228	16.77	41.82	19.30
MAUS-71	7.62	32.44	12.50
MAUS-81	12.50	29.57	9.64
Mean	11.65	35.08	14.13
CD at 5%	2.87	4.46	1.17

2.8 Viability in Storage

The problems of maintaining the soybean seed viability in storage have always been an important concern, and retention of high viability over a long period is necessary for crop production. Many factors determine the longevity of seeds during storage. These includes seed moisture content, temperature, relative humidity, initial viability, stage of maturity at harvest, storage gas and initial moisture content of seed entering into storage (Tatipata, 2009). Soybean seed loses its viability in a very short period of storage, even when stored in good nonporous container (Woodruff, 1998). The seeds with low moisture content and stored in any air tight containers could retain viability for a longer period of time.

Sealed plastic pot and polythene bag are more effective storage containers than cloth bag or jute bag and earthen pot, etc. (Rahman et al., 2010). Soybean seed is rapidly deteriorated by high temperature and high relative humidity during storage (McDonald, 1999). Several factors may affect the quality of seeds in storage, however, the most critical among them is high seed moisture content (O'Hare et al., 2001). High moisture content and presence of oxygen were the main causes for lipids autoxidation in soybean, leading to rapid seed deterioration and quality decline (Chang, 2004).

From the result of a study, it was concluded that soybean seed could be stored with above 80% germination for at least six months under a range of relative humidities (50 to 60%) if stored in polythene bag after drying to 8% initial seed moisture content (Ali et al., 2014) (Table 2.18).

Twenty six field emergence experiments were used to investigate the relationship between soybean (*Glycine max* (L.) Merrill) seed germination, vigor and field emergence. Each year for 10 years, standard

Table 2.18 Interaction effect of storage relative humidity, storage container and initial seed moisture content (SMC) on germination of soybean seed at different days after storage and field emergence in 2008 and 2009.

Storage RHx container x Initial SMC	Germination (%) 2008			Germination (%) 2009		
	60DAS	120DAS	180DAS	60DAS	120DAS	180DAS
H1C1M1	85.33	76.67	73.33	89.33	82.67	76.67
H1C1M2	72.00	62.00	34.00	72.67	55.33	41.33
H1C2M1	93.33	91.33	89.33	96.00	94.00	92.67
H1C2M2	84.00	75.33	48.00	90.00	73.33	57.33
H2C1M1	81.33	74.67	68.00	92.67	83.33	73.33
H2C1M2	70.00	54.67	26.67	63.33	43.33	30.67
H2C2M1	90.67	87.33	85.33	97.33	94.67	91.67
H2C2M2	76.67	62.00	44.00	77.33	60.67	46.00
H3C1M1	70.00	62.00	52.67	68.67	30.67	0.00
H3C1M2	56.00	26.00	0.00	66.67	32.00	0.00
H3C2M1	81.33	74.00	66.00	92.67	74.67	65.33
H3C2M2	69.33	48.00	0.00	66.00	39.33	4.00
H4C1M1	66.00	52.00	36.00	54.67	19.33	0.00
H4C1M2	48.67	20.00	0.00	46.00	10.00	0.00
H4C2M1	78.67	66.67	58.00	94.67	73.33	61.33
H4C2M2	62.67	38.00	0.00	62.00	23.33	2.67
F test	ns	**	**	**	**	**
CV(%)	5.67	7.84	4.04	6.46	6.41	8.41

DAS = Days after storage; CV = Coefficient of variation; ns = Non-significant,
** = Significant at 1% level.
Note: H1 = 50% relative humidity (RH), H2 = 60% RH, H3 = 70% RH, H4 = 80% RH,
C1 = Cloth bag, C2 = Polythene bag, M1 = 8% Seed moisture content (SMC), M2 = 12% SMC.

germination (SG), accelerated aging (AA) germination and cold test (CT) germination were measured on 12 to 52 seed lots of 5 to 21 cultivars and related to field emergence in two to four field plantings. A total of 272 seed lots were included, of which 146 were carryover seed. Standard germination was generally above 80% but AA and CT varied from > 80% to near zero. The field emergence index (FEI), calculated for each field emergence experiment to characterize seedbed conditions, varied from 108 to 44. The prediction accuracy varied from 0 to 100%. As the FEI decreased, prediction accuracy of all tests decreased. The prediction accuracy for SG with a critical level of 80% was near 100% only when the FEI index approached 100. The prediction accuracy for AA at critical levels of 80 or 90% remained near 100% until the FEI approached 80. Increasing the critical level for SG to 95% improved the predictive ability to nearly equal AA at a critical level of 80%. No test accurately predicted performance when the FEI < 80 is not desirable.

One of the major constraints in soybean cultivation is the non-availability of high vigor seeds at the time of sowing. Soybean seeds undergo rapid loss of vigour and viability during storage, which is more pronounced under sub-tropical conditions. Deterioration of seed during storage is manifested as a reduction in percent germination, while those seeds that do germinate produce weak seedlings, which ultimately affect the growth and yield of crop plant (Tekrony et al., 1993). Contents of soluble carbohydrates generally decline with aging of seed (Petruzelli and Taranto, 1989; Sharma et al., 2005) and this decline might result in limited availability of respiratory substrates for germination. Depletion of disaccharides may lessen the protective effects of sugars on structural integrity of membrane (Crowe et al., 1984). The lipid related changes of seeds during storage revealed a decline in phospholipids and polyunsaturated fatty acids leading to a marked decline in seed vigour (Priestley and Leopold, 1983). An increase in necrosis in cotyledons and substantial reduction in total germination was reported in soybean seeds stored at high temperature by Falivene et al. (1980) (Fig. 2.6).

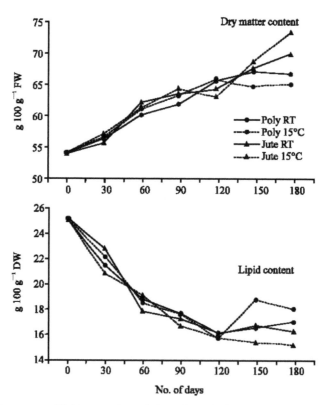

Fig. 2.6 Changes in dry matter and lipid content in cotyledons of germinating soybean seeds during storage. Each point represents the mean of three replications. CD (p > 0.05) for DOSx PMxT is 0.19 for lipid content (DOS: days of storage; PM: packing material; T: temperature).

The phospholipid content increased up to 90 DOS followed by a decrease with further increase in storage period up to 180 DOS, irrespective of the packing and temperature conditions (Table 2.19). The glycolipid content varied very little in germinating seeds stored in polythene or jute bags with the storage period, however, the decrease in glycolipid content was more at 15°C in both the packings as compared to RT. On germination, the free fatty acid content in cotyledons of germinating soybean seeds increased with the increase in storage period from 30 to 180 days in all the treatments. The sterol content increased up to 90 DOS and then decreased with the increase in storage period, irrespective of the packings and temperature conditions. The increase in sterol content during the initial 90 days of storage was notably more in the seeds stored at RT than those stored at 15°C in polythene bags.

Table 2.19 Variation of different lipids of soybean under storage.

	Polythene bags		Jute bags	
Days of storage (DOS)	**Room temperature**	**15°C**	**Room temperature**	**15°C**
Phospholipids (g 100 g^{-1} oil)				
30	0.9	0.8	0.8	0.9
60	1.1	1.1	1.2	1.1
90	1.3	1.3	1.3	1.3
120	1.1	0.9	0.9	1.0
150	0.8	0.8	0.6	0.6
180	0.6	0.6	0.6	0.5
Glycolipid content (g 100 g^{-1} oil)				
30	1.5	1.4	1.2	1.3
60	1.3	1.2	1.4	1.1
90	1.2	1.0	1.1	0.9
120	1.5	1.5	1.7	1.2
150	1.4	1.3	1.3	1.1
180	1.2	0.9	1.1	0.8
Free fatty acids (g 100 g^{-1} oil)				
30	1.1	1.4	1.2	1.4
60	1.0	1.1	1.0	1.1
90	1.4	1.4	1.3	1.3
120	1.8	1.8	1.7	1.7
150	2.2	1.9	1.9	2.0
180	2.1	1.9	1.9	2.6
Sterol (g 100 g^{-1} oil)				
30	8.2	7.5	7.6	7.8
60	8.8	8.2	8.8	7.9
90	9.7	9.3	9.6	9.5
120	9.1	8.9	9.1	9.0
150	7.5	6.8	6.5	6.7
180	6.4	5.9	5.7	5.3
CD (p < 0.05)	**Phospholipids**	**Glycolipids**	**Free fatty acids**	**Sterols**
DOS X PM	0.05	0.08	0.09	0.19
DOS X T	NS	0.08	0.09	0.19
PM X T	0.03	0.05	0.05	0.11
DOS X PM X T	0.07	0.12	0.13	0.28

In the studies, the data on lipid composition has revealed that lipid degradation marginally decreased triglyceride content during storage and increased free fatty acids, sterol and phospholipid content, suggesting that the lipase present in the seeds remains active and alters membrane integrity (Bernal Lugo and Leopold, 1992) (Table 2.19).

2.9 Seed Deterioration

Free radical-mediated phospholipid degradation is commonly thought to involve peroxidation of unsaturated fatty acids (Wilson and McDonald, 1986). The process can be initiated by the abstraction of hydrogen by a hydroxyl radical from a methylene group, forming a conjugated diene. The superoxide radical is too reactive and too polar to abstract hydrogen from lipids (Smirnoff, 1993). The conjugated diene reacts readily with oxygen to form a peroxyl radical (ROD"), which can abstract hydrogen from another unsaturated fatty acid, thus initiating a chain reaction of lipid peroxidation (Wilson and McDonald, 1986). These reactions result in the formation of lipid peroxides (ROOH), oxygenated fatty acids, and more free radicals. Hydroperoxides can fragment to form ethane and aldehydes (Smirnoff, 1993), including malondialdehyde. Aldehydes, unlike free radicals, are more stable and provide a mechanism for long distance detrimental effects (Esterbauer, 1982). They produce a variety of cytotoxic effects, among which is their reaction with sulfhydryl groups, leading to inactivation of proteins (Benedetti et al., 1980).

Metaions can influence peroxidation by reacting with lipid peroxides to form alkoxy (RO) and peroxy radicals, both of which can further the process of lipid peroxidation (Halliwell and Gutteridge, 1989).

Sites of attack would increase via the free radical chain reactions, damaging membranes in the process, thus leading to membrane leakage (Simon, 1974; Parish and Leopold, 1978).

In the presence of oxygen, aging of seed can lead to peroxidative changes in the polyunsaturated fatty acids (Stewart and Bewley, 1980; Wilson and McDonald, 1986) and this free radical-induced, non-enzymatic peroxidation has the potential to damage membranes, thereby causing the deterioration of the stored seeds (Sung and Jeng, 1994).

Alternatively, lipoxygenase (LOX) mediated pathway operating in many un-imbibed seeds, in which LOX is capable of catalyzing lipid peroxidation by acting on membrane phospholipid components, such as linolenic and linoleic acid, as its substrates (Priestley, 1986; Wang et al., 1990) also leads to the formation of hydroperoxides by addition of molecular oxygen to these fatty acids having cis, cis, 1,4 pentadiene motif (Feussner and Wasterneck, 2002) and these hydroperoxides produced act as a substrate for hydroperoxidelyase (HPL)—the second most important enzyme in LOX pathway, catalyzing the formation of aldehydes such as 3-2-nonenal and 3-2,6-2-nonadienal (Fauconnier et al., 1997) and ώ-oxo acids [9-oxo-nonanoic acid]. These medium chain aldehydes have been reported to be mainly responsible for the poor storability as well as poor seed germination potential of soybean (Gardner et al., 1990). Normal soybean seeds contain 3 lipooxygenase isozymes, LOX-1, LOX-2 and LOX-3, which differ in substrate specificity, optimum pH for catalytic activity, iso-electric point and thermal stability (Gardner et al., 1990; Matsui, 2006).

In soybean seeds, LOXs are abundant proteins that constitute 1–2% of the total protein content (Loiseau et al., 2001). The enzymes are involved in the production of volatile compounds (such as n-hexanal) associated with grassy-beany and rancid off-flavors in soybean and soy foods (Robinson et al., 1995). Foods made from soybean lacking LOXs generate less hexanal than those with normal LOX activities (King et al., 2001).

On the basis of reported storability and seed germination, thirteen soybean [*Glycine max* (L.) Merill] genotypes (grouped as six "good storers" and seven "poor storers") were selected in order to understand the relationship between lipid peroxidation, antioxidant activity and seed storability. Good storers possessed significantly high activity ($p < 0.05$) of LOX-1 and lower activity of LOX-2 as compared to poor storers.

Significant increase ($p < 0.05$) in HPL activity was observed in all poor storer genotypes and correlated with higher accumulation of lipid peroxides, total MDA and carbonyl content. Further results indicated that the good storers possessed high antioxidant activities, when analyzed through DPPH and CUPRAC method, than the poor storers; however, the activity of antioxidants enzymes, viz., SOD and catalase, remained unchanged in both the good and poor storers, with the exception of a good storer genotype-M1090.

Although no significant difference in the antioxidant compounds, like tocopherols and isoflavones, were found between the good and poor storers, an increased ascorbic acid content was observed in the good storers (8.5 mg/100 g to 14.74 mg/100 g) as compared to the poor storers (4.82 mg/100 g to 6.84 mg/100 g).

This study reflected the possible role of ascorbic acid, LOX-1, LOX-2 and HPL enzymes as potential indicators to determine the storability of soybean seeds and their potential to be used as the parameters to improve the nutritional quality of the soybean seeds.

2.10 Seed Priming

Under various conditions, the potential of seeds for rapid uniform emergence and development under various conditions is determined mostly by seed vigor trait (Paparella et al., 2015). Recent strategies for improvement of seed quality involved classical genetic, molecular biology and invigoration treatments, known as priming treatments. Seed priming was aimed primarily to control seed hydration by lowering external water potential, or shortening the hydration period, because most seeds are partially hydrated after priming process and reach a pre-germinate stage without radicle protrusion (Hilhorst et al., 2010). It was reported that primed seeds showed improved germination rate and uniformity under both optimal and adverse environments in wheat (Zhuo et al., 2009). The cellular mechanism of priming as it relates to improved stress tolerance in germinating seeds requires further study.

Currently, seed priming techniques include osmopriming (soaking seeds in osmotic solutions as PEG or in salt solutions), hydropriming (soaking seeds in predetermined amounts of distilled water or limiting imbibition periods), and hormone priming (seed are treated with plant growth regulators), which are more commonly studied in laboratory conditions, and thermopriming (a physical treatment achieved by pre-sowing of seeds at different temperatures that improve germination vigor under adverse environmental conditions) and matric priming (mixing seeds with organic or inorganic solid materials and water in definite proportions and, in some cases, adding chemical or biological agents) (Paparella et al., 2015; Jisha et al., 2013). Hydropriming and osmopriming with large-sized priming molecules cannot permeate cell wall/membrane, so water influx would be the only external factor affecting priming. The determination of suitable priming technique is dependent mainly on plant species, seed morphology and physiology. On the other hand, salts and hormone priming affect not only the seed hydration but also other germination-related processes due to absorption of exogenous ions/hormones, consequently confusing the effects of imbibition *versus* that of ions/hormones.

It is a technique for controlling seed slow absorption and post dehydration (Heydecker and Coolbear, 1977). According to, Khan (1992), osmotic conditioning, in its modern sense, aims to reduce the time of seedling emergence, as well as synchronize and improve the germination percentage by subjecting the seeds to a certain period of imbibitions using osmotic solutions. The beneficial effects of priming have also been demonstrated for many field crops, such as wheat, sugar beet, maize, soybean and sunflower (Khajeh-Hosseini et al., 2003; Sadeghian and Yavari, 2004). Park et al. (1997) reported that priming of aged seeds of soybean resulted in good germination and stand establishment in the field trials.

Seeds of soybean Cv. DS 2706 were primed with distilled water, KNO_3 (1%), NaCl (0.5%) and PEG 6000 (5%) for 8, 12, 24 and 48 hrs at 25°C. The experiment was carried out in complete randomized design, replicated four times with 17 priming treatments. Results of variance clearly showed the osmopriming and hydropriming had significant effect on seed germination and emergence. Mean comparison showed that the highest germination %, germination Index and vigor Index were achieved by priming with PEG6000 (5%) for 12 hours. Maximum seedling length and dry seedling weight was obtained by hydropriming for 24 hours and least mean germination time and highest energy of emergence was achieved by osmopriming by NaCl for 12 hours. Hydropriming and osmopriming both showed better characteristics than control. It can be surmised that osmopriming with PEG6000 (5%) for 12 hours is more suitable for the better germination in soybean. Results of investigation in this research showed that the osmopriming of soybean seeds can improve the efficiency of seeds with low vigor and cause quick homogenous establishment of seedling intact (Kujur and Lal, 2015) (Table 2.20).

Table 2.20 Some hydropriming and osmopriming effects of soybean seed germination.

Treatment	Germination (%)	Germination Index (%)	Mean germination time (%)	Energy of emergence (%)
Control (DW)	65	23	4	42
8 hrs soaking in DW	67	26	4	46
12 hrs soaking in DW	69	26	4	45
24 hrs soaking in DW	71	27	4	49
48 hrs soaking in DW	69	27	4	47
1% KNO_3 for 8 hrs	67	26	4	48
1% KNO_3 for 12 hrs	67	26	3	47
1% KNO_3 for 24 hrs	71	27	4	48
1% KNO_3 for 48 hrs	68	26	4	46
0.5% NaCl for 8 hrs	67	26	3	46
0.5% NaCl for 12 hrs	66	27	3	50
0.5% NaCl for 24 hrs	70	27	4	50
0.5% NaCl for 48 hrs	68	27	3	48
5% PEG_{6000} for 8 hrs	67	27	3	48
5% PEG_{6000} for 12 hrs	72	28	4	49
5% PEG_{6000} for 24 hrs	69	27	3	49
5% PEG_{6000} for 48 hrs	68	27	3	49
Mean	68	26	4	47
CD	1.57	1	0.16	3

Treatment	Seedling length (cm)	Dry seedling weight (mg)	Vigor index I	Seed vigor II
Control	26	109	1698	7101
8 hrs soaking in DW	27	123	1845	8288
12 hrs soaking in DW	28	128	1965	8866
24 hrs soaking in DW	31	132	2229	9338
48 hrs soaking in DW	29	128	1989	8847
1% KNO_3 for 8 hrs	29	120	1923	8040
1% KNO_3 for 12 hrs	29	123	1928	8228
1% KNO_3 for 24 hrs	29	123	2093	8746
1% KNO_3 for 48 hrs	29	124	1988	8480
0.5% NaCl for 8 hrs	29	125	1918	8360
0.5% NaCl for 12 hrs	29	125	1919	8312
0.5% NaCl for 24 hrs	29	126	2064	8801
0.5% NaCl for 48 hrs	28	123	1888	8381
5% PEG_{6000} for 8 hrs	29	127	1971	8509
5% PEG_{6000} for12 hrs	31	130	2259	9378
5% PEG_{6000} for 24 hrs	30	129	2102	8951
5% PEG_{6000} for 48 hrs	29	125	1982	8500
Mean	29	125	1986	8537
CD	0.26	1	52	224

According to Matthews (1980), slow asynchronous and unreliable germination and emergence, within germinable, low vigour seeds, arise due to seed ageing. Seed invigoration treatments have, therefore, been developed in order to improve the seed performance during germination and emergence. Most of these involve a period of controlled hydration of the seed to a point close to, but before, the emergence of the radicle, after which the seeds are dried back to their initial moisture content before sowing (Basu, 1994; Khan, 1992; Matthews and Powell, 1988). Such treatments include priming, in which hydration is controlled in an osmoticum, such as polyethylene glycol (PEG) or a salt solution (Heydecker and Coolbear, 1977), solid matrix priming, in which seeds imbibe in an inert medium held at a known matrix potential (Taylor et al., 1998), humidification, where the seeds are hydrated at a high relative humidity (Van Pijlen et al., 1996), and aerated hydration, in which the seeds imbibe in aerated water for a specified time (Thornton and Powell, 1992). These treatments, and others, have improved the rate, uniformity and reliability of germination and emergence in a range of crop species (Basu, 1994; Khan, 1992; Matthews and Powell, 1988).

In seed priming regime, seed water potential is at a level sufficient enough to initiate metabolic events in phase-II of germination process but which prevents radicle emergence (Simon, 1984). Germination response to priming are obtained approximately at seed moisture content of 30% and it increased linearly over the range of 45 to 50%, the upper limit depending on the species.

Basuchaudhuri (1990) reported that soybean seeds soaked in water recorded higher germination and indicated rapid translocation of nutrients after hydrolysis of the cotyledonary reserves to growing seedling.

Mewael et al. (2010) concluded that soybean seeds primed with $CaCl_2$ (0.5%), GA_3 (20 ppm) and KH_2PO_4 (50 ppm) were found to enhance seed quality.

While studying the effect of osmo-conditioning of seeds, such as soybean, peas and sweetcorn, with polyethylene glycol, Anwar Khan et al. (1978) obtained favourable results in establishing a uniform seedling stand, particularly at suboptimal temperature. PEG priming reduces the time of imbibition required for the onset of RNA and protein synthesis and polyribosome formation and increases the total amount of RNA and protein. The activity of enzymes, like acid phosphatase and esterase, increases by osmo-conditioning.

Osmo-conditioning leads to the complete disappearance of abscsisic acid. The mobilization of storage materials, such as sugars, fats and proteins, by activation or *de novo* synthesis of key enzymes may underlie the mechanism of osmo-conditioning.

The study with osmotic priming of soybean seeds with polyethylene glycol (PEG-6000) at -5 bars water potential increased the seed germination to the magnitude of 10, 8, 11 and 12.0%, with concomitant increase in the seed yield in all the four types of soils (Narasimha Prasad, 1994).

Basuchaudhuri (1990) reported that germination of soybean seeds soaked in KH_2PO_4 solution recorded higher germination percentage (100%). The value for proximity index showed that cotyledonary utilization was the highest with potassium dihydrogen phosphate.

Although priming has been found to improve both the rate and uniformity of germination and emergence in many species, little is known about the biochemical and molecular mechanisms. Some reported evidence on molecular studies indicated that the increased rate of metabolic processes is involved in germination when primed seeds are rehydrated. The beneficial effect of priming has been related to the physiological changes occurring in the partially hydrated embryos and on subsequent germination.

The improved performance of seedling after priming has been explained by the completion of DNA repair mechanisms during the priming period, qualitative and quantitative increase in protein content and rapid declining of reserve materials like phytate and micronutrients during germination of primed seeds (Coolbear and Grierson, 1979; Bray et al., 1989; Job et al., 1997). On-farm seed priming involves soaking the seed in water, surface drying and sowing on the same day. The rationale is that sowing of soaked seed decreases the time needed for germination and allows the seedling to escape deterioration caused by soil physical conditions. According to Khan (1992), osmotic conditioning, in its modern sense, aims to reduce the time of seedling emergence, as well as synchronize and improve the germination percentage, by subjecting the seeds to a certain period of imbibition using osmotic solutions. The seeds normally begin to take up water on contact with this solution and stop the process as soon as they become balanced with the water potential of the solution.

2.11 Effect of Seed Priming on Seed Biochemical and Molecular Content of Soybean

Seeds primed with GA_3 (50 ppm) showed significantly highest germination percentage, root length, shoot length, seedling vigour index and seedling dry weight, followed by KH_2PO_4 (0.5%). GA_3 (50 ppm) primed seeds showed significantly higher protein content and sugar content during germination process.

Enzymatic activity, such as amylase and invertase activity, were significantly higher in GA_3 (50 ppm) seeds, followed by KH_2PO_4 (0.5%) priming.

DNA content replicated at faster rate and found significantly higher in GA_3 (50 ppm) primed seeds, followed by KH_2PO_4, and the lowest DNA content was found in unprimed seeds.

The effect of priming in enhancing the seed quality was more prominent in low vigour seed lots.

Positive and significant correlation was observed between seed quality parameters and biochemical and molecular content of primed seeds.

2.12 Effect of Seed Priming on Storability of Soybean

- Seed quality parameters exhibited significantly decreasing trend throughout twelve months period of storage irrespective of seed priming treatments and seed quality. The moisture content of the seed was found to vary in concomitant with the fluctuation of relative humidity of ambient environment.

- Germination percentage, root length, shoot length, seedling dry weight and seedling vigour index showed significant difference due to seed quality levels. At the end of twelve months storage, higher seed quality parameters were recorded in the seed lot of higher quality over low quality seed lot.

- The rate of deterioration in primed low vigour seed lots was less as compared to primed high vigour seed lots.

- Seeds primed with GA_3 (50 ppm) showed significantly higher germination percentage soon after the priming was imposed, followed by KH_2PO_4 (0.5%) primed seed.

- GA_3 (50 ppm) primed seeds showed significantly higher results in biochemical parameters, such as total protein content and sugar content, up to eight months of storage. However, unprimed seeds recorded higher biochemical content at the end of storage period.

- GA_3 (50 ppm) primed seeds showed higher enzymatic activities, such as amylase activity and invertase activity, up to eight months of storage. However, unprimed seeds recorded higher enzymatic activity at the end of storage period.

- GA_3 (50 ppm) primed seeds significantly favoured germination, root length, shoot length, seedling dry weight, vigour index, and low electrical conductivity of seed leachate at the initial months of storage period, followed by KH_2PO_4 (0.5%) primed seeds.

- Higher germination percentage was recorded in hydro primed seeds over the PEG-6000 primed ones for first six months of storage, but these were later overtaken by PEG-6000 primed seeds after six months of storage.

- The higher quality seeds primed with GA_3 (50 ppm) were significantly privileged to have a better germination, root length, seedling dry weight, vigour index and low electrical conductivity at initial months of the storage period, followed by KH_2PO_4 (0.5%) primed seeds.

Significantly low performance was observed in low quality seeds which were hydroprimed.

From the study, it can be concluded that seed priming helps in enhancing the seed quality and storability of soybean seeds to some extent of storage period. Among the seed priming treatments, priming with GA_3 (50 ppm) and KH_2PO_4 were found to be good for enhancing seed quality parameters.

The priming of soybean seeds also helped in the increase of protein, sugar content, and total DNA content of seed. Priming also helped in enhancement of enzymatic activities during the germination process.

The starting values of germination ranged between 48% and 89%. Seeds were surface sterilized with 3% sodium hypochlorite and immersed in different primers: 1% potassium nitrate, 1% potassium chloride and 1% hydrogen peroxide. Untreated seeds were used as the control. The obtained results revealed that the effects of priming depended on soybean line and treatment, whereas the efficiency of this pre-sowing treatment was not affected by the starting value of seed germination. Some lines responded favourably to immersion, while in others priming had an inhibitory effect, causing a significant decrease in germination. There was an increase in germination up to 12% or a decrease up to 11%, depending on line and treatment. Lines that were positively affected by this method also exhibited increased values for other germination parameters: Mean germination time (MGT) and time to 50% germination (T50). Lower values of MGT and T50 were observed in lines which showed a negative response to priming and a decrease in germination, but also a more rapid radicle protrusion, as compared to the control (Miladinov et al., 2018).

References

Adams, C.A., R.W. Rinne and M.C. Fjerstad. 1980. Starch deposition and carbohydrase activities in developing and germinating soybean seeds. Ann. Bot. 45: 577–582.

Ali, M.R., M.M. Rahman and K.U. Ahammad. 2014. Effect of relative humidity, initial seed moisture content and storage container on soybean (*Glycine max* L. Meril) seed quality. Bangladesh J. Agril. Res. 39: 461–469.

Anwar Khan, A., Kar-Ling Tao, J.S. Knypl, B. Borkowska and E.P. Loyd. 1978. Osmotic conditioning of seeds: Physiological and biochemical changes. Acta Hort. 83: 267–279.

Andriotis, V.M.E. and J.D. Ross. 2003. Isolation and characterization of phytase from dormant *Corylus avellana* seeds. Phytochemistry 64: 689–699.

Azarkovich, M.I., M.I. Dmitrieva and A.M. Sobolev. 1999. Mobilization of protein and phytin in aleurone grains of germinating castor beans. Russian Journal of Plant Physiology 46: 349–356.

Bahin, E., C. Bailly, B. Sotta, I. Kranner, F. Corbineau and J. Leymarie. 2011. Crosstalk between reactive oxygen species and hormonal signaling pathways regulates grain dormancy in barley. Plant Cell and Environment 34: 980–993.

Bailly, C., A. Benamar, F. Corbineau and D. Come. 1998. Free radical scavenging as affected by accelerated ageing and subsequent priming in sunflower seeds. Physiologia Plantarum 104: 646–652.

Basu, R.N. 1994. An appraisal of research on wet and dry physiological seed treatments and their applicability with special reference to tropical and sub-tropical countries. Seed Sci. Technol. 22: 107–126.

Basuchaudhuri, P. 1990. Partitioning of assimiltes in soybean seedlings. Annals Agricultural Research 11(3-4): 300–304.

Bewley, J.D., K.J. Bradford, H.W.M. Hilhorst and H. Nonogaki. 2013. Seeds: Physiology of Development, Germination and Dormancy. 3rd Edition, Springer, New York.

Bejandi, T.K., M. Sedghi, R.S. Sharifi, A. Namvar and P. Molaei. 2009. Seed priming and sulfur effects on soybean cell membrane stability and yield in saline soil. Pesquisa Agropecu´aria Brasileira 44: 1114–1117.

Benedetti, A., M. Comporti and H. Esterbauer. 1980. Identification of 4-hydroxynonenal as a cytotoxic product originating from the peroxidation of liver microsomal lipids. Biochemica et Biophysica Acta 620: 281–296.

Bernal, L.I. and A.C. Leopold. 1992. Changes in soluble carbohydrates during seed storage. Plant Physiol. 98: 1207–1210.

Bewley, J.D. 1997. Seed germination and dormancy. Plant Cell 9: 1055–1066.

Bhardwaj, H.L., A.S. Bhagsari, J.M. Joshi, M. Rangappa, V.T. Sapra and M.S.S. Rao. 1999. Yield and quality of soymilk and tofu made from soybean genotypes grown at four locations. CropScience 39: 401–405.

Bhardwaj, H.L., A.A. Hamama, M. Rangappa, J.M. Joshi and V.T. Sapra. 2003. Effects of soybean genotype and growing location on oil and fatty acids in tofu. Plant Foods for Human Nutrition 58: 197–205.

Borek, S. and L. Ratajczak. 2010. Storage lipids as a source of carbon skeletons for asparagine synthesis in germinating seeds of yellow lupine (*Lupinus luteus* L.). Journal of Plant Physiology 167: 717–724.

Borowski, E. and S. Michaek. 2004. The effect of chilling temperature on germination and early growth of domestic and canadian soybean (*Glycine max* L. Merr) cultivars. Acta Sci. Pol. Hortorum cultus. 13: 31–43.

Brash, A.R. 1999. Lipoxygenases: Occurrence, functions, catalysis, and acquisition of substrate. J. Biol. Chem. 274: 23679–23682.

Bray, C.M., P.A. Davison, M. Ashraf and R.M. Taylor. 1989. Biochemical changes during osmo-priming of leek seeds. Ann. Bot. 63: 185–193.

Chang, S.K.C., Z.S. Liu, H.J. Hou and L.A. Wilson. 2004. Influence of storage on the characteristics of soybean, soymilk and tofu. Proc. VII-World Soybean Res. Con., IV-*In:* Soybean Proc. and Util. Con., III-Congresso Brasileiro de Soja Brazilian Soybean Congress, Foz do Iguassu, PR, Brazil, 29 February–5 March, 977–983.

Chen, H., J. Zhang, M.M. Neff, S.W. Hong, H. Zhang et al. 2008. Integration of light and abscisic acid signaling during seed germination and early seedling development. Proc. Natl. Acad. Sci. USA 105: 4495–4500.

Chien, C.T. and T.P. Lin. 1994. Mechanism of hydrogen peroxide in improving the germination of Cinnamomum camphora seed. Seed Science Technology 22: 231–236.

Colpas, F.T., E.O. Ono, J.D. Rodrigues and J.R.D. Passos. 2003. Effect of some phenolic compounds on soybean seed germination and on seed borne fungi. Braz. Arch. Biol. Tech. 46: 155–161.

Cookson, W.R., J.S. Rowarth and J.R. Sedcole. 2001. Seed vigour in perennial ryegrass (*Lolium perenne* L.): Effect and cause. Seed Science and Technology 29: 255–270.

Coolbear, P. and D. Grierson. 1979. Studies on the changes in the major nucleic acid components of tomato seeds (*Lycopersicon esculentum* Mill.) resulting from osmotic pre-sowing treatments. J. Exptl. Bot. 30: 1153–1162.

Crowe, L.M., R. Mourdian, J.H. Crowe, S.A. Jackson and C. Womersly. 1984. Effects of carbohydrates on membrane stability at lower water activities. Biochem. Biophys. Acta 769: 141–150.

Desikan, R., J.T. Hancock, J. Bright, H. Harrison, I. Weir, R. Hooley and S.J. Neill. 2005. A role for ETR1 in hydrogen peroxide signalling in stomatal guard cells.

Dierking, C. and K.D. Bilyeu. 2009. Raffinose and stachyose metabolism are not required for efficient soybean seed germination. J. Plant Physiol. 166: 1329–1335.

Dinkova, T.D., N.A. Marquez-Velazquez, R. Aguilar, P.E. Lazaro-Mixteco and E.S. de Jimenez. 2011. Tight translational control by the initiation factors eIF4E and eIF(iso)4E is required for maize seed germination. Seed Science Research 21: 85–93.

Donohue, K., R. Rubio de Casas, L. Burghardt, K. Kovach and C.G. Willis. 2010. Germination, postgermination adaptation, and species ecological ranges. Annual Review of Ecology, Evolution, and Systematics 41: 293–319.

Eastmond, P.J., V. Germain, P.R. Lange, J.H. Bryce, S.M. Smith et al. 2000. Post germinative growth and lipid catabolism in oilseeds lacking the glyoxylate cycle. Proc. Natl. Acad. Sci. USA 97: 5669–5674.

Edwards Jr., C.J. and E.E. Hartwig. 1971. Effect of seed size upon rate of germination in soybeans. Agronomy Journal 63: 429–430.

Esterbauer, H. 1982. Aldehydic products of lipid peroxidation. pp. 101–129. *In*: McBrien, D.C.H. and T.F. Slater (eds.). Free Radicals, Lipid Peroxidation and Cancer. Academic Press, New York.

Falivene, S.M.P., M.A.C. de Miranda and L.D.A. de Almeida. 1980. Temperature and the occurrence of cotyledons necrosis in soybean. Revista-Brasileira-Desementes 2: 43–51.

Fauconnier, M.L., A.G. Perez, C. Sanz and M. Marlier. 1997. Purification and characterization of tomato leaf (*Lycopersicon esculentum* Mill.) hydroperoxidelyase. J. Agric. Food Chem. 45: 4232–4236.

Feussner, I. and C. Wasterneck. 2002. The lipoxygenase pathway. Annu. Rev. Plant Biol. 53: 275–97.

Finch-Savage, W.E. and G. Leubner-Metzger. 2006. Seed dormancy and the control of germination. New Phytol. 171: 501–523.

Finkelstein, R.R. 2004. The role of hormones during seed development and germination. pp. 513–537. *In*: Davies, P.J. (ed.). Plant Hormones-Biosynthesis, Signal Transduction, Action! Dordrecht: Kluwer.

Fontaine, O., C. Huault, N. Pavis and J.P. Billard. 1994. Dormancy breakage of *Hordeum vulgare* seeds: Effects of hydrogen peroxide and scarification on glutathione level and glutathione reductase activity. Plant Physiology and Biochemistry 2: 677–683.

Gandhi, A.P. 2009. Quality of soybean and its food products, biotecnologia: Produção. International Food Research Journal 16: 11–19.

Gardner, H.W., D.L. Dornbosand and A.E. Desjardins. 1990. Hexanal, trans-2-hexenal, and trans-2-nonenal inhibit soybean (*Glycine max*) seed germination. J. Agric. Food Chem. 38: 1316–1320.

Grabe, D.F. 1965. Storage of Soybeans for Seed. Soy bean Digest. 26: 14–16.

Graham, I.A. 2008. Storage oil mobilization in seeds. Annu. Rev. Plant Biol. 59: 115–142.

Green, P.B. 1962. Mechanism for plant cellular morphogenesis. Science 138: 1404–1405.

Gubler, F., A.A. Millar and J.V. Jacobsen. 2005. Dormancy release, ABA and pre-harvest sprouting. Curr. Opin. in Plant Biol. 8: 183–187.

Halliwell, B. and J.M.C. Gutteridge. 1989. Free Radicals in Biology and Medicine. 2nd edition. Oxford University Press.

Harris, D., A.K. Pathan, P. Gothkar, A. Joshi, W. Chivasa and P. Nyamudeza. 2001. On-farm seed priming: Using participatory methods to revive and refine a key technology. Agricultural Systems 69: 151–164.

Heatherly, L.G. 2015. Soybean Seed Treatments and Inoculants. Mississippi Soybean Promotion Board.

Heydecker, W. and P. Coolbear. 1977. Seed treatments and performance survey and attempted prongnosis. Seed Science and Technology 5: 353–425.

Hilhorst, H.W.M., W.E. Finch-Savage, J. Buitink, W. Bolingue and G. Leubner-Metzger. 2010. Dormancy in plant seeds. pp. 43–67. *In*: Lubzens, E., J. Cerdà and M. Clarck (eds.). Dormancy and Resistance in Harsh Environments. Heideberg: Springer-Verlag.

Howell, K.A., A.H. Millar and J. Whelan. 2006. Ordered assembly of mitochondria during rice germination begins with pro-mitochondrial structures rich in components of the protein import apparatus. Plant Mol. Biol. 60: 201–223.

Huang, C.-H. and C.-M. Yang. 1995. Use of Weibull function to quantify temperature effect on soybean germination. Chinese Agron. J. 5: 25–34.

Hubel, F. and E. Beck. 1996. Maize root phytase. Purification, characterization, and localization of enzyme activity and its putative substrate. Plant Physiology 112: 1429–1436.

Hoy, D.J. and E.E. Gamble. 1987. Field performance in soybean with seeds of differing size and density. Crop Science 27: 121–126.

Ishibashi, Y., K. Yamamoto, T. Tawaratsumida, T. Yuasa and M. Iwaya-Inoue. 2008. Hydrogen peroxide scavenging regulates germination ability during wheat (*Triticum aestivum* L.) seed maturation. Plant Signaling and Behavior 3: 183–188.

Ishibashi, Y., T. Tawaratsumida, S.H. Zheng, T. Yuasa and M. Iwaya-Inoue. 2010a. NADPH oxidases act as key enzyme on germination and seedling growth in barley (*Hordeum vulgare* L.). Plant Production Science 13: 45–52.

Ishibashi, Y., T. Tawaratsumida, K. Kondo et al. 2012. Reactive oxygen species are involved in gibberellin/abscisic acid signalling in barley aleurone cells. Plant Physiology 158: 1705–1714.

Jacela, J.Y., M.D. de Rouchey Tokach, R.D. Goodband, J.L. Nelssen, D. Renter and S.S. Dritz. 2010. Feed additives for swine: Fact sheets-prebiotics and probiotis and phytogenics. Journal of Swine Health and Production 18: 87–91.

Jisha, K.C., K.J.T. Vijayakumari and J.T. Puthur. 2013. Seed priming for abiotic stress tolerance: An overview. Acta Physiologiae Plantarum 3: 1381–1396.

Job, C., A. Kersulec, L. Ravasio, S. Chareyre, R. Pepin and D. Job. 1997. The solubilisation of the basic subunit of sugar beet seed 11-S globulin during priming and early germination. Seed Sci. Res. 7: 225–243.

Johnson, D.R. and V.D. Luedders. 1974. Effect of planted seed size on emergence and yield of soybean (*Glycine max* (L.) Merr.). Agronomy Journal 66: 117–118.

Johnson, R.R. and L.M. Wax. 1978. Relationship of soybean germination and vigor tests to field performance. Agronomy Journal 70: 273–278.

Khan, A.A. 1992. Pre-plant physiological conditioning. Horti. Rew. 13: 131–181.

Khajeh-Hosseini, M., A.A. Powell and I.J. Bingham. 2003. The interaction between salinity stress and seed vigour during germination of soybean seeds. Seed Science and Technology 31: 715–725.

Kim, S.G., A.K. Lee, H.K. Yoon and C.M. Park. 2008. A membrane-bound NAC transcription factor NTL8 regulates gibberellic acid-mediated salt signalling in Arabidopsis seed germination. Plant J. 55: 77–88.

King, J.M., S.M. Chin, L.K. Svendsen, C.A. Reitmeier, L.A. Johnson and W.R. Fehr. 2001. Processing of lipoxygenase-free soybeans and evaluation in foods. J. Am. Oil Chemists Soc. 78: 353–360.

Koornneef, M., L. Bentsink and H. Hilhorst. 2002. Seed dormancy and germination. Current Opinion in Plant Biology 5: 33–36.

Kucera, B., M.A. Cohn and G. Leubner-Metzger. 2005. Plant hormone interactions during seed dormancy release and germination. Seed Science Research 15: 281–307.

Kujur, A.B. and G.M. Lal. 2015. Effect of hydropriming and osmopriming on germination behaviour and vigor of soybean (*Glycine max* L.) seeds. Agric. Sci. Digest. 35: 207–210.

Kumar, A. 2017. Germination behaviour of soybean varieties under different salinity stress. Int. J. App. Agric. Res. 12: 69–76.

Kuo, T.M., D.C. Doehlert and C.G. Crawford. 1990. Sugar metabolism in germinating soybean seeds. Plant Physiol. 93: 1514–1520.

Lazali, M., L. Louadj, G. Ounane, J. Abadie, L. Amenc, A. Bargaz, V. Lullien-Pellerin and J.J. Drevon. 2014. Localization of phytase transcripts in germinating seeds of the common bean (*Phaseolus vulgaris* L.). Planta 240: 471–478.

Lee, Y.P., K.H. Baek, H.S. Lee, S.S. Kwak, J.W. Bang and S.Y. Kwon. 2010. Tobacco seeds simultaneously over-expressing Cu/Zn-superoxide dismutase and ascorbate peroxidase display enhanced seed longevity and germination rates under stress conditions. Journal of Experimental Botany 61: 2499–2506.

Leymarie, J., G. Vitkauskaite, H.H. Hoang et al. 2012. Role of reactive oxygen species in the regulation of *Arabidopsis* seed dormancy. Plant and Cell Physiology 53: 96–106.

Liao Fang Lei, Jiang Wu, Zheng Yue Ping, Xu Hang Lin, Li Li Qing and Lu Hong Fei. 2011. Influences of temperature regime on germination of seed of wild soybean (*Glycine soja*). Agric. Sci. Technol. 12: 480–483.

Lien, D.T.P., T.M. Phuc, T.B. Tram and H.T. Toan. 2016. Time and temperature dependence of antioxidant activity from soybean seeds (*Glycine max* L. Merr) during germination. Int. J. Food Sci. Nutr. 1: 22–27.

Linkies, A., K. Graeber, C. Knight and G. Leubner-Metzger. 2010. The evolution of seeds. New Phytol. 186: 817–831.

Liu, Y., N. Ye, R. Liu, M. Chen and J. Zhang. 2010. H_2O_2 mediates the regulation of ABA catabolism and GA biosynthesis in Arabidopsis seed dormancy and germination. Journal of Experimental Botany 61: 2979–2990.

Loewus, F.A. and M.W. Louewus. 1983. Myo-Inositol: Its biosynthesis and metabolism. Ann. Rev. Plant Physiol. 34: 137–161.

Loiseau, J., B.L. Vu, M.-H. Macherel and Y.L. Deunff. 2001. Seed lipoxygenases: Occurrence and functions. Seed Sci. Res. 11: 199–211.

Lott, J.N.A., J.S. Greenwood and G.D. Batten. 1995. Mechanisms and regulation of mineral nutrient storage during seed development. pp. 215–235. *In*: Kigel, J. and G. Galili (eds.). Seed Development and Germination. New York: Marcel Dekker Inc.

Macovei, A., A. Balestrazzi, M. Confalonieri, M. Fae and D. Carbonera. 2011. New insights on the barrel medic MtOGG1 and MtFPG functions in relation to oxidative stress response in planta and during seed imbibition. Plant Physiol. Biochem. 49: 1040–1050.

Mandal, S.M., D. Chakraborty and K. Gupta. 2008. Seed size variation: Influence on germination and subsequent seedling performance in *Hyptis suaveolens* (*Lamiaceae*). Research Journal of Seed Science 1: 26–33.

Matsui, K. 2006. Green leaf volatiles: Hydroperoxidelyase pathway of oxylipin metabolism. Current Opinion in Plant Biol. 9: 274–80.

Matthews, S. 1980. Controlled deterioration: A new vigour test for crop seeds. pp. 513–526. *In*: Hebblethwaite, P.D. (ed.). Seed Production. London: Butterworths.

Matthews, S. and A.A. Powell. 1988. Seed treatments: Developments and prospects. Outlook on Agriculture 17: 97–103.

McDonald, M.B. 1999. Seed deterioration: Physiology, repair and assessment. Seed Sci. Technol. 27: 177–237.

Mewael, K.A., Ravi Hunje, R.V. Koti and N.K. Biradarpatil. 2010. Enhancement of seed quality in soybean following priming treatments. Karnataka J. Agric. Sci. 23(5): 787–789.

Miladinov, Z., S. Balesevic-Tubic, V. Dukic, A. Ilic, L. Cobanovic, G. Dozet and L. Merkulov-Popadic. 2018. Effect of priming on soybean seed germination parameters. Acta Agric. Serbica 23: 15–26.

Mittler, R., S. Vanderauwera, M. Gollery, F. Van Breusegem. 2004. Reactive oxygen gene network of plants. Trends in Plant Science 9: 490–498.

Molin, W.T., C.G. Wilkerson and R.C. Fites. 1980. Thiamine phosphorylation by thiamine pyrophosphotransferase during seed germination. Plant Physiol. 66: 313–315.

Mukherji, S., B. Dey, A.K. Paul and S.M. Sircar. 1971. Changes in phosphorus fractions and phytase activity of rice seeds during germination. Physiologia Plantarum 25: 94–97.

Müller, K., A. Linkies, R.A.M. Vreeburg, S.C. Fry, A. Krieger-Liszkay and G. Leubner-Metzger. 2009a. *In vivo* cell wall loosening by hydroxyl radicals during cress (*Lepidium sativum* L.) seed germination and elongation growth. Plant Physiology 150: 1855–1865.

Nakamura, M. 1974. Structure of starch granules. II. [beans, barley,soybeans]. Denpun Kagaku – J. Jap. Soc. Starch Sci. 21: 230–254.

Narasimha Prasad, K. 1994. Studies on certain aspects of seed management in soybean (*Glycine max* L.) Merrill. Ph. D. Thesis, Tamil Nadu Agric. Univ., Coimbatore (India).

Ogawa, M., K. Tanaka and Z. Kasai. 1979. Accumulation of phosphorus, magnesium, and potassium in developing rice grains: Followed by electron microprobe X-ray analysis focusing on the aleurone layer. Plant and Cell Physiology 20: 19–27.

Ogawa, M. 2003. Gibberellin biosynthesis and response during Arabidopsis seed germination. Plant Cell 15: 1591–1604.

O'Hare, T., J. Bagshaw and W.L.G. Johnson. 2001. Storage of oriental bunching onions. Post-harvest handling of fresh vegetables. *In*: Proc. of a Workshop held in Beijing, China, 9–11 May.

Okazaki, Y. and T. Katayama. 2005. Reassessment of the nutritional function of phytic acid, with special reference to myo-inositol function. Journal of Japan Society of Nutrition and Food Sciences 58: 151–156.

Oracz, K., H. El-Maarouf-Bouteau, J.M. Farrant et al., 2007. ROS production and protein oxidation as novel mechanism of seed dormancy alleviation. Plant Journal 50: 452–465.

Paparella, S., S.S. Araújo, G. Rossi, M. Wijayasinghe, D. Carbonera and A. Balestrazzi. 2015. Seed priming: State of the art and new perspectives. Plant Cell Reports 34: 1281–1293.

Paques, F.W. and G.A. Macedo. 2006. Lipases de Látex Vegetais: Propriedades e Aplicações Industriais: A review. Química Nova. 29(1): 93.

Parish, D.J. and A.C. Leopold. 1978. On the mechanism of aging in soy bean seeds. Plant Physiology 61: 365–368.

Park, J., Y.S. Kim, S.G. Kim, J.H. Jung, J.C. Woo et al. 2011. Integration of auxin and salt signals by the NAC transcription factor NTM2 during seed germination in *Arabidopsis*. Plant Physiol. 156: 537–549.

Parolin. 2001. Chemical damage induced by chilling in soybean seeds. Physiol. Plant. 111(4): 473–482.

Penfield, S., E.M. Josse, R. Kannangara, A.D. Gilday, K.J. Halliday et al. 2005. Cold and light control seed germination through the bHLH transcription factor SPATULA. Curr. Biol. 15: 1998–2006.

Pereira, E.P., G.M. Zanin and H.F. Castro. 2003. Immobilization and catalytic properties of lipase on chitosan for hydrolysis and etherification reactions. Brazilian Journal of Chemical Engineering 20(4): 343.

Peksen, E., A. Peksen, H. Bozoglu and A. Gulumser. 2004. Some seed traits and their relationships to seed germination and field emergence in pea (*Pisum sativum* L.). J. Agron. 3: 243–246.

Petruzelli, L. and G. Taranto. 1989. Wheat ageing: The contribution of embryonic and non-embryonic lesions to loss of seed viability. Physiol. Plant. 76: 289–294.

Pinfield-Wells, H., E.L. Rylott, A.D. Gilday, S. Graham, K. Job et al. 2005. Sucrose rescues seedling establishment but not germination of Arabidopsis mutants disrupted in peroxisomal fatty acid catabolism. Plant J. 43: 861–872.

Piskurewicz, U., V. Tureckova, E. Lacombe and L. Lopez-Molina. 2009. Far-red light inhibits germination through DELLA-dependent stimulation of ABA synthesis and ABI3 activity. EMBO J. 28: 2259–2271.

Posmyk, M.M., F. Corbineau, D. Vinel, C. Bailly and D. Côme. 2001. Osmoconditioning reduces physiological and biochemical damage induced by chilling in soybean seeds. Physiologia Plantarum 111: 473–482.

Pracharoenwattana, I. and S.M. Smith. 2008.When is a peroxisome not a peroxisome? Trends in Plant Science 13: 522–525.

Priestley, D.A. and A.C. Leopold. 1983. Lipid changes during natural aging of soybean seeds. Physiol. Plant 59: 467–470.

Priestley, D.A. 1986. Seed Aging. Implications for seed storage and persistence in soil. Cornell University Press, Ithaca.

Puntarulo, S., R.A. Sánchez and A. Boveris. 1988. Hydrogen peroxide metabolism in soybean embryonic axes at the onset of germination. Plant Physiology 86: 626–630.

Puntarulo, S., M. Galleano, R.A. Sánchez and A. Boveris. 1991. Superoxide anion and hydrogen peroxide metabolism in soybean embryonic axes during germination. Biochimica et Biophysica Acta 1074: 277–283.

Quettier, A.L. and P.J. Eastmond. 2009. Storage oil hydrolysis during early seedling growth. Plant Physiology and Biochemistry 47: 485.

Raboy, V. 2003. Myo-Inositol-1,2,3,4,5,6-hexakisphosphate. Phytochemistry 64: 1033–1043.

Rahman, M.M., B. Hagidok, M.M. Masood and M.N. Islam. 2010. Effect of storage container and relative humidity on the quality of wheat seed. Bangladesh J. Seed Sci. Technol. 14(1&2): 89–94.

Robinson, D.S., Z. Wu, C. Domoneyb and R. Casey. 1995. Lipoxygenases and the quality of foods. Food Chem. 54: 33–43.

Roy, S.K.S., A. Hamid, M.G. Miah and A. Hashem. 1996. Seed size variation and its effects on germination and seedling vigour in rice. Journal of Agronomy and Crop Science 176: 79–82.

Sadeghian, S.Y. and N. Yavari. 2004. Effect of water-deficit stress on germination and early seedling growth in sugar beet. Journal of Agronomy Crop Science 190: 138–144.

Schroeder, J.I., J.M. Kwak and G.J. Allen. 2001. Guard cell abscisic acid signalling and engineering drought hardiness in plants. Nature 410: 327–330.

Schulz, T.J. and K.D. Thelen. 2008. Soybean seed inoculant and fungicidal seed treatment effects on soybean. Crop Science 48: 1975–1983.

Sharma, S., P. Virdi, S. Gambhir and S.K. Munshi. 2005. Changes in soluble sugar content and antioxidant enzymes in soybean seeds stored under different storage conditions. Ind. J. Agric. Biochem. 18: 9–12.

Shu, K.,Y. Qi, Y. Meng, X. Luo, H. Shuai, W. Zhou, J. Ding, J. Du, F. Yang, Q. Wang, W. Liu, T. Yang, X. Wang, Y. Feng and W. Yang. 2017. Salt stress represses soybean seed germination by negatively regulating GA biosynthesis while positively mediating ABA biosynthesis. Front. Plant Sci. 8: 1372.

Shuai, H., Y. Meng, X. Luo, F. Chen, W. Zhou, Y. Dai, Y. Qi, J. Du, F. Yang, J. Liu, W. Yang and K. Shu. 2017. Exogenous auxin represses soybean seed germination through decreasing the gibberellins/abscisic acid(GA/ABA) ratio. Sci. Rep. 7: 12620.

Silva, L.G. and L.C. Trugo. 1996. Characterization of phytase activity in lupin seed. Journal of Food Biochemistry 20: 329–340.

Simon, E.W. 1984. Early events in germination. pp. 77–115. *In*: Murray, D.R. (ed.). Seed Physiology, Vol. 2, Germination and Reserve Mobilization. Academic Press, Orlando, FL.

Smith, T.J. and H.M. Camper Jr. 1975. Effect of seed size on soybean performance. Agronomy Journal 67: 681–684.

Smirnoff, N. 1993. The role of active oxygen in the response of plants to water deficit and desiccation. New Phytologist 125: 27–58.

Stewart, R.C. and D.J. Bewley. 1980. Lipid peroxidation associated with accelerated aging of soybean axes. Plant Physiol. 65: 245–248.

Stobbe, E., J. Moes, Y. Gan, H. Ngoma and L. Bourgeca. 2008. Seeds, Seed Vigor and Seeding Research Report, Department of Plant Science, NDSU Agriculture and University Extension, Fargo, ND, USA.

Sung, J.M. and T. Jeng. 1994. Lipid peroxidation and peroxide-scavenging enzymes associated with accelerated aging of peanut seed. Physiologia Plantarum 91: 51–55.

Tatipata, A. 2009. Effect of seed moisture content packaging and storage period on microchondria inner membrane of soybean seed. J. Agric. Technol. 5(1): 51–54.

Taylor, A.G., D.E. Klein and T.H. Whitlow. 1998. SMP: Solid matrix priming of seeds. Scientia Horticulturae, 37 : 1–11.

Tekrony, D.M., C. Nelson, D.B. Egli and G.M. White. 1993. Predicting soybean seed germination during warehouse storage. Seed Sci. Technol. 21: 127–137.

Theimer, R.R. and I. Rosnitschek. 1978. Development and intracellular localization of lipase activity in rapeseed (*Brassica napus* L.) cotyledons. Planta 139: 249–256.

Thompson, L.U. and L. Zhang. 1991. Phytic acid and minerals: Effect on early markers of risk for mammary and colon carcinogenesis. Carcinogenesis 12: 2041–2045.

Thornton, J.M. and A.A. Powell. 1992. Short-term aerated hydration for the improvement of seed quality in *Brassica oleracea* L. Annals of Applied Biology 127: 183–189.

Tian, X., S. Song and Y. Lei. 2008. Cell death and reactive oxygen species metabolism during accelerated ageing of soybean axes. Russian Journal of Plant Physiology 55: 33–40.

Van Pijlen, J.G., S.P.C. Groot, H.L. Kraak, J.H.W. Bergervoet and R.J. Bino. 1996. Effects of pre-storage hydration treatments on germination performance, moisture content, DNA synthesis and controlled deterioration tolerance of tomato (*Lycopersicon esculentum* Mill). Seed Sci. Res. 6: 57–63.

Vaughn, K.C. and L.P. Lehnen Jr. 1991. Mitotic disrupter herbicides. Weed Sci. 39: 450–457.

Wang, P. and C.P. Songm. 2008. Guard-cell signalling for hydrogen peroxide and abscisic acid. New Phytologist 178: 703–718.

Wilson, K.A. 2006. Mobilization of storage proteins in dicots. pp. 672–674. *In*: Black, M., J.D. Bewley and P. Halmer (eds.). Encyclopedia of Seeds. Science, Technology and Uses. CABI, Wallingford.

Wilson, D.O. and M.B. McDonald. 1986. The lipid peroxidation model of seed deterioration. Seed Sci. Technol. 14: 269–300.

Wilson, L.A., V.A. Birmingham, D.P. Moon and H.E. Snyder. 1978. Isolation and characterization of starch from mature soybeans. Cereal Chem. 55: 661–670.

Woodruff, M.J. 1998. Reports on the soybean: Its status, and potential for Bangladesh. Agro based Industries and Technology Development Project (ATDP). May 1998. Ministry of Agriculture and International Fertilizer Development Centre (IFDC).

Zhou, Y., P. Chu, H. Chen et al. 2012. Overexpression of *Nelumbo nucifera* metallothioneins 2a and 3 enhances seed germination vigor in Arabidopsis. Planta 235: 523–537.

Zhuo, J., W. Wang, Y. Lu, W. Sen and X. Wang. 2009. Osmopriming-regulated changes of plasma membrane composition and function were inhibited by phenylarsin oxide in soybean seeds. Journal of Integrative Plant Biology 9: 858–867.

Soybean (*Glycine max* L.) is an annual plant commercially grown primarily for oil and protein production. Soybean is generally a plant which grows 90–120 cm in height with first leaves simple, opposite and all other leaves alternate and trifoliate. Floral primodia are initiated within 3 weeks and flowering begins at 6–8 weeks. Pods are visible l0 days to 2 weeks after the onset of flowering.

Flowering continues for 3–4 weeks. Many stages of pod and seed development occur on the plant until near physiological maturity. Soybean stem growth and flowering habits are of two types: Indeterminate and determinate. The indeterminate type is characterized by the apical meristem continuing vegetative activity during most of the growing season, the inflorescences are auxiliary racemes and pods are produced. The determinate stem type is characterized by vegetative development which ceases when the apical meristem become an inflorescence, both auxiliary and terminal racemes exist and pods are borne rather uniformly along the stem. Extensive flower abortion (20–80%) can occur at any stage of development from time of bud initiation to seed development. The seed of soybean varies in shape but is generally oval and consists of an embryo enclosed by the seed coat.

Soybean maturity is classified in different groups (MGs) ranging from 000 for the very early maturing varieties to 9 for the later. Gradations within MGs are also commonly noted by adding a decimal to the MG number. A variety is classified to a specific MG according to the length of time from planting to maturity. This phenological attribute is determined by two abiotic factors: Photoperiod and temperature (Cober et al., 2001), and these factors can dictate the most suitable MG for a particular geographical location.

Producers are looking for ways to improve soybean yields and profitability and many are planting longer maturing soybean varieties as a way to reach these goals. The theory behind this strategy is that later maturing varieties will have a longer reproductive period and take full advantage of the growing season. However, planting later maturing varieties carries some risk. The most obvious risk is that the crop could be damaged by frost or freeze events, reducing yield and quality and increasing harvest delays. Additional risks include a greater potential for delayed wheat planting and early snow cover.

SOYGRO is a crop growth model which includes a development model designed to predict the duration of growth phase in soybean cultivars growing in various climates, using cultivar response to photoperiod and temperature. Average errors in predicted flowering date were 1–9 d, but when the crop was planted late and maturity occurred during cool late autumn weather the predicted maturity date was up to 19 d late.

3.1 Vegetative Growth Stages

Accurate identification of soybean growth stages is important to maximize grain yield and profitability, because most management decisions are based upon the growth stage of soybean plants within the fields. Two plant growth habits exist for soybean: Indeterminate and determinate. Indeterminate cultivars can produce both vegetative and reproductive growth simultaneously until about R5 growth stage. Most indeterminate cultivars are maturity groups 00 to IV. In contrast, determinate cultivars will complete most of their vegetative growth prior to initiating reproduction. Most determinate cultivars are maturity groups

V to IX. Soybean growth stages are divided into two phases: Vegetative (V) and reproductive (R). When identifying vegetative growth stages of soybean (Fig. 3.1), the primary consideration is the number of fully developed leaves on the main stem. A fully developed leaf is one that has all leaflets open, while an undeveloped leaf has leaflet edges that are still touching. Reproductive growth stages are identified by specific flower, pod, and seed characteristics. When classifying a field of soybean as a specific growth stage, at least 50 percent of the plants in the field must be at or beyond that growth stage. For example, a soybean field that has 10 percent of the plants with two fully developed trifoliate leaves (V2), 60 percent of the plants with three fully developed trifoliate leaves (V3), and 30 percent of the plants with four fully developed trifoliate leaves (V4) will be the V3 growth stage. This is because 70 percent of the field is or has already been the V3 growth stage.

Growth Stage (50% or more of field) Description/Importance Vegetative Stages

VE Emergence: The cotyledons and growing point are above the soil surface.
                Full-season soybean typically emerge 1 to 2 weeks after planting.
                Double-crop soybean typically emerge 5 days after planting.

VC Cotyledon: The two unifoliate leaves are fully developed (the leaf edges are no longer touching).
                Nitrogen (N)-fixing (Bradyrhizobium japonicum) root nodules may be visible, but not functional.

V1 1 Trifoliate Leaf: The first trifoliate leaf is fully developed.
                        N-fixing root nodules may be visible, but not functional.

V2 2 Trifoliate Leaves: Two trifoliate leaves are fully developed.
                          N-fixing root nodules are typically functional.

V3 3 Trifoliate Leaves: Three trifoliate leaves are fully developed.
                          Typically occurs 2–3 weeks after emergence for full-season soybean and 2 weeks after emergence for double-crop soybean.
                          This is final growth stage when several herbicides can be applied.

V4 to V(n) 4 Trifoliate Leaves to nth Trifoliate Leaves: Four to nth trifoliate leaves are fully developed.

                          The number of trifoliate leaves is determined by variety and environmental conditions.

Fig. 3.1 Different vegetative growth stages of soybean.

Indeterminate Varieties

- MG 000-IV
- Main stem node number stop at R5
- Potential number of main stem nodes produced by the plant is a function of the number of days between V1 and R5.5—A new node for every 3.7–5 days after V1 (linear)
- Planting date and maturity can restrict the final node number below the potential node number

Determinate Varieties

- MG V–IX
- Main stem node number accrual ceases abruptly at R1
- Leaves will continue to develop on branches (most yield)

When 5 trifoliate leaves have developed, the plant is at V5. Lateral branches may grow to compensate to some degree for low plant populations or wide row spacings. However, they can't compensate fully for under seeding.

At V5, plants reach 10" to 12". In the top stem, axillary buds develop; they will grow into flower clusters (racemes). The total number of nodes the plant can produce is set. If something damages the growing point, the axillary buds will branch off and grow profusely (Fig. 3.2).

If the plant breaks off below the cotyledon node, the plant will die.

The primary focus of the vegetative stage is height and leaf growth. The overall size of the plant and number of flower positions greatly depends on the length of the vegetative stage and the environmental conditions present (Scott and Aldrich, 1983).

Usually 0 to 6 branches per plant are typical, but all nodes have the potential to produce both branches and flowers (Shibles et al., 1975). The period of vegetative growth can vary depending on environmental factors, such as temperature and length of day (Carlson and Lersten, 2004).

Vegetative stages occur until the first flower blooms, after which reproductive stages begin. Even after reproductive stages begin, vegetative growth of indeterminate varieties will continue until seed enlargement begins (Howell, 1960).

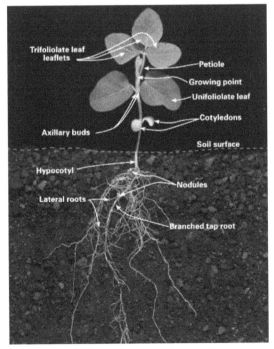

Fig. 3.2 Description of a soybean seedling.

During the vegetative growth stage, soybean seedlings are more tolerant of adverse environmental conditions, including cold temperatures, wind, hail and drought or flooding.

3.2 Seedling Growth

Early development after emergence is important for seedling establishment and early vegetative growth of soybean. The ability of seedlings to become autotrophic rapidly and to develop a large plant in a short period is ecologically advantageous (Nelson and Larson, 1984). Furthermore, rapid seedling growth results in extensive growth of root and shoot systems. Seedling vigor and rapid early vegetative growth may be more important under a low planting density and inferior environmental conditions, such as a limited plant-growing season and low temperatures in early spring. The advantage of seedling vigor is to enhance plant survival and develop the leaf area to intercept radiation, and also to enhance weed-suppressive ability, as has been documented in rice (Zhao et al., 2006) as well as in soybean (Jannink et al., 2000).

There are many reports on seed vigor and seedling emergence (Sadeghi et al., 2011) but less information is available on soybean seedling development.

Hydro- and halo-priming could increase the length and the dry weight of soybean seedlings produced from the seeds of all three vigor levels, except those from SD3 when pre-treated with distilled water (Fig. 3.3).

The seedlings from the large seeds were significantly heavier than those from the other seed size classes (Fig. 3.4). However, the deteriorating conditions could decrease the seedling dry weight in all four seed lots, especially in small soybean seeds.

The seedlings produced from large and medium seeds were longer and heavier than those from other two (mixed lot and small) size classes. Although it has been reported that smaller soybean seeds are more efficient than large ones regarding reserves remobilization (Edje and Burris, 1971; Barkke and Gradner, 1987), higher germination percentage (Larsen and Andreasen, 2004) and longer and heavier seedlings from large seeds are due to their high reserves (Helm and Spilde, 1990) which may lead to increase yield of some crops (Lowe and Rise, 1973). Subsequent seedling growth after seed germination

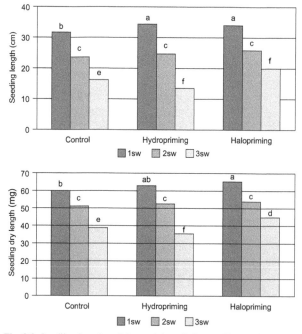

Fig. 3.3 Seedling length and dry weight as influenced by seed priming.

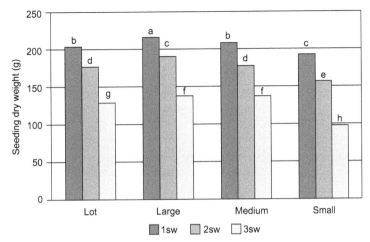

Fig. 3.4 Influence of seed size on seedling growth.

directly depends on the quantity and efficient mobilization of seed reserves (Westoby et al., 1992; Soltani et al., 2006).

Longer et al. (1986) reported that the seedlings emerging from large seeds face greater impedance from the soil crust, but a large seed size could accelerate seedling growth and early vegetative growth of soybean (Buris et al., 1973) due to the energy stored in the cotyledon. In soybean, the green cotyledons play an important role after emergence because the cotyledons not only have a storage-mobilizing function but also assimilate energy through photosynthesis (Hanley and May, 2006; Zhang et al., 2008). Zheng et al. (2011) insisted that the cotyledons provide a major proportion of the assimilation needed for seedling growth until the first true leaf becomes a significant exporter of photosynthate. Therefore, the dry matter accumulation during seedling growth is faster in soybean compared with some leguminous crops with small seeds (mung bean) or those that keep the cotyledons underground (Azuki bean, pigeon pea, etc.).

Twenty-seven soybean varieties originating from six countries were examined in 2009 and 2010. The pots were arranged in a completely random block design with 5 replications (10 pots per variety), and the seedlings were sampled at 14 and 28 days after sowing (DAS). The shoot dry weight at 14 and 28 DAS was highly correlated with seed size, cotyledon digestion, and leaf area. However, no positive correlation was found between shoot dry weight and photosynthetic rate at 28 DAS. Chamame, a Japanese cultivar, with the largest seed size grew rapidly, and showed the heaviest shoot dry weight, greatest cotyledon digestion, fast leaf expansion and high photosynthetic rate. However, Moyashimame, a medium-seed-size cultivar, also grew rapidly with a high photosynthetic rate. Some varieties, such as Tachinagaha (Japan), Hefeng (China), Parana and Pérola (Brazil), had a large or medium seed size, and high photosynthetic rate, but showed a relatively small leaf area and light shoot dry weight. These results suggested that big seeds with rapid cotyledon digestion developed a wider leaf area and, therefore, large dry matter production, indicating that the conversion of stored energy was more important than the leaf photosynthetic activity during early growth (Table 3.1 & Fig. 3.5).

3.3 Vegetative Growth

Hanway and Webber (1971) observed the differences among soybean varieties in the rates of dry matter accumulation and leaf area and concluded that yield differences were largely the result of the length of time of dry matter accumulated in seed. Basuchaudhuri and Munda (1987) observed that data on DM accumulation of shoot and pods, relative water content, leaf area density, chlorophyll content and peroxidase activity of leaves at different heights of a soybean canopy (about 70 cm height) at the pod formation stage indicated that most active region was around a height of 55 cm.

Table 3.1 Correlation coefficients among soybean growth characteristics in 2010.

	CD	RCD	PR1	LA1	SPAD1	DW1	PR2	LA2	SPAD2	DW2
SS	0.99**	−0.01	−0.05	0.48*	0.54**	0.77**	0.15	0.28	−0.10	0.56**
CD		0.02	−0.07	0.50**	0.52**	0.79**	0.17	0.32	−0.12	0.58**
RCD			−0.14	0.14	−0.12	0.24	0.02	0.35	0.01	0.10
PR1				−0.70**	0.49*	−0.46*	0.08	−0.40*	0.28	−0.20
LA1					−0.25	0.71**	−0.01	0.51**	−0.34	0.41*
SPAD1						0.18	0.16	−0.20	0.45*	0.10
DW1							0.14	0.48**	−0.16	0.60**
PR2								−0.17	0.28	0.16
LA2									−0.41*	0.75**
SPAD2										−0.18

* significant at the 0.05 probability level; ** significant at the 0.01 probability level.
SS, seed size; CD, cotyledon digestion; RCD, ratio of cotyledon digestion to seed size; PR1, apparent photosynthetic rate at 14 DAS; LA1, primary leaf area at 14 DAS; SPAD1, SPAD at 14 DAS; DW1, shoot dry weight at 14 DAS; PR2, apparent photosynthetic rate at 28 DAS; LA2, total leaf area at 28 DAS; SPAD2, SPAD at 28 DAS; DW2, shoot dry weight at 28 DAS.

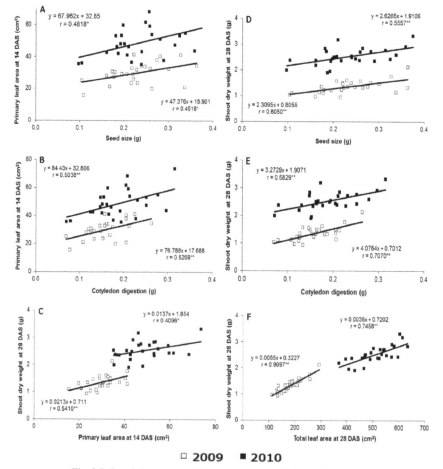

□ 2009   ■ 2010

Fig. 3.5 Correlations among different parameters during seedling stage.

The productivity of a crop is determined by parameters like Net assimilation rate (NAR), Crop growth rate (CGR), Relative growth rate (RGR), and partitioning of total photosynthates into economic and biological sinks. It is also suggested that competition exist between vegetative and reproductive growth during flowering and fruit set in soybean.

Buttery (1970) concluded that CGR increases during the first 50 to 60 days after soybean planting and then rapidly decreased. Koller et al. (1970) reported a rapid increase in crop growth rate (CGR) after pod formation resulted in increased accumulation of dry matter in reproductive organs of soybean.

Varietal differences for growth parameters among soybean cultivars were also observed. The differences in NAR, RGR and LAR among soybean varieties were reported. Sivakumar and Shaw (1978) observed that RGR and relative leaf growth rate declined with the growing age of the plant because the level of self-shading increased (Fig. 3.6).

Beaver et al. (1985) observed that the determinate growth type compensated for the reduced production of vegetative sinks on the main stem by partitioning assimilate to branch growth during early flowering and pod set rather than to reproductive sinks. Mehta and Sharma (1992) investigated in different diverse genotypes of soybean that none of the growth parameters contribute substantially to the final seed yield till 40 DAS. However, their cumulative contribution to yield became considerable under monoculture at 60 DAS.

Soybean seed size is representative of food or nutrient storage. Larger seeds have more nutrients stored in the cotyledons. There was a high positive correlation between primary leaf area (leaf area at 14 DAS) and shoot dry weight at 14 and 28 DAS. The primary leaf area was associated with cotyledon digestion rather than seed size (Table 3.1). This indicated that converted energy from the cotyledons contributes more to the leaf expansion, causing high photosynthetic activities and leading to more dry matter production. This was supported by the positive correlation between leaf area at 14 and 28 DAS (Table 3.1).

After seedling emergence, the green cotyledons and early developed leaves begin photosynthesis in order to support seedling growth. When the green cotyledons were cut partially, plant growth was suppressed in proportion to the degree of the cut compared with the intact plants (Ikeda and Kiso, 1981). It is certain that cotyledons play an important role in seedling development (Hanley and May, 2006; Zhang et al., 2008).

Ojima and Kawashima (1968) reported that the mean variation of photosynthetic rate in 38 varieties was around 20%. However, results showed a larger variation in 2009 (11.5–21.3 μmol CO_2 m^{-2}s^{-1}) than in 2010 (18.1–24.3 μmolCO_2 m^{-2}s^{-1}). The big variation in 2009 might be caused by the low air temperature, since most cultivars with low photosynthetic rate are from the tropical area. However, it was also found that the photosynthetic rate tended to correlate negatively with leaf area (Table 3.1), as reported by Kokubun et al. (1988).

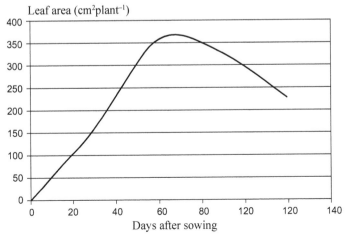

Fig. 3.6 Changes in leaf area with days after sowing in soybean.

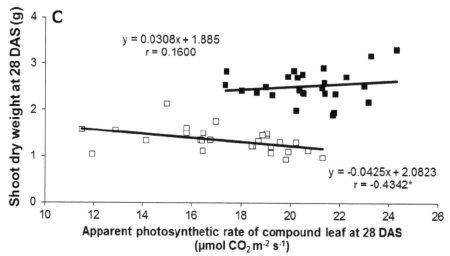

Fig. 3.7 Relationship between shoot dry weight and apparent photosynthetic rate of compound leaf of soybean seedling.

Although several studies have shown that leaf photosynthesis affects the plant growth (Hogan, 1988; Makino et al., 1997), results showed no positive correlation between photosynthetic rate and shoot dry matter except at 14 DAS in 2009 (Table 3.1 and Fig. 3.7), suggesting that the photosynthetic rate is not a main factor in the seedling development. Harris et al. (1986) argued that energy storage is a more important role of soybean cotyledons, and that their photosynthetic capacity is sufficient only to compensate for the respiratory losses of the seedling.

3.4 Formation of Nodes

Formation of nodes on the main stem was more rapid after emergence for the delayed planting than the earlier. Delayed planting tended to reduce the number of nodes produced on the main stem between R1 and R5 compared with the early planting and caused a lower number of nodes on the main stem at harvest (15.5) than the early planting (16.3). Differences in number of nodes produced during R1 and R5 result from differences in the rate of node production or variation in the length of the flower to pod setting period. The differences between the two planting dates were primarily because of differences in rate of node production and not in the length of the flower to pod setting period.

Soybean in the different management systems tended to have similar node number on the main stem and small differences were observed during the vegetative stages. From R3 to harvest, the most nodes on the main stem was found at the four management systems at Arlington, averaging 5% more nodes on the main stem at harvest than the management system at Hancock. Tillage system did not affect node number on the main stem at Arlington. However, soybean in the two irrigated systems at Arlington averaged 2% more nodes on the main stem than those in the non-irrigated systems, which is consistent with Korte et al. (1983). Momen et al. (1979) observed little effect on node number from irrigation.

Other comparisons of cultivars have shown that even though a cultivar produced the fewest nodes on the main stem, it may have produced the most nodes on the plant, and more extensive branching (Egli et al., 1985; Parvez et al., 1989).

Venkatareddy (1991) reported that shoot length in soybean genotypes was highest (20.10 cm) in KHSb2, followed by Monetta, medium length in PK-71 and Bragg, whereas the lowest (13.40 cm) in Hardee. Similarly, root length was the shortest (19.70 cm) in Hardee and PK-71 over Monetta, Bragg and KHSb2 but no genotype was recorded for long length. Whereas, the length of seedling was the highest (45.20 cm) in KHSb-2, followed by Monetta and Bragg, but it was the lowest (33.20 cm) in Hardee. No genotype was recorded for medium group of seedlings.

While the qualitative consequences of incorporating the long-juvenile gene into soybean are obvious, there has been little quantitative assessment of the response of commercial long-juvenile cultivars to environmental variables. Sinclair et al. (1991) included in their analysis of flowering date of soybean the long-juvenile line PI 159225, which has been a genetic source for the long-juvenile trait. They found the response to temperature and photoperiod in this cultivar was similar to other cultivars. The critical difference in the long-juvenile genotypes was that its overall maximum development rate to flowering was much less than all other cultivars. Subsequently, Sinclair and Hinson (1992) reported an analysis of the differences in flowering of 15 isoline pairs developed using PI 159925 as a parent. While the overall maximum development rate was decreased in 14 of the 15 lines, their analysis also indicated that both the temperature and photoperiod coefficients were systematically changed in the long-juvenile isolines.

Eight cultivars, including three long-juvenile cultivars, were sown each week throughout a year in a plastic greenhouse so that the plants would develop under differing temperature and photoperiod. While there were differences among cultivars in the cumulative temperature required for plant emergence and the rate of leaf appearance, these differences were not necessarily associated with the long-juvenile trait. An extended duration to flowering was confirmed for the three long-juvenile cultivars but this delay was not associated with any difference in sensitivity to temperature and photoperiod. The trait that distinguished the long-juvenile cultivars was a much lower maximum development rate towards flowering than that found in the other cultivars (Sinclair et al., 2005).

Mean number of days from sowing to flowering

Cultivars	Sow to flower (days)
Normal development	
Bragg	31.3
IAS-5	35.1
BR-15	42.7
Delayed flower	
Parana	35.3
OCEPAR-8	37.2
Long-juvenile	
OCEPAR-9	46.8
BR-27	51.7
Paranagoiara	54.6

Khan et al. (2003) observed that the determinate and indeterminate soybean types have different growth habits. Number of days to emergence, unifoliate and 6th trifoliate leaf formation were reduced in both cultivars with delay in sowing.

Highest plant density of 60 plants m^{-2} attained maximum plant height as compared with the lowest plant densities. The number of days to maturity declined with each successive planting date for both cultivars.

Investigation on the effect of cultivars, growth habit, and environmental conditions on the partitioning of assimilate between vegetative and reproductive plant parts in soybean [*Glycine max* (L.) Merr.] during flowering and fruit set in experiments included the following variables: Cultivars from maturity group II to V, including both indeterminate and determinate growth habits, optimum and delayed plantings, and moisture stress, data were collected from growth stage R1 to R6 on vegetative growth characteristics (nodes on main stem, total nodes, and plant height), pod development, and weight of vegetative and reproductive plant parts. Partitioning coefficients were calculated by dividing fruit weight by the total aboveground biomass. The nodes on the main stem and on the branches reached a maximum at R5 in all cases. Determinate cultivars produced fewer nodes on the main stem between R1 and R5; however, they produced more branch nodes and the total node production was greater than the indeterminate cultivars. The number of pods containing developing seed increased until approximately R6, suggesting that the production of new vegetative sinks (node production) in soybean stopped before the total fruit load was established. The partitioning coefficients increased rapidly from < 5% at growth stage R3 to ≥ 30% at

R6. The partitioning coefficient at any growth stage varied only slightly across the variables included in the experiments. This suggests that variation in fruit and seed number may be more closely related to variations in crop growth rate than to variations in the allocation of assimilate between vegetative and reproductive plant parts.

But the specific time between stages, number of leaves developed, and different varieties, seasons, locations, planting dates, and planting patterns ultimately determine the growth.

For example:

1. An early maturing variety may develop fewer leaves or progress through the different stages at a faster rate than indicated here, especially when planted late. A late maturing variety may develop more leaves or progress more slowly than indicated here.

2. The rate of plant development for any variety is directly related to temperature, so the length of time between the different stages will vary as the temperature varies both between and within the growing season.

3. Deficiencies of nutrients, moisture, or other stress conditions may lengthen the time between vegetative stages, but shorten the time between reproductive stages.

4. Soybeans planted at high densities tend to grow taller and produce fewer branches, pods, and seeds per plant than those planted at low densities. High density soybeans also will set pods higher off the ground and have a greater tendency to lodge.

3.5 Growth of Plant Parts

Rate of growth of the leaves, petioles, and stems closely follows that of the whole plant until the pods and beans begin growth, or at about R4. Shortly after R5.5, dry weight maximizes in these vegetative parts and begins to relocate to the rapidly developing beans. Leaf and petiole loss begins at V4 to V5 on the lowest node leaves and petioles and progresses very slowly upward until shortly after R6. At this time, leaf and petiole loss becomes rapid and continues until R8 when all leaves and petioles have generally fallen.

Root growth begins when the primary root emerges from the planted seed. Under favourable conditions, the primary root and several major laterals grow rapidly and may reach a depth of 0.8 to 1.0 m (2 .5 to 3.25 feet) by V6. During late vegetative and early flowering (V6 to R2), the root system is expanding at its fastest rate. Most of this growth occurs in the upper 30 cm (12 inches) of soil, if adequate soil moisture is available. Some roots may be within 2.5 cm (1 inch) of the surface. By R6, under favourable conditions, soybean roots may have reached depths greater than 1.8 m (6 feet) and have spread 25 to 50 cm (10 to 20 inches) laterally. Roots are growing very slowly at this stage, but some root growth continues until physiological maturity (R7).

Pantalone et al. (1996) related visual root scores to other measurable root characteristics in order to provide breeders with a means for rapid phenotypic evaluation of soybean roots. Root score, root surface, and root dry weight were measured at three soybean stages of development in 1992 and 1993. Nodule number and nodule dry weight were also measured during the second year. PI416937 had higher root score, root surface, nodule number and nodule dry weight than Lee 74. Root score was positively correlated with root surface, nodule number and nodule dry weight. Phenotypic root scores could be utilized effectively in selection programmers to rapidly evaluate large numbers of progeny in order to identify those with extensive fibrous root systems in the vegetative plant parts, but between R3 to R5.5, accumulation gradually shifts to the pods and beans. A portion of the nitrogen used by the soybean plant is made available by fixation of N from the air by *Rhizobium japonicum* bacteria in the root nodules. These bacteria infect the roots causing nodule production as early as the V1 stage. Throughout the V stages, the number of nodules increases along with the rate of N fixation. At about R2, the N-fixation rate increases dramatically, peaks at about R5.5, and drops rapidly thereafter.

The impact of starter nitrogen fertilizer on soybean root activity, leaf photosynthesis, grain yield and their relationship on field experiments were conducted in 2013 and 2014. Nitrogen was applied during

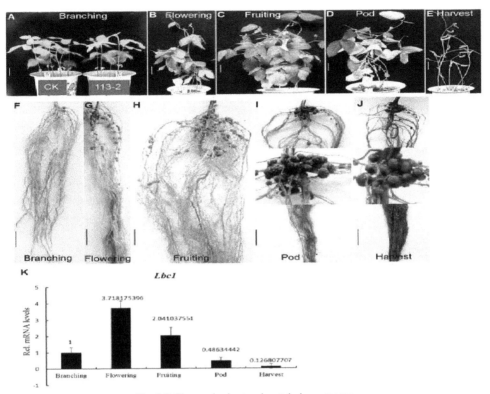

Fig. 3.8 Changes in shoot and root during ontogeny.

planting at rates of 0, 25, 50 and 75 kgN ha^{-1}. In both years, starter nitrogen fertilizer benefited root activity, leaf photosynthesis and, consequently, its yield. Statistically significant correlation was found among root activity, leaf photosynthetic rate, and grain yield at the developmental stage. The application of N25, N50 and N75 increased grain yield by 1.28%, 2.47% and 1.58% in 2013 and by 0.62%, 2.77% and 2.06% in 2014 compared to the N0 treatment. Maximum grain yields of 3238.91 kgha^{-1} in 2013 and 3086.87 kgha^{-1} in 2014 were recorded for N50 treatment. Grain yield was greater for 2013 than 2014, possibly due to more favourable environmental conditions. This indicated that applying nitrogen as starter is necessary to increase soybean yield in Sangjiang River Plain in China (Tables 3.2 & 3.3).

Table 3.2 Root dry weight (g/plant) on level of starter nitrogen.

Nitrogen	V4 (2013)	V4 (2014)	R2 (2013)	R2 (2014)	R4 (2013)	R4 (2014)
0 kg/ha	0.26	0.29	1.38	1.63	3.01	3.05
25 kg/ha	0.36	0.36	1.87	1.88	3.11	3.25
50 kg/ha	0.38	0.45	1.96	2.08	3.18	3.37
75 kg/ha	0.31	0.33	1.51	1.83	3.07	3.17

Table 3.3 Root activity (TTC reduction µg/g/h) on level of starter nitrogen.

Nitrogen	V4 (2013)	V4 (2014)	R2 (2013)	R2 (2014)	R4 (2013)	R4 (2014)
0 kg/ha	78.3	72.5	129.8	123.3	105.4	99.2
25 kg/ha	80.3	79.6	131.7	127.8	110.0	102.3
50 kg/ha	96.6	94.1	148.4	143.1	125.5	113.2
75 kg/ha	83.8	84.2	134.5	133.2	111.4	108.4

The rate of increase in soybean plant dry weight is very slow at first but gradually increases through the V stages and R1 as more leaves develop and ground cover increases. At about R2, the daily rate of dry weight accumulation by the whole plant is essentially constant until it gradually decreases during the late seed-filling period (shortly after R6) and measurably stops shortly after R6.5 (Fig. 3.9).

Egli et al. (1990) reported that the seed size had no effect on specific growth rate or NAR. Seed vigor had no effect on SGR, NAR or seedling weight at 20 DAE. In a 2nd set of experiments, individual seed conductivity of seeds from high and low vigor seed lots were related to seedling DW accumulation. The data indicated that seed vigor has no effect on the ability of the seedling to accumulate DW if there is no injury to the cotyledons. It is reported that erect plant type in soybean was superior than spreading type for almost all the growth and physiological parameters, viz., fresh weight, number of root nodule, LAI, NAR and CGR significantly increased with 0 to 80 kg P_2O_5 ha^{-1}.

Soybean (*Glycine max* L. cv Enrei) seeds were sown in each seedling case for 10 days and roots, hypocotyls, and leaves were collected. Proteins were extracted and analyzed by nano-liquid chromatography mass spectrometry. Mole percent abundance was calculated using emPAI values. To determine the functional role of the proteins identified in the MS analysis, functional categorization was performed using MapMan bin codes. Visualization of protein abundance ratio was performed using MapMan software. Enzyme activity and quantitative reverse transcription-polymerase chain reaction analyses were performed.

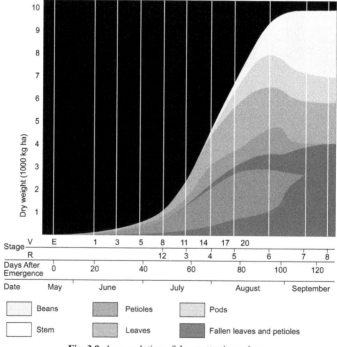

Fig. 3.9 Accumulation of dry matter in soybean.

3.6 Proteins of Vegetative Stage

A total of 357, 360 and 392 proteins were identified in root, hypocotyl, and leaf of vegetative stage soybean, respectively. Proteins related to stress, cell organization, transport, signalling, and mitochondrial electron transport chain decreased in root, hypocotyl, and leaf. Proteins related to protein metabolism, glycolysis, and cell wall were comparable in root and hypocotyl; however, in leaf, glycolysis and cell wall related proteins were decreased. Aldehyde dehydrogenase was decreased in abundance and activity in hypocotyl and leaf as compared to root. Major latex proteins 43 and 423 changed in abundance in

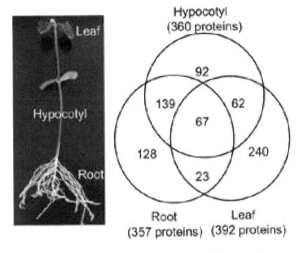

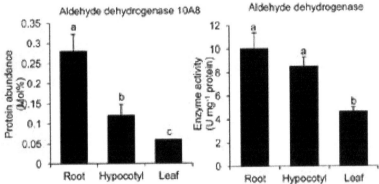

Fig. 3.10 Abundance of proteins in seedling parts of soybean.

an organ-specific manner. The mRNA expression level of major latex proteins exhibited a differential expression in the hypocotyl of soybean during flooding stress and recovery. Results suggest that aldehyde dehydrogenase and major latex proteins play key roles in the growth of soybean in an organ-specific way.

3.7 Temperature

The crop growth is considered optimum in the temperature range of 26.5 to 30°C, but growth ceases at temperatures below 10°C. Agrometeorological conditions cause wide fluctuations in growth, development and yield of soybean crop. Brown (1960) proposed the "soybean development unit" based on temperature, which could be used to predict soybean maturity. Whigham and Minor (1978) have studied the influence of temperature at different growth stages of soybean. Reduced moisture availability affects several physiological functions of soybean plants. Insufficiency of water during the pod development can limit the yield severely. Providing support irrigation at flowering and pod development has been useful in attaining better yield (Wang et al., 1980). Jeyaraman et al. (1990) studied the influence of weather parameters prevailing during critical growth stages and whole cropping period on the seed yield of soybean. Samui et al. (2002) have studied inter relationship of various energy balance components during the growth of soybean.

Soybean plants are also sensitive to chilling in the juvenile period; temperatures above zero but below 10°C cause damage to soybean plants, resulting from temperature-induced physiological and biochemical changes (Wolfe, 1991; McKersie and Leshem, 1994). A distinct increase in proline content belongs to the most frequently encountered changes and it has been observed in soybean seedlings by Heerden et al. (2002) and Yadegari et al. (2007).

High temperature causes loss of cell water content, which leads to the cell size and ultimately the growth being reduced (Rodriguez et al., 2005). Reduction in net assimilation rate (NAR) is also another reason for reduced relative growth rate (RGR) under HT, which was confirmed in maize and millet (Wahid, 2007) and sugarcane (Srivastava et al., 2012). The morphological symptoms of heat stress include scorching and sunburns of leaves and twigs, branches and stems, leaf senescence and abscission, shoot and root growth inhibition, fruit discoloration and damage (Rodriguez et al., 2005).

Tan et al. (2011) noted that soybean (*Glycine max*), when exposed at 38/28°C (day/night), for 14 days at flowering stage decreased the leaf Pn and stomatal conductance (*gs*), increased thicknesses of the palisade and spongy layers, damaged plasma membrane, chloroplast membrane, and thylakoid membranes, distorted mitochondrial membranes, cristae and matrix.

Chilling stress also causes the leakage of intracellular electrolytes from tissues as a result of the loss of cytoplasmic membrane integrity (Baczek-Kwinta et al., 2004; Borowski and Blamowski, 2009). The destructive effect of chilling on the membranes is even greater in the light than in the dark (Szalai et al., 1996). This is undoubtedly associated with the negative effect of stress also on the process of photosynthesis, which has been found both in soybean (Heerden et al., 2003a, b) and in other species (Lu-Cun Fu et al., 1994; Starck et al., 2000).

Similarly to other types of environmental stress, chilling induces the production of H_2O_2 and other reactive oxygen species (ROS) in plants. Under such conditions, plants activate the enzymatic system that prevents ROS accumulation. One of the elements of this system is catalase (CAT EC 1.11.1.6). Increased catalase activity under chilling conditions has been observed in germinating soybean seeds by Posmyk et al. (2001), while in seedlings of soybean and other species by Prasad et al. (1994) and Posmyk et al. (2005). Increased synthesis of proline and other substances serving as osmoprotectants and antioxidants (ROS) under the influence of chilling requires large energy inputs from plants, which causes the inhibition of growth and development under stress conditions (Borowski and Blamowski, 2009).

Another experiment, carried out using pot cultures, investigated the response of 2-week soybean plants of the same cultivars to a 6-day chilling period. The following temperatures were used: 25/20°C (control), 25/0°C, 10/0°C (day/night). Both experiments tested the response of 6 domestic soybean cultivars ('Aldana', 'Jutro', 'Progres', 'Mazowia', 'Nawiko', and 'Augusta') and 2 Canadian cultivars ('OAC Vision', 'Dorothea') to chilling. The obtained results showed that a temperature of 10°C used during germination (28/10°C), and even to a larger extent during imbibition and germination (10/10°C), clearly reduced the speed of germination, percentage of germinated seeds, and radical length relative to the control, but it increased catalase activity in sprouts. A chilling temperature of 25/0°C and 10/0°C (day/night) significantly increased leaf electrolyte leakage, free proline content and catalase activity relative to the control, but it decreased the photosynthetic rate and total plant leaf area. Seeds and seedlings of cvs. 'Jutro' and 'Nawiko' were generally the least sensitive to chilling, while 'Aldana' and 'Dorothea' were the most sensitive.

Table 3.4 Effect of short-term chilling on the rate of photosynthesis in plants of 8 soybean cultivars (μmol CO_2 m^{-2} s^{-1}).

Temperature °C	Cultivars								
Day/Night	Aldana	Jutro	Progres	Mazowia	Nawiko	Augusta	OAC Vision	Dorothea	Mean
25/20	4.68	5.92	4.95	4.72	6.02	4.85	4.95	4.87	5.12
25/0	3.89	4.78	3.98	3.63	4.45	3.64	4.03	4.21	4.08
10/0	2.71	3.16	2.85	2.14	3.32	2.28	2.32	2.11	2.61
Mean	3.76	4.62	3.93	3.50	4.60	3.59	3.77	3.73	
LSD0.05: for temperature – 0.35; for cultivars – 0.72; for interaction – 1.40.									

Table 3.5 Effect of short-term chilling on catalase activity in leaves of soybean seedlings of selected cultivars ($\mu g. \ g^{-1}$ FW).

Temperature °C	Cultivars				
Imbibition/Germination	Jutro	Nawiko	Aldana	Dorothea	Mean
25/20 (control)	20.4	22.6	21.9	19.8	21.2
25/0	61.6	62.9	54.8	52.3	57.9
10/0	87.6	89.5	73.5	76.8	81.8
Mean	56.5	58.3	50.1	49.6	
LSD0.05: for temperature – 22.3; for cultivars – n.s.; for interaction – n.s.					

Table 3.6 Effect of short-term chilling on total leaf area in plants of 8 soybean cultivars (dm^2 $plant^{-1}$).

Temperature °C	Cultivars								
Day/Night	Aldana	Jutro	Progres	Mazowia	Nawiko	Augusta	OAC Vision	Dorothea	Mean
25/20	1.60	2.38	1.95	1.85	2.45	1.74	1.88	1.94	1.97
25/0	1.28	1.97	1.57	1.48	2.07	1.39	1.52	1.40	1.58
10/0	0.87	1.55	1.21	1.18	1.59	1.12	1.17	1.11	1.22
Mean	1.25	1.97	1.58	1.50	2.04	1.42	1.52	1.48	
LSD0.05: for temperature – 0.17; for cultivars – 0.37; for interaction – n.s.									

3.8 Water

The water requirements of soybean vary with soil, climatic conditions, growth duration, and yield level of cultivars. Water use for soybeans can vary from 450 to 825 mm, where the growing season ranges from 100 days at low altitude to 190 days in higher altitudes (Doorenbos and Pruit, 1977). Generally, soybean water use is low during the germination and seedling stages; the water use is especially high during the reproductive stages (R1–R6) and less during the maturation stages (Fig. 3.11). In parallel to this pattern of water use, irrigation for soybeans was often carried out when the soil water depletion reaches 80% in the vegetative stage, 45% in early to peak flowering, 30% in late flowering to early pod development, and 80% in late pod to maturity. General field observation of drought stressed soybeans indicates that the amount of soil water available to the plants throughout all developmental stages exerts a major influence on plant growth. According to Brady et al. (1974), the best yield and most efficient water use are generally obtained when the available soil water in the root zone is not depleted by more than 50–60%. Thus, sufficient water supply, especially during the early reproductive stages is essential for soybean production under water-limited conditions.

In a study, ET data available on soybean crop for Bhopal during 1991–95 have been utilized. With regard to the water requirement of the crop, the life span of soybean has been divided into five important growth stages, viz., seedling up to 2 weeks after sowing (WAS), vegetative (3–8 WAS), flowering (9–10 WAS), pod development (11–13 WAS), and maturity (14–15 WAS). In this report, consumptive use of water (ET), Water Use Efficiency (WUE), Heat Units (HU), Heat Use Efficiency (HUE) and crop coefficient (Kc) for different growth stages of the crop have been computed and discussed.

The study revealed that, on average, soybean crop consumed about 450 mm of water. The average WUE was found to be 3.23 kg/ha/mm. It was also observed that WUE does not depend only on the total amount of water consumed by the crop but also indicates the importance of its distribution during various growth stages. On average, the crop consumed nearly 7%, 36%, 24%, 25% and 8% of water during seedling, vegetative, flowering, pod development and maturity stages, respectively. The crop consumed maximum amount of water during vegetative stage. However, the average weekly ET rate was found to be highest during flowering stage (nearly 52 mm). Average heat unit requirement of soybean was found

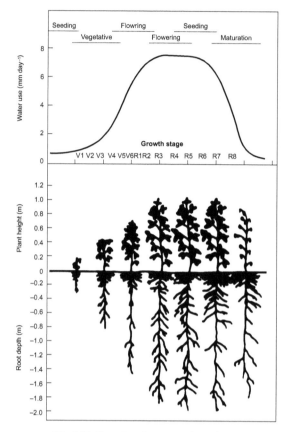

Fig. 3.11 Growth and water use in soybean.

to be 1694 degree-days. Maximum heat units were required during vegetative stage (638 degree days), followed by pod development stage (358 degree days). The average HUE was found to be 0.86 kg/ha/ degree days. Crop coefficient (Kc) values varied in the range 0.30–0.45, 0.55–0.90, 1.00–1.15, 0.85–0.70 and 0.55–0.40 during seedling, vegetative, flowering, pod development and maturity stages, respectively. The crop coefficient values attained the peak during the flowering stage (Tables 3.7 & 3.8).

Table 3.7 Agrometeorological parameters and crop yield of Soybean at Bhopal.

Station Year Crop Variety	Crop duration (Weeks)	Crop yield (kg/ha)	ET* (mm)	WUE* (kg/ha/mm)	HU	HUE*
					(degree days)	
Bhopal 1991 PK472	14	1200	414	2.90	1490	0.81
Bhopal 1992 Punjab 1	14	1450	442	3.28	1509	0.96
Bhopal 1993 Punjab 1	15	1860	484	3.84	1639	1.13
Bhopal 1994 Punjab 1	14	1575	438	3.60	2150	0.73
Bhopal 1995 Punjab 1	15	1150	458	2.51	1680	0.68
Mean - - -		1447	447	3.23	1694	0.86

* ET – Evapotranspiration, * WUE – Water use efficiency, * HUE – Heat use efficiency.

Table 3.8 Phase wise water requirement (ET, percent ET* and weekly ET) and growing degree days (HU, HU percent* and weekly HU) of Soybean crop at Bhopal.

Seedling		Vegetative		Flowering		Pod development		Maturity	
ET (mm)	Weekly ET (mm)	ET (mm)	Weekly ET (mm)	ET (mm)	Weekly ET (mm)	ET (mm)	Weekly ET (mm)	ET (mm)	Weekly ET (mm)
32 (7.2)	16	160 (35.8)	27	105 (23.5)	52	113 (25.2)	38	37 (8.3)	19
HU	Weekly HU	HU	Weekly HU	HU	Weekly HU	HU	Weekly HU	HU	Weekly HU
(degree days)									
241 (14.2)	120	638 (37.7)	106	234 (13.8)	117	358 (21.1)	119	223 (13.2)	112

ET – Evapotranspiration, HU – Heat units, Figures in parenthesis are percentages. Average WUE.

3.9 Light

The study aims to determine the effect on various intensities of light on the response of vegetative growth of local soybean to three vegetative stages. Greenhouse experiments with 15 experimental units were prepared in a completely randomized design with treatments consisting of: Without shade, 50, 60, 70, and 80% shade. The observed variables are plant height, number of leaves, number of branches, and stem diameter, root length, wet weight, and dry weight of stove.

The results showed that shading percentage had an effect on plant response on all growth variables until the end of vegetative-3 stage. The higher the shade percentage, the more increase the growth of plant height and root length, on the contrary, decrease the growth of the number of leaves, in effect on the decrease of leaf number, branch number, stem diameter, and wet weight and dry weight of plant.

Luquez et al. (1997) compared wild type soybean variety (Clark) and chlorophyll deficient line homozygous for the recessive allele Y9 under watered or water stress condition and found that matured leaves of Y9 had a 65 percent lower chlorophyll content than wild type. However, net photosynthetic rate (PN) of Y9 leaves was only 20 percent lower than wild type. Transpiration rates (E) were significantly higher in the leaves of Y9 than in the wild type; the higher E of Y9 correlated with increase in stomatal conductance.

Koesmaryono et al. (1998) investigated diurnal variation of net photosynthetic rate (PN) and transpiration rate (E) in six week old soybean (cv. Fukuyutaka) under condition of 100, 50 or 25 percent of full sun irradiance (I-100, I-50, I-25 plants) and found that, in the morning, photosynthetic rate and transpiration rate in plant at irradiances, grew. However, during the afternoon, all plants were tested under full irradiance. At low level of irradiance, low PN and E was observed. Stomatal conductance was considerably lower in I 25 plants only.

Determinate cultivar Delta Pine 36 (Maturity Group VI) was planted at an optimal planting date during 1995 and 1996 at low (80000 plants ha^{-1}), medium (145000 plants ha^{-1}), and high (390000 plants ha^{-1}) plant populations on a Commerce silt loam near Baton Rouge, LA (30'N). Yield was unaffected by plant population'. Equilibration of CGR for low vs. higher plant populations near R1 was achieved through greater NAR for the low plant population during the vegetative period, created by greater light interception efficiency (LIE, light interception per unit leaf area). Although NAR equilibrated to minimal levels across plant populations near Rl, low population maintained CGR parity with higher populations until R5 through greater relative leaf area expansion rate (RLAER) during the late vegetative and early reproductive periods. Higher relative leaf area expansion rate for low vs. higher plant populations resulted from increased partitioning of dry matter into branches, probably induced by greater red/far red light ratios within the canopy. In conclusion, a possible genetic characteristic conducive to low optimal plant population is greater partitioning of dry matter into branches (Fig. 3.12).

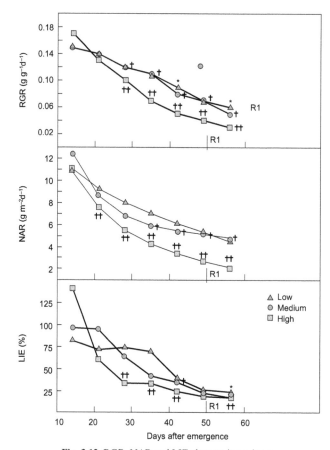

Fig. 3.12 RGR, NAR and LIE changes in soybean.

3.10 Photoperiod

Soybeans were subjected to three photoperiod treatments: (1) short days-(SD) 9-h solar radiation; (2) long days-(LD) 9-h solar radiation with a 3-h night interruption; and (3) 10 short days plus long days (SD + LD). Short inductive photoperiods of SD and SD + LD promoted a rapid transition of axillary and terminal shoot meristems to the reproductive condition. Terminal shoot meristems were transformed to floral racemes, and vegetative structure differentiation stopped, coincident with the beginning of branch elongation, indicating a release from apical dominance. Long photoperiods of SD + LD and LD promoted elongation of internodes and leaf expansion. Unlike SD plants, SD + LD plants developed all of the differentiated but not elongated apical internodes, and terminal racemes became visible. In the non-induced LD plants, inter node elongation was promoted, and vegetative growth of terminal shoot meristems was prolonged. The latter effect may be caused by (1) a delay in flower induction and (2) independent behavior of terminal meristems with respect to the reproductive condition of the axillary meristems. Photoperiod may regulate endogenous hormonal levels, and correlative controls of the vegetative growth are postulated.

Toledo et al. (1993) observed that the soybean is a short day plant highly influenced by photoperiod. Response to day length is determined by the cultivars, and the genetic control of flowering and growth is distinct and independent for long and short day conditions. Parental final plant height, days to flowering, trifoliolate leaf number and average length of the internodes and the genetic mechanisms controlling photoperiodic response of these traits.

3.11 Carbon Dioxide

Seed yield increases of soybeans in response to increases in CO_2 concentration of 180 to 200 $\mu mol.mol^{-1}$ above the current ambient concentration applied using free air carbon dioxide enrichment (FACE) systems have ranged from about 0% to 45% in different cultivars (Bishop et al., 2015; Bunce, 2014; Bunce, 2015; Bunce, 2016). Reasons for the wide range of responses among cultivars at the same location remain unclear. Delay in the transition from vegetative to reproductive growth in response to elevated CO_2 varied among cultivars and was highly correlated with the seed yield increase in both indoor chambers and field FACE systems (Bunce, 2015). Delayed transition to reproductive growth increased main stem and axillary node number, providing more sites for pods, and increasing seed yield. However, the cultivars compared in that study varied in maturity group, and it is not known whether variation exists within a soybean maturity group in effects of elevated CO_2 on the duration of vegetative growth, or whether any such variation would be correlated with yield increase. Soybean cultivars used in North America have been assigned to "maturity groups" in order to specify the latitudinal band best suited to that cultivar.

Prior experiments in indoor chambers and in the field using free-air carbon dioxide enrichment (FACE) systems indicated variation among soybean cultivars in whether and how much elevated CO_2 prolonged vegetative development. However, the cultivars tested differed in maturity group, and it is not known whether variation exists in CO_2 effects on the duration of vegetative growth within a maturity group. In these experiments, a total of five soybean cultivars of maturity group IV were grown at ambient and elevated CO_2 in the field in Maryland, USA using FACE systems, over three years. The time of first flowering, the time of the first open flowers at the apex of the main stem, the total number of main stem nodes at maturity, and seed yield were recorded. In each year of the study, there were cultivars in which elevated CO_2 did not affect the duration of vegetative growth or the main stem node number, and other cultivars in which elevated CO_2 prolonged vegetative growth and increased the number of main stem nodes and seed yield at maturity. The stimulation in yield by elevated CO_2 was highly correlated with the increase in the number of main stem nodes, indicating that CO_2 effects on the duration of vegetative growth may be important in adapting soybean to higher atmospheric CO_2 (Fig. 3.13).

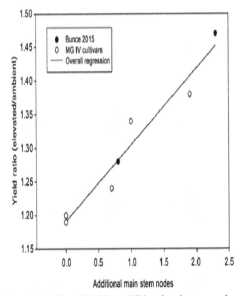

Fig. 3.13 Effect of yield on additional main stem nodes.

3.12 Waterlogging

A major agronomic problem in the southeastern USA is low yield of late-planted soybean (*Glycine max* (L.) Merr.). This problem is aggravated by the adverse effect of waterlogging on crop growth.

Experiments were undertaken to identify soybean growth stages sensitive to waterlogging, identify yield components and physiological parameters explaining yield losses induced by waterlogging, and determine the extent of yield losses induced by waterlogging under natural field conditions. Greenhouse and Field studies were conducted during 1993 and 1994 near Baton Rouge, LA, on a Commerce silt loam. Waterlogging tolerance was assessed in cultivar Centennial (Maturity Group VI) at three vegetative and five reproductive growth stages by maintaining the water level at the soil surface in a greenhouse study. Using the same cultivar, it evaluated the effect of drainage in the field for late-planted soybean. Rain episodes determined the timing of waterlogging, redox potential and oxygen concentration of the soil were used to quantify the intensity of waterlogging stress.

Results of the greenhouse study indicated that the early vegetative period (V2) and the early reproductive stages (R1, R3, and R5) were most sensitive to waterlogging. Three to 5 cm of rain per day falling on poorly drained soil was sufficient to reduce crop growth rate, resulting in a yield decline from 2453 to 1550 kg ha[-1]. Yield loss in both field and greenhouse studies was induced primarily by decreased pod production, resulting from fewer pods per reproductive node. Thus, waterlogging was determined to be an important stress for late-planted soybean in high rainfall areas, such as the Gulf Coast Region (Fig. 3.14).

In summary, waterlogging influenced yield through yield formation mechanisms similar to those shown by other stresses (non-optimal row spacing, reduced light, partial defoliation) whose main effect is to reduce crop growth rate. It appears that, regardless of the type of stress that influences crop growth rate, effects on yield components are similar.

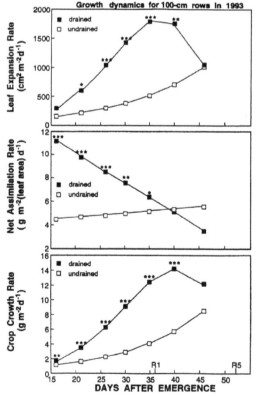

Fig. 3.14 Growth dynamics of soybean under waterlogging.

Waterlogging can also effect soybean when soybean was flood irrigated and water ponds on the land for more than two days (Griffin and Saxton, 1988; Scott et al., 1989). Research has generally indicated that soybean is more sensitive to waterlogging during the early reproductive period [R1(first flower) to R5 (seed initiation)] stages, according to Fehr and Caviness (1977) vs. the vegetative period (emergence to R1) (Griffin and Saxton, 1988; Linkemer et al., 1998). Inadequate oxygen supply for root respiration is the main cause of reduced yield under waterlogged conditions (Grable, 1966; Russell et al., 1990).

This results in disruption of root growth and function, nodulation, nitrogen fixation, photosynthesis, crop growth rate and stomatal conductance (Bacanamwo and Purcell, 1999a, b). Since all these secondary factors are interrelated, it is difficult to identify the main factor resulting in yield loss. Although changes in soil chemistry and plant hormones are recognized (Reid and Bradford, 1985), an immediate response to waterlogging-induced hypoxia is reduced uptake and concentration of nitrogen, induced partly by reduced nodulation (Minchin and Pate, 1975), reduced rate of nitrogen fixation (Bacanamwo and Purcell, 1999a) and denitrification of soil nitrate (Ponnamperuma, 1972). Nitrogen is considered the most important plant nutrient for crop dry matter accumulation because of its influence on enzymes regulating photosynthesis (Hay and Porter, 2006). If sustained, nitrogen deficiency can restrict growth so that leaf area index will be reduced below levels required for optimal light interception (95%) (Hay and Porter, 2006) and crop growth rate. However, waterlogging is also recognized as being responsible for reducing the availability of other minerals to crop plants (Kozlowski and Pallardy, 1985).

Table 3.9 Yield for soybean cultivar Hartz 5000 flooded over two years and grown near Baton rouge, LA, 1999 and 2000.

Waterlogging treatment	Yield (kg.ha^{-1})
Control (no waterlogging)	4153
1 wk waterlogging at V4	3454
1 wk waterlogging at R1	3618
1 wk waterlogging at R3	3485
1 wk waterlogging at R5	3380

3.13 Salinity

Soil degradation caused by salinization and sodification is of universal concern. Nearly one billion hectares of soil around the world had some degree of salinization and sodification problem (FAO, 1992). About 2.78 million hectares of land is classified as unsuitable for agriculture due to salinization and sodification. This problem manifests itself especially in arid and semi-arid areas with poorly drained soils because of continual addition of salts with irrigation practices (Ayars and Tanji, 1999). High soil salinity causes nutrient imbalances, result in the accumulation of elements toxic to plants, and reduces water infiltration if the level of one salt element—sodium—is high. Salinity affects plant growth through ionic and osmotic effects. Sometimes, these effects are distinct from each other, other times they overlap each other.

Results have indicated that salinity affects growth and development of plants through osmotic and ionic stresses. Due to the accumulated salts in soil, plants wilt apparently under salt stress conditions, while soil salts, such as Na$^+$ and Cl$^-$, disrupt normal growth and development of plants (Farhoudi et al., 2007). The difference in a plant response to a given level of salinity is dependent on the concentration and composition of ions in solution as well as the genotype that is exposed to the salinity (Cramer, 1992). Seed germination is usually the most critical stage in seedling establishment, determining successful crop production (Almansouri et al., 2001).

Soybean is an important agricultural crop and has, among its genotypes, a relatively wide variation of salt tolerance. As measured by vegetative growth and yield, however, the emergence or failure of a high emergence ratio and seedling establishment on saline soils can have significant economic implications in areas where soil salinity is a potential problem for soybean.

Fifteen soybean genotypes were tested in sand culture experiment. The seeds were irrigated with saline waters of different EC levels (0, 3, 6, 7.2, 10, 12, 14 dSm^{-1}). Length and dry weight of root and shoot as well as PR were evaluated under salinity at 7 DAS. Salinity significantly reduced dry matter accumulation in both roots and shoots in all the cultivars, though declension was more pronounced in PS 1347 and PS 1024. Shoot growth was affected more adversely than root growth. Cultivars showed a wide range of variation in their salinity tolerance as mediated by PR (percent reduction in seedling dry weight over control) and SSI (salinity susceptibility index). PK 1029 and PK 416 exhibited higher levels of tolerance to salinity compared to the other cultivars (Table 3.10).

The growth of root and shoot is the most important parameter for salt tolerance because roots are in direct contact with the soil and absorb water from the soil and shoot supply it to the rest of the plant. For this reason, root and shoot length provides an important clue to the response of plants to salt stress (Jamil et al., 2005). Salt stress inhibited the root and shoot length of all the soybean genotypes as the level of salinity increased, however shoot length was more affected than root length. Similar results were reported by (Farhoudi et al., 2011) who demonstrated that root growth is less inhibited than the shoots in most of the crops. These results differ from those reported by Jamil et al. (2007). They found that decrease in the length of root was more prominent than in the shoot.

Table 3.10 Effect of salinity on germination and seedling growth of soybean after 7 days of germination.

Variety	Salinity (dSm^{-1})	% germ.	Length (cm)		Dry wt. (mg. plant^{-1})		Seedling Height (cm)	Seedling dry wt. (mg.plant^{-1})	SSI
PK327			Shoot	Root	Shoot	Root			
	0	92	13.6	6.00	18.00	8.00	19.60	23.00	
	3.0	80	9.00	5.20	14.00	4.46	14.20	18.46	0.511
	6.0	68	3.30	2.20	6.00	1.40	5.50	7.40	1.050
	7.2	56	2.50	1.20	4.20	0.80	3.70	5.00	1.030
	10.0	32	1.50	0.00	3.20	0.00	1.50	3.20	1.010
	12.0	24	0.80	0.00	2.80	0.00	0.80	2.80	0.987
	14.0	10	0.60	0.00	2.40	0.00	0.60	2.40	0.965
	CD 5% P	1.98	0.45	0.75	0.77	0.69	0.54	0.74	

3.14 Drought

An experiment was conducted to investigate the drought tolerance in 11 genotypes of soybean at vegetative growth stage. The plants were cultivated in the form of completely randomized blocks design as a factorial with two factors of drought stress and genotype. The factor of genotype had 11 levels and the factor of drought stress has 4 levels including control, –3, –7 and –11 bar soil water potential. The experiment was done in greenhouse of Gorgan University. The water requirement of plants for each irrigation was calculated with daily scaling of pots and soil moisture curve drawing. The results indicated shoot height, leaf area, dry matter of roots and total plant were significantly reduced under drought stress. The transpiration rate decreased under drought stress, consequently, reduction of leaf area (and above ground plant) was more severe than root dry matter. Maintenance of root growth may result in the higher water uptake and better adaptation to drought stress. Proline concentration increased in plants tissues under drought stress. The higher proline concentration was observed in the more drought tolerant genotypes. However, no significant dependency was found between proline concentration and other parameters. There was a positive correlation between identified parameters except proline concentration. Data indicated that genotypes of sahar, Hill, Dair and Gorgan-3 had the highest drought tolerance. In spite of the low value of all identified parameters in Habit genotype, this genotype was less affected than other genotypes by drought stress and indicated the least of sensitivity index. The most sensitive genotypes were Williams, LBK and BP.

3.15 Nutritional Requirement

The nutritional requirements of soybean are moderately high in comparison with other grains and the crop does best in soils of medium to high fertility and with a favourable soil P^H of around 6.5 while making good use of applied fertilizers. Therefore, even when the best soybean varieties and cultural practices are used it could result in the crop not reaching full potential unless soil fertility is properly managed. In addition, soybean plant require a loose, well-drained loamy soil that allows for good aeration and water-holding capacity.

The dry mass, total nitrogen and the natural abundance of 15N were measured in stems plus leaves, pods and seeds during all stages of growth of a crop of irrigated Forrest soybeans. The sources of nitrogen accumulated in the seed and the magnitudes of N_2 fixation were calculated. During early growth, the proportion (PN2) of plant nitrogen arising from N_2 fixation reflected residual effects of different rates of inoculation imposed two years previously and increased from 26–42% at 58 days to 49–57% at 71 days. Growth and accumulation of nitrogen in stems plus leaves ceased at about 108 days; growth of and nitrogen accumulation by seeds continued from 100 days to 142 days. During the most rapid phase of seed growth (100–125 days), 75–85% of seed nitrogen arose from N_2 fixation. After net accumulation of plant nitrogen ceased (125 days), continued accumulation of seed nitrogen was derived mainly by relocation from pods and leaves but some N_2 fixation continued. During ripening (142–167 days), there was loss of seed mass and nitrogen. The amounts of N_2 fixed during seed development were substantial. After attainment of the maximum levels of nitrogen in the supporting shoots, N_2 fixation totalling almost twice that achieved in the first 108 days of growth was directed exclusively to developing seeds during a period of only 34 days.

A study showed that the soybean plant has a strong tap-root system and is able to use nutrients on the subsoil very effectively. However, research on soil testing reported that about 28% of the soil samples for soybean cultivation tested low to very low in available P and K. It also reported that, approximately 60% of P and 50% of K taken up by a soybean plant is removed from the field when the seeds are harvested. As such, if availability is low, a band application of 4.5 to 14 kg ha^{-1} is beneficial and if the soil is low in K, 2.5 kg ha^{-1} is required to raise the soil K test by 1 mg kg^{-1} with the recommended fertilizer source being potassium sulphate.

Some researchers have shown that a response to P fertilization should not be expected if the soil test for P is higher than 20–25 mg kg^{-1} (measured by bray-1 producer). As with phosphate use, the response to potash fertilization should not be expected if the soil test for K is higher than 120 mg kg^{-1}. Phosphate and potassium play an important role in the growth and development of soybean. They increase frost and disease tolerance, palatability, storage quality and yield (Manitoba soil fertility Guide, 2004). Phosphate in particular enhances photosynthesis rates, enzymatic activity, root development, uptake and transfer of other nutrients and seed germination. Phosphorus deficiency is reported to reduce nodule formation and growth, while an adequate supply leads to good development of nodules.

However, very high soil phosphate value may depress seed protein and oil content while yielding will be low if available phosphorus is less than 30 kg P ha^{-1}. Similarly, K regulates several plant processes, including water and nutrient transport across cell walls, and regulation of water vapour and carbon dioxide (CO_2) exchange through stomata. Potassium deficiency is reported to cause stunted growth and chlorosis. Since soybean are very efficient in deriving most of their nitrogen needs through N-fixation, no additional N fertilizer application is required. For this reason, the application of other essential nutrient elements, like P and K, for soybean is often overlooked, with the myth being that soybean don't need fertilizer management at all.

3.16 Plant Growth Regulators

Gibberellins (GAs) play an essential role in many aspects of plant growth and development, such as seed germination (Maske et al., 1997), stem elongation and flower development (Yamaguchi and Kamiya, 2000). Gibberellins regulate plant growth by affecting stem elongation; leaves show weak responses to GA. GA_3 stimulated extensive growth and chlorophyll development in soybean hypocotyls *in vitro*.

GA_3 has a profound effect upon the antioxidant potentials and it caused a significant enhancement in the production of antioxidant compounds when compared to control. GA_3 1 mg/L in soaking solution was proposed as the optimum concentration to apply for the soaking process.

To better understand the role of these compounds, a pot experiment was carried out in order to study effects of GA_3 and cytokinin on the vegetative growth of the soybean. GA_3 (50 mg L^{-1}) was applied as seed treatment, leaving plants with water application as control. GA_3 (100 mg L^{-1}) and cytokinin (30 mg L^{-1}) were sprayed on leaves at the physiological stage V3/V4, and, 15 days later, cytokinin (30 mg L^{-1}) also, as foliar spray. Seed treatment decreased plant emergence and initial soybean root growth, but, as the season progressed, differences in root growth disappeared; plants were shorter, and presented a decrease in the number of nodes, in stem diameter, in leaf area and in dry matter yield. Conversely, foliar application of GA_3 led to an increase in plant height, first node height and stem diameter. Leaf area and dry matter production also increased as a result of GA_3 foliar application. There was no effect of exogenous gibberellin and cytokinin on the number of soybean leaves, number of stem branches and root dry matter. Joint application of gibberellin and cytokinin tended to inhibit gibberellin effects. Cytokinin applied to leaves during soybean vegetative growth was not effective in modifying any of the evaluated plant growth variables.

Table 3.11 Results of plant emerged, plant height, first node height, stem diameter, leaf area, total dry matter and length root of soybean plants cv. IAC 17 under seed treatment with GA3 and leaf application of GA3 and CK.

Treatment	Plant height (cm)	F.N.H.[1]	Stem diameter (mm)	Leaf area (cm^2 plant^{-1})	T.D.M.[2] (g.plant^{-1})	Root length (m)
First Evaluation						
No GA3	8.4 a*	17.1 a	2.2 a	237 a	1.35 a	184 a
With GA3	5.8 b	17.2 a	2.1 a	237 a	1.24 a	94 b
VC %	8.91	12.12	14.08	15.07	19.45	33.62
Second Evaluation						
No GA3	66.5 bc	5.9 b	9.3 a	1951 a	14.06 a	199a
With GA3	98.5 a	6.1 b	9.1 a	1984 a	14.52 a	255a
GA3 + CK	101.4 a	5.5 b	9.5 a	2183 a	16.41 a	266 a
No GA3	55.0 c	8.6 b	7.2 b	1328 a	8.09 a	190 a
With GA3	104.0 a	11.5 a	8.6 ab	2166 a	16.22 a	294 a
GA3 + CK	86.6 ab	9.0 ab	7.6 b	1476 a	10.18 a	199 a
VC %	12.73	23.41	7.57	20.86	28.79	31.52
Third Evaluation						
No GA3	108.8 cde	5.1 a	13.3 a	4957 a	40.73 a	278 a
With GA3	139.8 abcd	6.5 a	12.8 a	5070 a	39.79 a	289 a
GA3 + CK	148.3 ab	5.6 a	13.6 a	5010 a	48.99 a	382 a
No GA3 CK	113.3 cde	5.6 a	12.8 a	5141 a	41.26 a	307 a
With GA3 CK	155.0 a	5.3 a	14.1 a	6094 a	48.90 a	393 a
GA3 + CK CK	152.3 a	7.4 a	13.4 a	5759 a	46.65 a	357 a
No GA3	90.8 e	9.1 a	11.8 a	4404 a	37.83 a	315 a
With GA3	156.0 a	10.0 a	13.4 a	5326 a	45.54 a	295 a
GA3 + CK	142.5 abc	8.1 a	13.1 a	5122 a	44.06 a	302 a
No GA3 CK	104.0 de	7.4 a	12.9 a	5189 a	39.31 a	315 a
With GA3 CK	146.0 abc	8.8 a	13.5 a	5858 a	47.86 a	416 a
GA3 + CK CK	143.3 abc	8.4 a	13.0 a	5240 a	41.08 a	313 a
VC %	11.52	28.27	11.30	20.47	17.40	35.25

* Averages followed by the same letter do not differ; Tukey test, $P = 0.05$. [1] First node height. [2] Total dry matter.

Plant growth substances are well-known to improve the source-sink connection and encourage the translocation of photoassimilates thereby helping in effective flower formation, fruit and seed development and ultimately increase the yield of crops. A pot experiment was conducted at Sher-e-Bangla Agricultural University, Dhaka, Bangladesh during November, 2013 to March, 2014 in a Randomized Complete Block Design (RCBD) based on five replications with a view to find out the influence of different plant growth regulators and their stages of application on the growth and yield of soybean cv. BARI Soybean-6. Application of plant growth regulators at different stages of plant showed significant effect on plant height, number of branches plant^{-1}, chlorophyll content (SPAD value), average length of internode, dry weight plant^{-1} and seed yield of soybean. Results showed that application of GA$_3$ at vegetative stage produced the tallest plant (61.16 cm) and longest average length of internode (8.79 cm) and spray at flower initiation stage provided maximum SPAD value (50.38). Kinetin at vegetative stage gave the highest dry weight (23.68 g plant^{-1}) of soybean. Results also revealed that salicylic acid applied at flower and pod initiation stage gave the highest number of branches plant^{-1} (11.00) and seed yield (6.38 g plant^{-1}), compared to other growth regulators. So salicylic acid acts an important role for increasing soybean yield, when it was applied at flower and pod initiation stage.

Results of the experiment conducted by Vello and Castro (1987) in soybean cv. Davis revealed that pre-flowering foliar spray of GA$_3$ (100 ppm) significantly increased the plant height (119.7 cm); while, CCC (2000 ppm) reduced it to the extent of 94.8 cm. Bruce (1990) treated determinate soybean with GA$_3$ and ethephon, GA$_3$ treatment mainly produced positive effects on number of pods, height and nodes/branch: while, ethephon decreased yield and 100-seed weight. Shimano and Matsumoto (1991) conducted pot trials to study the effect of 0.01 to 1000 ppm GA$_3$ applied at different growth stage on internodal elongation of soybean and found that GA$_3$ (10–100 ppm) increased the 4th internodal length compared with the control, GA$_3$ was most effective in promoting internodal elongation when applied at floral initiation stage.

Leite et al. (2003) observed that foliar application of GA$_3$ increased the plant height, first nodal elongation and stem diameter in soybean. Leaf area and dry matter production also increased; however, there was no effect on number of leaves, stem branches and root dry matter.

Kothule et al. (2003) reported that plant growth substances of different concentrations, i.e., GA, NAA, CCC and salicylic acid each @ 100 and 200 ppm and urea @ 1 and 2% when applied exogenously as foliar spray improved morphological characters, viz., plant height, number of branches, leaf area, total dry matter of plant and reduced the number of days to 50 per cent flowering in soybean. GA @ 200 ppm was found most effective in increasing plant height. While, it was revealed that TIBA (30, 40 or 50 gm/l) application at V5 phenological stage in soybean (cv. Pintado) was effective in reducing plant height without affecting parameters related to productivity.

Tanner and Ahmed (1974) observed that in soybean, total dry matter production and rate of total dry matter production were not affected by TIBA, however, both seed yield and rate of reproductive dry matter production were greater with treated than with control plants. Similar results were obtained by Singh and Sarkar (1976).

Zaidi and Singh (1995) conducted an experiment where soybean seeds were soaked for 4 hours in distilled water or 100 or 200 ppm GA$_3$/IAA. They were then sown in pots and exposed to salinities of 0.8, 10 or 20 dS/m. The detrimental effects of salinity on dry matter production and distribution were eliminated by pre soaking in GA$_3$ or IAA.

Gulluoglu et al. (2006) observed that the application of plant growth regulators GA$_3$, Atonik, Cytozyme, Maxicrop reduced the effect of heat stress on both main and double cropped soybean and increased biomass yield under extended heat and dry conditions. Pankaj Kumar et al. (2006) studied the influence of plant growth regulators on determinate (JS-335) and semi-determinate (MACS-124) soybean genotypes and revealed that the growth retardants TIBA, mepiquat chloride and cycocel increased total dry matter production and BMD in both the soybean genotypes. They were more beneficial in terms of translocation of photo-assimilates towards developing reproductive parts as compared to growth promoter, kinetin and control.

Ravichanadran and Ramaswami (1991) studied the source-sink relationship in soybean as influenced by TIBA. They found that pre-flowering application of TIBA (50 ppm) decreased LAI but increased the dry matter production, CGR and NAR.

Deotale and Sorte (1996) found that among various concentrations of TIBA and B-9 (50, 100, 150, 200 and 250 ppm), TIBA (100 ppm) showed stimulatory effect on CGR, NAR and leaf nitrogen content which ultimately increased the grain yield in soybean. While, Maske et al. (1998) found that GA_3 was relatively more effective than NAA in increasing CGR at 30–45 and 45–60 DAS and accelerating the yield contributing factors in soybean.

Jadhav (2000) stated that the application of increasing concentrations of GA_3 and NAA increase the morphological and physiological parameters like CGR, RGR, NAR and LAR in soybean which inturn led to the increased yield and yield attributes. Similarly, Sarkar et al. (2002) showed that double spraying of GA_3 and IAA (100 ppm) at 20 and 42 days after sowing increased LAI, CGR and NAR in soybean (cv. BS-3). The foliar spray of GA_3 (100 ppm) at 30 DAS had the most regulatory effect to enhance root, stem, leaf and total dry matter, LAI, CGR, RGR and NAR in soybean (cv. PB-1) (Rahman et al., 2004).

Aqueous solutions of SA, applied as a spray to the shoots of soybean (*Glycine max* (L.) Merr. cv. Cajeme), significantly increased the growth of shoots and roots as measured after seven days of treatment. Shoot spraying of SA had no significant effect on photosynthetic rate. Growth increases were obtained in plants cultivated either in the greenhouse or in the field; SA-induced increases in root growth of up to 100% were measured in the field.

The idea that apically derived auxin inhibits shoot branching by inhibiting the activity of axillary buds was first proposed 70 years ago, but it soon became clear that its mechanism of action was complex and indirect. Recent advances in the study of axillary bud development and of auxin signal transduction are allowing a better understanding of the role of auxin in controlling shoot branching. These studies have identified a new role for auxin early in bud development as well as some of the second messengers involved in mediating the branch-inhibiting effects of auxin.

Soybean genotypes grown in sub-tropical climate may exhibit lodging. The plant lodging is influenced by soil type and fertility level, sowing date, latitude and altitude of the location, plant population and conditions of crop development. Plant regulators and herbicides are able to avoid or reduce plant lodging. This study verified the effects of the growth regulators TIBA and daminozide on vegetative growth and yield of soybean cultivar CD 214 RR. The experiment was carried out at a field in randomized block design with four replications in a factorial scheme. The A factor was represented by the combination of regulators TIBA and daminozide and its concentrations, and the Factor B was seven times of evaluation of injury and plant height or eight times of evaluation of lodging. In the range of doses used, the application of daminozide resulted in greater injury to soybean plants than TIBA. The smaller plant height was achieved by the application of 6 g ha^{-1} of TIBA and 1200 g ha^{-1} of daminozide. Treatments with daminozide (100 g ha^{-1}) and TIBA (10 g ha^{-1}) stood out due to the reduced lodging of soybean plants. Grain weight increased linearly when the levels of TIBA increased. There was a negative correlation between lodging and grain yield and a positive correlation between plant height and lodging. There was also a negative correlation between injury caused by the application of plant regulators and lodging.

The effect of exogenous naphthaleneacetic acid (NAA) on the internal levels of indole-3-acetic acid (IAA) in rooting hypocotyls of *Glycine max* was studied. The hypocotyls of NAA-treated cuttings grew significantly higher numbers of adventitious roots with an increase in endogenous IAA levels that corresponded with a decrease in IAA oxidase activity (32 kDa) examined. Moreover, a decline of peroxidase activity was accompanied by a decrease of lignin content during root formation. Caffeic acid and ferulic acid, two critical phenolic compounds for lignin synthesis, accumulated in NAA-treated tissues. Consequently, the increased IAA levels with a decrease of IAA oxidase activity accompanied a lower lignin content and a reduced peroxidise activity in NAA-treated tissues suggests that the induction of adventitious roots by NAA in soybean cuttings may be due to the higher IAA levels accumulated in tissues.

Deotale et al. (1995) studied the effect of GA and NAA on growth parameter of soybean and obtained highest values for plant height, number of leaves per plant, number of branches per plant, leaf area, dry matter, days to maturity and seed yield with 100 mg L^{-1} NAA. Maximum number of seeds per pod and grain yield was obtained when NAA was applied 15 days after emergence stage (Tables 3.12 & 3.13).

The intensity and quality (red to far-red (R/Fr) ratio) of light directly affect growth of plant under shading. Gibberellins (GAs) and auxin [indole-3-acetic acid (IAA)] play important roles in mediating the shading adaptive responses of plants. Thus, the intensity and quality of the uncoupling light from shading were assessed to identify the influence of each component on the morphology and matter distribution of the leaf, stem, and petiole. This assessment was based on the changes in endogenous Gibberellin 1 (GA1) and IAA levels. Soybean plants were grown in a growth chamber with four treatments [normal (N), N+Fr, low (L), and L+Fr light]. Results revealed that the reductions in photosynthetically active radiation (PAR) and R/Fr ratio equally increased height and stem mass fractions (SMFs) of the soybean seedling. The light intensity significantly influenced the dry mass per unit area and mass fraction of soybean leaves, whereas the light quality regulated the petiole elongation and mass fraction. Low R/Fr ratio (high Fr light) increased the soybean biomass by improving the photosynthetic assimilation rate and quantum yield of photosystem II. In addition, the IAA and GA1 levels in the leaf, stem, and petiole did not reflect the growth response trends of each tissue toward light intensity and quality; however, trends of the IAA-to-GA1 content ratios were similar to those of the growth and matter allocation of each soybean tissue under different light environments. Therefore, the response of growth and matter allocation of soybean to light intensity and quality may be regulated by the IAA-to-GA1 content ratio in the tissues of the soybean plant.

Hormones are critical to maintain branches; shoot branching is highly regulated by endogenous and environmental cues (Umehara et al., 2008). Auxin and cytokinin have long been known to have an important involvement in controlling shoot branching (Leyser, 2003) since auxin suppresses axillary bud outgrowth and cytokinin promotes axillary bud outgrowth (Shimizu-Sato et al., 2009). Hormones always

Table 3.12 Effect of GA and IAA on morphological characters of soybean.

Conc. (ppm)	Plant height (cm) at 100 DAS	Branches per plant at 100 DAS	Leaves per plant at 80 DAS
GA$_3$			
Control	25.11c	2.11c	6.67b
100	57.10a	6.22a	11.89a
200	50.67b	4.56b	11.56a
IAA			
Control	25.51c	2.11c	6.89b
100	47.67a	3.78b	8.22ab
200	42.33b	4.78a	9.22a

Values with different letter(s) within a column differ significantly at 5% probability (LSD).

Table 3.13 Effect of time of application of GA and IAA on morphological characters of soybean.

Treatment	Plant height (cm) at 100 DAS	Branches per plant at 100 DAS	Leaves per plant at 80 DAS
GA$_3$			
T1	43.56ab	3.89b	8.44
T2	48.33a	5.11a	10.89
T3	41.39b	3.89b	9.78
IAA			
T1	37.89	3.22b	7.67a
T2	40.00	4.00a	8.89ab
T3	37.62	3.44ab	7.78b

T1 = spray at 20 DAS, T2 = double spray at 20 and 42 DAS, T3 = spray at 42 DAS.

have advantages to maintain yield and ameliorate the optimum yield (Khan et al., 2007). Drought and shade stress have effects on auxin, cytokinins and abscisic acid concentrations (Davies, 2010).

Multiple frequently interactive stress factors naturally influence plant due to global change. The leaf's hormone concentrations, main-stem and branch yield response to the combination of shade and drought were studied in a greenhouse experiment during 2009 and 2010 seasons. Pot experiments were conducted under shade of maize (LI) and normal irradiance (HI). Shade stress was removed once maize was harvested. Manipulative progressive soil drying period at branching stages under good soil conditions (HW) and water stress treatment (LW) were applied in 2010, while well-watered (WW) and moderate drought (MD) were applied in 2009. Under shade stress, seedling height and first internode length increased, stem diameter decreased, abscisic acid (ABA) and zeatin (ZT) concentration decreased, while indole acetic acid (IAA) and gibberellins 3 (GA$_3$) concentration increased. More also, branch numbers, pod number of branches and seed number of branches increased. Branch yield did not reduce significantly under shade stress, which was related to the decrease of ABA and IAA. Based on the results, soybean yield decreased under shade and drought stresses was mainly due to the yield reduction of the main-stem under drought (Table 3.14).

The finding of research, the branch pod and seed number did not reduce significantly under drought and shade stresses, but rather they increased under drought stress oppositely. Such results may be related to the growth stage; the stresses happened in V5 stage. Soybean had better morphological and physiological elasticity and decrease of ABA for shaded soybean and reduction of IAA for water deficit soybean. Yield decrease of shade and drought stresses was mainly due to yield reduction of the main-stem. Hence, soybean genotype which has better main-stem yield stability may be suggested under the environment of low irradiance and soil moisture.

The levels of different cytokinins, indole-3-acetic acid (IAA) and abscisic acid (ABA) in roots of *Glycine max* [L.] Merr. cv. Bragg and its supernodulating mutant nts382 were compared. Forty-eight hours after inoculation with Bradyrhizobium, quantitative and qualitative differences were found in the root's endogenous hormone status between cultivar Bragg and the mutant nts382. The six quantified cytokinins, ranking similarly in each genotype, were present at higher concentrations (30 ± 196% on average for isopentenyl adenosine and dihydrozeatin riboside, respectively) in mutant roots. By contrast, the ABA content was 2-fold higher in Bragg, while the basal levels of IAA [0.53 lmol (g DW)$^{-1}$, on average] were similar in both genotypes. In 1 mM NO$_3$-fed Bragg roots 48 h post-inoculation, IAA, ABA and the cytokinins isopentenyl adenine, and isopentenyl adenosine quantitatively increased with respect to uninoculated controls. However, only the two cytokinins increased in the mutant. High NO$_3$ (8 mM) markedly reduced root auxin concentration, and neither genotypic differences nor the inoculation-induced increase in auxin concentration in Bragg was observed under these conditions. Cytokinins and ABA, on the other hand, were little affected by 8 mM NO$_3$. Root IAA/cytokinin and ABA/cytokinin ratios were always higher in Bragg relative to the mutant, and responded to inoculation (mainly in Bragg)

Table 3.14 Effects of drought and shade on plant hormone of soybean.

Treatment			ABA (ng g^{-1}FW)	IAA (ng g^{-1}FW)	GA$_3$ (ng g^{-1}FW)	ZT (ng g^{-1}FW)
2009	HI	WW	26.62c	47.36c	227.29bc	52.66a
		MD	358.55a	22.15d	205.64c	38.08b
	LI	WW	14.55c	117.07a	411.81a	49.01a
		MD	285.79b	78.08b	278.74b	33.46b
2010	HI	HW	24.50c	79.60b	128.29c	37.45a
		LW	207.41a	61.86b	104.01c	30.86ab
	LI	HW	15.57c	136.06a	300.26a	32.87a
		LW	125.12b	76.55b	274.58b	24.61b

HI, normal irradiance; LI, low irradiance, under the shade of light screen or corn; WW, represents well water (75 ± 2% of the soil field capacity); MD, moderate drought (45 ± 2% of the soil field capacity); LW, low-water treatment; HW represents high water treatment. Within columns means in the same season followed by the same small letters are not significant at the 0.05 and 0.01 levels of probability according to LSD test, respectively.

Table 3.15 Effect of NO_3 and inoculation on IAA/cytokinin and ABA/cytokinin ratios in roots of soybean cv. Bragg and its supernodulating mutant nts382.

Ratio	NO_3 (mM)	Inoculation (+/–)	Bragg (pmol.pmol^{-1}.10^{-3})	Nts382 (pmol.pmol^{-1}.10^{-3})
IAA/Cytokinin	1	–	1.22	0.60
	1	+	1.01	0.54
	8	–	0.36	0.20
	8	+	0.42	0.22
ABA/Cytokinin	1	–	3.56	1.01
	1	+	2.62	0.84
	8	–	3.05	1.44
	8	+	2.03	1.09

and nitrate (both genotypes). The overall results are consistent with the auxin-burst-control hypothesis for the explanation of auto regulation and supernodulation in soybean. However, they are still inconclusive with respect to the inhibitory effect of NO_3 (Table 3.15).

Cinnamic acid is a known allelochemical that affects seed germination and plant root growth and therefore influences several metabolic processes. In a work, it was evaluated its effects on growth, indole-3-acetic acid (IAA) oxidase and cinnamate 4-hydroxylase (C_4H) activities and lignin monomer composition in soybean (*Glycine max*) roots. The results revealed that exogenously applied cinnamic acid inhibited root growth and increased IAA oxidase and C_4H activities. The allelochemical increased the total lignin content, thus altering the sum and ratios of the *p*-hydroxyphenyl (H), guaiacyl (G), and syringyl (S) lignin monomers. When applied alone or with cinnamic acid, piperonylic acid (PIP, a *quasi*-irreversible inhibitor of C_4H) reduced C_4H activity, lignin and the H, G, S monomer content compared to the cinnamic acid treatment. Taken together, these results indicate that exogenously applied cinnamic acid can be channeled into the phenylpropanoid pathway via the C_4H reaction, resulting in an increase in H lignin. In conjunction with enhanced IAA oxidase activity, these metabolic responses lead to the stiffening of the cell wall and are followed by a reduction in soybean root growth.

References

Almansouri, M., J.M. Kinet and S. Lutts. 2001. Effects of salt and osmotic stresses on germination in Durum wheat (*Triticum durum* Desf.). Plant Soil 231: 243–254.

Ayars, J.E. and K.K. Tanji. 1999. Effects of drainage on water quality in arid and semi-arid lands. pp. 831–867. *In*: Skaggs, R.V. and J. Van Schifgaarde (eds.). Agricultural Drainage. ASA-CSSA-SSSA, Madison.

Bacanamwo, M. and L.C. Purcell. 1999a. Soybean dry matter and N accumulation responses to flooding stress, N sources, and hypoxia. J. Exp. Bot. 50: 689–696.

Bacanamwo, M. and L.C. Purcell. 1999b. Soybean root morphological and anatomical traits associated with acclimation to flooding. Crop Sci. 39: 143–149.

Baczek-Kwinta, R., A. Hyrlicka, J. Maslak, A. Oleksiewicz and B. Serek, 2004. Porównanie reakcji bazylii wlasciwej i melisy lekarskiej na rózne stresy srodowiskowe. Zesz. Probl. Post. Nauk Roln. 496: 537–544.

Barkke, M.P. and F.P. Gradner. 1987. Juvenile growth in pigeon pea, soybean, and cowpea in relation to seed and seedling characteristics. Crop Sci. 27(2): 311–316.

Basuchaudhari, P. and G.S. Munda. 1987. Characterizing active physiological region in soybean (*Glycine max* (L.) Merrill) at pod formation stage. Legume Res. 10(1): 27–28.

Beaver, J.S., R.L. Cooper and R.J. Martin. 1985. Dry matter accumulation and seed yield of determinate and indeterminate soybeans. Agron. J. 77: 675–679.

Bernstein, L. and H.E. Hayward. 1958. Physiology of salt tolerance. Annu. Rev. Plant Physiol. 9: 25–46.

Bishop, K.A., A.M. Betzelberger, S.P. Long and E.A. Ainsworth. 2015. Is there potential to adapt soybean (*Glycine max* Merr.) to future [CO_2]? An analysis of the yield response of 18 genotypes in free-air CO_2 enrichment. Plant, Cell & Environment 38: 1765–1774.

Borowski, E. and Z.K. Blamowski. 2009. The effects of triacontanol 'TRIA' and Asahi SL on the development and metabolic activity of sweet basil (*Ocimum basilicum* L.) plants treated with chilling. Folia Hort. 21/1: 39–48.

Brady, R.A., L.R. Stone, C.D. Nickell and W.L. Powers. 1974. Water conservation through proper timing of soybean irrigation. J. Soil Water Conserv. 29: 266–268.

Brown, D.M. 1960. Soybean ecology I. Development temperature relationships from controlled environmental studies. Agron. J. 52: 493–496.

Bruce, A.P. 1990. The use of plant growth regulators to enhance yield and production efficiency of soybean. Dissertation Abstract. Intl. B. Science and Engg. 51(6): 2678.

Bunce, J.A. 2014. Limitations to soybean photosynthesis at elevated carbon dioxide in free-air enrichment and open top chamber systems. Plant Science 226: 131–135.

Bunce, J.A. 2015. Elevated carbon dioxide effects on reproductive phenology and seed yield among soybean cultivars. Crop Science 55: 339–343.

Bunce, J.A. 2016. Responses of soybeans and wheat to elevated CO_2 in free-air and open top chamber systems. Field Crops Research 186: 78–85.

Burris, J.S., O.T. Edje and A.H. Wahab. 1973. Effects of seed size on seedling performance in soybean: II. Seedling growth and photosynthesis and field performance. Crop Sci. 13: 207–210.

Buttery, B.R. 1970. Effect of variation in leaf area index on growth of soybean and maize. Crop Sci. 10: 9–10.

Cober, E.R., D.W. Stewart and H.D. Voldeng. 2001. Photoperiod and temperature responses in early-maturing, near-isogenic soybean lines. Crop Sci. 41: 721–727.

Carlson, J.B. and N.R. Lersten. 2004. Reproductive morphology. pp. 59–95. *In*: Boerma, H.R. and J.E. Specht (eds.). Soybeans: Improvement, Production, and Uses. 3rd ed. ASA-CSSA-SSSA Publications, Madison, WI.

Cramer, G.R. 1992. Response of a Na-excluding cultivar and a Na-including cultivar to varying Na/Ca. J. Exp. Bot. 43: 857–864.

Davies, P.J. 2010. The plant hormones: Their nature, occurrence, and functions. Plant Horm. pp. 1–15.

Deotale, R.D., D.S. Katekhaye, N.V. Sorte, J.S. Raut and V.J. Golliwar. 1995. Effect of TIBA and B-9 on morpho-physiological characters of soybean. J. Soil and Crops 4(2): 172–176.

Deotale, R.D. and N.D. Sorte. 1996. Effect of TIBA and B-9 on growth parameters, biochemical aspects and yield of soybean. J. Soils Crops 6(1): 89–93.

Doorenbos, J. and W.O. Pruit. 1977. Guidelines for predicting crop water requirement. FAO Irrigation and Drainage Paper 24, Food and Agricultural Organisation of the United Nation.

Edje, O.T. and J.S. Burris. 1971. Effects of soybean seed vigour on field performance. Agron. J. 63: 536–538.

Egli, D.B., D.M. TeKrony and R.A. Wiralaga. 1990. Effect of seed vigour and size on seedling growth. Journal of Seed Technology 14: 1–12.

FAO. 1992. The Use of Saline Water for Crop Production. Irrigation and Drainage Paper, 48. Rome.

Farhoudi, R., F. Sharifzadeh, M. Makkizadeh and M. Kochakpour. 2007. The effects of NaCl priming on salt tolerance in canola (*Brassica napus*) seedlings grown under saline conditions. Seed Sci. Technol. 35: 754–759.

Farhoudi, R. and M.M. Tafti. 2011. Effect of salt stress on seedlings growth and ions homoestasis of soybean (Glycine max) cultivars. Adv. Environ. Biol. 5(8): 2522–2526.

Fehr, W.R. and C.E. Caviness. 1977. Stages of soybean development. Iowa State University Cooperative Extension Service, Special Report 80.

Grable, A.R. 1966. Soil aeration and plant growth. Advances in Agronomy 18: 57±106.

Griffin, J.L. and A.M. Saxton. 1988. Response of solid-seeded soybean to flood irrigation. II. Flood duration. Agronomy Journal 80: 885–888.

Gulluoglu, L., H. Arioglu and M. Arslan. 2006. Effects of some plant growth regulators and nutrient complexes on pod shattering and yield losses of soybean under hot and dry conditions. Asian Journal of Plant Sciences 5(2): 368–372.

Hanley, M.E. and O.C. May. 2006. Cotyledon damage at the seedling stage affects growth and flowering potential in mature plants. New Phytol. 169: 243–250.

Hanway, J.J. and C.R. Weber. 1971b. Dry matter accumulation in eight soybean (*Glycine max* (L.) Merrill) varieties. Agron. J. 63: 227–232.

Harris, M., R.O. Mackender and D.L. Smith. 1986. Photosynthesis of cotyledons of soybean seedlings. New Phytol. 104: 319–329.

Hay, R.K.M. and J.R. Porter. 2006. The Physiology of Crop Yield. Second edition. Blackwell Publishing, Oxford, 314 pp.

Helm, J.L. and L.A. Spilde. 1990. Selecting quality seed of cereal grains. NDSU Extension Service, North Dakota State University of Agriculture and Applied Science, and U.S. Department of Agriculture Cooperating 701: 231–788.

Heerden, P.D.R. and G.H.J. Krüger. 2002. Separately and simultaneously induced dark chilling and drought stress effect on photosynthesis, proline accumulation and antioxidant metabolism in soybean. J. Plant Physiol. 159: 1077–1086.

Heerden, P.D.R., M.M. Tsimilli, G.H.J. Krüger and R.J. Strasser. 2003a. Dark chilling effects on soybean genotypes during vegetative development; parallel studies of CO_2 assimilation, chlorophyll a kinetics O-J-I-P and nitrogen fixation. Physiol. Plant. 117(4): 476–491.

Heerden, P.D.R., G.H.J. Krüger, J.E. Loveland, M.A.J. Parry and C.H. Foyer. 2003b. Dark chilling imposes metabolic restrictions on photosynthesis in soybean. Plant Cell Environ. 26: 323–337.

Hogan, K.P. 1988. Photosynthesis in two neotropical palm species. Funct. Ecol. 2: 371–377.

Ikeda, K. and M. Kiso. 1981. On the role of cotyledon in early growth of soybean plants. Rep. Tokai Br. Crop Sci. Soc. Japan 91: 11–14.

Jadhav, B.P. 2000. Influence of plant growth regulators on growth and yield of soybean genotypes. M.Sc. (Agri) Thesis (unpub.), MPKV, Rahuri.

Jannink, J.-L., J.H. Orf, N.R. Jordan and R.G. Shaw. 2000. Index selection for weed suppressive ability in soybean. Crop Sci. 40: 1087–1094.

Jamil, M., C.C. Lee, S. Ur Rehman, D.B. Lee, M. Ashraf and E.S. Rha. 2005. Salinity (NaCl) tolerance of Brassica species at germination and early seedling growth. EJEAF Che 4: 970–976.

Jamil, M., S. Rehman, K.J. Lee, J.M. Kim, H.S. Kim and E.S. Rha. 2007. Salinity reduced growth PS II photochemistry and chlorophyll content in radish. Sci. Agric. 64: 1–10.

Jeyaraman, S., S. Subramanian and S.R. Sree Rangaswamy. 1990. Influence of weather parameters at the crop growth stages on seed yield of soybean. Mausam 41: 575–578.

Khajeh-Hosseini, M., A.A. Powell and I.J. Bingham. 2003. The interaction between salinity stress and seed vigour during germination of soybean seeds. Seed Sci. Technol. 31: 715–725.

Khafagi, O.A., S.M. Khalaf and W.I. El-Lawendry. 1986. Effect of GA3 and CCC on germination and growth of soybean, common bean, cowpea and pigeon pea plants grown under different levels of salinity. Annals of Agricultural Science 24: 1965–1982.

Khan, R., M.M.A. Khan, M. Singh, S. Nasir, M. Naeem, M.H. Siddiqui and F. Mohammad. 2007. Gibberellic acid and triacontanol can ameliorate the opium yield and morphine production in opium poppy (Papaver somniferum L.). Acta Agric. Scand B-S P. 57: 307–312.

Khan, A.Z., P. Shah, S.K. Khalil and F.H. Taj. 2003. Influence of planting date and plant density on morphological traits of determinate and indeterminate soybean cultivars under temperate environment. Pakistan Journal of Agronomy 2(3): 146–152.

Koesmaryono, Y., H. Sugimoto, D. Ito, T. Haseba and T. Sato. 1998. Photosynthetic and transpiration rates of soybean as affected by different irradiances during growth. Photosynthetica 35(4): 573.

Kokubun, M., H. Mochida and Y. Asahi. 1988. Soybean cultivar difference in leaf photosynthetic rate and its relation to seed yield. Jpn. J. Crop. Sci. 57: 743–748.

Koller, M.R. 1971. Analysis of growth within distinct strata of the soybean community. Crop Sci. 11: 400–402.

Kondetti, P., N. Jawali, S.K. Apte and M.G. Shitole. 2012. Salt tolerance in Indian soybean (*Glycine max* L. Merill) varieties at germination and early seedling growth. Ann. Biol. Res. 3(3): 1489–1498.

Korte, L.L., J.H. Williams, J.E. Specht and R.C. Sorensen. 1983. Irrigation of soybean genotypes during reproductive ontogeny. I. Agronomic responses. Crop Science 23(3): 521–527.

Kothule, V.G., R.K. Bhalerao and T.H. Rathod. 2003. Effect of growth regulators on yield attributes, yield and correlation coefficient in soybean. Ann. Plant physiol. 17(2): 140–142.

Kozlowski, T.T. and S.G. Pallardy. 1985. Effect of flooding on water, carbohydrate and mineral relations. pp. 165–193. *In*: Kozlowski, T.T. (ed.). Flooding and Plant Growth. Orlando, FL, Academic Press.

Larsen, S.U. and F. Andreasen. 2004. Light and heavy seeds differ in germination percentage and mean germination thermal time. Crop Sci. 44: 1710–1720.

Leite, V.M., C.R. Rosolem and J.D. Rodrigues. 2003. Gibberellin and cytokinin effects on soybean growth. Sci. agric. (Piracicaba, Braz.) 60(3): 55–59.

Leyser, O. 2003. Regulation of shoot branching by auxin. Trends Plant Sci. 8: 541–545.

Linkemer, G., J.E. Board and M.E. Musgrave. 1998. Waterlogging effects on growth and yield components in late-planted soybean. Crop Sci. 38: 1576–1584.

Longer, D.E., E.J. Lorenz and J.T. Cothren. 1986. The influence of seed size on soybean [*Glycine max* (L.) Merrill] emergence under simulated soil crust conditions. Field Crops Res. 14: 371–375.

Lowe, L.B. and S.K. Rise. 1973. Endosperm protein of wheat seed as a determinate of seedling growth. Plant Physiol. 51: 57–60.

Lu Cun Fu, Pen Gui Ying, Lu Gf and Pen Gy. 1994. Effects of light on photosynthesis of alfalfa under cold stress. Grass. China 5: 15–18.

Luquez, V.M., J.J. Guiamet and E.R. Motaldi. 1997. Net photosynthetic and transpiration rate in a chlorophyll deficient isoline of soybean under well-watered and drought condition. Photosynthetica 34(1): 125–131.

Maguire, J.D. 1962. Speed of germination—Aid in selection and evaluation for seedling emergence and vigor. Crop Science 2: 176–177.

Makino, A., T. Sato, H. Nakano and T. Mae. 1997. Leaf photosynthesis, plant growth and nitrogen allocation in rice under different irradiances. Planta 203: 390–398.

Maske, V.G., R.D. Dotale, P.N. Sorte, B.D. Tale and C.N. Chore. 1997. Germination, root and shoot studies in soybean as influenced by GA3 and NAA. Journal of Soils and Crops 7: 147–149.

McKersie, B.D. and Y.Y. Lesham. 1994. Stress and stress coping in cultivated plants. Dordrecht, Boston, London, Kluwer Academic Publishers, p. 79–103.

Minchin, F.R. and J.S. Pate. 1975. Effects of water, aeration and salt regime on nitrogen fixation in a nodulated legume: Definition of an optimum root environment. J. Exp. Bot. 26: 60–69.

Momen, N.M., R.E. Shae and O. Arjamand. 1979. Moisture stress effect on the components of two soybean cultivars. Agron. J. 17(1): 86–87.

Nelson, C.J. and K.L. Larson. 1984. Seedling growth. pp. 93–129. *In*: Tesar, M.B. (ed.). Physiological Basis of Crop Growth and Development. American Society of Agronomy, Crop Science Society of America, Inc. Madison, WI.

Ojima, M. and R. Kawashima. 1968. Studies on the seed production of soybean: 5 Varietal differences in photosynthetic rate of soybean. Proc. Crop Sci. Soc. Japan 37: 667–675.

Pankaj Kumar, S.M. Hiremath and M.B. Chetti. 2006. Influence of growth regulators on dry matter production, distribution and shelling percentage in determinate and semi determinate soybean genotypes. Legume Res. 29(3): 191–195.

Pantalone, V.R., J.W. Burton and T.E. Carter Jr. 1996. Soybean fibrous root heritability and genotypic correlations with agronomic and seed quality traits. Crop Sci. 36: 1120–1125.

Parvez, A., F. Gardner and K. Boote. 1989. Determinate- and indeterminate-type soybean cultivar responses to pattern, density, and planting date. Crop Sci. 29: 150–157.

Ponnamperuma, F.N. 1972. The chemistry of submerged soils. Advances in Agronomy 24: 29±95.

Posmyk, M.M., F. Corbineau, D. Vinel, C. Bailly and D. Come. 2001. Osmoconditioning reduces physiological and biochemical damage induced by chilling in soybean seeds. Physiol. Plant. 111(4): 473–482.

Posmyk, M.M., C. Bailly, K. Szafranska, K.M. Janas and F. Corbineau. 2005. Antioxidant enzymes and isoflavonoids in chilled soybean (*Glycine max* (L.) Merr.) seedlings. J. Plant Physiol. 162: 403–412.

Prasad, T.K., M.D. Anderson, B.A. Martin and C.R. Stewart. 1994. Evidence for chilling-induced oxidative stress in maize seedlings and a regulatory role for hydrogen peroxide. Plant Cell. 6(1): 65–74.

Rahman, M.S., M.I. Nashirul, A. Tahar and M.A. Karim. 2004. Influence of GA_3 and MH and their time of spray on morphology, yield contributing characters and yield of soybean. Asian J. Pl. Sci. 3(5): 602–609.

Ramana, S., Ajay, A.B. Singh and R.B.R. Yadava. 2003. Role of cotyledons in regulation of physio-biochemical components of soybean (*Glycine max* L.). Indian J. Agric. Res. 37: 204–208.

Ravinchandran, V.K. and C. Ramaswami. 1991. Source and sink relationship in soybean as influenced by TIBA. Indian J. Pl. Physiol. 34(1): 80–83.

Reid, D.M. and K.J. Bradford. 1985. Effects of flooding on hormone relations. pp. 195–219. *In*: Kozlowski, T.T. (ed.). Flooding and Plant Growth. Orlando, FL, Academic Press.

Rodríguez, M., E. Canales and O. Borrás-Hidalgo. 2005. Molecular aspects of abiotic stress in plants. Biotechnol. Appl. 22: 1–10.

Russell, D.A., D.M.L. Wong and M.M. Sachs. 1990. The anaerobic response of soybean. Plant Physiology 92: 401±407.

Sadeghi, H., F. Khazaei, L. Yari and S. Sheidaei. 2011. Effect of seed osmopriming on seed germination behavior and vigor of soybean (*Glycine max* L.). ARPN Journal of Agricultural and Biological Science 6: 39–43.

Sadeghian, S.Y. and N. Yavari. 2004. Effects of water deficit stress on germination and seedling growth in sugar beet. J. Agron. Crop Sci. 190(2): 138–144. United State Salinity Laboratory Staff, 1954.

Sarkar, P.K., M.S. Haque and M.A. Karim. 2002. Effects of GA_3 and IAA and their frequency of application on morphology, yield contributing characters and yield of soybean. Pak. J. Agron. 1: 119–122.

Samui, R.P., S.S. Mondal and A.K. Dhotre. 2002. Comparative studies on energy balance components and their inter relations on soybean crops and their soil in kharif season at Pune situated in semi-arid tract. Mausam 53: 319–328.

Scott, W.O. and S.R. Aldrich (eds.). 1983. Modern Soybean Production, 2nd ed. S & A Publications, Champaign, IL.

Shibles, R., I.C. Anderson and A.H. Gibson. 1975. Soybean. pp. 151–189. *In*: Evans, L.T. (ed.). Crop Physiology Some Case Histories. 1st ed. Cambridge University Press, London.

Shimano, I. and S. Matsumoto. 1991. Effect of gibberellic acid on internode elongation. Japanese J. Crop Sci. 60(1): 15–19.

Shimizu-Sato, S., M. Tanaka and H. Mori. 2009. Auxin–cytokinin interactions in the control of shoot branching. Plant Mol. Biol. 69: 429–435.

Sinclair, T.R., S. Kitani, K. Hinson, J. Bruniard and T. Horie. 1991. Soybean flowering date: Linear and logistic models based on temperature and photoperiod. Crop Sci. 31: 786–790.

Sinclair, T.R. and K. Hinson. 1992. Soybean flowering in response to the long-juvenile trait. Crop Sci. 32: 1242–1248.

Singh, B.B. and S.K. Sarkar. 1976. Effect of growth retardants on growth, flowering, productivity and chemical composition of soybean. Haryana J. Hort. Sci. 5: 195–202.

Sivakumar, M.V.K. and R.H. Shaw. 1978. Methods of growth analysis in field grown soya beans (*Glycine max* (L.) Merril). Ann. Bot. 42: 213–222.

Soltani, A., M. Gholipoor and E. Zeinali. 2006. Seed reserve utilization and seedling growth of wheat as affected by drought and salinity. J. Environ. Exp. Bot. 55: 195–200.

Starck, Z., B. Niemyska, J. Bogdan and R.N. Akour Tawalbeh. 2000. Response of tomato plant to chilling stress in associated with nutrient or phosphorus starvation. Plant Soil 226: 99–106.

Srivastava, S., A.D. Pathak, P.S. Gupta, A.K. Shrivastava and A.K. Srivastava. 2012. Hydrogen peroxide-scavenging enzymes impart tolerance to high temperature induced oxidative stress in sugarcane. J. Environ. Biol. 33: 657–661.

Szalai, G., T. Janda, E. Paldi and Z. Szigeti. 1996. Role of light in the development of post-chilling symptoms in maize. J. Plant Physiol. 148: 378–383.

Tan, D.X., R. Hardeland, L.C. Manchester, A. Korkmaz, S. Ma, S. Rosales-Corral and R.J. Reiter. 2012. Functional roles of melatonin in plants, and perspectives in nutritional and agricultural science. Journal of Experimental Botany 63: 577–597.

Tanner, J.W. and S. Ahmed. 1974. Growth analysis of soybeans treated with TIBA. Crop Sci. 14: 371–374.

Toledo, J.F.F., M.F. Oliveira, A.C. Tsutida and R.A.S. Kiihl. 1993. Genetic analysis of growth of determinate soybean genotypes under three photoperiods. Rev. Bras. Genet. 16: 713–748.

Umehara, M., A. Hanada, S. Yoshida, K. Akiyama, T. Arite, N. Takeda-Kamiya, H. Magome, Y. Kamiya, K. Shirasu and K. Yoneyama. 2008. Inhibition of shoot branching by new terpenoid plant hormones. Nature 455: 195–200.

Vello, N.A. and P.R.C. Castro. 1987. Action of growth regulators on development of soybean cultivar 'Davis', *Anais da Escola. Superiourde Agriculture.* Luiz de quieroz. 38(1): 269–279.

Venkatareddy, D.M. 1991. Investigations on seed technology of soybean, Ph.D. Thesis, Univ. Agric. Sci., Bangalore, Karnataka (India).

Wahid, A. 2007. Physiological implications of metabolites biosynthesis in net assimilation and heat stress tolerance of sugarcane (*Saccharum officinarum*) sprouts. J. Plant Res. 120: 219–228.

Wang, S.R., G.M. Wang, E.F. Quenroz and C.M. Mesquita. 1980. Research on drought resistance and irrigation of soybean in Parana Brazil In irrigated soybean production in arid and semi-arid regions. W.H. Judy and J.A. Jacobs (eds.). International Soybean Programme 20: 92–96.

Wang, Z., Y.R. Reddy and B. Quebedeaux. 1997. Growth and photosynthetic responses of soybean to short-term cold temperature. Environ. Exp. Bot. 37: 13–24.

Westoby, M., E. Jurado and M. Leishman. 1992. Comparative evolutionary ecology of seed size. Trends Ecol. Evolution 7(11): 368–372.

Whigham, D.K. and H.C. Minor. 1978. Agronomic characteristics and environmental stress. pp. 77–118. *In*: Normal, A.G. (ed.). Soybean Physiology, Agronomy and Utilization. Academic Press, Inc., New York.

Wolfe, D.W. 1991. Low temperature effects on early vegetative growth, leaf gas exchange and water potential of chilling-sensitive and chilling-tolerant crop species. Ann. Bot. 67: 205–212.

Yadegari, L.Z., R. Heidari and J. Carapetian. 2007. The influence of cold acclimation on proline, malondialdehyde (MDA), total protein and pigments contents in soybean (*Glycine max*) seedlings. J. Biol. Sci. 7(8): 1436–1441.

Yamaguchi, S. and Y. Kamiya. 2000. Gibberellin biosynthesis: Its regulation by endogenous and environmental signals. Plant and Cell Physiology 41: 251–257.

Zaidi, P.H, and B.B. Singh. 1995. Effect of growth regulators on IAA-oxidase and peroxidase activity in soybean under salinity. Indian. J. of Plant Physiol. 35(1): 123–131.

Zhang Jing Xian, Cui Si Ping, Li Jun Ming, Wei Jian Kun and M.B. Kirkham. 1995. Protoplasmic factors, antioxidant responses and chilling resistance in maize. Plant Physiol. Biochem. 33: 567–575.

Zhang, H., D. Zhou, C. Matthew, P. Wang and W. Zheng. 2008. Photosynthetic contribution of cotyledons to early seedling development in *Cynoglossum divaricatum* and *Amaranthus retroflexus*. New Zeal. J. Bot. 46: 39–48.

Zhao, D.L., G.N. Atlin, L. Bastiaans and J.H.J. Spiertz. 2006. Cultivar Weed competitiveness in aerobic rice: Heritability, correlated traits, and the potential for indirect selection in weed-free environments. Crop Sci. 46: 372–380.

Zheng, W., P. Wang, H. Zhang and D. Zhou. 2011. Photosynthetic characteristics of the cotyledon and first true leaf of castor (*Ricinus communis* L.). Aust. J. Crop Sci. 5: 702–708.

CHAPTER 4
Reproductive Development

Flower induction is a phase change from vegetative to floral production. Flowering commences from "floral induction signal" which induces "floral evocation". This is followed by "floral initiation", "flower development" and eventually "anthesis". For many species, flowering must occur at appropriate seasons for floral induction and following reproductive development (Fig. 4.1).

Floral development is controlled by both internal and external cues. Plants have developed sophisticated mechanisms with complex genetic network in regulation of flowering. Four flowering pathways have been determined, as reviewed by various authors. These include photoperiod, autonomous, vernalization and gibberellin-induced pathways. There are a common set of genes which define the signalling pathways in flowering among plant species, while photoperiod and vernalization and/or their interactions are reported as main external factors influencing flowering responses and behaviours.

Floral development includes initiation of floral primordial and subsequent anthesis as discrete events, although in many investigations only anthesis is considered. For 'Ransom' soybean (*Glycine max* (L.) Merrill) grown at day/night temperatures of 18/14, 22/18, 26/22, 30/26, and 34/30°C and exposed to photoperiods of 10, 12, 14, 15 and 16 h, time of anthesis ranged from less than 21 days after exposure at the shorter photoperiods and warmer temperatures to more than 60 days at longer photoperiods and cooler temperatures. For all temperature regimes, however, floral primordial were initiated under shorter photoperiods within 3 to 5 days after exposure and after not more than 7 to 10 days exposure to longer photoperiods. Once initiation had begun, time required for differentiation of individual floral primordia and the duration of leaf initiation at shoot apices increased with increasing length of photoperiod. While production of nodes ceased abruptly under photoperiods of 10 and 12 h, new nodes continued to be formed concurrently with initiation of axillary floral primordial under photoperiods of 14, 15 and 16 h. The vegetative condition at the main stem shoot apex was prolonged under the three longer photoperiods and is indicative of the existence of an intermediate apex under these conditions. The results indicate

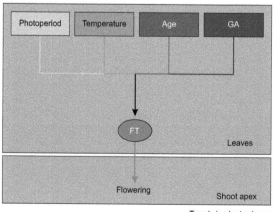

Trends in plant science

Fig. 4.1 Flow chart to flowering.

that initiation and anthesis are controlled independently rather than collectively by photoperiod, and that floral initiation has two independent steps—one for the first-initiated flower in the axil of a main stem leaf and a second for the transformation of the terminal shoot apex from vegetative to reproductive condition (Thomas and Raper, 1983).

4.1 Photoperiod

A photoperiod sensitive (sensitive) soybean [*Glycine max* (L.) Merr.] line and photoperiod in sensitive (insensitive) line were used to determine the critical seedling stage before possible floral induction and the length of photo induction required for anthesis. Seedlings of each entry were grown in pots and subjected to either a 10 h or 16 h photoperiod. Beginning 3 days after emergence, two pots per entry in the 10 h room were exchanged daily with two pots from the 16 h photoperiod. No detectable differences between treatments were observed in the flowering time of the insensitive line. However, the sensitive line exhibited the following: (i) The number of days to first flowering was not affected by the transfer from a 16 h to a 10 h photoperiod up to 9 days after emergence; (ii) Plants moved from a 10 h to a 16 h photoperiod before 36 days after emergence did not flower, indicating that induction was completed at 36 days after emergence; (iii) The earliest anthesis occurred 46 days after emergence.

Therefore, it was concluded that the induction period was 27 short days (10 hours) and that anthesis occurred 10 days after the completion of induction. The critical time to begin induction was 9 days after emergence for the sensitive line (Shanmugasundaran and Tsou, 1978).

Flowering process governs seed set and, thus, affects agricultural productivity. Soybean, a major legume crop, requires short-day photoperiod conditions for flowering. While leaf-derived signal(s) are essential for the photoperiod-induced floral initiation process at the shoot apical meristem, molecular events associated with early floral transition stages in either leaves or shoot apical meristems are not well understood. To provide novel insights into the molecular basis of floral initiation, RNA-Seq was used to characterize the soybean transcriptome of leaf and micro-dissected shoot apical meristem at different time points after short-day treatment. Shoot apical meristem expressed a higher number of transcripts in comparison to that of leaf, highlighting greater diversity and abundance of transcripts expressed in the shoot apical meristem. A total of 2951 shoot apical meristem and 13,609 leaf sequences with significant profile changes were identified during the time course examined. Most changes in mRNA level occurred after 1 short-day treatment. Transcripts involved in mediating responses to stimulus, including hormones, or in various metabolic processes represent the top enriched GO functional category for the SAM and leaf dataset, respectively. Transcripts associated with protein degradation were also significantly changing in leaf and SAM, implying their involvement in triggering the developmental switch. RNA-Seq analysis of shoot apical meristem and leaf from soybean undergoing floral transition reveal major reprogramming events in leaves and the SAM that point toward hormones gibberellins (GA) and cytokinin as key regulators in the production of systemic flowering signal(s) in leaves. These hormones may form part of the systemic signals in addition to the established florigen, FLOWERING LOCUS T (FT). Further, evidence is emerging that the conversion of shoot apical meristem to inflorescence meristem is linked with the interplay of auxin, cytokinin and GA, creating a low cytokinin and high GA environment (Wong et al., 2013).

As a paleopolyploid, soybean has undergone at least two major genome duplication and subsequent diploidization events, resulting in a complex genome with homeolog expected for most genes (Shoemaker et al., 2006). It is a short-day plant, grown broadly across the latitude but with each cultivar having a narrow range of north to south adaptation. This geographic adaptation of soybean is likely a result of genetic diversity associated with a large number of genes and quantitative trait loci regulating flowering behavior (Kong et al., 2010).

The floral initiation process is regulated by complex networks incorporating endogenous as well as exogenous cues in order to ensure the reproduction process occurring under optimal conditions. Studies carried out using *Arabidopsis thaliana*, a facultative long-day plant, have revealed the involvement of about 180 genes in controlling flowering time and a proportion of these genes occur in a network of six major flowering regulatory pathways (Fornara et al., 2010).

The photoperiod and vernalization pathways regulate flowering in response to either seasonal changes in day length or temperature while the ambient temperature pathways do so under the influence of daily growth temperature. The three remaining pathways are more responsive to internal developmental cues and these involve the age, autonomous and gibberellins (GA) pathways. Central to these pathways are three floral pathway integrators: *FLOWERING LOCUS T (FT)*, *SUPPRESSOR OF OVEREXPRESSION OF CONSTANS1 (SOC1/AGL20)* and *LEAFY (LFY)* that are proposed to integrate signals from these multiple pathways and coordinate floral developmental program in the shoot apical meristem (SAM).

FT is the mobile flowering signal produced in leaves that travels to the SAM (Corbesier et al., 2007) and forms a complex with the bZIP transcription factor, FD. The FT/FD complex then induces the expression of SOC1, the first floral gene activated in the SAM after exposure to long-days converting the SAM into an inflorescence meristem (IM) (Borner et al., 2000; Samach et al., 2000). SOC1 activates *LFY* and, similar to the FT/FD complex, can also induce the expression of floral meristem identity genes, such as *APETALA1 (AP1)*, triggering a developmental program culminating in the formation of flowers. Although the function of FT is conserved across different species (Kong et al., 2010; Lin et al., 2007; Laurie et al., 2011), the fact that *ft* mutants are only late-flowering suggests additional factors could eventually override the mutation.

Counterparts of *Arabidopsis* flowering time genes are beginning to be studied in soybean (Jung et al., 2012) and the functional conservation of these orthologs has been demonstrated, albeit with some intriguing variations. For example, while GmFT2b and GmFT5b are reported to have florigen-like functions, like the *Arabidopsis* FT, they are repressed by the GmPHYA1 and GmPHYA2 under long-days and, hence, inhibit the flowering process (Kong et al., 2010). This is in stark contrast to *Arabidopsis*, whereby PHYA plays a promotive role together with CRY2 resulting in the stabilization of CONSTANS (CO) (Valverde et al., 2004). Furthermore, unlike *Arabidopsis*, it is the GmCYR1a and not GmCRY2a that play a role in promoting flowering (Zhang et al., 2008).

In the same way, a variety of soybean called Biloxi, which normally flowers very late in the summer, could be made to flower in midsummer if it received only seven hours light each day. If Biloxi soybeans are grown in the winter in a warm glasshouse they will flower profusely, but if the length of winter days is extended by leaving an electric light on near the plants from about 5 p.m. to midnight they will remain vegetative. Conversely, species which normally produce their flowers in mid-summer can be made to flower even in mid-winter by extending the hours of daylight with artificial lighting.

Seedlings will not flower even when they are placed in favourable day-lengths. Before a plant can respond to a photoperiodic stimulus, it must have attained a certain stage of development, and this stage has been termed 'ripeness-to-flower'. Just what is involved in the change from seedling to 'ripeness-to-flower' is not clearly understood, but it is in some ways similar to the change from the sexually-immature juvenile to the sexually-mature adult of animal species.

The first visible indication of flowering is a change in the shape of the apical meristem of the shoot, but this can be seen only by careful examination under a microscope. During vegetative growth, the meristem, from which new leaves and stem tissues arise, is usually conical in shape and quite small, often only 0.1 mm in diameter. With the onset of flowering, however, each meristem becomes flattened into a broad disc which may give rise to a single flower, e.g., poppy, or to many flowers, e.g., capitulum of daisy.

During the later stages of flower development, some of the younger internodes of the stem elongate, raising the flowers above the vegetative body of the plant, and the leaves in this part of the stem are usually smaller and of simpler shape than those farther down the shoot.

Within certain limits, photoperiodic induction produces a quantitative flowering response. Although two short-days are sufficient to initiate flowering in Biloxi soybeans, heavier flowering follows further short-day treatment up to the point where all the vegetative apices have been converted to flowering apices.

Evidently, in short-day plants, florigen is formed only in leaves which are exposed to short-days. It travels through leaf tissue receiving short-days or continuous darkness but will either not pass through leaves exposed to long days or is inactivated in them. The inhibition exerted by parts of a leaf receiving long days is only apparent when the long-day part of the leaf is between the short-day part and the terminal bud of the shoot.

Before short-day plants will flower they must be subjected to a day-length which does not exceed a certain critical value and each short-day must be followed by a night of relatively long duration. Short-day plants will not normally flower if kept in continuous darkness and it seems to be the regular rhythm of short days and long nights which causes them to flower. The necessity for light suggests that photosynthesis or one of the intermediate reactions of photosynthesis may be involved in flowering (reaction I). The relationship between photosynthesis and flowering is also suggested by the following facts: Both processes require carbon dioxide to be present in the air and both proceed only in light of high intensity. Furthermore, some plants, such as potato, which have large photosynthetic reserves, are able to flower even when kept in darkness, and in some other short-day plants injected sugars can substitute for the light period.

For flowering to occur, each favourable light period must be followed by a long period of uninterrupted darkness. From this we may infer that the substance formed in the light is utilised during the darkness. This dark reaction (reaction II) becomes effective in causing flowering only when the dark period exceeds the critical night-length for the species. Evidently the substance formed in the dark reaction is synthesised only slowly in the leaves each night, therefore, a fairly prolonged period of darkness is necessary. This is because lengthening the days beyond the critical day-length by means of supplementary artificial illumination even of low intensity prevents flowering. The dark reaction is photosensitive, only occurring in the absence of light. Thus, short-day plants will not flower even in short days if the night is interrupted by a brief period of light (reaction X). This light break need last only about one minute and be of low intensity illumination to destroy the beneficial effects of a long night. Furthermore, a light break is most effective in preventing flowering when it is given at about the middle of the dark period. Earlier or later interruption of the dark period is less effective, and if, as previously suggested, the synthesis of the dark reaction product occurs fairly slowly, it can explain this result, and also gain considerable insight into the mechanism of floral induction. A light break early in the night would destroy only the small amount of dark product already formed in the leaves; the ensuing period of darkness would still be long enough to allow more of this substance to be synthesised and flowering would occur. By the middle of the dark period, a large amount of dark product would have been synthesised and this would all be destroyed by a light break at this time. The ensuing dark period would be too short to allow enough dark product to form, so flowering would not occur. However, towards the end of the dark period, most of the light product would have been converted to dark product and most of this would already have been translocated out of the leaves, so that a light break at this time would destroy only the small amount of dark product which still remained in the leaves and would not prevent flowering. Light therefore seems to have two opposing actions, high light intensities providing precursors and promoting flowering, low intensity light breaks destroying photosensitive products of the dark reaction and thus inhibiting flowering.

The substance formed in the dark reaction is translocated from the leaves, where it is synthesised, to the apices, where it causes the initiation of flowers. During this time, a further reaction must occur in order to render the flowering hormone light-stable (reaction III), otherwise plants requiring more than one photoperiodic cycle for induction would always fail to flower because the hormone would be destroyed during each light period. It is thought that the hormone is light-sensitive only in the leaves and that in the stem it is converted to a light-stable form. Once the hormone reaches the apex it evidently begins to be synthesised in meristematic tissue, and its further synthesis there is independent of length of day, for once flowering has commenced it will continue under all conditions of day-length and after all photoperiodically induced leaves have been removed from the plant.

The sequence of photoperiodic induction in short-day plants can be written as follows:—

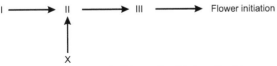

Reaction I = High intensity light reaction (photosynthesis).
     II = Photosensitive reaction occurring only in mature leaves in darkness.
   III = Photostabilisation of flower hormone in stem tissue.
    X = Antagonistic affect of long days, low intensity supplementary illumination, or light breaks on reaction II.

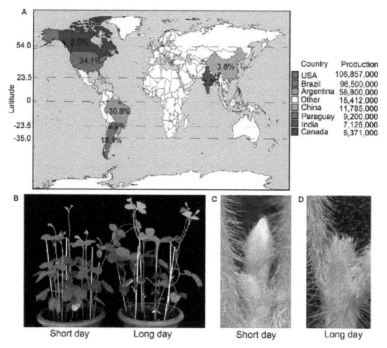

Fig. 4.2 Short-day and long-day effects in soybean.

4.2 Flowers

The papilionaceous flower consists of a tubular calyx of five sepals, a corolla of five petals (one banner, two wings and two keels), one pistil and nine fused stamens with a single separate posterior stamen (Fig. 4.3). The stamens form a ring at the base of the stigma and elongate one day before pollination, at which time, the elevated anthers form a ring around the stigma. The pod is straight or slightly curved, varies in length from two to seven centimetres, and consists of two halves of a single carpel, which are joined by a dorsal and ventral suture. The shape of the seed, usually oval, can vary amongst cultivars from almost spherical to elongate and flattened.

Soybean seed number is determined by the number of flowers produced, the number of pods retained on the plant, and the number of seeds per pod. Because flowers can be produced on all stem and branch

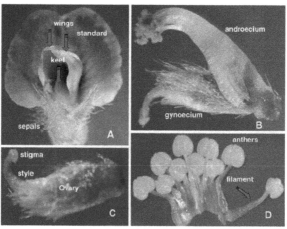

Fig. 4.3 Flower of a soybean plant.

nodes, flower number is highly influenced by the amount of branching. The number of branches and branch length are amazingly flexible and respond to stand density and plant spacing.

Unlike corn, soybean plants produce "complete flowers". Complete means that they contain all four basic flower parts: Sepals, petals, stamens, and pistil (Fig. 4.3). Soybean flower structure ensures that they are highly self pollinated. The two keel petals enclose the sexual parts making it nearly impossible for wind or insects to carry pollen into the flower. The 10 stamens (male parts) are closely situated near the pistil (female structure) so that pollen grains produced in the anthers (part of stamen) are deposited directly onto the stigma (part of pistil) (Fig. 4.3). More than 98% of soybean pods result from self pollination.

Shortly after pollen grains land on the stigma, pollen tubes emerge from the pollen grains and penetrate the stigma. Pollen tubes elongate through the short style. The style tissue provides nourishment and water to the growing pollen tube. It also provides direction, so that the pollen tube's journey ends in the correct place—inside the ovary (Fig. 4.4).

Each ovary contains two to four ovules. The ovary wall will become the pod wall and the ovules will become seeds. As with all agronomic plants, soybean flowers undergo double fertilization. Three nuclei (plural of nucleus) move into the pollen tube. One of the three nuclei directs pollen tube growth and will not be involved in fertilization. The other two nuclei travel down the pollen tube and enter into the ovule once the pollen tube completes its journey. One male nucleus combines with the female gamete to form the embryo within the seed. The other male gamete joins with two female nuclei to form the endosperm. Mature soybean seeds contain almost no endosperm. The large cotyledons accomplish the food storage function usually associated with endosperm.

Each ovule in an ovary requires a separate pollen tube for fertilization. If an ovary contains three ovules, at least three pollen tubes must enter the ovary if all three ovules are to be fertilized. For corn, the number of female flowers that become fertilized is an important determinant of seed number, and that number is highly influenced by weather. Fertilization of soybean flowers is nearly 100%. Reasons for high success rate are: Many pollen grains are produced in the 10 stamens, no pollen grains are lost by wind, weather has little effect on maturity sync of stamens and pistil, pollen tubes must travel a short distance from stigma to ovary, and the flower petals cover the pistil, which reduces dehydration.

Soybean plants bear flowers on inflorescences called racemes. Racemes have multiple flowers attached by a short stalk (pedicel) to a central, unbranched axis (rachis). After flowers are fertilized, the rachis elongates and separates the developing pods. Flowers in a raceme are fertilized in a specific pattern, starting with the flower nearest the plant and proceeding up the rachis to the last flower. It may take 4 to 10 days for all flowers to open on a single raceme. There are three buds at each leaf axil that can produce racemes, so flowers may continue to open at a node for two or more weeks.

As stated before, nearly 100% of soybean flowers are fertilized. So, technically, all of the reproductive structures that abscise are pods. However, many of the structures that abscise are very small pods that may have petals still attached. Not all flowers have an equal chance of remaining on the plant. Flowers produced on nodes near the bottom of the canopy are more likely to abscise than flowers located in the upper one-third (Table 4.1 and Fig. 4.5).

Fig. 4.4 Photograph of reproductive parts of soybean flower.

Table 4.1 Variation in flower color and leaf shape and cultivars.

Violate	White	Broad leaves	Narrow leaves	Small leaves	4-5 leaflets
DSb 30-2	PS 1569	RVS 2010-1	PS 1569	JS 20-94	MACS 1491
KDS 975	RVS2010-1	DSb 30-2	AMS 115	KDS 975	KDS 975
MAUS 710	RSC 10-30	RSC 10-30	JS 20-94	SL 1074	KDS 775
KDS 775	VLS 91	MACS 1491	NRC 117	JS 97-52(c)	NRC122
RKS 18(c)	NRC 118	NRC 117	AMS 115		
MACS 1491	VLS 90	TS 72	Himso 1686		
JS 20-94	NRC 119	DS 3103	PS1570		
NRC 117	NRC 120	MACS 1480	DS3104		
AMS 115	VLB 202	AMS 100-1	SL 1074		
TS 72	JS 20-116	KDS 775	NRC120		
MACS 1488	AMS 100-1	NRC 119	VLB 202		
RVS2010-2	Himso1686	DSb 29	NRC 122		
AMS 1001	KDS 754	DS 3104			

Fig. 4.5 Violet and white flowers and primary (A), secondary (B), tertiary (C) raceme in soybean.

4.3 Pod Development

The reproductive growth period is usually represented by flowering, pod and seed development. Flowering may become visible at 25 days or may be delayed until 50 days when certain genotypes and environments interact (Whigham, 1983). The soybean flower stigma is receptive to pollen approximately 24 hours before anthesis and remains receptive 48 hours after anthesis. The anthers mature in the bud and directly pollinate the stigma of the same flower. As a result, soybeans exhibit a high percentage of self fertilisation and cross-pollination is usually less than one percent (Abernethy et al., 1977).

Flowering may occur over 4–6 weeks, depending on the environment and cultivar. After fertilisation of the flower, the pods develop slowly for the first few days, then the rate of development increases until the pod reaches maximum length after 15–20 days (Whigham, 1983). Peterson et al. (1992) proposed a flower and pod staging system for soybean based on the morphological characteristics of the flower a few days prior to and following anthesis (Figs. 4.6 and 4.7, Table 4.2). Studies of Westgate and Peterson (1993) have shown that this system is useful for categorising morphological changes resulting from drought stress and for quantifying the impact of drought stress on pod set.

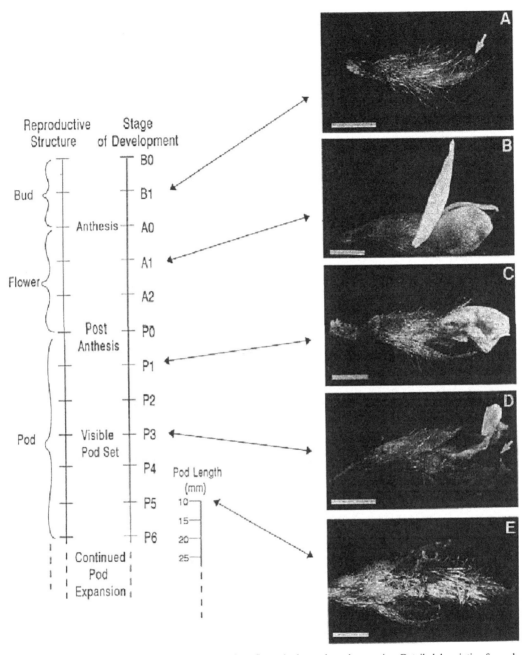

Fig. 4.6 Stages of soybean reproductive development from flower bud to early pod expansion. Detailed description for each stage is shown in Table 4.2. Adapted from Westgate and Peterson (1993).

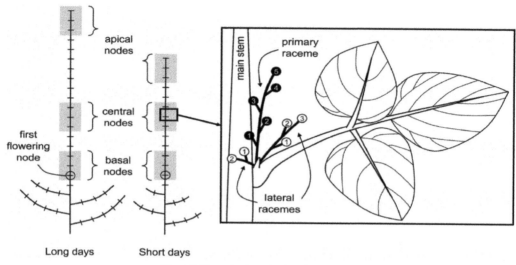

Fig. 4.7 Flowering in main shoot in soybean.

Table 4.2 A flower and pod staging system of soybean (Morphological description).

Stage	DAA	Flower	Pistil or pod	Ovules
B0	–2	No visible corolla	Pre-pollination	Maturing embryo sac
B1	–2	Corolla visible, but not fully extended beyond calyx lobes	Pre-pollination	Mature embryo sac, central cell filled with multi-grained amyloplasts
B2	0	Banner petal fully extended	Early pollination and pollen germination	Mature embryo sac, degenerate synergid in some ovules
A0	0	Partial opening of banner petal	Pollination completed	Fertilisation beginning
A1	0	Banner petal completely reflexed, full anthesis	Pollen tube growth	Fertilisation completed; zygote formation
A2	+1	Banner petal collapsed; appears 'hooded'	--	Zygote to 2-celled proembryos; starch grains disappearing in central cell
P0	+2	Margins of banner petal slightly wilted and rolled inward. Small brownish spots visible on banner and/or keel petals	--	2- to 4-celled proembryos; acellular endosperm
P1	+2–3	More spots discoloration and wilting along petal edges; part of petals withered	--	4- to 16-celled proembryos; parietal free nuclei in central cell
P2	+3–4	Completely withered petals, no turgid petal tissues visible	Pistil elongation and extension; early pod set	16-celled proembryos; ecellular endosperm
P3	+4–5	Completely withered corolla	Stigma of pistil visible beyond corolla. Pod set	Upto 32-celled proembryos with early suspensors, cellularization of endosperm beginning
P4	+4–6	Withered corolla may be torn away or abscised from receptacle by expanding pod	Visible pod swelling and extension; 7–10 mm length	Globular embryos with early suspensors, cellularization of emdosperm beginning
P5	+5–8	Withered corolla attached at base of pod or shed	Pod 1–2 cm in length	Embryos with developed suspensors surrounded by cellular endosperm

Adapted from Peterson et al. (1992).

The initiation of flower was started from 54 DAS irrespective to the application of GA+ABA. Number of flowers produced increased gradually up to 64 DAS, then increased sharply up to 72 DAS, and then slowed down both in plants under control and treated (Fig. 4.8). The lower most node (1st) produced 8–9 flower per plant and attained to pick in 4th node ascendingly where plant under control produced 13 and 16 flowers, respectively, in plant under controlled and GABA treated (Fig. 4.9). From 5th to 7th node, the number of flower per node decreased gradually and then get zigzag trend pattern up to end. Flower abscission was found to be started from 60 DAS (Fig. 4.10). From the beginning, the floral abscission trend in plants under control revealed steady which slowed down after 70 DAS (Islam et al., 2010).

Four hormone groups, namely cytokinins (CKs), ABA, gibberellins (GA) and auxins (IAA) are important in reproductive development of crops. In normal seed development, a CKs peak shortly after anthesis is followed by a GA peak, and then by an auxin peak. ABA increases steadily during seed development.

It is well known that phytohormones play important roles in regulating crop reproductive development by affecting both sink strength and assimilate partitioning from source to sink organs (de Bruijn et al., 1993; Emery et al., 2000; Wang et al., 2001). It is also becoming clear that phytohormones may exert

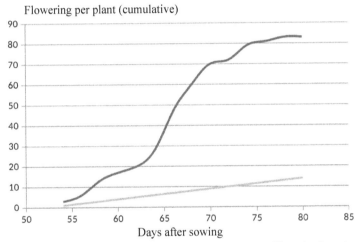

Fig. 4.8 Curve showing changes of flowers per plant after on-set of flowering in soybean.

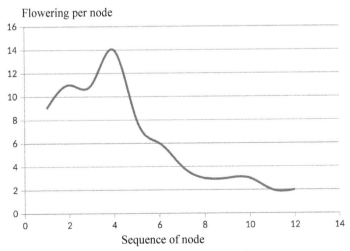

Fig. 4.9 Variations in flowers in nodes of soybean.

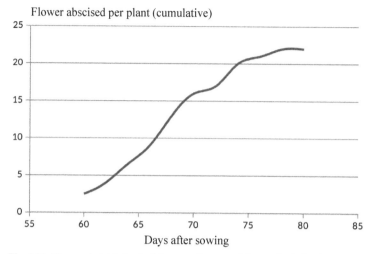

Fig. 4.10 Changes in total abscised flowers after start of flower abscission in soybean.

distinct effects at different stages during ovary development (Setter and Flannigan, 2001). During early reproductive development, it is generally found that large concentrations of ABA inhibit cell division whilst large concentrations of CKs promote cell division in the young ovary of crop plants, including soybean (Emery et al., 2000; Setter and Flannigan, 2001; Yang et al., 2002a; Kokubon and Honda, 2000).

A NPK factorial experiment was conducted during two successive cropping seasons to study direct and residual effects of various treatments on the flower initiation and yield of soybeans. Plants were slower to reach the maximum flowering stage in the year of fertilizer application than in the following year. During the first year, potassium and phosphorus delayed the flower forming process, whereas potassium alone hastened flowering in the second year, unless associated with a high dosage of nitrogen. All the treatments gave far greater yields in the first season, suggesting that fertilizers should be applied during every cropping season. Nitrogen and phosphorus at 20 pounds of N and P_2O_5, per acre gave the biggest yields, but potassium failed to show any response. The residual responses to potassium and phosphorus during the second year were significant but not to nitrogen. A description of reproductive stages and influence of sowing date on days to different reproductive stages are given in Tables 4.3 and 4.4.

Table 4.3 Description of reproductive stages.

Stage no.	Stage title	Description
R1	Beginning bloom	One open flower at any node on the main stem
R2	Full bloom	Open flower at one of the two uppermost nodes on the main stem with a fully developed leaf
R3	Beginning pod	Pod 3/16 inch long at one of the four uppermost nodes on the main stem with a fully developed leaf
R4	Full pod	Pod ¾ inch long at one of the four uppermost nodes on the main stem with a fully developed leaf
R5	Beginning seed	Seed 1/8 inch long in a pod at one of the four uppermost nodes in the main stem with a fully developed leaf
R6	Full seed	Pod containing a green seed that fills the pod cavity at one of the four uppermost nodes on the main stem with a fully developed leaf
R7	Beginning maturity	One normal pod on the main stem has reached its mature pod color
R8	Full maturity	95% of pods have reached their mature pod color; 5–10 days of drying weather are required after R8 before the soybeans have less than 15% moisture

Table 4.4 Sowing date and days from sowing to emergence (VE), flowering (R1), beginning of pod setting (R3) and beginning of seed filling (R5) across the four experiments.

Expt.	Treatment+	MG	Sowing date	VE	R1	R3	R5
1	HT	2.1	Feb	6c	33g	41f	52g
		3.0		6c	37f	57e	70e
	MT	2.1		9b	49c	59e	67f
		3.0		9b	53e	67d	82d
	IT	2.1		17a	57d	67d	81d
		3.0		18a	67b	79b	95b
	LT	2.1		17a	61c	73c	89c
		3.0		18a	73a	85a	101a
2	SD1	2.1	April 23	12b	53a	66a	81ab
		3.0		12b	55a	66a	84a
	SD2	2.1	April 30	15a	52a	65a	78bc
		3.0		15a	54a	65a	84a
	SD3	2.1	May 13	12b	47b	56b	72de
		3.0		12b	45b	56b	79cd
	SD4	2.1	May 26	7d	40c	47e	63f
		3.0		7d	40c	51c,d	71e
	SD5	2.1	June 8	8c	39c	50de	64f
		3.0		8c	39c	55c	66f
	SD6	2.1	June 19	5e	36d	48e	59g
		3.0		5e	39c	50de	61g
3	HT	2.1	Sept 16	5d	33d	40d	51d
		3.0		5d	37c	55c	71c
	MT	2.1		9d	58b	64b	76b
		3.0		9c	66a	72a	83a
	IT	2.1		11b	nr	nr	nr
		3.0		11b	nr	nr	nr
	LT	2.1		13a	nr	nr	nr
		3.0		13a	nr	nr	nr
4	F1	2.7	April 25	13d	65a	79a	100a
	F2	2.4	May 2	19a	57b	67c	89c
	F3	2.4	May 13	16b	52c	71b	99b
	F4	3.1	May 18	14c	45d	63d	84d

nr: no record because experiments were terminated due to severe infestation of powdery mildew around 85 days after sowing.
Different letters indicate significant ($P < 0.05$) differences between treatments within each experiment for the variables identified at the top of each column.
HT: high temperature; MT: moderate temperature; IT: increasing temperature; LT: low temperature; SD: sowing date; F: farmer; MG: maturity group.

4.4 Abortion of Flowers

Soybean plants produce an abundance of floral buds, but a large proportion of the ovaries are aborted prior to developing into mature pods (Wiebold et al., 1981; Dybing et al., 1986). It has been reported that 40–80% of the flowers and pods initiated eventually abort under conventional cultivation (Wiebold et al., 1981). Abscission of flowers and young pods occurs mostly following flowering, after pollination and fertilisation have completed (Brun and Betts, 1984). Therefore, pod number is primarily determined during early stage of pod development (within 5 days after anthesis) (Dybing et al., 1986). The individual seed weight is a product of the rate and the duration of seed filling (Munier-Jolain et al., 1998), it is generally determined during seed filling after the pod number had been fixed (Westgate and Grant, 1989; Desclaux et al., 2000; Brevedan and Egli, 2003).

It is well recognised that drought stress during flowering and early pod development is the major cause of pod abortion in soybeans (Momen et al., 1979; Boyer, 1983; Westgate and Peterson, 1993; Saitoh et al., 1999). A brief soil water deficit during this period can decrease pod set up to 70% (Momen et al., 1979; Andriani et al., 1991). This is because the reproductive development during this phase involves several processes that are extremely vulnerable to a change in plant water status (Saini, 1997).

According to Peterson et al. (1992), it is clear that, during flowering and early pod growth, there is a transition from cell division in the pre-embryo and free nuclear division in the endosperm to rapid cellularization of the endosperm and further differentiation of the embryo. Successful passing of this transition is essential for continuous growth and setting of pods (Westgate and Peterson, 1993). The processes involved in this transition, such as cell division and expansion, are very sensitive to changes in plant water status; a high water potential of the flowers or pods is vital for setting of the pods (Westgate and Peterson, 1993). Studies have shown that soil water deficits have a direct effect on flower water status and flower function in soybean. Drought stress imposed prior to or soon after flowering can significantly decrease flower water potential (Westgate and Peterson, 1993) and impair ovary function (Kokubun et al., 2001).

Similar findings have also been reported in other crops, like maize, where a low water potential in the style could inhibit silk elongation, arrest flower development, and disrupt ovary metabolism (Westgate and Boyer, 1985). However, to little is known about the physiological and biochemical reasons for pod abortion in soybeans grown under drought stress during early reproductive development.

The characteristics of pod set and seed growth as affected by raceme order were investigated in order to determine the yield-determining process of soybean. Observations of racemes were made on the 4th, 7th, 10th, 13th and terminal nodes of the main stem of three cultivars (Indeterminate type: Harosoy; Determinate types: Enrei, Tamahomare). Pod-setting ratio and dry seed weight reduced in higher-order racemes, while the number of seeds in a pod did not vary among raceme orders. The lower seed weight of higher-order racemes was due to the short seed growth period. Compound leaves of secondary racemes enhanced the seed growth but not the pod-setting. With the highest node of determinate types, seed dry weight and rate of dry matter accumulation (RDA) of primary racemes exceeded those of terminal racemes, suggesting that more competition for assimilates among racemes occurred in terminal racemes. The results indicate that, regardless of growth habits, the lower the raceme order, the higher the number of pods and the pod-setting ratio. Seeds derived from lower-order racemes accounted for the majority of the yield. Hence, pod-setting and seed growth of lower-order racemes are more important than those of higher-order racemes in determining soybean yield.

Water stress caused by restriction of watering for three days during the pre-anthesis stage significantly increased the abortion of the basal flowers, which are destined to develop into pods under optimal irrigation. The experiment also revealed that the pistils of well-watered plants, whether pollinated with water-stressed or unstressed pollen, produced pods at a considerable rate, whereas only a small percentage of water-stressed pistils developed into pods, even when crossed with unstressed pollen.

Soybean (*Glycine max* [L.] Merr.) ovary growth was measured from anthesis to 6 days after anthesis (DAA) to establish a timetable of biochemical events that might be useful in identifying processes that initiate abscission. Two procedures were developed to provide samples with either high or low percent pod set for; IX93-100, a semi determinate line having long racemes. Characteristics measured were fresh and dry weight, soluble and insoluble protein, soluble carbohydrate, starch, RNA, and DNA. Setting ovaries grew more rapidly than abscising ovaries. Since there was a daily increase in ovary weight in both groups, all measured characteristics showed daily increases when expressed on per ovary basis. Statistically significant differences between groups were detected between 2 and 5 DAA for most characteristics. When chemical composition was expressed on concentration basis, starch level was significantly higher in setting ovaries at 5 and 6 DAA. Regression analysis showed that these deviations between setting and abscising samples started between anthesis and 1 DAA. It was concluded that processes leading to eventual shedding of fertilized ovaries (called flower abortion in soybeans) commence soon after anthesis of the shed flower, and that setting and abscising ovaries do not differ in protein, soluble carbohydrate, starch, or nucleic acid content when abscission processes begin.

Soybean (*Glycine max* L. Merr.) yield is determined by the number of seeds per unit area and individual seed weight. The seed number depends upon the number of floral buds that initiate pods and attain maturity. Soybean plants produce an abundance of floral buds, but a large proportion of them abscise during development to mature pods (van Schaik and Probst, 1958; Kato, 1964; Bevedan et al., 1978; Wiebold et al., 1981). A reduction in this abscission might increase pod and seed number, and can thereby lead to an increased yield. The magnitude of abscission varied with the position on the plant, being higher in the branches, the lower part of the main stem and the top nodes of the main stem (Hansen and Shibles, 1978; Wiebold et al., 1981; Hein-dl and Brun, 1984; Gai et al., 1984). Within individual racemes, the distal floral positions on the rachis showed a higher abscission percentage than the proximal positions (Huff and Dybing, 1980; Spollen et al., 1986a; Dybing et al., 1986; Carlson et al., 1987; Wiebold, 1990; Wiebold and Panciera, 1990). However, the physiological mechanism controlling reproductive abortion remains unclear. Possible physiological factors affecting abortion include quantity of certain plant hormones (Huff and Dybing, 1980; Beckmann, 1981; Heindl et al., 1982; Spollen et al., 1986b; Carlson et al., 1987; Yarrow et al., 1988), deficiency in or competition for carbohydrates and nutrients (Brevedan et al., 1978; Brun and Betts, 1984; Antos and Wiebold, 1984; Heitholt et al., 1986a, b) and the quality and quantity of light in the plant canopy (Heindl and Brun, 1983; Brun et al., 1985; Myers et al., 1987). The application of 6-benzylaminopurine to racemes reduced abortion and increased pod-set probability (Crosby et al., 1981; Carlson et al., 1987; Dyer et al., 1987; Peterson et al., 1990; Mosjidis et al., 1993; Reese et al., 1995). Endogenous cytokinins detected in the root pressure exudate showed a maximum concentration during a period of 0 to 9 days after the initial flowering, when most flowers are destined to either initiate pods or abort (Heindl et al., 1982; Carlson et al., 1987). These lines of evidence suggest that cytokinins produced in roots and translocated to flowers might be involved in the reduction of flower abortion or enhancement of pod initiation. However, data on the ontogenetic changes in the cytokinin contents of racemes are not available, therefore, whether the cytokinin contents in each portion of a raceme relate to the pod-set probability or not remains unknown.

The critical role of phytohormones in the formation and abortion of reproductive organs in soybean was clearly recognized when Huff and Dybing (1980) observed that extracts from flowers and young pods applied to growing flowers accelerated flower abortion. They then applied a lanolin paste containing either indoleacetic acid (IAA), giberellin (GA) or 6-benzylaminopurine (BA) to the growing raceme, and found that IAA enhanced the abortion rate, as did the extract, whereas GA and BA did not. These results indicated that IAA plays a crucial role in increasing the abortion rate, although there was a conflicting report indicating that IAA delays the abortion (Oberholster et al., 1991).

However, whether intra-raceme variation in the pod-set probability relates to endogenous cytokinin levels remains unknown. To address this question, intra-raceme variation in cytokinin content and pod-set probability was investigated. A soybean genotype IX93-100, which has long racemes, was grown in an environmentally controlled chamber (30/20°C day/night temperature, 15 h day length, 600 μmol m^{-2}s^{-1} photosynthetic photon flux density). Flowers, which were divided into three floral positions (proximal, middle, distal) on individual racemes, were sampled at intervals after anthesis. The cytokinins in the samples were identified by gas chromatography-mass spectrometry (GC-MS) and further quantified by enzyme immunoassay (EIA). The GC-MS analysis revealed that cis-zeatin riboside (c-ZR) and isopentenyl-adenosine (iPA) were predominant forms of cytokinin in soybean racemes. The total amount of these cytokinins in racemes, which was monitored by EIA, peaked one to two weeks after the first flowering on a raceme, when pod development was initiated. Within individual racemes, the total cytokinin concentrations were greater at more proximal floral positions, as was the probability of pod set. Removal of proximal flowers at anthesis enhanced both cytokinin concentrations and pod set at middle positions on the raceme. Thus, pod-set probability was significantly associated with the cytokinin concentration at different floral positions within individual soybean racemes (Fig. 4.11).

The ability of soybean [*Glycine max* (L.) Merr.] to adjust its pod load to environmental conditions is an important, but not well understood, part of the yield production process. To better understand this process at the whole plant level, a single-node model (SOYPOD) was extended to predict pod set of whole plants. Pod survival in SOYPODP was determined by comparing assimilate from photosynthesis (calculated from solar radiation) to pod requirements during initial pod growth. Measured profiles of

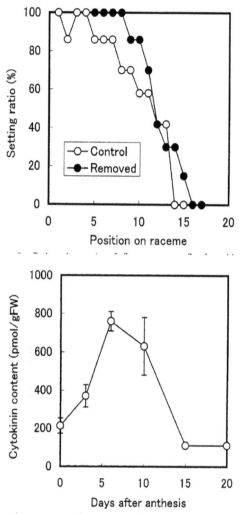

Fig. 4.11 Pod setting percent on the raceme and cytokinin content changes with DAA.

flowers per node and pod and seed growth rates were used as input. Pods that do not receive adequate assimilate for a specified number of consecutive days do not survive. When inter-nodal movement of assimilate was prohibited and the nodal assimilate production profile matched the input flower profile, pod set on the stem matched measured profiles of two field-grown cultivars. In this form, SOYPODP accurately mimicked known responses of pod number to variation in assimilate supply, nodes per plant and individual seed growth rate. Competition for assimilate at a node-controlled pod set, so increasing uniformity of pod development (more flowers or a shorter flowering period) increased pod set. Increasing nodal variation in the beginning of pod development reduced the sensitivity of the plant to short-term fluctuations in assimilate supply. Variation in uniformity of pod development at a node was responsible for increases in pod set with no change in assimilate availability. The value of these increases is not clear, but it seems unlikely that they would affect yield.

The nodes on the soybean plant will produce a cluster of flowers. Soybean flowers self-pollinate, so pod set can occur shortly after the appearance of the flower. During this early reproductive stage, as many as 60 to 75% of the flowers and/or small pods may be shed by the plant. Flower and small pod abortion is a natural part of soybean growth and development. When a flower or pod is shed from the plant, the soybean plant is adjusting to its environment. The amount of plant photosynthate available will dictate the number of pods that will reach maturity. If the plant produces more pods than the production of photosynthate, flowers and/or pods will be shed. If the plant produces sufficient photosynthate, the plant may continue to flower and set additional pods. Fields that have severe early flower and small pod shed may initiate more flowering and possibly set new pods if environmental conditions improve to increase photosynthate production; however, yield potential may already be compromised by early stress.

Heat and/or drought stress during the R3 to R6 (beginning pod to full seed) growth stages may increase flower and small pod abortion. Results from a study conducted by Mann and Jaworski (1970) showed that pod formation was severely limited at temperatures above 40°C (104°F). Pods will typically not abort once a plant reaches R6 stage. During seed filling stages, remember that any yield loss encountered is likely realized through reduced seed size due to unfavourable environmental conditions, not necessarily by pod shed.

4.5 Pod Setting

High soybean yield requires an aggressive pod and seed set; however, a moderate amount of early flower or pod abortion will not necessarily hurt yields. The soybean plant is amazingly adaptable and will simply produce more and larger seed if enough photosynthates are available. The final yield is determined by the genetics of the variety and the end result of the environment. Little can be done to prevent soybean plants from shedding flowers or pods in hot weather, except reducing other environmental stresses as much as possible. Stresses to the soybean plant should be closely monitored and controlled if possible. Insect or disease stress and nutrient availability are factors a grower can usually control. When irrigation is available, the application of water may help alleviate heat stress.

Soybean (*Glycine max* (L.) Merr. cv. Enrei) plants were grown in pots (15-L volume) placed in a greenhouse with ventilation. At the time when the first flower opened, pots were transferred to growth chambers with natural lighting under day temperature of 30°C and night temperatures of 20, 25 or 30°C. The numbers of flowers opened and pods set each day were recorded and the seed yield and yield components were investigated after harvest. The increase in night temperature decreased the seed size and increased the number of flowers and pods. As a result, the seed weight per plant was unaffected by night temperature. However, high night temperatures increased the number of flowers on the secondary and tertiary racemes. These flowers opened after the 18th day of the flowering period and showed a high rate of pod setting. These results suggest that a high night temperature stimulated flower opening and pod setting in the secondary and tertiary racemes. The increases in the numbers of flowers and pods could serve to moderate the reduction of seed yield caused by a high night temperature (Fig. 4.12).

High temperatures, due to global warming, is an increasing environmental stress that influences soybean (*Glycine max* (L.) Merr.) growth and yield. Breeding tolerant cultivars for high temperature conditions is, therefore, of high importance. This experiment was conducted in order to evaluate the varietal differences of soybean photosynthetic apparatus and agricultural characteristics, such as flowering, pod number, and yield, under high temperature conditions. Seven cultivars were selected from the world soybean core collection, which were derived from the Gene bank Project, National Agriculture and Food Research Organization in Japan, and 2 Japanese cultivars were used and grown in greenhouses. The high temperature (HT) treatment delayed the beginning pod and full maturity stages by 1 to 10 days and −1 to 17 days, respectively. Flower number per plant had a tendency to become larger in the HT treatment when compared to that of the control. In the HT treatment, pod setting rates decreased and then seed number decreased, which resulted in a decrease in yield, but flower number increased. The decrease in pod number depended mainly on the decrease in pod setting rate. The actual quantum yield of Photosystem II (PSII) (Φ_{PSII}) was not different between the control and HT treatment, indicating the HT treatment did not reduce the efficiency of electron transport in the PSII for any of the cultivars. The maximum quantum yield of PSII (Fv/Fm) did not show a significant difference between the control and the treatment groups. Every plot was more than 0.79, and assumed no photoinhibition occurred in the HT treatment. The degree of heat dissipation in PSII was similar in both the control and the treatment groups and among the cultivars. The CO_2 assimilation rate (A_N) had a close relationship with stomatal conductance (g_s) in the control and treatments groups, indicating that a cultivar with a high stomatal conductance had a tendency for high CO_2 assimilation rate. The other photosynthetic characteristics

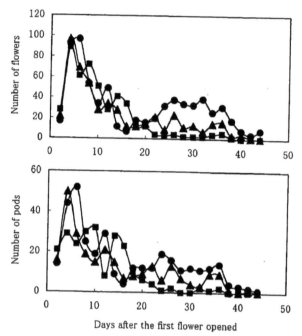

Fig. 4.12 Numbers of flowers opened and pods set, and the rate of pod set at different periods during flowering period in 1998.

did not show a relationship with A_N. There was no significant decrease in Fv/Fm in the HT treatment when compared to that of the control in this experiment. The photosynthetic apparatus may have not been damaged by the HT treatment in this experiment. A higher transpiration ability in soybean may be associated with a higher adaptability for high temperature conditions.

Table 4.5 shows flower number, pod number, and pod setting rate the flower number per plant tended to be higher in the HT group than in the control. Kongnamul Kong, Heukdaelip, and Uronkon showed significantly more flower numbers in the treated group than in the control.

Abscission of flowers and immature pods are significant yield-limiting factors in soybean. The abscission is caused by the lack of assimilates or nutrients, and by environmental stresses, such as deficiency of soil moisture, chilling or heat. Although Ueki and Igawa (1958) reported that high night temperature reduced the rate of pod setting, there are few reports supporting an enhancing effect of night temperature on the abscission of flowers and pods (Sato and Ikeda, 1979; Seddigh and Jolliff, 1984; Gibson and Mullen, 1996). The results of this study also suggested that a high night temperature had no influence on the rate of pod setting, although it stimulated the flowering of secondary and

Table 4.5 Soybean flowering and pod setting as influenced by high temperature.

Cultivar	Flower Number (plant-1)			Pod Number (plant-1)			Pod Setting Rate (%)		
	Control	HT		Control	HT		Control	HT	
Enrei	105c	121d	ns	49c	43cd	ns	49a	36a	ns
Tachinagaha	112c	132d	ns	22d	38cd	ns	20de	28ab	ns
Chunhoku 2	289a	360ab	ns	145a	115a	ns	50a	33a	*
Shirosota	174bc	198cd	ns	55c	55bcd	ns	32c	28ab	ns
Chieneum Kong	280a	335ab	ns	81b	64bc	ns	30cd	19bc	*
Kongnamul Kong	216ab	275bc	*	89b	92ab	ns	41ab	34a	ns
Heukdaelip	159bc	293abc	*	54c	21d	ns	36bc	7d	*

* significant at 5% level, ns not significant.

Table 4.6 Influence of high night temperature on flowering, pod setting and rate of pod set of soybean.

Temperature	Days after the first flower opened					
(Day/night)	1–10	11–20	21–30	31–40	41–50	Total
30/20 (°C)						
No. of flowers (plant-I)	78.8 ± 16.6	29.0 ± 8.6	3.3 ± 2.8	3.5 ± 2.4	0.3 ± 0.5	114.8 ± 25.8
No. of pods (plant-1)	35.0 ± 3.6	18.3 ± 4.0	1.3 ± 1.9	1.3 ± 1:0	0 ± 0.5	5.8 ± 3.9
Rate of pod set (%)	44.4	62.9	38.5	35.7 0	48.6	
30/25 (°C)						
No. of flowers (plant -I)	71.5 ± 7.1	26.5 ± 9.0	17.0 ± 5.0	11.0 ± 0.8	0 ± 0	126.0 ± 13.3
No. of pods (plant-t)	33.5 ± 2.6	15.3 ± 2.6	9.5 ± 4.4	5.5 ± 3.5	0 ± 0	63.8 ± 2.8
Rate of pod set (%)	46.9	57.5	55.9	50.0	50.6	
30/30 (°C)						
No. of flowers (plant-1)	80.0 ± 5.2	21.8 ± 6.8	41.5 ± 9.9	26.0 ± 6.2	2.5 ± 3.3	171.8 ± 14.5
No. of pods (plant-I)	42.8 ± 6.8	14 ± 5.2	17.3 ± 3.9	9.5 ± 5.5	0 ± 0	84.3 ± 5.7
Rate of pod set (%)	53.4	64.4				

The values are the mean of four plants with standard deviation.

tertiary racemes. Of great interest is how the assimilates are supplied to facilitate flowering and pod setting without the increase of photosynthetic rate at high night temperature (Gibson and Mullen, 1996) (Table 4.6).

Exogenous application of cytokinin to raceme tissues of soybean (*Glycine max* (L.) Merr.) has been shown to stimulate flower production and to prevent flower abortion. The effects of these hormone applications have been ascertained for treated tissues, but the effects of cytokinins on total seed yields in treated plants have not been evaluated. The objectives were to examine the effects of systemic cytokinin applications on soybean yields using an experimental line of soybeans, SD-87001, that has been shown to be highly sensitive to exogenous cytokinin application. Soybeans were grown hydroponically or in pots in the greenhouse, and 6-benzylaminopurine (BA) was introduced into the xylem stream through a cotton wick for 2 weeks during anthesis. After the plants had matured, the number of pods, seeds per pod, and the total seed weight per plant were measured. In the greenhouse, application of 3.4×10^{-7} moles of BA resulted in a 79% increase in seed yield compared with controls. Results of field trials showed much greater variability within treatments, with consistent, but non-significant increases in seed number and total yields of about 3%. Data suggest that cytokinin levels play a significant role in determining total yield in soybeans, and that increasing cytokinin concentrations in certain environments may result in increased total seed production.

The results showed that severe drought stress significantly decreased pod set up to 40% and the critical stage for pod abortion was 3–5 days after anthesis (DAA), when cell division was active in the ovaries. Drought at later stages, when pod filling had begun, reduced seed size but had no significant effect on pod set. Pod water potential decreased by drought, however pod turgor was maintained at similar level to the well-watered controls. ABA concentration increased significantly in the xylem sap, leaves, and pods of drought stressed plants. Xylem-borne ABA and leaf ABA were seemingly the source of ABA accumulated in the drought-stressed pods.

Carbohydrate metabolism was disrupted by drought stress in both leaves and floral organs. In leaves, drought stress decreased photosynthetic rate, starch and sucrose concentrations but increased hexoses (glucose + fructose) concentrations, indicating a source limitation. In flowers and pods, drought stress increased sucrose and hexoses concentrations but decreased starch concentration, soluble invertase activity, and hexoses to sucrose ratio, indicating that the capacity of the pods to utilise the incoming sucrose was impaired by drought stress. As a consequence of both source and sink restrictions, non-structural carbohydrate (sucrose + hexoses + starch) accumulated in the pods was significantly reduced under drought stress.

The soil water thresholds for reduction in pod growth and pod set were 0.43 and 0.30 of FTSW, respectively. Pod growth was reduced before a significant decrease of pod water potential was detected, and the decrease of pod fresh weight was closely correlated with increasing xylem sap ABA concentration, implying that root signal and not pod water potential controlled pod growth during soil drying. Pod set began to decrease only when pod water potential had decreased and photosynthetic rate and pod fresh weight had decreased by 40% and 30%, respectively. Pod ABA concentration had increased 1.5-fold compared to the well-watered controls. Below the threshold water potential, pod set decreased further and correlated positively with photosynthetic rate and pod fresh weight, whilst it correlated negatively with pod ABA concentration.

Manipulation studies showed that application of 0.1 mM ABA on the canopy decreased gas exchange rates and pod set in well-watered soybeans. In drought-stressed plants, ABA treatment induced stomatal closure during early stage of soil drying, leading to higher leaf water potential, which maintained greater gas exchange rates, resulting in an increased pod set compared to the plants without ABA application. Application of 1 mM 6-benzylaminopurine (BA, an artificial cytokinin) on the canopy increased gas exchange rates and pod set in well-watered plants, but decreased leaf water potential, gas exchange rates and slightly decreased pod set in drought-stressed plants. In ABA- and BA-treated plants, pod set was linearly correlated with the leaf photosynthetic rate, implying that the two hormones exert their roles in altering pod set partly by modifying photosynthate availability.

To elucidate the performance of flowering and pod set under water stress at different growth stages, cultivars Enrei (determinate) and Touzan 69 (indeterminate) were grown in Wagner pots under vinyl-house conditions during the year 1996. Water stress prolonged the flowering duration in each water stressed plot. Although, in Enrei, water stress at the pre-flowering stage decreased the number of nodes, flowers and pod set on the basal racemes, in Touzan 69, less flowers were compensated for by an increase in pod set on the upper racemes. In both cultivars, water stress at flowering and young pod stages depressed the number of flowers as well as pod set on the basal racemes, which ultimately resulted in less pods and yield. The later the stage of water stress treatment, the smaller the seed size, so yield was reduced more at the seed growth stage in both cultivars. The results revealed that water stress during the development of floral organs increased flower and pod abortion, and furthermore, many seeds were aborted due to water stress after the beginning of flowering. From late flowering to young pod stages, soybean cultivars are susceptible to sink abortion, and the seed growth stage is most critical for yield determination under water stress (Saitoh et al., 1999) (Fig. 4.13).

In soybean, long days during post-flowering increase seed number. This positive photoperiodic effect on seed number has been previously associated with increments in the amount of radiation accumulated during the crop cycle because long days extend the duration of the crop cycle. However, evidence of

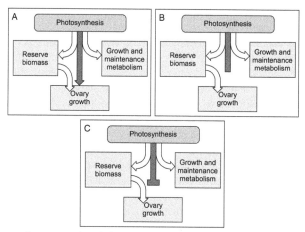

Fig. 4.13 Schematic diagram of ovary growth and photosynthesis-photoassimilate reserve relationships in well-watered (A), mild water stressed (B) and severe water stressed (C) soybean.

intra-nodal processes independent of the availability of assimilates suggests that photoperiodic effects at the node level might also contribute to pod set. This work aims to identify the main mechanisms responsible for the increase in pod number per node in response to long days; including the dynamics of flowering, pod development, growth and set at the node level. Long days increased pods per node on the main stems, by increasing pods on lateral racemes (usually dominated positions) at some main stem nodes. Long days lengthened the flowering period and thereby increased the number of opened flowers on lateral racemes. The flowering period was prolonged under long days because effective seed filling was delayed on primary racemes (dominant positions). Long days also delayed the development of flowers into pods with filling seeds, delaying the initiation of pod elongation without modifying pod elongation rate. The embryo development matched the external pod length irrespective of the pod's chronological age. These results suggest that long days during post-flowering enhance pod number per node through a relief of the competition between pods of different hierarchy within the node. The photoperiodic effect on the development of dominant pods, delaying their elongation and, therefore, postponing their active growth, extends flowering and allows pod set at positions that are usually dominated.

The distribution of flower and pod production during flowering may be an important determinant of pod and seed number in grain crops. It was characterized the dynamics of small pod production and survival to maturity on indeterminate and determinate soybean [*Glycine max* (L.) Merrill] cultivars growing in the field or greenhouse. Two soybean cultivars (maturity group IV, indeterminate and determinate) were grown in the field near Lexington, KY (38°N latitude) in 2001 and 2002 in 0.76 cm rows using late May and late June (2002 only) planting dates, and normal (24 plants m^{-2}) and low (9 plants m^{-2}, 2002 only) plant populations. Cultivar Elgin 87 (indeterminate, maturity group II) was grown in a greenhouse in 3.0 L pots with one plant per pot. All unmarked pods that were ≥ 10 mm long were marked with acrylic paint at the base of the pod at 3-day intervals. Paint colour was changed at each marking to provide a temporal profile of pod production and pod survival. The pod production (marked pods) period was longer in the indeterminate cultivar (nearly 50 days after R1) than the determinate cultivar (≤ 40 days after R1). Delayed planting shortened the pod-production period, but a two- to three-fold difference in pods per plant, created by changing plant population, did not affect it. The temporal distribution of small pods that survived to maturity (full sized pods with at least one normal seed) closely followed the distribution of pod production in all experiments. Some surviving pods initiated growth after the beginning of seed filling (i.e., between growth stage R5 and R6), but most of the pods were initiated in a much shorter interval (up to 84% were initiated in < 40% of the period) before R5. Abortion of pods > 10 mm long was relatively low (20–30%), so production of a pod ≥ 10 mm long seems to be a key event in the pod set process. The average length of the pod set period at individual nodes on the main stem was larger for the determinate cultivar (14 days) than for the indeterminate (9 days), so the longer total period in the indeterminate cultivar resulted from the delay in initiating pod production at the upper nodes on the main stem. Temporal profiles of pod production and pod set seem to be more sensitive to changes in flower and main stem node production than to changes in photosynthesis per plant (created by varying plant population). These results provide some of the information needed to integrate time into models predicting pod and seed number (Fig. 4.14).

Using the same plant material (IX93-100) grown in pots and in the field, it was examined changes in the concentrations of endogenous auxin and cytokinin within racemes and the effects of application of the two hormones on pod set. The auxin (IAA) concentration in racemes was high for a long period from pre-anthesis to ca. 10 days following the anthesis (DAA) of the first flower on a raceme, but the cytokinin concentration remained elevated for a shorter period, with a peak at 9 DAA (Nonokawa et al., 2007) (Fig. 4.11). The two phytohormones are located primarily at different positions within a raceme; the IAA concentration was higher in distal portion of racemes, whereas the cytokinin concentration was higher in basal portions of racemes. IAA application to racemes reduced the number of flowers and pods throughout the reproductive stage. In contrast, the effect of cytokinin (BA) application varied depending on the growth stage: application of BA at around 7 DAA significantly increased the pod-set percentage, while at other stages, BA application reduced pod set. Thus, the concentrations of the two endogenous hormones changed in a different manner, with cytokinins exerting a positive effect, and auxin exerting a negative effect on pod set, depending on the growth stage.

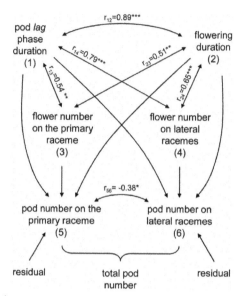

Fig. 4.14 Correlation amongst parameters related to flower number and the total pod number.

Environmental conditions prevailing during the reproductive period are important determinants of soybean yield and yield components (Board and Harville, 1996; Liu et al., 2010; Liu et al., 2013). Board and Harvill (1996) suggested that intensity and quality of solar radiation intercepted by the canopy influenced yield by changing survival rate of flowers in soybean plant. Liu et al. (2010) stated that light enrichment and shading significantly decreased and increased abortive rate of flower resulting in change of pod number per plant. Umezaki and Yoshida (1992) indicated that plant height of soybean will increase with the decrease of light intensity.

Much of the soybean yield variation is related to changes in flower number that survived to mature. From 32 to 81% of the flowers of field-grown soybean do not develop into mature pods (Wiebold et al., 1981).

Flowering in soybean is a dynamic system in which flower survival may depend on where a flower is located and when it is initiated. Egli and Bruening (2006b) stated that the temporal distribution of flower and pod production plays an important role in determining pod or seed number at maturity. Both of flowering and pod production periods at individual nodes continue for 30 days or more, and they are nearly the same length (Huff and Dybing, 1980; Gai et al., 1984; Egli and Bruening, 2006b). The timing of a flower initiate during the bloom or seed-filling period was important and late developing flowers may abort because large rapidly growing pods and seeds from early flowers consume most of the assimilate (Bruening and Egli, 2000).

Most flowers are produced in a much shorter time compared to total reproductive period (Kuroda et al., 1998). Flowering on whole plants includes inter- and intra nodal variation, and there is 15–50 days interval between the first and last flowers (Gai et al., 1984; Constable and Ross, 1988; Dybing, 1994; Zheng et al., 2002). The node location of a flower in soybean main axis determines, in part, when it develops. This time of a flower developing is a critical factor in determining it survival or abscission, because at individual node, the abortion of early developing flowers always had lower than those developing later (Brun and Betts, 1984; Heitholt et al., 1986).

Soybean plants have highly asynchronous flowering characteristic and the degree of asynchrony may determine, in part, soybean growth character (determinate and indeterminate). This phenomenon often occurs in soybean in a certain node of the main stem; some pods were filling however flowering still continues. Huff and Dybing (1980) indicated that flowering at individual nodes usually starts at the base of the primary raceme and continues upward with about 1-day intervals between flowers. Flowering on the higher order racemes or sub-branches usually starts after the primary raceme (Munier-Jolain

Table 4.7 Effect of shading on flower, pod number, flower abscission and yield in three soybeans.

Cultivar	Treatment	Flowers (no. plant^{-1})	Pods (no. plant^{-1})	Abscission (%)	Yield (g.plant^{-1})
Hai339	CK	147.3a	41.8a	72a	24.8a
	Shade	89.4b	23.3b	74a	16.0b
Heinong35	CK	114.9a	48.4a	58a	17.2a
	Shade	78.9b	26.4b	67b	9.9b
Kennong18	CK	147.5a	52.7a	64a	17.0a
	Shade	97.9b	26.0b	73b	8.2b

Different letters within the row represent significantly different from CK (natural light) and shading under same cultivar (P < 0.05).

et al., 1994; Saitoh et al., 1998; Egli and Bruening, 2002; Zheng et al., 2002). In soybean plants, when the lowermost nodes start filling seeds, the uppermost nodes are still in the process of producing flowers. Whether from the single plant level or an individual node, the abscission of late developing flowers mostly are inevitable when a large number of filling pods exist.

A 2-year field experiment was conducted under natural light and shading conditions to examine the responses of spatial distribution of flower and flower abscission in three soybeans, characteristics of flowering progress were also discussed. The results showed that responses to shading occurred proportionately across the main axis node positions despite the differences in the time of development of flower and pod between the high and low node positions. Reproductive organ of middle node was more sensitive than that of low and high node in single plant. Compared with that under natural light, shading increased flowering time 7 days for H339,3 days for HN35 and 1 day for KN18 (Table 4.7). Flowering process showed four significant stages: Early-bloom phase (4–6 days), full-bloom phase (about 15 days), slow-bloom phase (about 10 days) and final-bloom phase (about 9–16 days). Soybean has the characteristic of excessive flower production. Data may give a hint that flower number produced per plant isn't the most significant factor determining the final pod number survived per plant. Excessive flower produced per plant is may be just a precondition as reproductive prosperity in soybean.

References

Abernethy, R.H., R.G. Palmer, R. Shibles and I.C. Anderson. 1977. Histological observations on abscising and retained soybean flowers. Can. J. Plant Sci. 57: 713–716.

Andriani, J.M., F.H. Andrade, E.E. Suero and J.L. Dardanelli. 1991. Water deficits during reproductive growth of soybeans. 1. Their effects on dry-matter accumulation, seed yield and its components. Agronomie 11: 737–746.

Antos, M. and W.J. Wiebold. 1984. Abscission, total soluble sugars, and starch profiles within a soybean canopy. Agron. J. 76: 715–719.

Beckmann, K.A. 1981. The internal control of flower abscission on soybeans. PhD thesis, Purdue University, West Lafayette, IN.

Board, J.E. and B.G. Harville. 1996. Growth dynamics during the vegetative period affects yield of narrow-row, late-planted soybean. Agronomy Journal 88: 567–572.

Boyer, J.S. 1983. Environmental stress and crop yields. pp. 3–7. *In:* Raper, C.D. and P.J. Kramer (eds.). Crop Reaction to Water and Temperature Stresses in Humid, Temperate Climates. Westview Press, Boulder, CO.

Brevedan, R.E., D.B. Egli and J.E. Legget. 1978. Influence of N nutrition on flower and pod abortion and yield of soybeans. Agron. J. 70: 81–84.

Brevedan, R.E. and D.B. Egli. 2003. Short periods of water stress during seed filling, leaf senescence, and yield of soybean. Crop Sci. 43: 2083–2088.

Bruening, W.P. and D.B. Egli. 2000. Leaf starch accumulation and seed set at phloem-isolated nodes in soybean. Field Crops Research 68: 113–120.

Brun, W.A. and K.J. Betts. 1984. Source/sink relations of abscising and nonabscising soybean flowers. Plant Physiol. 75: 187–191.

Brun, W.A., J.C. Heindl and K.J. Betts. 1985. The physiology of reproductive abscission in soybeans. pp. 866–874. *In:* Shibles, R. (ed.). World Soybean Conference III: Proceedings. Westview Press, Boul-der.

Carlson, D.R., D.J. Dyer, C.D. Cotterman and R.C. Durley. 1987. The physiological basis for cytokinin induced increases in pod set in IX93-100 soybeans. Plant Physiol. 84: 233–239.

Constable, G.A. and I.A. Ross. 1988. Variability of soybean phenology response to temperature, day length and rate of change in day length. Field Crops Res. 18: 57–69.

Corbesier, L., C. Vincent, S.H. Jang, F. Fornara, Q.Z. Fan et al. 2007. FT protein movement contributes to long-distance signaling in floral induction of Arabidopsis. Science 316: 1030–1033.

Crosby, K.E., L.H. Aung and G.R. Buss. 1981. Influence of 6-benzylaminopurine on fruit-set and seed development in two soybean, *Glycine max* (L.) Merr. genotypes. Plant Physiol. 68: 985–988.

de Bruijn, S.M., E.A.M. Koot-Gronsveld and D. Vreugdenhil. 1993. Abscisic acid and assimilate partitioning to developing seeds. III. Does abscisic acid influence sugar release from attached empty seed coat in an ABA-deficient *Pisum sativum* mutant? J. Exp. Bot. 44: 1735–1738.

Desclaux, D., T.-T. Huynh and P. Roumet. 2000. Identification of soybean plant characteristics that indicate the timing of drought stress. Crop Sci. 40: 716–722.

Dybing, C.D., H. Ghiasi and C. Paech. 1986. Biochemical characterization of soybean ovary growth from anthesis to abscission of aborting ovaries. Plant Physiol. 81: 1069–1074.

Dyer, D.J., D.R. Carlson, C.D. Cotterman, J.A. Sikorski and S.L. Ditson. 1987. Soybean pod set enhancement with synthetic cytokinin analogs. Plant Physiol. 84: 240–243.

Egli, D.B. and W.P. Bruening. 2001. Source-sink relationships, seed sucrose levels and seed growth rates in soybean. Ann. Bot. 88: 235–242.

Emery, R.J.N., Q.F. Ma and C.A. Atkins. 2000. The forms and sources of cytokinins in developing white lupine seeds and fruits. Plant Physiol. 123: 1593–1604.

Fornara, F., A. de Montaigu and G. Coupland. 2010. SnapShot: Control of flowering in Arabidopsis. Cell 141.

Gai, J., R.G. Palmer and W.R. Fehr. 1984. Bloom and pod set in determinate and indeterminate soybeans in China. Agron. J. 76: 979–984.

Gibson, L.R. and R.E. Mullen. 1996. Influence of day and night temperature on soybean seed yield. Crop Sci. 36: 98–104.

Hansen, W.R. and R. Shibles. 1978. Seasonal log of the flowering and podding activity of field-grown soybeans. Agron. J. 70: 47–50.

Heindl, J.C., D.R. Carlson, W.A. Brun and M.L. Brenner. 1982. Ontogenetic variation of four cytokinis in soybean root pressure exudate. Plant Physiol. 70: 1619–1625.

Heindl, J.C. and W.A. Brun. 1983. Light and shade effects on abscission and 14C-photoassimilate partitioning among reproductive structures in soybean. Plant Physiol. 73: 434–439.

Heindl, J.C. and W.A. Brun. 1984. Patterns of reproductive abscission, seed yield, and yield components in soybean. Crop Sci. 24: 542–545.

Heitholt, J.J., D.B. Egli and J.E. Legget. 1986a. Characteristics of reproductive abortion in soybean. Crop Sci. 26: 589–595.

Heitholt, J.J., D.B. Egli, J.E. Legget and C.T. MacKown. 1986b. Role of assimilate and carbon-14 photosynthate partitioning in soybean reproductive abortion. Crop Sci. 26: 999–1004.

Huff, A. and C.D. Dybing. 1980. Factors affecting shedding of flowers in soybean (*Glycine max* (L.) Merrill). J. Exp. Bot. 31: 751–762.

Jung, C-H., C.E. Wong, M.B. Singh and P.L. Bhalla. 2012. Comparative genomic analysis of soybean flowering genes. PLoS ONE 7(6): e38250.

Kato, I. 1964. Histological and embryological studies on fallen flowers, pods and abortive seeds in soybean, *Glycine max* (L.). Tokai-Kinki Natl. Agr. Exp. Stn. Bull.

Kokubun, M. and I. Honda. 2000. Intra-raceme variation in pod set probability is associated with cytokinin content in soybeans. Plant Product. Sci. 3: 354–359.

Kokubun, M., S. Shimada and M. Takahashi. 2001. Flower abortion caused by preanthesis water deficit is not attributed to impairment of pollen in soybean. Crop Science 41: 1517–1521.

Kong, F., B. Liu, Z. Xia, S. Sato, B.M. Kim et al. 2010. Two co-ordinately regulated homologs of FLOWERING LOCUS T are involved in the control of photoperiodic flowering in soybean. Plant Physiol. 154: 1220–1231.

Kuroda, T., K. Saitoh, T. Mahmood and K. Yanagawa. 1998. Differences in flowering habit between determinate and indeterminate types of soybean. Plant Production Science 1: 18–24.

Laurie, R.E., P. Diwadkar, M. Jaudal, L.L. Zhang, V. Hecht et al. 2011. The medicago FLOWERING LOCUS T homolog, MtFTa1, is a key regulator of flowering time. Plant Physiology 156: 2207–2224.

Lin, M-K., H. Belanger, Y.-J. Lee, E. Varkonyi-Gasic, K.-I. Taoka et al. 2007. FLOWERING LOCUS T protein may act as the long-distance florigenic signal in the cucurbits. Plant Cell 19: 1488–1506.

Liu, B., S. Watanabe, T. Uchiyama, F. Kong, A. Kanazawa, Z. Xia, A. Nagamatsu, M. Arai et al. 2010. The soybean stem growth habit gene Dt1 is an ortholog of Arabidopsis TERMINALFLOWER1. Plant Physiol. 153: 198–210.

Liu, Y., X. Li, K. Li, H. Liu and C. Lin. 2013. Multiple CIBs form heterodimers to mediate CRY2-dependent regulation of flowering-time in *Arabidopsis*. PLoS Genet. 9: e1003861.

Mann, J.D. and E.G. Jaworski. 1970. Comparison of stresses which may limit soybean yields. Crop Sci. 10: 620–624.

Momen, N.N., R.E. Carlson, R.H. Shaw and O. Arjmond. 1979. Moisture-stress effects on the yield components of two soybean cultivars. Agron. J. 71: 86–90.

Mosjidis, C.O., C.M. Perterson, B. Truelove and R.R. Dute. 1993. Stimulation of pod and ovule growth of soybean, *Glycine max* (L.) Merr. by 6-benzylaminopurine. Annals Bot. 71: 193–199.

Munier-Jolain, N.G., N.M. Munier-Jolain, R. Roche, B. Ney and C. Duthion. 1998. Seed growth rate in grain legumes–I. Effect of photoassimilate availability on seed growth rate. J. Exp. Bot. 49: 1963–1969.

Myers, R.L., W.A. Brun and M.L. Brenner. 1987. Effect of raceme-localized supplemental light on soybean reproductive abscission. Crop Sci. 27: 273–277.

Nonokawa, K. et al. 2007. Roles of auxin and cytokinin in soybean pod setting. Plant Production Science, Tokyo, 10: 199–206.

Nagel, L., R. Brewster, W.E. Riedell and R.N. Reese. 2001. Cytokinin regulation of flower and pod set in soybean (*Glycine max* (L.) Merr.). Ann. Bot. 88: 27–31.

Oberholster, S.D., C.M. Perterson and R.R. Dute. 1991. Pedicel abscission of soybean: Cytological and ultrastructural changes induced by auxin and ethephon. Can. J. Bot. 69: 2177–2186.

Peterson, C.M., J.C. Williams and A. Kuang. 1990. Increased pod set of determinate cultivars of soybean, *Glycine max*, with ~ 6-benzylaminopurine. Bot. Gaz. 151: 322–330.

Peterson, C.M., C.O.H. Mosjidis, R.R. Dute and M.E. Westgate. 1992. A flower and pod staging system for soybean. Ann. Bot. 69: 59–67.

Reese, R.N., C.D. Dybing, C.A. White, S.M. Page and J.E. Larson. 1995. Expression of vegetative storage protein (VSP-(3)) in soybean raceme tissues in response to flower set. J. Exp. Bot. 46: 957–964.

Saini, H.S. 1997. Effects of water stress on male gametophyte development in plants. Sex. Plant Reprod. 10: 67–73.

Saitoh, K., T. Mahmood and T. Kuroda. 1999. Effect of moisture at different growth stages on flowering and pod set in determinate and indeterminate soybean cultivars. Japan. J. Crop Sci. 68: 537–544.

Samach, A., H. Onouchi, S.E. Gold, G.S. Ditta, Z. Schwarz-Sommer et al. 2000. Distinct roles of CONSTANS target genes in reproductive development of *Arabidopsis*. Science 288: 1613–1616.

Sato, K. and T. Ikeda. 1979. The growth responses of soybean plant to photoperiod and temperature IV. The effect of temperature during the ripening period on the yield and characters of seeds. Jpn. J. Crop Sci. 48: 283–290.

Seddigh, M. and G.D. Jolliff. 1984. Night temperature effects on morphology, phenology, yield components of indeterminate field-grown soybean. Agron. J. 76: 824–828.

Setter, T.L. and B.A. Flannigan. 2001. Water deficit inhibits cell division and expression of transcripts involved in cell proliferation and endoreduplication in maize endosperm. J. Exp. Bot. 52: 1401–1408.

Shanmugasundaram, S. and S. Tsou. 1978. Photoperiod and critical duration for flower induction in soybean. Crop Science 18: 598–601.

Shoemaker, R.C., J. Schlueter and J.J. Doyle. 2006. Paleopolyploidy and gene duplication in soybean and other legumes. Curr. Opin. Plant Biol. 9: 104–109.

Spollen, W.G., W.J. Wiebold and D.S. Glenn. 1986a. Intraraceme competition in field-grown soybean. Agron. J. 78: 280–283.

Spollen, W.G., W.J. Wiebold and D.S. Glenn. 1986b. Effect of intraraceme competition on carbon-14-labeled assimilate and abscisic acid in soybean. Crop Science 26: 1216–1219.

Thomas, J.F. and C.D. Raper, Jr. 1983. Photoperiod and temperature regulation of floral initiation and anthesis in soya bean. Annals of Botany 51: 481–489.

Ueki, K. and M. Igawa. 1958. The influence of high night temperature on the growth and fruiting in soybean. Tech. Bull. Fac. Agr. Kagawa Univ. 9: 111–118.

Umezaki, T. and T. Yoshida. 1992. Effect of shading on the internode elongation of late maturing soybean. J. Fac. Agric. Kyushu Univ. 36: 267–272.

Valverde, F., A. Mouradov, W. Soppe, D. Ravenscroft, A. Samach et al. 2004. Photoreceptor regulation of CONSTANS protein in photoperiodic flowering. Science 303: 1003–1006.

van Schaik, P.H. and A.H. Probst. 1958. The inheritance of inflorescence type, peduncle length, flowers per node, and percent flower shedding in soybeans. Agron. J. 50: 98–102.

Wang, D., M.C. Shannon and C.M. Grieve. 2001. Salinity reduces radiation absorption and use efficiency in soybean. Field Crop Res. 69: 267–277.

Westgate, M.E. and J.S. Boyer. 1985. Carbohydrate reserves and reproductive development at low leaf water potentials in maize. Crop Sci. 25: 762–769.

Westgate, M.E. and T.G.L. Grant. 1989. Water deficits and reproduction in maize. Responses of the reproductive tissue to water deficits at anthesis and mid-grain fill. Plant Physiol. 91: 862–867.

Westgate, M.E. and C.M. Peterson. 1993. Flower and pod development in water–deficient soybean (*Glycine max* L. Merr.). J. Exp. Bot. 44: 109–117.

Whigham, D.K. 1983. Soybean. pp. 205–225. *In*: International Rice Research Institute (ed.). Potential Productivity of Field Crops under Different Environments. IRR, Los Banos, Philippines.

Wiebold, W.J., D.A. Ashley and H.R. Boerma. 1981. Reproductive abscission levels and patterns for eleven determinate soybean cultivars. Agron. J. 73: 43–46.

Wiebold, W.J. and M.T. Panciera. 1990. Vasculature of soybean racemes with altered intraraceme competition. Crop Sci. 30: 1089–1093.

Wong, C.E., M.B. Singh and P.L. Bhalla. 2013. The dynamics of soybean leaf and shoot apical meristem transcriptome undergoing floral initiation process. PLoS ONE 8(6): e65319.

Yang, J.C., J.H. Zhang, Z.L. Huang, Z.Q. Wang, Q.S. Zhu and L.J. Liu. 2002a. Correlation of cytokinin levels in the endosperm and roots with cell number and cell division activity during endosperm development in rice. Ann. Bot. 90: 369–377.

Yarrow, G.L., W.A. Brun and M.L. Brenner. 1988. Effect of shading individual soybean reproductive structures on their abscisic acid content, metabolism, and partitioning. Plant Physiol. 86: 71–75.

Zhang, Q.Z., H.Y. Li, R. Li, R.B. Hu, C.M. Fan et al. 2008. Association of the circadian rhythmic expression of GmCRY1a with a latitudinal cline in photoperiodic flowering of soybean. Proc. Natl. Acad. Sci. USA 105: 21028–21033.

Zheng, S., H. Nakamoto, K. Yoshikawa, T. Furuya and M. Fukuyama. 2002. Influence of high night temperature on flowering and pod setting in soybean. Plant Prod. Sci. 5: 215–218.

CHAPTER 5
Pod Growth and Yield

The period of reproductive growth (from the start of flowering to maturation) in soybean is very long compared with that of cereals. The pod shells in soybean develop after flowering, whereas the husks in rice or wheat, whose function is the same as that of pod shell in soybean, are formed before flowering. Therefore, the seed filling begins about four weeks after flowering in soybean (Konno, 1976), but about one week after flowering in rice or wheat (Sofield et al., 1977; Chowdhury and Wardlaw, 1978). However, in some legumes, such as mung bean, cowpea and common bean, the period from flowering to maturation is also shorter than that of the soybean (Egli, 1998).

Several studies documented that the pod growth is very slow after flowering in soybean. Zheng et al. (2003) found that the lag period of pod growth is much longer in the flowers that opened early. On the other hand, Saitoh et al. (1998) indicated that the delay of pod growth is much longer in the low order racemes (early opening flowers) and is reduced in high order racemes (late opening flowers). It is suggested that the timing of pod growth plays a role in the regulation of synchronous pod maturation in soybean.

5.1 Pod Growth

The pod growth was obviously slow in soybean compared with that of the other three crops. Soybean pods reached the maximum length about 25 days after flower opening, against 7 days in mung bean, and 10 days in azuki bean and common bean (Fig. 5.1).

Figure 5.2 shows the pod growth after flowering on individual raceme. The flowers opened first on primary raceme at 42 DAS (days after sowing), then followed by terminal raceme (5 days late), sub-branch raceme (12 days late), secondary raceme (14 days late) and tertiary raceme (22 days late). However, the delay observed at flowering time seemed to be shortened at the phase of pod elongation.

In order to evaluate the developmental difference in pod growth among the raceme orders, in Table 5.1, a linear equation on individual raceme between the pod length and DAS during the linear phase of pod elongation based on Fig. 5.2 was described. As a result, the DAS for the pod reached at 20 cm in length was 59.2 on primary raceme, but was 72.7 on tertiary raceme. The difference between two raceme

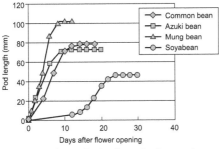

Fig. 5.1 Pod growth after flowering in some beans.

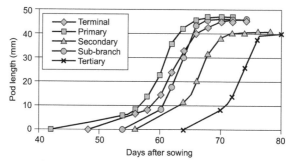

Fig. 5.2 Pod growth of soybean in different racemes.

Table 5.1 Linear equation between pod length (Y) and days after sowing (X) on the individual raceme.

Raceme	Equation	R2	X (Y = 20)	Delay (days)
Primary	Y = 4,9633X − 274.05	0.985	59.2	0.0
Terminal	Y = 4,3177X − 244.31	0.999	61.2	2.0
Sub-branch	Y = 5,7542X − 337.66	0.991	62.2	3.0
Secondary	Y = 4,5563X − 279.80	0.988	65.8	6.6
Tertiary	Y = 4,9500X − 340.10	0.971	72.7	13.5

orders was 13.5 days and it was shorter than the difference at flowering time (22 days). The same results were also observed on terminal, sub-branch and secondary racemes.

By the same procedure, the seed growth on individual raceme is shown in Fig. 5.3 and the linear equation between seed weight and DAS is shown in Table 5.2. The seed growth on individual raceme, especially on primary, terminal, sub-branch and secondary racemes came about almost at the same time, but is a little late on tertiary raceme. Therefore, the delay of seed growth on the racemes that opened the flower late was shorter than those of pod growth, such as it was reduced to 3.1 against 6.6 in the pod

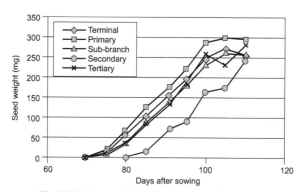

Fig. 5.3 Seed growth in different racemes of soybean.

Table 5.2 Linear equation between seed weight (Y) and days after sowing (X) on the individual raceme.

Raceme	Equation	R2	X (Y = 100)	Delay (days)
Primary	Y = 10,545X − 773.8	0.995	82.7	0.0
Terminal	Y = 9,3753X − 691.63	0.998	84.4	1.7
Sub-branch	Y = 8,9927X − 673.22	0.989	86.0	3.3
Secondary	Y = 9,8645X − 745.91	0.977	85.8	3.1
Tertiary	Y = 8,7954X − 732.92	0.976	94.7	12.0

growth on secondary raceme and was reduced to 12.0 against 13.5 in the pod growth on tertiary raceme. However, the overtaking of pod and seed growth in late opening flower was more dramatic during pod elongation more than seed growth (Zheng et al., 2004).

Since more than 2/3 of vegetative organs develop after the start of flowering in soybean (Baba et al., 2003), slow development in the pods set on the flowers that opened early is considered to be caused by the competition for assimilate between reproductive and vegetative organs after the start of flowering (Brun and Betts, 1984). It also indicates that the vegetative growth is more preceded than reproductive growth at that stage. Zheng et al. (2003) had found that the pod growth is stimulated by short photoperiod after flower opened but is not by sink removal (removing all flowers and stem tops except the target flowers) and benzyl adenine application. However, what factors limit the assimilate proportion between vegetative and reproductive organs is unknown.

Soybean flowers continue to open for a long period but the pods mature simultaneously. The developmental differences of pods and seeds between early- and late-opened flowers are not well known. Soybean plants were grown in a greenhouse in 2001 and in the field in 2003 in Kyushu University, Fukuoka, Japan. Soybean pods grew very slowly after flower opened whereas the pods of other three legumes (common bean, azuki bean and mung bean) started to elongate immediately after flower opened. In soybean, the pods that opened earlier started to elongate later, leading to uniform development of the pods that opened at different dates. The similar phenomenon was observed in seed growth but it seemed to be more pronounced in pod elongation than in seed growth.

During the early reproductive stages, a soybean plant may abort flowers and small pods as it adjusts to the surrounding environment. The percentage of flower and/or pod shed is determined by how much fruit the plant can support to full development. While some flower and small pod shed is a normal occurrence, unfavorable growing conditions, such as high temperatures or drought, may cause the plant to abort many more flowers and/or small pods than usual. The loss of flowers and small pods can be alarming to many producers who may already be concerned for their crops struggling with unfavorable growing conditions.

Flower and small pod abortion is a natural part of soybean growth and development. When a flower or pod is shed from the plant, the soybean plant is adjusting to its environment. The amount of plant photosynthate available will dictate the number of pods that will reach maturity. If the plant produces more pods than the production of photosynthate, flowers and/or pods will be shed. If the plant produces sufficient photosynthate, the plant may continue to flower and set additional pods. Fields that have severe early flower and small pod shed may initiate more flowering and possibly set new pods if environmental conditions improve and increase photosynthate production; however, yield potential may already be compromised by early stress.

Heat and/or drought stress during the R3 to R6 (beginning pod to full seed) growth stages may increase flower and small pod abortion. Results from a study conducted by Mann and Jaworski (1970) showed that pod formation was severely limited at temperatures above 40°C (104°F). Pods will typically not abort once a plant reaches R6 stage. During seed filling stages, any yield loss encountered is likely realized through reduced seed size due to unfavorable environmental conditions and not necessarily by pod shed.

High soybean yield requires an aggressive pod and seed set; however, a moderate amount of early flower or pod abortion will not necessarily hurt yields. The soybean plant is amazingly adaptable and will simply produce more and larger seed if enough photosynthates are available. The final yield is determined by the genetics of the variety and the end result of the environment.

Prevention of the abortion might result in an increase in the number of pods and seeds, and thereby lead to an increase in grain yield (Kokubun, 2011). Kato (1964) observed that abortion occurred most frequently at initial stages of pro embryo development after fertilization. The abortion was amplified by unfavourable environments, including water deficiency (Kato, 1964; Saitoh et al., 1999), suboptimum solar radiation and temperature (Kurosaki et al., 2003). Previous studies suggested two putative factors controlling the abortion: Availability of photosynthate or nutrients for pod development (Heitholt et al., 1986a, b) and availability of certain hormones (Nonokawa et al., 2007; Kokubun, 2011). Within individual racemes, the pod-set percentage in basal flowers was considerably higher than that in distal ones (Kokubun and Honda, 2000), which appeared to be associated with the intra-raceme variation of endogenous cytokinin content (Kokubun and Honda, 2000). In addition, the application of synthetic cytokinin to racemes enhanced the pod formation (Nonokawa et al., 2007). These lines of evidence suggest that cytokinin plays a promotive role in pod setting in soybean.

The application of synthetic cytokinin (6-benzylaminopurine, BA) to racemes of soybean genotype IX93-100 at 7 days after anthesis (DAA) enhanced pod-set percentage of the florets at the 5th position and above (numbered from the base on rachis). The endogenous cytokinin (*trans* zeatin riboside) content of individual florets was measured at the 1st, 3rd, 5th and 7th position every 3 days after anthesis. Cytokinin was detected only from the florets at 9 DAA, and the content was higher in the more proximal florets while it became negligible in the 7th floret. These results suggest that an increase in the amount of cytokinin in individual florets might enhance the pod setting of the florets positioned at the middle or distal part within the raceme.

Previous investigations have shown the feasibility of increasing pod number on legumes by the application of 6-benzylaminopurine (BA) directly to the raceme. These investigations were designed to determine what reproductive parameter was affected by cytokinin application, and if these applications were overcoming a deficiency in root-produced cytokinins during late flowering. Five individual main stem racemes on greenhouse grown soybeans (*Glycine max* L. Merr.) were treated with 2 millimolar BA. A single application of BA when pods appeared at 25 to 50% of the proximal floral positions resulted in a 58% increase in pod set due primarily to a 33% reduction in floral abscission. Applications of BA at later intervals also resulted in significant reductions in total abscission. When three applications of BA were imposed on the upper five nodes of field grown soybeans, total pod number and seed weight were significantly increased in this section of the canopy by 27 and 18%, respectively. Throughout the flowering period, root pressure exudate was sampled for the subsequent separation and quantification of zeatin, dihydrozeatin, zeatin riboside, dihydrozeatin riboside, and isopentenyladenine. Total cytokinin flux peaked from 0 to 9 days after flowering began, and then dropped to one-half of this level by 15 days post anthesis. The probability that a flower would initiate a pod was directly related to the concentration of total cytokinins present in the exudate when the flower opened.

Pod size of soybean is an important factor in the determination of seed weight. However, little is known about pod growth of soybean. Brassinosteroid, a group of phytohormones, regulate the pod growth of faba bean. Therefore, the role of brassinosteroid in pod growth of soybean was investigated. So, a measurement of pod length and cell number and cell area in pods treated with a brassinosteroid biosynthesis inhibitor was undertaken. The inhibitor suppressed pod growth through the reduction of cell area. Then, pod morphology and the expression of brassinosteroid biosynthesis (*GmCYP450 85A1, 2* and *3*) and response (*GmBZR1, GmBES1* and *GmBRU1*) genes in the pods of two cultivars that differ in pod size were examined. The difference in pod size was attributable to cell area, and the expression of brassinosteroid biosynthesis and response genes in pods was higher in the cultivar that has large pods. These results suggest that pod size of soybean is regulated through cell hypertrophy caused by brassinosteroid.

Soybean (*Glycine max* [L.] Merr.) ovary growth was measured from anthesis to 6 days after anthesis (DAA) in order to establish a timetable of biochemical events that might be useful in identifying processes that initiate abscission. Two procedures were developed in order to provide samples with either high or low percent pod set for; IX93-100,' a semi determinate line having long racemes. Characteristics measured were fresh and dry weight, soluble and insoluble protein, soluble carbohydrate, starch, RNA, and DNA. Setting ovaries grew more rapidly than abscising ovaries. Since there was a daily increase in ovary weight in both groups, all measured characteristics showed daily increases when expressed on per ovary basis. Statistically significant differences between groups were detected between 2 and 5 DAA for most characteristics. When chemical composition was expressed on concentration basis, starch level was significantly higher in setting ovaries at 5 and 6 DAA. Regression analysis showed that these deviations between setting and abscising samples started between anthesis and 1 DAA. It was concluded that processes leading to eventual shedding of fertilized ovaries (called flower abortion in soybeans) commence soon after anthesis of the shed flower, and that setting and abscising ovaries do not differ in protein, soluble carbohydrate, starch, or nucleic acid content when abscission processes begin (Dybing et al., 1986).

The pedicels of fully opened flowers on terminal racemes of field-grown IX93-100 soybean plants were treated three times with 200 mg kg^{-1} BAP in lanolin over a 6-d period. Racemes were then excised and ^{32}P uptake was recorded for each flower position within a raceme; histological features of pedicels and ovules also were determined. Application of BAP increased pod and ovule length, width and weight at all four distal nodes (D, D-1, D-2, D-3) relative to controls treated with lanolin. Length and width of parietal endosperm cells were smaller in BAP-treated ovules at the most proximal node being studied (D-3), and greater numbers of parietal endosperm cells were observed at D-1 and D-3 nodes when compared to lanolin controls. Smaller amounts of starch were found in suspensor cells, endosperm, and integuments of lanolin-treated ovules, and starch depletion over time was observed within starch sheaths of pedicels from lanolin-treated pods when compared to BAP-treated tissues. BAP-treated racemes had more ^{32}P uptake at the four most distal nodes. A higher rate of uptake (cpm mg^{-1} f. wt) was evident in ovules than in ovary tissues. These results suggest that for racemes otherwise destined to abscise, application of BAP promotes pod set and growth by stimulating ovule development.

5.2 Pod Growth Period

The delay of pod growth seems to lengthen the reproductive growth period. On the other hand, Saitoh et al. (1998) indicated that the delay of pod growth is much longer in the low order racemes (early opening flowers) and is reduced in high order racemes (late opening flowers). It is suggested that the timing of pod formation plays a role in the regulation of synchronous pod maturation in soybean (Saitoh et al., 1998). However, there are no further investigations focused on the delay of pod growth in soybean. The understanding of this function may be helpful to understand the mechanism of pod set, seed filling and yield.

It is well known that the vegetative growth continues for a while even after the start of flowering in soybean. Thus, the competition for assimilate supply between reproductive and vegetative growth after the start of flowering could inhibit the pod growth (Brun and Betts, 1984; Baba et al., 2003). Furthermore, the pod set could be stimulated by cytokinin (Peterson et al., 1990), and the pod filling duration is shortened by a short photoperiod (Thomas and Raper, 1976; Raper and Thomas, 1978).

Pod growth in soybean (*Glycine max* (L.) Merr.) begins several days after flower opening, compared with more immediate growth in other beans. Some investigated the relationship between genotype, raceme order of pod set, assimilate supply or photoperiod and the length of lag period of pod growth (LP, days from flower opening to the time when pod length reaches 10 mm). Soybean (five cultivars) plants were grown in a greenhouse and in the field in 2001. The lengths of pods developed from 20 flowers which opened on the same day and set on the same raceme order, were measured every other day. The length of LP varied with the cultivar from 5 to 16 days and it was longer in late cultivars. The LP in the primary raceme (early flowers) was 15 days, but that in the secondary raceme (late flowers) was 8 days. Both late sowing and short photoperiod (10 h) after the start of flowering shortened the LP by up to 7 days in Enrei and 5 days in Fukuyutaka. However, neither sink (except the target racemes) removal nor BA application to the target racemes at the start of flowering affected the length of LP, even though these treatments were expected to stimulate pod growth (Fig. 5.4 and Table 5.3).

Efforts to determine the mechanistic relationships between fruit and canopy development are complicated by difficulties in designing non-destructive treatments that modify plant-sink development. Pod growth was physically restricted by placing plastic straws, referred to as plastic pod-restriction devices, over 0, 50, or 100% of the soybean (*Glycine max* L. Merr.) pods in greenhouse experiments. The objective was to determine how decreases in pod growth influenced whole-plant growth and development. Both the rate and final accumulation of seed dry matter were decreased by restricting pod growth. Conversely, restricting pod growth increased seed number due to increased production and decreased abscission of fruits. Plant dry matter and N accumulations during the linear seed-fill period were not affected by restricting pod growth. This resulted from proportional increases in partitioning of assimilates into stems and leaves. Thus, decreases in reproductive growth apparently did not cause feedback inhibition of photosynthesis. Although leaf abscission was delayed by restricting pod growth, dry matter and N accumulation were affected only slightly. In general, restricting pod growth influenced

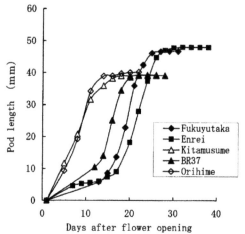

Fig. 5.4 Varietal differences in pod growth.

Table 5.3 Lag period of pod growth in different order racemes in two soybean cultivars.

Raceme	Sown on 8 June		Sown on 9 July		Sown on 1 august	
	Flg.date[#]	Lagperiod (d)	Flg.date[#]	Lagperiod (d)	Flg.date[#]	Lagperiod (d)
Enrei						
Primary	0	17.0 ± 2.49	0	10.0 ± 1.89	0	4.7 ± 0.70
Terminal	4	14.1 ± 1.14	4	7.8 ± 0.94	2	4.7 ± 0.74
Secondary*	8	13.3 ± 1.10	10	5.8 ± 1.47	–	–
Secondary	11	9.8 ± 1.67	14	7.7 ± 1.38	–	–
Tertiary	21	8.2 ± 1.13	–	–	–	–
LSD (0.05)	–	2.45	–	1.39	–	0.55
Fukuyutaka						
Primary	0	+	0	15.6 ± 1.55	0	9.3 ± 1.0
Terminal	4	+	6	12.6 ± 2.76	4	8.0 ± 1.0
Secondary*	13	13.3 ± 1.24	13	8.2 ± 1.03	–	–
Secondary	13	15.7 ± 2.14	15	7.5 ± 2.07	–	–
Tertiary	–	–	24	9.0 ± 0.00	–	–
LSD (0.05)	–	1.46	–	2.94	–	0.89

* Secondary raceme with a compound leaf; + The flower opened but no pod set.
\# Days after the first flower opened on the plant; – No flower opened.

plant development and assimilate allocation in a similar manner as physical removal of fruits. The inverse relationship between the rate of dry matter accumulation in seed and pod and seed number per plant indicated limited assimilate availability.

5.3 Seed Growth Rate and Duration

Seed growth rate and duration of growth were studied for different soybean [*Glycine max* (L.) Merr.] cultivars and isolines differing in maturity and stem termination behavior under field conditions. Various methods for estimating rate and duration were compared, and such estimates were compared with yield, along with mature seed weight, seed number, and days from planting to maturity. Three estimates of seed filling period were all highly correlated with each other, but the final seed weight, divided by its growth rate during the linear phase of dry matter accumulation, or the 'effective filling period', correlated best with yield. Two estimates of seed growth rate were also highly correlated with each other but not with yield. Allelles that delayed maturity did not generally increase the seed filling period, and, in some cases, caused slight reductions. Seeds are the primary sinks for photosynthates during reproductive growth. Variation in light intercepted during and after seed initiation has been found to be a major environmental determinant of soybean (*Glycine max* (L.) Merrill) seed size. The influence of light enrichment and shading on seed growth rate, effective filling, cotyledon cell number, cell volume and endogenous ABA concentrations of cotyledons/testas during seed filling of soybean was investigated. Evans, an indeterminate Group 0 soybean, was subjected to light reduction and enrichment treatments from the beginning of pod formation until final harvest for two years in Massachusetts. Higher rates of seed growth, greater seed dry weight, and higher cotyledon cell number were all observed with light enrichment. There was a reduction in seed growth rate and cotyledon cell number, along with a significant lowering of endogenous ABA levels in testa and cotyledon with shade. The level of ABA in cotyledon during seed development was significantly correlated with seed growth rates only under shade treatments. Both the growth rates and seed filling duration were influenced by variation in light interception by the soybean canopy. The effects of varying light treatment on seed size, within one genotype, were most likely due to the differences in seed growth rate and cotyledon cell number (Fig. 5.5).

The influence of 6-benzylaminopurine (BA) on the premature abscission of developing soybean, *Glycine max* (L.) Merr. fruits of 2 genotypes was studied. BA was applied during the critical period of fruit-setting. The tested concentration range of BA was from 1 micro molar to 5 milli molar; 2 milli molar was optimal. Spray application of 2 milli molar BA to terminal inflorescences at the R3 developmental stage of field-grown soybeans significantly increased fruit-set and seed yield of the Shore genotype during three growing seasons. In contrast, the Essex genotype gave significant responses two out of three seasons. The response of Shore was generally more pronounced than that of Essex. The apical fruits on the inflorescences gave the greatest response to BA. Seed weight increase was apparent 3–4 weeks after BA treatment (Crosby et al., 1981).

Soybeans (*Glycine max* L. Merrill) were grown in the greenhouse and in the field in order to investigate the effect of variations in the assimilate supply during the linear phase of seed development on the rate and duration of growth of individual seeds. Increased assimilate supplies, created by partial fruit removal, increased rates of dry matter accumulation, duration of seed growth, and final seed size (weight per seed). Reductions in the supply of assimilate to the developing seed, created by shading (60%) the plants during the linear phase of seed development, lowered seed growth rate but did not affect final seed size because of a longer duration of seed growth. Nitrogen stress during seed development, created by removing N from the nutrient medium, did not affect seed growth rate but shortened the duration of seed growth and reduced final seed size. This indicates that the growth characteristics of soybean seed are influenced by the supply of assimilate to the seed during the linear phase.

The growth characteristics of soybean (*Glycine max* [L.] Merr.) embryos in culture and seeds *in situ* were found to be similar, but developmental differences were observed. Embryos placed in culture when very small (< 2 milligrams dry weight) failed to attain the maximal growth rates attained by embryos which were more mature when placed in culture. When nutrient levels were maintained in the culture medium, embryos continued to grow indefinitely, reaching dry weights far in excess of seeds matured *in situ*. Apparently, maternal factors were important in early and late development during the determination

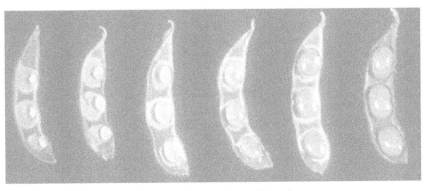

Fig. 5.5 Soybean seed growth in pod.

of maximum growth rate and the cessation of growth. Embryo growth rate was not affected by substituting glucose plus fructose for sucrose in the medium, nor by hormone treatments, including abscisic acid. Glutamine was found to give substantially better growth than glutamate, however. Contrary to prior reports, the response of soybean embryo growth rate to moderate temperatures was found to be primarily an artifact of the effect of irradiance on media temperature; however, at higher temperatures, the maternal plant is unable to support the rapid growth rates that the embryo is capable of attaining under conditions of unlimited irradiance. Across seven genotypes, the correlation coefficient between seed growth rate *in situ* and embryo growth rate *in vitro* was 0.94, indicating that all of the variability of *in situ* seed growth rate between cultivars could be attributed to inherent growth rate differences associated with the embryos. The response to temperature was very similar for both embryos in culture and seeds *in situ* at temperatures below 30°C. Beyond that temperature, embryo growth rate continued to increase, while seed growth rate did not. The implication is that *in situ* seed growth rate is determined by the inherent growth potential of the embryo at low assimilate supply.

Ultrastructural studies of the soybean (*Glycine max* [L.] Merrill) fruit (Thorne, 1981), along with kinetic studies of sucrose uptake by the fruit (Thorne, 1980), have revealed the route of assimilate imported by the fruit and possible physiological roles that different parts of the fruit play in assimilate transport. Assimilate transported from the leaf canopy enters the fruit through the vascular bundles of the pod from which it moves directly into the seed coat. In the seed coat, assimilate is rapidly unloaded into the free space of the innermost cell layers, where it diffuses away from the vascular bundles toward the inner epidermis of the seed coat, prior to uptake by the embryo. No direct vascular connection exists between the seed coat and embryo (Carlson, 1975).

The potential contribution of photosynthesis to embryo metabolism depends on the amount of light transmitted to the embryo. The estimated nm that is the amount of light with wavelengths between 400 and 700 incident on developing embryos by measuring light transmission through the relevant plant tissues (Fig. 5.6). Measurements in the field after canopy closure and flowering but before seed maturation

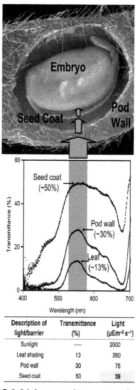

Fig. 5.6 Light transmittance to embryo.

showed that shading by leaves reduces the light levels by 80–94%, depending on the location of a fruit within the canopy. Of the light reaching the fruit, approximately 30% is transmitted through the wall to the seed coat and the seed coat transmits approximately 50% of this to the embryo. At a latitude of 43° at noon in a typical soybean field it was calculated that the light received in μE by most embryos is in the range of $5-30^{-2}\text{sec}^{-1}$. Light is preferentially absorbed by plant tissues at higher and lower wavelengths, so that the penetrating light is enriched in the range nm. Embryos used in subsequent experiments were at a 525–575 post-mitotic stage where storage deposition was close to linear (Satterlee and Koller, 1984; Egli, 1998). Such pods are generally located higher on the plant relative to the m μE canopy, resulting in higher light intensities. Therefore, a $30-35^{-2}\text{sec}^{-1}$ m μE light level was used for most cultures, with others at ~ 5 or $\sim 100^{-2}\text{sec}^{-1}$ for comparison, with a range of possible *in planta* conditions (Allen et al., 2009).

During the growing season, sunlight in soybean growing regions can average 1000^{-2}sec^{-1}m μE for daylight hours and reach 2000^{-2}sec^{-1}m μE. Transmission spectra for a soybean leaf, pod, and seed coat suggest the embryo may receive up to 39^{-2}sec^{-1}m μE of light. Field measurements in central Michigan suggest values closer to $5-30^{-2}\text{sec}^{-1}$, depending upon planting density, pod location on the plant, and sunlight levels at the time of measurement. Spectra represent an average of two measurements over different areas of tissue; measurements on other parts of each tissue consistently gave similar values.

The relationships between various carbohydrate pools of the soybean (*Glycine max* [L.] Merrill) fruit and growth rate of seeds were evaluated. Plants during mid pod-fill were subjected to various CO_2 concentrations or light intensities for 7 days in order to generate different rates of seed growth. Dry matter accumulation rates of seeds and pod wall, along with glucose, sucrose, and starch concentrations in the pod wall, seed coat, and embryo were measured in three-seeded fruits located from nodes six through ten. Seed growth rates ranged from 4 to 37 milligrams day^{-1} fruit^{-1}. When seed growth rates were greater than 12 milligrams day^{-1} fruit^{-1}, sucrose concentration remained relatively constant in the pod wall (1.5 milligrams–100 milligrams dry weight^{-1}), seed coat (8.5 milligrams–100 milligrams dry weight^{-1}), and embryo (5.0 milligrams–100 milligrams dry weight^{-1}). However, sucrose concentrations decreased in all three parts of the fruit as growth rate of the seeds fell below 12 milligrams day^{-1} fruit^{-1}. This relationship suggests that, at high seed growth rates, flux of sucrose through the sucrose pools of the fruit was more important than pool size for growth. Starch concentration in the pod wall remained relatively constant (2 milligrams–100 milligrams dry weight^{-1}) at higher rates of seed growth but decreased as seed growth rates fell below 12 milligrams day^{-1} fruit^{-1}. This suggests that pod wall starch may buffer seed growth under conditions of limiting assimilate availability. There was no indication that carbohydrate pools of the fruit were a limitation to transport or growth processes of the soybean fruit (Fader and Koller, 1985) (Fig. 5.7).

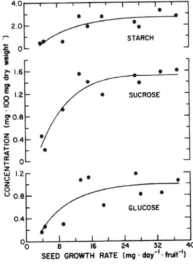

Fig. 5.7 Carbohydrate changes with seed growth rate in soybean.

Root-originated chemical signals have been shown to regulate the response of vegetative shoot to drought in soybeans (*Glycine max* L. Merr.). However, their roles in the growth of soybean reproductive structures under drought stress have been investigated. To explore this, a glasshouse experiment, in which potted soybeans were either well-watered (WW) or subjected to six levels of drought stress, was conducted. Irrigation was withheld in pots at six different dates before anthesis in order to induce drought of different severity (D1–D6) at sampling, viz., 4 days after anthesis (DAA). Root water potential, leaf water potential, pod water potential, xylem sap [ABA], pod fresh weight (FW), and pod set percentage were determined. Soil water status in the pot was expressed as the fraction of transpirable soil water (FTSW). Pod FW started to decrease at FTSW ¼ 0.43–0.02, when pod water potential was similar to that in the WW plants, while root water potential had decreased to –0.15 MPa and xylem sap [ABA] had increased 9-fold as compared with the WW plants. Pod set started to decrease at FTSW ¼ 0.30–0.01, and coincided with the decrease in pod water potential. Pod set started to decrease only when pod FW had decreased ca. 30%. Based on the results, a potential role of drought-induced increase in xylem sap ABA in affecting pod growth was suggested. It was proposed that a low pod water potential, which might have led to disruptions in metabolic activities in the pods, is important in determining pod abortion.

Generally, soybean seed yield depends mostly on pod number per area. However, over 50% of soybean flowers and young pods abort and don't make mature pods and seeds. Under bad weather and growing conditions, the percentage of flower and pod abortion increases. The average number of seeds in a pod is relatively constant, although the seed number per pod differs from 1 to 4 in soybean. Average seed weight is affected by growing conditions in late growth stages. The low average yield compared with potential productivity (10 t ha^{-1}) may be due to several factors that interfere with maximum growth. First, soybean plants are very susceptible to physical, chemical and biological conditions of the soil as well as climatic conditions. In this row, the highest pod number was about 180 and the lowest was only 18, and the average pod number was about 100 pods per a plant. As seen, a plant with many pods tended to be next to a plant with low pod number. This may be mainly due to competition for solar radiation, and the plant growth is easily depressed by shading caused by the adjacent bigger plant.

Analysis of characteristics of soybean canopy under humid subtropics of high altitude of north-eastern hill region of India indicated that total dry matter production showed a quadratic pattern of change in respect of height. Stem dry weight was maximum in lower stratum and then linearly to the top of the canopy. However, pod weight increased slightly in the middle over lower stratum and thereafter decreased sharply at the top of the canopy. However, leaf weight as well as leaf area density increased slowly in the lower and middle strata with a high value in the top stratum. Number of seeds per plant, seed weight per plant and seed-pod ratio were higher but 100 seed weight was lower in the middle stratum. High seed yield is associated with high leaf area index, seeds per m^2 and moderate 100 seed weight and dry matter production (Table 5.4) (Basuchaudhuri, 1987).

Individual seed weight, germination, and seedling growth rate were strongly correlated when reduced by water and high AT stress. Severe stress during seed fill caused soybean plants to exceed their capacity to buffer seed number, shifting seed weight distributions towards a larger proportion of small seed, resulting in poor seed lot germination and vigor (Table 5.5).

Table 5.4 Yield and yield attributes in soybean genotypes grown at high altitude (1800 msl).

Variety	Leaf area index	Dry matter Production (g.m^{-2})	Seeds per m^2 (10^3)	100 seed weight (g)	Seed yield (g.m^{-2})	Harvest index
Bragg	2.0	700	3.35	8.6	287.5	0.40
Ankur	3.9	1354	3.47	10.6	367.5	0.27
Shilajit	5.5	1289	4.07	9.8	400.0	0.31
NP4	4.7	1487	3.40	13.2	450.0	0.30
JS 72-20	5.5	2106	3.90	13.0	480.0	0.23
PK 71-21	2.8	1090	3.37	14.3	465.2	0.42
Mean	4.0	1338	3.59	11.6	408.4	0.32
CD0.05	1.6	490	0.38	2.4	73.0	0.08

Spacing 40 × 10 cm.

Table 5.5 Yield of soybean plants exposed to three water stress levels at an optimum (27–29°C) and high air temperature (AT).

Water stress	Experiment I SDD yield (g seed plant⁻¹)		Experiment II SDD yield (g seed plant⁻¹)	
	29°C		27°C	
None	26	34.3	39	38.0
Moderate	41	24.0	70	31.4
Severe	70	18.3	99	23.7
LSD0.05z		2.9		2.8
	35°C		33°C	
None	–19	24.4	–24	29.4
Moderate	15	14.8	04	26.4
Severe	47	8.7	35	17.0
LSD0.05z		2.9		2.8
LSD0.05v		4.1		3.8

z: Comparison of water-stress means within a temperature.
v: Comparison of AT means within a water stress level.

The growth rate and duration of an individual seed are important parameters in the yield production process of a grain crop. Experiments were conducted in the CSIRO phytotron at Canberra, Australia to investigate the effect of temperature on the rate and duration of seed growth and associated plant characteristics in soybeans [*Glycine max* (L.) Merr.]. Soybeans ('Fiskeby V') were grown at 24/19°C until beginning seed growth and shifted to 18/13, 24/19, 27/22, 30/25, and 33/28°C (day/night temperature, 8-hour day) until maturity. Seed growth rates (SGR) were estimated by harvesting seeds at 5-day intervals, and from the growth rate of excised cotyledons cultured in nutrient media for 96 hours. The SGR of seed developing on the plant increased from 6.1 to 7.9 mg seed⁻¹ day⁻¹ as air temperatures increased from 18/13 to 27/22°C, but there was no further change as the temperature increased to 33/28°C. Excised cotyledons showed a similar growth rate not affected by temperatures of 24/19 to 30/25°C but was reduced by 3 days at 33/28°C, and this was associated with accelerated leaf senescence, as shown by leaf yellowing and reductions in CO_2 exchange rate. Final seed size was reduced from 200 to 151 mg seed⁻¹ at 18/13 and 33/28°C. The data suggest that SGR and duration are relatively insensitive to temperatures ranging from 24/19 to 30/25°C when these are imposed after flowering and pod development. The reduction in the duration of seed growth at high temperature may be one mechanism by which high temperatures reduce yield. Exposing the plants to high temperatures (33/28°C) during the period of flowering and pod set reduced SGR (36%) regardless of the temperature during the subsequent seed growth period, suggesting that seed growth is sensitive to temperature levels at these early stages. Seed respiration on a dry weight basis ($mgCO_2h^{-1} g^{-1}$) was not affected by temperature. The reduction in the duration of seed growth at high temperatures reduce yield.

Temperature and photoperiod are the two major environmental variables that affect the length of the preflowering period (PFP) in many crops. Unlike photoperiod, which affect the duration of the photo-induction phase in photoperiod sensitive cultivars (Collinson et al., 1992; Yin et al., 1997) whereas, temperature affects the duration of the entire PFP in all cultivars of rice (Yoshida, 1981). Temperature is the primary environmental factor controlling developmental rate and often determines the rate of seed germination, seedling establishment, flowering, fruit maturation and plant growth. Temperature could affect seed growth rate directly by affecting seed metabolism or by affecting the supply of assimilate to the seed (Egli and Wardlaw, 1980; Jones et al., 1981; Donovan et al., 1983). High temperature reduces seed growth rate in many crops and reduces plant growth and yield in many environments. Unfavorable environmental conditions (temperature, photoperiod, rainfall and relative humidity) during seed growth and development in the field can reduce germination and subsequently reduce yield. Length of photoperiod strongly influences the morphology of soybean plant by causing changes in the time of flowering, maturity and dry matter production. Soybean cultivars do not have the same critical day length and usually experience gradually warming temperatures and lengthening days during vegetative and reproductive periods when grown as a summer crop. The timing of reproductive events in the crop and especially the duration of the preflowering period are modulated strongly by photoperiod and air temperature (Summerfield et al., 1986). Significant differences in dry matter accumulation were found between determinate and indeterminate soybean (Beaver and Cooper, 1982) and between soybean isolines (Wilcox, 1985). Germination in developing soybean seed has been an important topic of investigation, because germination and development are distinct physiological stages in the life cycle of the plant. It is apparent from these studies that time from anthesis and the degree of desiccation play a crucial role in determining the ability of seeds to germinate and their subsequent growth. Appropriate date of sowing is not only important for proper emergence but also important to have the crop in the field when environmental conditions are conducive for proper growth and development. The aim of the present investigation was to characterize the pattern of seed development in precociously mature soybean seeds planted on different dates during their progression from germination to seedling growth and maturation and the effects that temperature has on yield.

Fresh Seed Weight

Data on fresh seed weight from R5 to R7 of the two soybean varieties planted on four dates are presented in Fig. 5.8 for 2000 and Fig. 5.9 for 2001. Planting date significantly affected the fresh seed weight of soybean varieties. Maximum fresh seed weight attained in both varieties decreased as sowing was delayed from May to August, except in Williams 82, whereas a slight increase in the maximum fresh seed weight was attained at the last sowing date in 2000. The mean fresh seed growth rates calculated from the regression equations (Table 5.6) shows that May planted crop of Epps has higher mean seed growth rate in 2000 and 2001. A steady decrease in means fresh seed growth rate was observed as planting was delayed and the minimum mean fresh seed growth rates were noted in plots planted in August of both years. Minimum fresh seed weight from late planted crop may be due to reduction in seed filling duration and slower seed growth rate which may be due to decrease in photoperiod and mean daily light integral. While in Williams 82, a somewhat inverse relationship was found between delay in date of sowing and mean seed growth rate as August planted crop had higher mean seed growth rate as compared to May planted crop in both years. The decrease in seed developmental rate with delay planting of Epps cultivar may be due to decrease in temperature and photoperiod.

In Williams 82, the slight increase in seed development rate with delay in sowing may be due to sink capacity and sink strength of the seed due to its indeterminate nature and were differently affected by reduction in temperature in delayed planting. The differences among the seed growth rates trend of determinate and indeterminate varieties may be the patterns of seed development and its assimilates requirements.

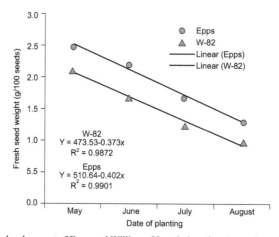

Fig. 5.8 Fresh seed weight development of Epps and Williams 82 varieties of soybean planted on four dates during 2000.

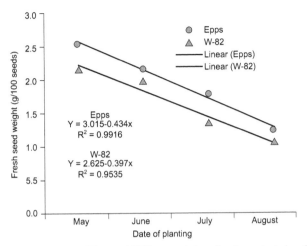

Fig. 5.9 Fresh seed weight development of Epps and Williams varieties of soybean planted on four dates during 2001.

Table 5.6 Mean fresh seed growth rate (g/day/100 seeds) of soybean varieties as affected by planting dates during 2000 and 2001.

Date of sowing	2000		2001	
	Epps	**Williams 82**	**Epps**	**Williams 82**
D1	0.80	0.57	0.67	0.41
D2	0.73	0.67	0.59	0.45
D3	0.67	0.66	0.52	0.61
D4	0.56	0.68	0.51	0.64

Fresh Seed Moisture

Data regarding fresh seed moisture (%) of the two soybean cultivars sown on four planting dates are reported in 2000 and 2001, which indicated fresh seed moisture (%) of soybean cultivars was initially the same in all seed samples from various planting dates. Fresh seed moisture content decreased as the seed developed and the trend of decrease was about the same in all planting dates. Mean rate of change in moisture (%) in fresh seed calculated from the regression equations reveals that Epps planted in May has higher mean rate of change in moisture (%) in both years as the crop stage advanced. A decrease in the rate of change in fresh seed moisture (%) was observed as planting was delayed up to the first week of August during both years. Slow rate of change in fresh seed moisture (%) from late planted crop may be due to slower seed growth rate which may be due to decrease in temperature and photoperiod. While, during both years, a somewhat inverse relationship was found in Williams 82 as August planted crop had higher rate of change in fresh seed moisture (%) as compared to May planted crop (Egli and TeKrony, 1997; Westgate, 1994).

Dry Seed Weight

Data on dry seed weight of Epps and Williams 82 sown on four planting dates are presented in (Table 5.7). The table shows that planting dates significantly affected the dry matter accumulation in soybean cultivars. Dry seed weight decreased in both varieties as sowing was delayed from May to August during both years. The mean dry seed accumulation rate indicates that May planted crop had higher dry matter accumulation rate as compared to delayed planted crop in August. A steady decrease in dry matter accumulation rate was observed as planting was delayed and the rate of decrease was more in

Table 5.7 Mean dry seed accumulation rate (g/day/100 seeds) of soybean varieties as affected by planting dates during 2000 and 2001.

Date of sowing	2000		2001	
	Epps	Williams 82	Epps	Williams 82
D1	0.34	0.37	0.31	0.38
D2	0.30	0.30	0.30	0.32
D3	0.28	0.28	0.23	0.24
D4	0.24	0.23	0.20	0.17

2001 than 2000. Minimum dry seed weight from late planted crop may be due to decrease in photoperiod and temperature which resulted in minimizing the seed filling duration and its growth.

Seed growth rate and duration of growth were studied for different soybean [*Glycine max* (L.) Merr.] cultivars and isolines differing in maturity and stem termination behavior under field conditions. Various methods for estimating rate and duration were compared, and such estimates were compared with yield, along with mature seed weight, seed number, and days from planting to maturity. Three estimates of seed filling period were all highly correlated with each other, but the final seed weight, divided by its growth rate during the linear phase of dry matter accumulation, or the 'effective filling period', correlated best with yield. Two estimates of seed growth rate were also highly correlated with each other but not with yield. Allelles that delayed maturity did not generally increase the seed filling period, and, in some cases, caused slight reductions. None of the genes studied affected seed growth rate or final seed size (Table 5.8).

Short-day photoperiods can increase the partitioning of assimilates to filling seeds of soybean (*Glycine max* L. Merr.), resulting in higher seed growth rates. The plant growth substance ABA has been implicated in the regulation of assimilate transfer within filling soybean seeds. Thus, it is hypothesized that an increased concentration of endogenous ABA in seeds may enhance sucrose accumulation and seed growth rate of soybeans exposed to short-day photoperiods. Plants of cv. Hood 75 were grown in a greenhouse under an 8-h short-day photoperiod (SD) until 11 d after anthesis (DAA) of the first flower, when half of the plants were transferred to a night-interruption (NI) treatment (3 h of low-intensity light inserted into the middle of the dark period). Plants remaining in SD throughout seed development had seed growth rates 43% higher than that of plants shifted to NI (7.6 mg seed^{-1} d^{-1} vs. 5.3 mg seed^{-1} d^{-1}). On

Table 5.8 The rate and duration of seed fill and seed size of cultivars in 1983.

Maturity group (MG)	Days from R5 to R7	Effective filling period (days)	Seed growth rate (mg.seed^{-1} day^{-1})	Final seed size (mg.seed^{-1})
MGII				
Mean	32.9	25.9	6.0	154.8
Range	26.0–40.0	19.2–30.7	4.4–9.5	88.4–213.5
CV	5.8	10.8	11.0	9.8
LSD0.05	3.1	4.7	1.1	25.3
N	18	17	17	17
MGIII				
Mean	33.1	27.6	5.5	147.7
Range	24.7–45.3	21.4–35.9	2.8–7.4	64.4–183.9
CV	4.2	20.5	16.3	7.9
LSD0.05	2.3	9.5	1.5	19.5
N	17	15	15	15
MGIV				
Mean	33.6	23.3	5.7	133.3
Range	25.3–45.7	14.7–31.4	3.4–7.8	49.8–206.2
CV	6.1	15.2	13.8	6.0
LSD0.05	3.4	5.9	1.3	13.7
N	18	11	11	11

a tissue-water basis, the concentration of ABA in SD seeds increased rapidly from 7.6 μmol l^{-1} at 11 DAA to 65.2 μmol l^{-1} at 18 DAA, but then declined to 6.6 μmol l^{-1} by 39 DAA. In contrast, the concentration of ABA increased more slowly in NI seeds, reaching only 47.4 μmol l^{-1} by 18 DAA, peaking at 57.0 μmol l^{-1} on 25 DAA, and declining to 10.2 μmol l^{-1} by 39 DAA. The concentration of sucrose in SD embryos peaked at 73.5 mmol l^{-1} on 25 DAA and remained relatively constant for the remainder of the seed-filling period. In NI, the concentration of sucrose reached only 38.3 mmol l^{-1} by 25 DAA, and peaked at 61.5 μmol l^{-1} on 32 DAA. Thus, in both SD and NI, sucrose accumulated in embryos only after the peak in ABA concentration, suggesting that ABA may have stimulated sucrose movement to the seeds. The earlier accumulation of ABA and sucrose in SD suggests that ABA may have increased assimilate availability during the critical cell-division period, thus regulating cotyledon cell number and subsequent seed growth rate for the remainder of the seed-filling period.

Light interception and leaf area criteria to maintain optimum soybean [*Glycine max* (L.) Merr.], yield during the last half of the seed filling period have been developed.

Planting dates were 3 June 1997 and 10 June 1997. The experimental treatments consisted of a factorial design of 10 soybean cultivars and three levels of defoliation treatments: A nondefoliated control. Partial defoliation at the temporal midpoint of seed filling, and total defoliation at the temporal three-fourths point of seed filling. Leaf area index (LAI) and canopy light interception (LI) were measured after defoliation treatments. Grain yield was determined at the maturity by machine harvest. Partial defoliation at mid-seed filling significantly (P < 0.05) reduced LAI, LI, and yield in 8 out of 10 cultivars in at least 1 yr of the study. Total defoliation at the three-fourths point of seed filling also reduced yields for almost every cultivar-year treatment combination. Results tended to support the original criteria related to LI and leaf area criteria for maintenance of optimal yield.

Field studies were conducted in two years using 638 F2 and 1185 F3 lines of selected 16 F1 and 15 F2 parent lines (< 80 pods plant^{-1}) to evaluate pod number and CO_2 exchange rate (CER) as selection criteria. Pod and seed number, and seed weight of individual lines were observed during harvesting time, and CER of Seed filling and yield of soybean under water and radiation deficits were investigated during 2011 and 2012. Treatments were irrigations (I1, I2, I3 and I4 for irrigation after 60, 90, 120 and 150 mm evaporation from class A pan, respectively) in main plots and light interceptions (L1: 100%, L2: 65% and L3: 25% sunlight) in sub-plots. Seeds per plant under I1 and I2 decreased, but under I3 and I4 increased as a result of radiation deficit. Maximum seed weight and seed filling duration of plants under 25% light interception (L3) were higher than those under full sunlight (L1) and 65% light interception (L2). In contrast, plants under full sunlight had the highest seed filling rate, particularly under water stress. Seed filling duration under severe light deficit (L3) was about 9 days longer than that under full sunlight (L1), leading to 15.8% enhancement in maximum seed weight. Decreasing seed yield of soybean under well watering and mild water stress and improving it under moderate and severe water deficit due to low solar radiation are directly related with changes in seed filling duration and consequently in seed weight and number of seeds per plant under these conditions (Table 5.9; Figs. 5.10 and 5.11).

Table 5.9 Means of maximum seed weight, seed filling duration and rate of soybean under irrigation and radiation treatments in 2011 and 2012.

Treatment	Seeds/plant	Max. seed weight (mg)	Seed filling duration (day)	Seed filling rate (mg.day^{-1})
Year				
2011	22.55b	130.66a	46.11a	2.83b
2012	26.91a	131.01a	28.96b	4.52a
Light interception				
L1	25.4a	125.15b	32.82b	3.81a
L2	25.34a	122.45b	37.67ab	3.25b
L3	23.45a	144.91a	42.11a	3.44b

Different letters in each column indicate significant difference at $P \leq 0.05$.
L1, L2, L3: 100%, 65% and 25% light interception, respectively.

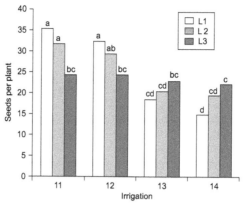

Fig. 5.10 Seeds per plant as varied with irrigation and light interception.

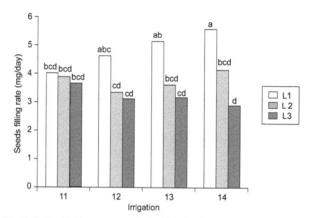

Fig. 5.11 Seed filling rate as varied with irrigation and light interception.

The effect of N supply on soybean (*Glycine max* L. Merrill) seed growth was investigated using an *in vitro* liquid culture system. Sucrose was maintained at 200 mM and N was supplied by asparagine and methionine in a 6.25:1 molar ratio. Media N concentrations from zero to 270 mM had little effect on cultured cotyledon dry matter accumulation rate for 7 or 14 d, but rates approached zero after 21 d when there was no N in the media. Only 17 mM N was required for maximum cotyledon growth rate up to 21 d. Cotyledon N accumulation and concentration increased in direct proportion to the N concentration in the media. The N concentration in cotyledons from a high protein genotype was higher than a normal genotype at all media N levels (0–270 mM). Soluble sugar and oil concentrations in the cotyledons were highest at zero media N and decreased as media N increased. These data suggest that the concept of seed N demand, which is thought to cause senescence in soybean, is incorrect. Soybean seeds can accumulate dry matter without accumulating N and apparently need only minimal supplies of N (17 mM) in order to maintain the metabolic enzymes necessary to sustain dry matter accumulation. Genetic differences in seed protein concentration seem to be regulated by the cotyledons, not the supply of N (Fig. 5.12).

Duncan (1986) suggested that light intercepted during and after seed initiation is a major determinant of yield. Mathew et al. (2000) indicated that light enrichment initiated at early flowering can modify seed size, with some internal control moderating the final size of most seeds.

Genetic differences in the seed growth rate of soybean were related to the number of cells in the cotyledons (Egli et al., 1980). There was a positive correlation between soybean cotyledon cell number and the ability of the seed to accumulate dry matter (Guldan and Brun, 1985). However, within the same soybean genotype, seed size may be influenced more by cotyledon cell size than by cell number (Hirshfield et al., 1992). There is evidence that plant hormones are involved in determining both sink

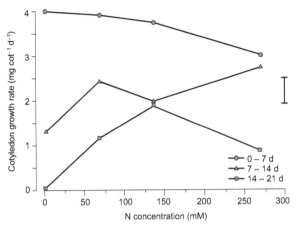

Fig. 5.12 *In vitro* cotyledon growth rate with N concentration.

size and capacity (Liu, 1993; Liu and Herbert, 2000). Abscisic acid (ABA), earlier considered an inhibitory substance, is now recognized as a naturally occurring plant hormone of major importance in the coordination of plant growth and development in response to the environment and the regulation of transport and storage of assimilates during grain development (Xiong and Zhu, 2003). Raschke and Hedrich (1985) reported that ABA reduced photosynthate production directly by a reduction in ribulose-1,5-bisphosphate carboxylase activity in soybean. Increased ABA in leaves caused stomatal closure, which resulted in a decline in photosynthesis due to low intracellular CO_2 levels (Fisher et al., 1986). ABA levels have been found to rise sharply and then fall during seed development of soybean (Liu et al., 2000). Soybean embryos require ABA for their continued development and for the accumulation of storage proteins (Tian and Brown, 2000).

Schussler et al. (1984) found a high concentration of ABA in large-seeded soybean genotypes compared to small and medium-sized seeds. He proposed that ABA may be involved in the stimulation of rapid unloading of sucrose into the testa of soybean and the ABA in cotyledons may enhance sucrose uptake by the cotyledons. A rapid increase in the fresh and dry weight of soybean seed was found to be correlated with a peak in the rate of ABA accumulation in the developing seed (Quebedeaux et al., 1976). Similarly, a decrease in the rate of dry weight accumulation was associated with a sharp decline in ABA concentration. Seeds are the primary sinks for photosynthates during reproductive growth.

Variation in light intercepted during and after seed initiation has been found to be a major environmental determinant of soybean [*Glycine max* (L.) Merrill] seed size. The influence of light enrichment and shading on seed growth rate, effective filling, cotyledon cell number, cell volume and endogenous ABA concentrations of cotyledons/testas during seed filling of soybean was investigated.

Evans, an indeterminate Group 0 soybean, was subjected to light reduction and enrichment treatments from the beginning of pod formation until final harvest for two years in Massachusetts. Higher rates of seed growth, greater seed dry weight, and higher cotyledon cell number were all observed with light enrichment. There was a reduction in seed growth rate and cotyledon cell number, along with a significant lowering of endogenous ABA levels in testa and cotyledon with shade. The level of ABA in cotyledon during seed development was significantly correlated with seed growth rates only under shade treatments. Both the growth rates and seed filling duration were influenced by variation in light interception by the soybean canopy. The effects of varying light treatment on seed size, within one genotype, were most likely due to the differences in seed growth rate and cotyledon cell number (Tables 5.10 and 5.11).

5.4 Seeds Per Unit Area

Randomly selected 32 F2 and 30 F3 lines were measured at initial seed filling stage, the selection of F2 lines based on pod number to generate F3 lines increased the average seed yield by 39%, and pod number

Table 5.10 Seed growth characteristics of soybean in response to light treatments across years.

Treatment	Seed growth rate (mg/seed/day)		Effective filling period (days)	
	1997	**1998**	**1997**	**1998**
CK	10.86b	13.82b	25.8b	23.5b
LE	13.04a	17.25a	25.5b	21.6b
S	8.34c	8.35c	28.9a	26.3a

Data points are means of 4 replicates; CK, LE and S are control, light enrichment and shade treatments, respectively; means within each column followed by the same letter are not significantly different ($P < 0.05$) using Duncan's Multiple Range Test.

Table 5.11 Cotyledon cell characteristics of soybean in response to light treatments across years.

Treatment	Cotyledon cell number (x 10^6)		Cotyledon cell volume (μlX10^5)		Growth rate/cell (ng/day)		Cotyledon cell weight (ng)	
	1997	**1998**	**1997**	**1998**	**1997**	**1998**	**1997**	**1998**
CK	8.32a	8.51a	14.21a	16.01a	1.42a	1.60b	17.61a	21.52a
LE	9.99a	10.12a	13.46a	16.82a	1.43a	1.86a	17.14a	21.64a
S	5.82b	6.93b	15.91a	17.54a	1.56a	1.02c	19.47a	23.28a

Data points are means of 4 replicates; CK, LE and S are control, light enrichment and shade treatments, respectively; means within each column followed by the same letter are not significantly different ($P < 0.05$) using Duncan's Multiple Range Test.

by 77% in F3 lines compared with F2 lines. A close relationship was found between seed weight and pod or seed number per plant. Net CER responded sensitively to a reduction of light in a short-term and showed 78% of F2 lines and all F3 lines with maximum CER (Pmax) $\geq$ 20 μmolCO$_2$.m^{-2}.s^{-1}. The ratio of pod number per plant and Pmax varied between lines and was used to group lines resulting in close relationships between Pmax and pod number. It is concluded that the use of pod number and CER (Pmax) as selection criteria offers an alternative approach to soybean breeding for high yield (Sitompul et al., 2015) (Fig. 5.13).

The number of seeds per unit area is an important yield component in soybean [*Glycine max* (L.) Merr.]; however, the mechanisms responsible for the regulation of this yield component are not well understood. Field experiments were conducted at Lexington, KY (3 yr), and at Taian, China (1 yr), in order to investigate the relationship between net canopy photosynthesis and seeds per unit area using genotypes with differences in individual seed growth rates (SGR). At Lexington, shades (30 and 63% reduction in insolation) were placed over plots from growth stage R1 until maturity in order to create differences in canopy photosynthesis. Planting dates (early and late) and row spacing (wide and narrow) were used at Taian to create differences in canopy photosynthesis. Crop growth rate (CGR) was measured between growth stage R1 and R5 as an estimate of net canopy photosynthesis. Yield, seeds per m^2, and SGR were also measured. Within each genotype, there was a linear relationship between CGR and seeds per m across treatments and years. Within an experiment, seeds per m^2 at a constant CGR was inversely related to genotypic differences in SGR. A partitioning coefficient was estimated by dividing the total sink demand (seeds per m^2 X SGR) by CGR. There were no apparent genotypic differences in partitioning coefficient; however, partitioning coefficient decreased linearly as CGR increased. The data suggest that the model proposed by Charles-Edwards, which describes seeds per m^2 as a direct function of canopy photosynthesis and a partitioning coefficient and in inverse function of assimilate flux to individual seeds, accurately describes the regulation of seeds per m^2 in soybean.

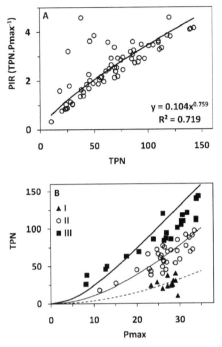

Fig. 5.13 Relationship between PIR and TPN for all selected F2 and F3 lines (A), and TPN and Pmax for each of groups (I, II and III) (B).

5.5 Yield Attributes and Yield

Kumudini et al. (2002) compared the new and old variety of soybean, the results indicated that the new cultivar of soybean had high quality due to the long durability of leaf at pods filling level and the escalation of dry materials at this level. The results of several studies also demonstrated that soybean with high-level of yield is reachable through high harvest index and the allocation of most of the photosynthetic materials into reproductive organs, whereas increasing surface of leaf until graining has a contradictory relation with seed yield (Kumudini et al., 2002). In this vein, Jian Jin et al. (2010) studied 41 varieties of soybean and found out that the duration (from sheathing to graining) was overriding to produce outstanding qualified soybean. Khan and Hatam (2000) illustrated that most of morphological attributions had meaningful and positive correlation with seed yield. Masudi et al. (2009) also reported that bush weight, numbers of seed and in bush had higher correlation with soybean yield. On the contrary, in a study by Bangar et al. (2003), it was found that soybean yield had significance relation with weight of 100 grains, numbers of days from germination to 50% flowering, and time of cultivation. Studies by Henrico et al. (2004) and Akhtar and Sneller (1996) indicated that the number of seed had meaningful correlation with seed yield per plant; this attribution had the highest direct impact on yield. Rezaizad (1999) investigated the existence of relations between seed yield and its components and he explored the idea that number of seed per plant, biological yield, and numbers of pod per plant had the most correlation with seed yield. Zhao et al. (1991) employed data factor analysis method in 12 important agricultural attributions by 16 soybean genotypes in China. These attributions were classified into four groups. The first variable was number of seed per plant and numbers of pod per plant. The second variable was consisted of plant height, number of node, height of the first pod from land and day numbers which have been needed to flower. The third variable included number of pod per plant, hundred seed weight, and weight of seed per plant. The fourth variable was the number of branches. Motivated by previous research, it was found that cultivar of soybean had a significant impact on soybean seed yield. It is also notable that various sorts released different yields accordance to environmental conditions and their respective adaptabilities to those conditions. Thus, recent studies tried to gain desired effects through modifying agricultural variables, including history of planting, model of planting, etc.

Previous research (Brun, 1978) has indicated that yield is more influenced by changes in source strength during the Rl to R7 period than by the emergence to Rl period (stages according to Fehr and Caviness, 1977). Jiang and Egli (1993) cited several studies (Egli and Yu, 1991; Johnston et al., 1969; Schou et al., 1978; Shibles et al., 1975), indicating that seed number (per unit ground area) was responsive to altered environmental conditions during flowering and pod set. Because seed per pod is genetically influenced (Shibles et al., 1975), seed number (per unit ground area) is usually highly correlated with pod number (per unit ground area). However, yield component mechanisms through which source strength influences pod number have not been completely identified. In response to a light interception (LI) gradient during R1 to R7, pod number was regulated by pods per reproductive node (Board and Harville, 1993), thus demonstrating source restriction during pod formation (R3 to R5). In their study, reproductive node number was constant across treatments.

Pod number is determined by pods per reproductive node and reproductive node number. The first parameter is determined by the difference between total pods per reproductive node initiated (pods at least 0.5 cm long) (Fehr and Caviness, 1977) and those aborted. Reproductive node number is determined by total node number and the percentage of these nodes becoming reproductive (percentage reproductive nodes). Although main stem node number in determinate cultivars is determined during the vegetative period, all other yield components contributing to pod number are formed mainly during R1 to R5 and possibly into the late reproductive period (R5 to R7).

Seed priming has presented promising and even surprising results for many seeds, including the legume seeds (Bradford, 1986). The advantage of seed priming in reducing the germination time and improving emergence uniformity is well established under laboratory conditions. The direct benefits of seed priming in all crops included: Faster emergence, better, more and uniform stands, less need to re-sow, more vigorous plants, better drought tolerance, earlier flowering, earlier harvest and higher grain yield. The indirect benefits reported were: Earlier sowing of crops, earlier harvesting of crops and increased willingness to use fertilizer because of reduced risk of crop failure. Park et al. (1997) reported that priming aged seeds of soybean resulted in good germination and stand establishment in the field trials. Seed pretreatment with PEG-6000 increased seed germination and vigor index (Gong Ping et al., 2000; Finch-Savage et al., 2004). This crop is, therefore, exposed many times to moisture and nutrient stresses during or immediately after germination. Considering the beneficial effects of seed priming on moisture use efficiency and seed quality parameters like germination and vigour which help in maintenance of optimum plant population it is recommended to obtain expected yield level.

The results exhibited that seeds primed with GA3 @ 100 ppm (T4), 0.5% KNO_3 (T5) recorded significantly higher germination percentage, i.e., 87.33% and 87.00%, respectively, over the untreated control T1 (83.00%). The treatments (T4) GA3 100 ppm 12 hr (77.19), (T8) Hydration with IAA @ 80 ppm 12 hr (76.55) and (T3) Hydration with $CaCl_2$ (2.0%) 12 hr (76.22) maintained the optimum plant stand at harvest over untreated control (T1). This may likely contribute for boosting up economic yield in soybean cultivar, JS-9305. The seed priming significantly influenced the seed yield and yield contributing characters of soybean. The seed priming treatments (T4) GA3 100 ppm 12 hr, (T7) hydration with water + Bavistin 3.0 g/kg were found to be effective for improvement in dry matter content of seedling (g) in soybean variety JS-9305. The treatments (T4) 100 ppm GA3 12 hr (2078.00 g/plot), (T8) hydration with IAA 80 ppm 12 hr (2008.67 g/plot), (T5) Hydration with 0.5 per cent KNO_3 12 hr (1991.00 g/plant) seed yield per plot, respectively, over the untreated control T1 (1647.67 g/plot) showing to the corresponding favourable improvement in number of pods per plant, number of seeds per pod, test weight (g), seed yield per plot (gm), seed yield per Ha (q), biological yield (g) and numerical harvest index (%) (Table 5.12).

Soybean {*Glycine max* (L.) Merr.} yield is more restricted by assimilatory capacity (source strength) during the reproductive (R1 to R7) compared with vegetative period (emergence to R1). Although pod number (per unit ground area) is recognized as an important factor affecting yield, the period in which this yield component is source restricted has not been clearly identified. Therefore, a study was undertaken in order to determine this period by identifying yield components through which source strength affects pod number and finding when these yield components are determined. Field studies were conducted

Table 5.12 Effect of seed priming on seed yield (q/ha) and harvest index.

Treatment	Seed yield (q/ha)	Harvest index (%)
Untreated (control)	22.88	36.61
Hydration with distilled water (12 h)	25.17	38.01
Hydration with CaCl$_2$ (2.0%) (12 h)	26.95	39.98
Hydration with 100 ppm GA$_3$ (12 h)	28.86	40.88
Hydration with 0.5% KNO$_3$ (12 h)	27.65	40.60
Hydration with 0.5% KCl (12 h)	25.32	38.03
Hydration (12 h) + Bavistin@3.0 g/kg of seed	27.56	39.24
Hydration with 80 ppm IAA (12 h)	27.90	40.63
Mean	26.54	39.25
SE±	0.43	1.01
CD (0.05)	1.30	NS

during 1991 and 1992 with 'Centennial' soybean at Baton Rouge, LA, on a Moon silty clay soil (fine-silty, mixed, nonacid, thermic Typic Fluvaquent). Treatments were partial defoliations designed to create light interception differences during Rl to R7. Yield was reduced 23% by defoliation. Harvest index, which was 0.59 for the control, ranged from 0.56 to 0.60 for the defoliation treatments. Source strength influenced pod number (per unit ground area) through branch dry matter, branch number, branch node number, and pods per reproductive node on the whole plant. Pods per reproductive node were regulated by pod initiation (pods at least 0.5 cm long) and/or abortion of initiated pods. Determination of final pod number occurred at 10 to 12 d after R5. In conclusion, pod number was source restricted from Rl to 10 to 12 d after R5. Stresses that restrict source strength during this period should be avoided in order to optimize pod number and yield (Board and Tan, 1995) (Tables 5.13, 5.14 and 5.15).

While early planting is a prudent management practice to increase soybean yield, logistical, equipment, environment, and labor challenges can delay planting. However, when early planting is possible, soybean is exposed to a greater risk of a spring killing frost, early season insects and seedling diseases, and damaging rainfall events that may result in suboptimal stand. In such years, replanting may be necessary. Furthermore, the climate variability that is affecting state and regional soybean yields (Mourtzinis et al., 2015) may also cause more frequent replanting situations. Proper replanting methods and optimal final plant stands (> 247,000 plants ha^{-1}) have been determined by Gaspar and Conley (2015), and yet, the proper MG to use in replant or late planting scenarios are unclear. Recent research conducted in the Midsouth, where a wide range of PDs and MGs are possible, has heavily investigated the MG × PD interaction to maximize yield (Salmeron et al., 2014; 2016).

Across the 12 environments in a study, soybean seed yield was positively correlated with both protein and oil contents. The stronger correlation with oil ($r = 0.40$) compared to protein ($r = 0.20$) suggest that oil content may be a more important factor affecting seed yield. Yield was positively correlated with all examined constituents, apart from linoleic and linolenic fatty acids and sucrose. Protein and oil exhibited a negative correlation ($r = -0.25$), as has been reported by Carrera et al. (2011). The sum of essential and non-essential amino acids was each positively correlated with protein content, $r = 0.77$ and $r = 0.93$, respectively.

The sum of essential amino acids as a percentage of the 18 amino acids was negatively correlated ($r = -0.70$) with protein, whereas, for non-essentials, the correlation was positive ($r = 0.83$). Across the examined region, large yield variability was observed due to PD and MG combinations (Fig. 5.14). The figure shows the yield response for several PD × MG combinations and shows that the greatest seed yield resulted from early planting (late April–early May) and MG 2. This result agrees with the highest-yielding MG identified by Mourtzinis and Conley (2017) for the same region. In that study, MG 1.4 to 2.2 resulted in the highest yields in Spooner and Arlington, WI, the northernmost and southernmost sites of the study, respectively. For MG 2, a large yield difference (~ 1200 kg ha^{-1}) was observed between

Table 5.13 Seed yield and harvest index for control and defoliation treatments (DF).

Defoliation Treatment	Seed yield (kg.ha⁻¹)			Harvest index (%)
	1991	**1992**	**Mean**	
Control	3592	2644	3118	59
DF-R3	2831**	2559	2695**	60
DF-R4	2734**	2306*	2520**	57**
DF-R5	2307**	2235*	2271**	56**
DF-R6.5	2095**	2082**	2091**	56**

* significant at 5% level; ** significant at 1% level.

Table 5.14 Yield components for control and defoliation treatments.

Defoliation treatment	Pod number (no.m⁻¹)	Seed per pod (no.)	Seed number (no.m⁻¹)	Seed size (g per 100 seed)
Control	1593	2.14	3396	11.66
DF-R3	1351**	2.12	2870**	11.48
DF-R4	1195**	2.10	2515**	11.25
DF-R5	1084**	2.02	2266**	11.04*
DF-R6.5	1162**	2.02	2360**	10.55**

* significant at 5% level; ** significant at 1% level.

Table 5.15 Effects of defoliation on reduction in yield due to seed characters.

Defoliation treatment	Reduction in yield due to	
	Reduced seed no. (%)	Reduced seed size
Control	–	–
DF-R3	100	0
DF-R4	100	0
DF-R5	86	14
DF-R6.5	76	24

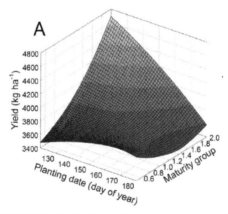

Fig. 5.14 Yield, planting date and maturity group relationships.

early and late PDs. Similar yield losses due to delayed planting have been reported by other studies in the region (Gaspar and Conley, 2015). For shorter season MGs, such as a 0.5, PD had little effect on yield, but the maximum was only 75% of the late MG's planted early (Fig. 5.14). There are situations in which early planting is not possible due to weather constraints, but there must be large economic benefits associated with agronomic practices that may delay soybean planting, such as fall planted cover crops, since early planting with a longer MG has no additional input costs associated with the increased yield potential. Maturity group selection and PD are simple and yet important management decisions that should not be separate across the examined region.

Planting date and MG selection were important factors for soybean seed protein and oil content and their sum. Late planting of all MGs resulted in the greatest protein concentrations. The opposite response was observed for oil content, in that early planting resulted in the greatest oil amount. Both protein and oil were affected mainly by PD and, to a lesser degree, by MG. The observed trend of higher protein and lower oil due to late planting is consistent with a previous study (Helms et al., 1990). Because of the inverse relationship of protein and oil, the response curve for the sum was curvilinear across PD and showed a maximum at mid-May PD's. These results suggest that a single combination of management practices with the goal of maximizing soybean yield, protein, and oil content may be difficult to attain. Planting a MG 2 in early May was found to maximize seed yield and oil content, but it resulted in the lowest observed protein content.

Across all environments included in the study, and depending on the PD, T5 to T8 ranged between 14 and 22°C. The negative correlation between T5 to T8 and PD ($r = -0.39$, $P < 0.0001$) indicates that earlier planting resulted in warmer average air temperatures between R5 and R8. It appears that these warmer temperatures favored yields of later-maturing soybean (Fig. 5.15) whereas, it slightly suppressed yields of MG 0.5 to 1. Temperature variability, which was introduced to this planting date and MG decisions can greatly affect yield and composition, and therefore, significantly increase or suppress overall farm profitability. The results show that the combination of early planting (late April–early May) and using the longest maturity group (MG 2) had the highest yield, oil, and oleic acid potential across the examined region. However, if a seed with high protein content is the overarching goal, a compromise in lower seed yield may be necessary.

For soybean, yield components which have the potential to influence yield are seed number per area, seed size (grams per seed), seed per pod (no.), pod number per area, pod per reproductive node (no.), and reproductive node number per area (*a* reproductive node is a node that contains at least one viable pod having at least one seed) (Kahlon et al., 2011). Similar to many other environmental stresses in soybean culture (Egli, 1998; Jiang and Egli, 1995), nonoptimal planting date, sub-optimal plant population, and row spacings, reduce yield through reductions in seed number, number per area (Kahlon and Board, 2012). Seed number per area is largely determined between emergence and shortly after R5 stages (Fehr and Caviness, 1977; Piigeaire et al., 1986). The R5 stage is the initiation of seed filling and is indicated by at least one pod among the top four main stem nodes that has one seed at least 0.3 cm long. Because seed number per area, pod number per area, and reproductive node number per area are all strongly linked to TDM by R5 (Board and Modali, 2005), CGR during the emergence stage to R5 period is very

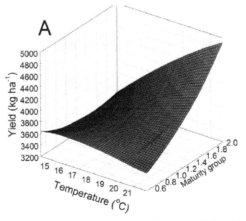

Fig. 5.15 Yield, temperature and maturity group relationships.

important for final yield determination. Crop growth rate during this period has long been recognized as being regulated by *LI* (Shibles and Weber, 1966; Board et al., 1990a; Sinclair and Horie, 1989). Light interception is largely controlled by LAI and can also be influenced by light interception efficiency (light interception/leaf area index), especially during the early vegetative period (Board, 2000). Both LAI and light interception efficiency are influenced by row spacing, plant population, maturity group, and planting date. Greater LI in narrow vs. wide row spacing starts occurring as early as three weeks after emergence and continues throughout this stage up to R5 period (Board et al., 1996). Greater LAI in late versus early maturity groups is usually achieved because of longer periods to R5 (i.e., longer periods for vegetative growth and leaf expansion) (Board et al., 1996; Board and Kahlon, 2012a).

Increased plant population creates greater LI compared with low plant population occurring during the early stages of the growing season. However, if environmental conditions are optimal, low plant populations can achieve yields equivalent to optimal plant populations (Board, 2000). Late planting, with its restricted emergence to R5 period, tend to have lower leaf area index and LI compared to normal planting dates simply because of less time available for leaf area development (Board and Settimi, 1986). Late planting in the south-eastern USA reduces crop yield by restricting the emergence to R5 period because of the shortening day lengths commencing in late June/early July (Board and Settimi, 1986). Essentially, this is due to limited time available to the crop to optimize TDM by R5, and consequently produces a suboptimal number of pods and seed per area, resulting in lower yields (Board et al., 1990b; Kahlon and Board, 2011; Board and Kahlon, 2012b).

Increased light interception *(LI)*, along with concomitant increases in crop growth rate *(CGR)*, is the main factor that explains how cultural factors, such as row spacing, plant population, and planting date, affect soybean yield. Leaf area index *(LAI)*, LI, and CGR are interrelated in a "virtuous spiral", where increased *LAI* leads to greater LI, resulting in a higher *CGR* and more total dry matter per area (TDM). This increases *LAI*, thus accelerating the entire physiological process to a higher level. A greater understanding of this complex growth dynamic process could be achieved through use of cluster analysis and principle components analysis (PCA). Cluster analysis involves grouping of similar objects in such a way that objects in the same cluster are similar to each other and dissimilar to objects in other clusters. PCA is a technique used to reduce a large set of variables to a few meaningful ones. Seasonal relative leaf area index (RLAI), relative light interception (RLI), and relative total dry matter (RTDM) response curves were determined from the data by a stepwise regression analysis, in which these parameters were regressed against relative days after emergence (RDAE). Greatest levels of RLAI, RLI and RTDM were observed in soybean planted early on narrow row spacings.

Soybean [*Glycine max* (L.) Merr.] yield is often associated with the number of pods per unit area, which can be construed to suggest that the number of pod bearing structures (nodes) plays a role in determining yield. This hypothesis was tested in a 2-yr field experiment at Lexington, KY (38 degrees N latitude). Four cultivars that varied in relative maturity (0.7 to 5.3) and high and low populations (one cultivar) were used to create variation in nodes per square meter. The experiments were planted on 25 May 2010 and 31 May 2011 in 38-cm rows (19 cm for the high population) and the target populations were approximately 43 (four cultivars), 26 (low population), and 78 (high) plants m(-2). Irrigation was used to minimize water stress. There were four replications of each treatment in a randomized complete block design. A 30-cm section of bordered row was harvested at maturity in order to determine flower scars, pods, and nodes. The photosynthetically active radiation interception was roughly 90% or above at R1 for all treatments and there was little difference in incident solar radiation or temperature among treatments during flowering and pod set. There was a two-fold variation in nodes per square meter in both years. Pods per square meter increased in concert with nodes per square meter up to approximately 70% of the maximum node number; above that level there was no change in pods per square meter. At low node levels, pods per square meter seems to be limited by the number of nodes, but at higher node levels, there was no relationship between nodes and pods.

The growth and seed yields of 2 Japanese and 3 Chinese cultivars of soybeans cultivated in 2002–2005 using a drip irrigation system in the arid area of Xinjiang, China, were analyzed with respect to growth parameters and air temperature. Seed yield was very high in 2002, 2003 and 2004, but relatively low in 2005. The variation among years in seed yield clearly depended on pod number.

The mean leaf area index (LAI) and crop growth rate (CGR) in 2005 was lower than those in the other 3 years. CGR showed significant positive correlations with mean LAI at the early growing stages, and with net assimilation rate (NAR) at the later growing stages. The increasing rate of pod number (IRP) was positively correlated with the mean LAI and CGR at the pod setting period, suggesting that an adequate supply of photosynthates would be required for pod setting. It was concluded that excellent growth in the years with high yields was supported by the large LAI before the pod setting periods and by high NAR and vigorous pod growth at the latter half of the growing season.

The result shows that growth rate was slow during vegetative phase in all genotypes. A relatively smaller portion of total dry mass (TDM) was produced before flower initiation and the bulk of it after anthesis. Maximum absolute growth rate (AGR) was observed during the pod filling stage in all genotypes due to maximum leaf area (LA) development and leaf area index (LAI) at this stage. Plant characters, like LAI and AGR, contributed to higher TDM production. Results indicate that a high-yielding soybean genotype should possess larger LAI, higher TDM production ability and higher AGR at all growth stages. A major agronomic problem in the south-eastern USA is low yield of late-planted soybean [*Glycine max* (L.) Merr.]. This problem is aggravated by the adverse effect of waterlogging on crop growth. To identify soybean growth stages sensitive to waterlogging, yield components and physiological parameters explaining yield losses induced by waterlogging, and determine the extent of yield losses induced by waterlogging under natural field conditions, greenhouse and field studies were conducted during 1993 and 1994 near Baton Rouge, LA, (30° N Lat) on a Commerce silt loam. Waterlogging tolerance was assessed in cultivar Centennial (Maturity Group VI) at three vegetative and five reproductive growth stages by maintaining the water level at the soil surface in a greenhouse study. Using the same cultivar, the effect of drainage in the field for late-planted soybean was evaluated. Rain episodes determined the timing of waterlogging; redox potential and oxygen concentration of the soil were used to quantify the intensity of waterlogging stress.

Results of the greenhouse study indicated that the early vegetative period (V2) and the early reproductive stages (R1, R3, and R5) were most sensitive to waterlogging. Three to 5 cm of rain per day falling on poorly drained soil was sufficient to reduce crop growth rate, resulting in a yield decline from 2453 to 1550 kg ha⁻¹. Yield loss in both field and greenhouse studies was induced primarily by decreased pod production resulting from there being fewer pods per reproductive node. In conclusion, waterlogging was determined to be an important stress for late-planted soybean in high rainfall areas, such as the Gulf Coast Region (Linkemer et al., 1998) (Table 5.16).

Greater understanding of how soybean [*Glycine max* (L.) Merr.] yield compensation occurs across plant populations would aid research aimed at reducing optimal plant population. It is interesting to determine how net assimilation rate (NAR) and leaf area index (LAI) contribute to crop growth rate (CGR) equilibration across low, medium, and high plant populations during the vegetative (emergence-R1) and early reproductive periods (Rl–R5). Determinate cultivar, Delta Pine 36ffi (Maturity Group VI), was planted at an optimal planting date during 1995 and 1996 at low (80000 plants ha⁻¹), medium (145000 plants ha⁻¹), and high (390000 plans ha⁻¹) plant populations on a Commerce silt loam near Baton Rouge, LA (30'N). Yield was unaffected by plant population. Equilibration of CGR for low vs. higher plant populations near R1 was achieved through greater NAR for the low plant population during the vegetative period, created by greater light interception efficiency (LIE, tight interception per unit leaf area). Although NAR equilibrated to minimal levels across plant populations near Rl, low population maintained CGR parity with higher populations until R5 through greater relative leaf area expansion rate

Table 5.16 Plot yield, sample yield, seed size, seed number, pods per reproductive node and total dry matter (R7) for Centennial soybean plant. Data are averaged across row spacing and years.

Site	Plot yield (kg.ha⁻¹)	Sample yield (g.m⁻²)	Seed size (g/100 seeds)	Seed no. (m⁻²)	Pods/rep. Node (no.)	Total DM (R7) (g.m⁻²)
Drained	2452	295	13.43	2249	2.43	493
Undrained	1550	219	11.88	1842	1.90	362
LSD0.05	255	38	0.28	263	0.14	63

Table 5.17 Yield, agronomic data, and whole plant yield components for soybean grown at low, medium, and high plant populations, arranged across 1995 and 1996. Baton Rouge, LA.

Plant population	Yield (kg.ha⁻¹)	Plant height (cm)	Days to R7 (d)	HI (%)	No. of seeds per m²	Seed size (g/100 seeds)	Seed per pod (no.)	No. of pods per m²
Low	4066a	77a	129a	51.6a	2459a	14.88a	1.84a	1336a
Medium	4153a	88a	128a	50.6a	2505a	14.92a	1.81a	1384
High	3961a	103a	129a	49.3a	2386a	14.97a	1.82a	1311a

Low population: 80000 plants ha⁻¹; medium population: 145000 plants ha⁻¹; high population: 390,000 plants ha⁻¹. Means in individual column followed by different letters are significantly different according to DMRT (P < 0.05).

(RLAER) during the late vegetative and early reproductive periods. Higher relative leaf area expansion rate for low vs. higher plant populations resulted from increased partitioning of dry matter into branches, probably induced by greater red/far red light ratios within the canopy. In conclusion, a possible genetic characteristic conducive to low optimal plant population is the greater partitioning of dry matter into branches (Table 5.17).

Reasons for the gradual genetic yield improvement (10–30 kg ha⁻¹yr⁻¹) reported for soybean [*Glycine max* (L.) Merr.] during decades of cultivar development are not clearly understood. Identification of mechanisms for the yield improvement would aid in providing indirect selection criteria for streamlining cultivar development. A study was undertaken in order identify yield components responsible for yield improvement in 18 public southern cultivars released between 1953 and 1999. The study was done at the Ben Hur Research Farm near Baton Rouge, LA (30° N Lat) during 2007 and 2008, plus a validation study in 2009. Experimental design was a randomized complete block with four replications and one factor (cultivar). In the 2007–2008 study, 18 cultivars released across the 1953–1999 period were selected. Three old and three new cultivars were used for the 2009 validation study. Data were obtained on yield, seed m⁻², seed size, seed per pod, pod m⁻², pod per reproductive node (a reproductive node is one having at least one pod having at least one seed), reproductive node m⁻², percent reproductive nodes and nodes per m⁻². Data were analyzed by ANOVAR and mean separation. Regression and path analyses were also done between yield and yield components, year of release and yield components, and among yield components themselves. Results of the 2007–2008 study indicated that yield differences were sequentially controlled by node m⁻², reproductive node m⁻², pod m⁻², and seed m⁻². However, node m⁻² was not as accurate at distinguishing low and high-yielding cultivars as the other three yield components and its role in yield formation was not substantiated in the validation study. A possible indirect selection criterion for yield during cultivar development is reproductive node m⁻².

A 2-year field experiment was conducted under light enrichment and shading conditions in order to examine the responses of seed yield and yield components distribution across main axis in soybean. The results showed that the maximum increase in seed yield per plant by light enrichment occurred at 27 plants.m⁻², while the most significant reduction in seed yield per plant by shading occurred at 54 plants.m⁻². Light enrichment beginning at early flowering stage decreased seed size on average by 7%, while shading increased seed size on average by 9% over densities and cultivars, resulting in a lesser extent compensation in seed yield decrement. Responses to light enrichment and shading occurred proportionately across the main axis node positions despite the differences in the time (15–20 days) of development of yield components between the high and low node positions (Liu et al., 2010). Variation intensity of seed size of three soybeans was dissimilar as a result of changes in the environment during the reproductive period. The small-seed cultivar had the greatest stability in single seed size across the main axis, followed by moderate-seed cultivar, while large-seed cultivar was the least stable. Although maximum seed size may be determined by genetic potential in soybean plants, our results suggested that seed size can still be modified by environmental conditions, and the impact can be expressed through some internal control moderating the final size of most seeds in main stem and in all pods. It indicates that, through redistributing the available resources across main stem to components, soybean plants showed the mechanism, in an attempt to maintain or improve yield in a constantly changing environment.

Table 5.18 Effects of light enrichment and shading on yield and yield components of Hai339 at three densities.

Yield component		2007			2008	
		D14	D27	D54	D27	D40
Yield (g/plant)	LE	36.0a	21.8a	9.5a	24.3a	17.2a
	CK	22.9b	12.7b	8.1a	12.5b	10.0b
	S	14.4c	7.0c	3.9b	8.7c	4.9c
Pods (No./plant)	LE	59.3a	40.2a	15.9a	46.8a	28.2a
	CK	39.3b	23.5b	13.3b	21.4b	17.2b
	S	21.5c	12.3c	7.73c	14.4c	8.4c
Seeds/pod (No.)	LE	2.36a	2.05a	2.22a	1.96a	2.21a
	CK	2.17b	1.93a	2.14ab	2.04a	1.94b
	S	2.28b	1.98a	2.01b	1.94a	1.80c
Seed size (mg/seed)	LE	255b	263b	271a	265c	276c
	CK	273b	287a	290a	287b	298b
	S	302a	292a	239b	313a	322a

Values followed by different letters within the row are significantly different from different light treatments under the same density within a year (P < 0.05). D14, D27 and D54 are 14 plants/m², 27 plants/m² and 54 plants/m², respectively. LE, CK and S are light enrichment, natural light and shade treatments, respectively.

Environmental conditions prevailing during the reproductive period, especially intensity and quality of solar radiation intercepted by the canopy, are important determinants of soybean yield and yield components (Board et al., 1996). Increased seed yield of soybean through narrow rows, can be attributed to increased light interception during reproductive period (Board et al., 1996). Light enrichment initiated at late vegetative or early flowering stages increased plentiful pod number, resulting in a 144% to 252% increase in seed yield (Mathew et al., 2000). Shading resulted in lengthening of internodes, decreasing of the number of pods and seeds per plant, the seeds yield per plant, the aerial part biomass per plant significantly (Li et al., 2006). In addition, influence of shading on seeds yield per area depends on duration of shading (Jiang and Egli, 1995). Adjusting planting density is an important tool to optimize crop growth and the time required for maturity.

Soybeans can tolerate moisture stress relatively well during the vegetative stages. Stress at this time reduces shoot growth, but not root growth. These conditions diminish water use by the plants and increase their ability to extract water from the soil. Planting early is a recommended tactic for mitigating the adverse effects of moisture stress. This is because the plants will have deeper roots for extracting soil moisture and produce a larger crop canopy, which shades the soil and reduces soil moisture losses due to evaporation. This tactic seems to be working in 2018 as the early–planted fields are tolerating the moisture and high-temperature stress better than the late-planted fields.

Soybean production has dominated the temperate regions due to cool to moderately warm climates. The increasing demand for soybeans in economically developing regions necessitates the rapid increase of production in the tropics. Soybean production under high temperature environments has also increased in the temperate regions due to global warming. The favorable temperature for soybean range is 16–28°C for the whole growing season (McBlain et al., 1987), the ranges of 15–22°C, 20–22°C, and 15–22°C are the optimum temperatures for the emergence, flowering and maturity stages, respectively (Liu et al., 2008), and there is a maximum of 27°C for the seed-filling period (Thomas et al., 2010). Concerns have emerged that global climate change may impact soybean production (Prasad et al., 2006; 2017). A recent analysis of long-term data revealed that growing season temperatures had a negative impact on soybean yields and caused a 17% reduction for every 1°C rise (Lobell and Asner, 2003). Moreover, studies using a temperature gradient chamber (TGC) and a model-aided analysis on farmers' yields have demonstrated significant negative effects of increased temperature on soybean yield under a temperate climate (Tacarindua et al., 2012; 2013), and reductions in both reproductive development and biomass production were associated. However, the variability of soybean genotypes in response to high temperature has been very limited (Chebrolu et al., 2016; Mochizuki et al., 2005), which is crucially important for soybean breeding for better adaptations to high temperature. The tropical environment with relatively higher temperatures has the potential to be employed to study plant development and agronomic performance under high temperatures. Unlike temperate regions, the mean air temperature in the tropics is relatively consistent and between 20 and 30°C every month of the year; the differences in monthly mean temperature between cool and hot months are rarely more than 7°C, and in some regions, the difference is only 2°C (Monteith, 1977). Additionally, the day-length is approximately 12 h throughout the year at the equator (Monteith, 1977), while in the temperate latitudes, it ranges from 11.2 to 14.4 h at latitude 35° in Kyoto, Japan. The large seasonal change of environmental factors and associated environmental stresses in the temperate region can have considerable seasonal effects on crop performances, to which the responsiveness of plant production processes is different depending on developmental stage. Therefore, the climate in the tropics, such as that in Indonesia, may be suitable for detecting cultivar differences in soybean adaptation to a high temperature environment.

Very high soil temperatures (33°C) can cause decreased nodulation and nitrogen fixation to occur in soybeans. High soil temperatures are most likely to have occurred in later planted beans due to the reduced canopy cover and coarse-textured soils. Producers can evaluate nodulation by digging up some plants and inspecting the nodules. Well-nodulated soybeans should have seven to 14 nodules on the tap root at flowering. Nodules that are actively fixing nitrogen will be pink to red when cut open.

During the R3 growth stage (one pod 3/16 inches long on the upper four nodes on the main stem having an unrolled leaf), drought-stressed soybean plants will abort both flowers and pods. Temperatures above 35°C have been shown to significantly decrease pod set. Leaf loss can also occur in severely stressed plants. If the moisture stress ends, soybeans will produce new flowers and pods up to the R5 stage (beginning seed).

Soybean yield losses will be the greatest when moisture stress occurs between the middle of the R4 growth stage (beginning seed swell in any pod on the plant) and the middle of the R5 growth stage (full seed in any pod on the plant). Stress at this time reduces the number of pods per plant as the plants are no longer able to produce new blossoms and pods. This is the major source of the lost yield. However, the number of seeds per pod and the size of the seed can also be reduced at this time. Leaf loss will continue in severely stressed plants.

Moisture stress occurring at the R6 growth stage (one pod having green seed that completely fills the pod on one of the upper four nodes on the main stem having an unrolled leaf) is largely due to a reduction in seed size. A reduction in pod number per plant can occur, but is much less likely after R6. Stress occurring after the R7 growth stage (one normal pod on the main stem has reached its mature color) does not affect yield.

Most of Indonesia's dry land is covered by acid soil, which leads to the decreasing potential yield of the crops. In different areas, soybean potential yield depends on the different soil pH and the availability of the soil. With this in view, the research was to study the potential yield of soybean promising lines in acid soil of Central Lampung, Indonesia. Ten promising lines and two check varieties (Tanggamus and Wilis) were grown in acid soil with pH 4.7. The results showed that the highest seed yield was shown by SC5P2P3.5.4.1-5 with 2.51 t/ha. Other soybean promising lines with seed yield over 2 t/ha were SJ-5/Msr.99.5.4.5-1-6-1 and the check variety Tanggamus. The highest yield of SC5P2P3.5.4.1-5 was caused by the high number of filled pods and the large size of seed.

Morphological adjustments are sometimes effective means to avoid drought stress. A number of root-related traits have been proposed as indicators of drought tolerance in soybean (Liu et al., 2005). Root distribution, which is measured in terms of horizontal and vertical root length density or dry matter in soil of different depth (Benjamin and Nielsen, 2006), will change in drought tolerant soybean cultivars under drought stress (Tzenova et al., 2008). It was reported that, under seasonal drought, there is a low root density in the dry surface soil but a high root density in the deeper region of the soil where the water content is higher. Moreover, using data from drought tolerant soybean cultivars, it was found that there is a positive correlation between drought tolerance and dry root weight/plant weight; total root length/ plant weight, and root volume/plant weight (Liu et al., 2005).

Root to shoot ratio increases under water deficit conditions. It has been proposed that the cessation of shoot but not root growth can be explained by the higher sensitivity to water deficit of shoot than root. The differential growth is closely related to the differential change in cell wall composition, which involves the thickening of shoot cell wall and relaxing of the expansion of root cell wall by certain catalytic enzymes and stiffening agents (Wu and Cosgrove, 2000).

There are only limited reports on related studies in soybean. The study on GmRD22 from soybean suggested a relationship between osmotic stress and cell wall metabolism. GmRD22 is a BURP-domain containing protein localized in the apoplast, which may play a role in stress tolerance by regulating lignin content of cell wall under stress, presumably through interacting with cell wall peroxidases (Wang et al., 2012).

The adjustments of leaf morphology may play a role in drought tolerance. Some cultivars take advantage from the maintenance of leaf area which provides a possible benefit for the growth of soybean plant after the stress is relieved (Manavalan et al., 2009). Under stress, drought tolerant soybean cultivars exhibited a larger leaf area when compared with less tolerant cultivars (Stolf-Moreira et al., 2010). This phenomenon was associated with the larger extent of reduction in stomatal conductance and yet a smaller extent of reduction in photosynthetic rate in the tolerant cultivar (Stolf-Moreira et al., 2010). In this case, the drought tolerant cultivar may benefit from the reduction of water loss while minimizing the cost of reduction of photosynthesis.

Seed weight can be evaluated using 100-seed weight or seed weight distribution. To eliminate the effects of the large measurement errors on the weight of a single seed, the weights of batches of 100 seeds are measured instead. Despite the general decreasing trend of seed weight under drought, the seed weight may not reduce uniformly as a function of drought intensity. Therefore, seed weight distribution has become another parameter employed to evaluate the effect of drought on seed weight, through the assessment of weight of seeds of different sizes. Dornbos and Mullen (1991) reported that, under severe drought, the proportion of seeds of diameter larger than 4.8 mm was reduced by 30%–40% while the proportion of seeds of diameter smaller than 3.2 mm was increased by 3%–15%. Under drought, soybean plants continued to produce heavy seeds. However, a greater portion of seeds were of low weight.

The causes of soybean yield improvement included changes in environmental conditions, genetic improvement, management practices, and the interactions among these factors. Increases in atmospheric CO_2 and O_3 concentration, air temperature, and climate variability affected yields in the previous decades (Southworth et al., 2002). Under elevated CO_2, total biomass in soybean increased proportionally more than yield, with positive changes in the photosynthetic rate and leaf area, and negative changes for harvest index (HI), stomata conductance, and Rubisco activity (Ainsworth et al., 2002). Kucharik and Serbin (2008) showed that increasing summer temperature could potentially decrease soybean yield in the United States by 16%, whereas increased precipitation might produce a counter effect, improving yield by 5 to 10%. Traits that contribute to improved soybean yield include longer reproductive or seed-filling periods (Shen and Liu, 2015), decreased lodging (Specht and Williams, 1984), and improved disease resistance (Foulkes et al., 2009). Changes in management practices that increased soybean yields are related to narrow rows (Heatherly and Elmore, 2004), improvement of weed control (Bradley and Sweets, 2008), conservation tillage and reduction in harvest losses (Heatherly and Elmore, 2004), and early sowing (Sacks and Kucharik, 2011). Early sowing can increase yield by lengthening both the vegetative and seed-filling phases (Egli and Cornelius, 2009).

The largest reported soybean grain yield is approximately three-fold more than the highest reported U.S. average yield. An understanding of yield determination is needed to identify avenues for increasing yield and for defining the yield potential of soybean. To illustrate physiological traits important for yield determination, we used a framework that models yield as the product of seed number (seed m^{-2}) and individual seed mass (mass seed). Developmentally, seed m^{-2} is determined first and is proportional to the biomass accumulation rate (BAR, gm^{-2} d^{-1}) and the fraction of assimilate allocated to reproductive structures (Fig. 5.16). Seed m^{-2} is inversely proportional to the individual seed growth rate (ISGR, mgseed^{-1} d^{-1}) where the ISGR represents the minimum amount of assimilate necessary to prevent a flower or pod from aborting. Hence, seed m^{-2} can be increased by optimizing conditions for crop growth

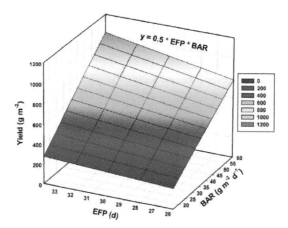

Fig. 5.16 Yield, EFP and BAR relationship in soybean.

(e.g., radiation interception, stress-free environment, high soil fertility levels) and having a low ISGR. Determination of mass seed occurs later during ontogeny than seed m^{-2} and can be expressed as the product of the ISGR and the effective seed filling period (EFP, d). Variation among genotypes for ISGR is quite large and is generally not affected greatly by the environment. There is also genotypic variation in the EFP, but the EFP is decreased by a variety of biotic and abiotic stresses. Analysis indicates that reaching the potential yield of soybean depends upon high BAR and extending the EFP, and a key factor affecting both of these variables is ensuring non-limiting crop nutrition, especially nitrogen. Strategies for increasing soybean maximum yield include early planting (which extends the EFP), optimizing crop nutrition, minimizing biotic and abiotic stresses, and developing breeding programs tailored for high yield environments. Characterizing physiological traits important for yield with genetic markers offers tools for combining favorable traits for high-yield environments (Van Roekel et al., 2015).

References

Akhtar, M. and C.H. Sneller. 1996. Yield and yield components of early maturing soybean genotypes in the Hid–South. Crop Sci. 36: 866–882.

Allen, D.K., J.B. Ohlrogge and Y. Shachar-Hill. 2009. The role of light in soybean seed filling metabolism. Plant J. 58: 220–234.

Ainsworth, E.A., P.A. Davey, C.J. Benacchi, O.C. Dermody, E.A. Heaton, D.J. Moore et al. 2002. A meta-analysis of elevated [CO$_2$] effects on soybean (*Glycine max*) physiology, growth and yield. GCB Bioenergy 8: 695–709.

Baba, A., S.H. Zheng, R. Matsunaga, T. Furuya and M. Fukuyama. 2003. Characteristics of dry matter production in Sachiyutaka, a new soybean cultivar for southwest of japan. Jpn. J. Crop Sci. 72: 384–389.

Bangar, N.D., G.D. Mukheka, M. Akhtar, D.B. Lad and D.G. Mukheka. 2003. Genetic variability, correlation and regression studies in soybean. J. Maharashtra Agri. Universities 28: 320–3.

Basuchaudhuri, P. 1987. Above-ground characteristics of soybean crop. Annal Agricultural Research 8: 135–140.

Beaver, J.S. and R.L. Cooper. 1982. Dry matter accumulation patterns and seed yield components of two indeterminate soybean cultivars. Agron. J. 74: 380–383.

Benjamin, J.G. and D.C. Nielsen. 2006. Water deficit effects on root distribution of soybean, field pea and chickpea. Field Crops Research 97(2-3): 248–253.

Board, J.E. and J.R. Settimi. 1986. Photoperiods effect before and after flowering on branch development in determinate soybean. Agron. J. 178: 996–1002.

Board, J.E., B.G. Harville and A.M. Saxton. 1990a. Narrow-row seed-yield enhancement in determinate soybean. Agron. J. 82: 64–68.

Board, J.E., B.G. Harville and A.M. Saxton. 1990b. Branch dry weight in relation to yield increases in narrow-row soybean. Agron. J. 82: 540–545.

Board, J.E. and B.G. Harville. 1993. Soybean yield component responses to a light interception gradient during the reproductive period. Crop Sci. 33: 772–777.

Board, J.E. and Q. Tan. 1995. Assimilatory capacity effects on soybean yield components and pod number. Crop Science 35: 846–851.

Board, J.E., W. Zhang and B.G. Harville. 1996. Yield rankings for soybean cultivars grown in narrow and wide rows with late planting dates. Agron. J. 88: 240–245.

Board, J.E. 2000. Light interception efficiency and light quality affect yield compensation of soybean at low plant population. Crop Sci. 40: 1285–1294.

Board, J.E. and H. Modali. 2005. Dry matter accumulation predictors for optimal yield in soybean. Crop Sci. 45: 1790–1799.

Board, J.E. and C.S. Kahlon. 2011. Soybean yield formation: What controls it and how it can be improved. pp. 1–36. *In*: El-Shemy, H.A. (ed.). Soybean Physiology and Biochemistry. Intech Publ. InTech pen Access, Rijeka, Croatia.

Board, J.E. and C.S. Kahlon. 2012a. A proposed method for stress analysis and yield prediction in soybean using light interception and developmental timing. Crop Management 11(1): 22.

Board, J.E. and C.S. Kahlon. 2012b. Contribution of remobilized total dry matter to soybean yield. Journal of Crop Improvement 26(5): 641–654.

Bradford, K.J. 1986. Manipulation of seed water relations via osmotic priming to improve germination under stress conditions. Horticultural Science 21: 1105–1112.

Bradley, K.W. and L.E. Sweets. 2008. Influence of glyphosate and fungicide coapplications on weed control, spray penetration, soybean response, and yield in glyphosate-resistant soybean. Agron. J. 100: 1360–1365.

Brun, W.A. 1978. Assimilation. pp. 45–76. *In*: Norman, A.G. (ed.). Soybean Physiology, Agronomy and Utilization. Academic Press, New York.

Brun, W.A. and K.J. Betts. 1984. Source/sink relations of abscising and non-abscising soybean flowers. Plant Physiol. 75: 187–191.

Carlson, J.B. 1975. Morphology. pp. 17–95. *In*: Caldwell, B.E. (ed.). Soybeans: Improvement, Production, and Uses. American Society of Agronomy, Madison, WI.

Carrea, C., M.J. Martinez, J. Dardanelli and M. Balzarini. 2011. Environmental variation and correlation of seed components in nontransgenic soybeans: Protein, oil, unsaturated fatty acids, tocopherols and isoflavones. Crop Science 51: 800–809.

Chebrolu, K.K., F.B. Frischi, S. Ye, H.B. Krishnan, J.R. Smith and J.D. Gillman. 2016. Impact of heat stress during seed development on soybean seed metablome. Metabolomics 12: 266.

Chowdhury, S.I. and I.F. Wardlaw. 1978. The effect of temperature on kernel development in cereals. Aust. J Agric. Res. 29: 205–223.

Collinson, S.T., R.H. Ellis, R.J. Summerfield and E.H. Roberts. 1992. Durations of the photoperiod-sensitive and photoperiod-insensitive phases of development to flowering in four cultivars of rice (*Oryza sativa* L.). Ann. Bot. 70: 339–346.

Crosby, K.E., L.H. Aung and G.R. Buss. 1981. Influence of 6-benzylaminopurine on fruit-set and seed development in two soybean, *Glycine max* (L.) Merr. genotypes. Plant Physiol. 68(5): 985–988.

Donovan, G.R., J.W. Lee, T.J. Longhurt and P. Martin. 1983. Effect of temperature on grain growth and protein accumulation in cultured wheat ears. Aust. J. Plant Physiol. 10: 445–450.

Dornbos, D.L. and R.E. Mullen. 1991. Influence of stress during soybean seed fill on seed weight, germination, and seedling growth rate. Canadian Journal of Plant Science 71(2): 373–383.

Duncan, W.G. 1986. Planting patterns and soybean yields. Crop Sci. 26: 584–588.

Dybing, C.D., H. Ghiasi and C. Paech. 1986. Biochemical characterization of soybean ovary growth from anthesis to abscission of aborting ovaries. Plant Physiol. 81: 1069–1074.

Egli, D.B. and D.M. TeKrony. 1997. Species differences in seed water status during seed maturation and germination. Seed Sci. Res. 7: 3–11.

Egli, D.B. and I.W. Wardlaw. 1980. Temperature response of seed growth characteristics of soybean. Agron. J. 72: 560–564.

Egli, D.B., J.E. Leggett and A. Cheniae. 1980. Carbohydrate levels in soybean leaves during reproductive growth. Crop Sci. 20: 468–473.

Egli, D.B. and Z. Yu. l991. Crop growth rate and seeds per unit area in soybean. Crop Sci. 31: 439–442.

Egli, D.B. 1998. Seed Biology and the Yield of Grain Crops. CABI Press, London, pp. 1–178.

Egli, D.B. and P.L. Cornelius. 2009. A regional analysis of the response of soybean yield to planting date. Agron. J. 101: 330–335.

Fader, G.M. and H.R. Koller. 1985. Seed growth rate and carbohydrate pool sizes of soybean fruit. Plant Physiol. 79: 663–666.

Fehr, W.R. and C.E. Caviness. 1977. Stages of soybean development. Iowa State University. Coop. Ext. Ser. Spec. Rep. 80.

Finch-Savage, W.E. 2004. The use of population-based threshold models to describe and predict the effects of seedbed environment on germination and seedling emergence of crops. pp. 51–95. *In*: Benech-Arnold, R.L. and R.L. Sánchez (eds.). Seed Physiology: Applications to Agriculture. New York, NY, USA: Haworth Press.

Fisher, E., K. Raschke and M. Stitt. 1986. Effect of abscisic acid on photosynthesis in whole leaves: Changes in CO_2 assimilation, levels of carbon-reduction-cycle intermediates, and activity of ribulose-1,5 bispphosphate carboxylase. Planta 169: 536–545.

Foulkes, M.J., M.P. Reynolds and R. Sylvester-Bradley. 2009. Genetic improvement of grain crops: Yield potential. pp. 355–385. *In*: Sadras, V.O. and D.F. Calderini (eds.). Crop Physiology: Applications for Genetic Improvement and Agronomy. Elsevier, Burlington, MA.

Gaspar, A.P. and S.P. Conley. 2015. Responses of canopy reflectance, light interception, and soybean seed yield to replanting suboptimal stands. Crop Sci. 55(1): 377–385.

GongPing, G.U., W.U. GuoRong, L. ChangMei and Z. ChangFang. 2000. Effects of PEG priming on vigor index and activated oxygen metabolism in soybean seedlings. Chinese J. of Oil Crop Science 22: 26–30.

Guldan, S.J. and W.A. Brun. 1985. Relationship of cotyledon cell number and seed respiration to soybean seed growth. Crop Sci. 25: 815–819.

Heatherly, L.G. and R.W. Elmore. 2004. Managing inputs for peak production. pp. 451–532. *In*: Soybeans: Improvement, Production, and Uses. Agron. Monogr. SV-16. ASA, CSSA, and SSSA, Madison, WI.

Heitholt, J.J., D.B. Egli and J.E. Legget. 1986a. Characteristics of reproductive abortion in soybean. Crop Sci. 26: 589–595.

Heitholt, J.J., D.B. Egli, J.E. Legget and C.T. MacKown. 1986b. Role of assimilate and carbon-14 photosynthate partitioning in soybean reproductive abortion. Crop Sci. 26: 999–1004.

Helms, T.C., C.R. Hurburgh Jr., R.L. Lussenden and D.A. Whited. 1990. Economic analysis of increased protein and decreased yield due to delayed planting of soybean. J. Prod. Agric. 3: 367–371.

Henrico, S.B., G.P. Claudio, R. Pinto and D. Destro. 2004. Path analysis under multicollineararity in soybean. Brazilian Archives Biol. and Tech. 47: 669–676.

Hirshfield, K.M., R.L. Flannery and J. Dale. 1992. Cotyledon cell number and cell size in relation to seed size and seed yield of soybean. Plant Physiol. 31: 395–403.

Jiang, H. and D.B. Egli. 1993. Shade induced changes in flower and pod number and flower and fruit abscission in soybean. Agronomy Journal 85: 221–225.

Jiang, H. and D.B. Egli. 1995. Soybean seed number and crop growth rate during flowering. Agronomy Journal 87: 264–267.

Jian Jin, A., A. Xiaobing Liu, A. Guanghua Wang, A. Liang Mi, B. Zhongbao Shen, B. Xueli Chen, J. Stephen and C. Herbert. 2010. Agronomic and physiological contributions to the yield improvement of soybean cultivars released from 1950 to 2006 in Northeast China. Field Crops Res. 115: 116–123.

Johnston, T.J., J.W. Pendleton, D.B. Peters and D.R. Hicks. 1969. Influence of supplemental light on apparent photosynthesis, yield, and yield components of soybean. Crop Sci. 9: 577–581.

Jones, R.J., B.G. Gengenbach and V.B. Cardwell. 1981. Temperature effect on in vitro kernel development in maize. Crop Sci. 21: 761–766.

Kahlon, C.S., J.E. Board and M.S. Kang. 2011. An analysis of yield component changes for new vs. old soybean cultivars. Agron. J. 103: 13–22.

Kato, I. 1964. Histological and embryological studies on fallen flowers, pods and abortive seeds in soybean, *Glycine max* (L.). Tokai-Kinki Natl. Agr. Exp. Stn. Bull.

Kahlon, C.S. and J.E. Board. 2012. Growth dynamic factors explaining yield improvement in new versus old soybean cultivars. J. Crop Improv. 26(2): 282–299.

Khan, A. and M. Hatam. 2000. Heritability and interrelationship among yield determining components of soybean varieties. Pakistan. J. Biological Sci. 116: 5–8.

Kokubun, M. and I. Honda. 2000. Intra-raceme variation in pod set probability is associated with cytokinin content in soybeans. Plant Product. Sci. 3: 354–359.

Kokubun, M. 2011. Physiological Mechanisms Regulating Flower Abortion in Soybean. pp. 541–554. *In*: Ng, T.B. (ed.). Soybean Biochemistry, Chemistry and Physiology. Janeza Trdine, Rijeka, Croatia.

Konno, S. 1976. Physiological study on the mechanism of seed production of soybean. Bull. Natl. Inst. Agric. Sci. D27: 139–295.

Kucharik, C.J. and S.P. Serbin. 2008. Impacts of recent climate change on Wisconsin corn and soybean yield trends. Environ. Res. Lett. 3: 034003.

Kurosaki, H., S. Yumoto and I. Matsukawa. 2003. Pod setting pattern during and after low temperature and the mechanism of cold-weather tolerance at the flowering stage in soybeans. Plant Prod. Sci. 6: 247–254.

Kumudini, S., D.J. Hume and G. Chu. 2002. Genetic improvements in short season soybean, nitrogen accumulation, remobilization and partitioning. Crop Sci. 24: 141–145.

Li, C.Y., Z.D. Sun, H.Z. Chen and S.Z. Yang. 2006. Influence of shading stress during different growth stage on yield and main characters of soybean. Southwest China Journal of Agricultural Sciences 19: 265–269 (In Chinese).

Linkemer, G., J.E. Board and M.E. Musgrave. 1998. Waterlogging effect on growth and yield components of late-planted soybean. Crop Sci. 38: 1576–1584.

Liu, X.B. 1993. Hormonal regulation of photoassimilates partitioning in crops. Crops Res. 3: 33–35 (In Chinese).

Liu, X.B. and S.J. Herbert. 2000. Some aspects of yield physiology research in soybean. J. Northeast Agr. Univ. (Engl. ed.) 7: 171–178.

Liu, X.B., J. Jin, G.H. Wang and Q.Y. Zhang. 2000. Endogenous hormone activities during seed development of soybean genotypes differing in protein content. Soybean Sci. 19: 238–242 (In Chinese).

Liu, X.B., J. Jin, G.H. Wang and S.J. Herbert. 2008. Soybean yield physiology and development of high-yielding practices in Northeast China. Field Crops Res. 105: 157–171.

Liu, Y., J.Y. Gai, H.N. Lu, Y.J. Wang and S.Y. Chen. 2005. Identification of drought tolerant germplasm and inheritance and QTL mapping of related root traits in soybean (*Glycine max* (L.) Merr.). Acta Genetica Sinica 32(8): 855–863.

Liu, B., X.B. Liu, C. Wang, Jian Jin, S. Herbert and M. Hashemi. 2010. Responses of soybean yield and yield components to light enrichment and plant density. Int. J. Plant Prod. 4: 1735.

Lobell, D.B. and G.P. Asner. 2003. Climate and management contributions to recent trends in US agricultural yields. Science 299: 1032–1032.

Manavalan, L.P., S.K. Guttikonda, L.S.P. Tran and H.T. Nguyen. 2009. Physiological and molecular approaches to improve drought resistance in soybean. Plant and Cell Physiology 50(7): 1260–1276.

Mann, J.D. and E.G. Jaworski. 1970. Comparison of stresses which may limit soybean yields. Crop Sci. 10: 620–624.

Masudi, B., M.R. Bihamta, H.R. Babai and S.A. Peighambari. 2009. Evaluation of genetic variation for agronomic, morphological and phonological traits in soybean. Seed and Plant 24(3): 413–427 (In Persian).

Mathew, J.P., S.J. Herbert, S.H. Zhang, F.A.A. Rautenkranz and G.V. Litchfield. 2000. Differential response of soybean yield components to the timing of light enrichment. Agronomy Journal 92: 1156–1161.

McBlain, B.A., J.D. Hesketh and R.L. Bernard. 1987. Genetic effects on reproductive phenology in soybean isolines differing in maturity genes. Can. J. Plant Sci. 67: 105–115.

Mochizuki, A., T. Shiraiwa, H. Nakagawa and T. Horie. 2005. The effect of temperature during the reproductive period on the development of reproductive organs and occurrence of delayed senescence in soybean. Jap. J. Crop Sci. 74: 339.

Monteith, J.L. 1977. Climate. pp. 1–25. *In*: Alvin, P.T. and T.T. Kozlowski (eds.). Ecophysiology of Tropical Crops. New York, NY: Academic Press.

Mosjidis, C.O., C.M. Perterson, B. Truelove and R.R. Dute. 1993. Stimulation of pod and ovule growth of soybean, *Glycine max* (L.) Merr. by 6-benzylaminopurine. Annals Bot. 71: 193–199.

Mourtzinis, S., J.E. Specht, L.E. Lindsey, W.J. Wiebold, J. Ross, E.D. Nafziger et al. 2015. Climate-induced reduction in US-wide soybean yields under-pinned by region- and in-season specific responses. Nature Plants 1: Article no. 14026.

Mourtzinis, S. and S.P. Conley. 2017. Delineating soybean maturity groups across the US. Agron. J. 109: 1–7.

Nonokawa, K., M. Kokubun, T. Nakajima, T. Nakamura and R. Yoshida. 2007. Role of auxin and cytokinin in soybean pod setting. Plant Production Science 10: 199–206.

Park, C.S., G.D. Marx, Y.S. Moon, D. Wiesenborn, K.C. Chang and V.L. Hofman. 1997. Alternative uses of sunflower. pp. 765–807. *In*: Schneiter, A.A. (ed.). Sunflower Technology and Production. Agronomy Monograph No. 35. ASA, CSSA, SSSA, Madison, Wisconsin.

Peterson, C.M., J.C. Williams and A. Kuang. 1990. Increased pod set of determinate cultivars of soybean, *Glycine max*, with 6-benzylaminopurine. Bot. Gaz. 151: 322–330.

Pigeaire, A., C. Duthion and O. Turc. 1986. Characterization of the final stage in seed abortion in indeterminate soybean, white lupin and pea. Agronomie 6(4): 371–378.

Quebedeaux, B., P.B. Sweetser and J.C. Rowell. 1976. Abscisic acid levels in soybean reproductive structures during development. Plant Physiol. 58: 363–366.

Raper, C.D. and J.F. Thomas. 1978. Photoperiodic alteration of dry matter partitioning and seed yield in soybean. Crop Sci. 18: 654–656.

Raschke, K. and R. Hedrich. 1985. Simultaneous and independent effects of abscisic acid on stomata and photosynthetic apparatus in whole leaves. Planta 163: 105–118.

Rezaizad, A. 1999. An investigation on genetic diversity in soybean cultivars. M.Sc. Thesis. Department of Agronomy and Plant Breeding, College of Agriculture, Tehran University, Karadj, Iran. 120 pp (In Persian).

Sacks, W.J. and C.J. Kucharik. 2011. Crop management and phenology trends in the U.S. Corn Belt: Impacts on yields, evapotranspiration and energy balance. Agric. For. Meteorol. 151: 882–894. doi:10.1016/j.agrformet.2011.02.010.

Saitoh, K., S. Isobe and T. Kuroda. 1998. Pod elongation and seed growth as influenced by nodal position on stem and raceme order in a determinate type of soybean cultivar. Jpn. J. Crop Sci. 67: 523–528.

Saitoh, K., S. Isobe and T. Kuroda. 1999. Intraraceme variation in the numbers of flowers and pod set in field-grown soybean. Japanese Journal of Crop Science 68: 396–400.

Salmerón, M., E.E. Gbur, F.M. Bourland, N.W. Buehring, L. Earnest, F.B. Fritschi et al. 2014. Soybean maturity group choices for early and late plantings in the mid-south. Agron. J. 106: 1893–1901.

Salmerón, M., E.E. Gbur, F.M. Bourland, N.W. Buehring, L. Earnest, F.B. Fritschi et al. 2016. Yield response to planting date amount soybean maturity groups for irrigated production in the US Midsouth. Crop Sci. 56: 747–759.

Schou, J.B., D.L. Jeffers and J.G. Streeter. 1978. Effects of reflectors, black boards, or shades applied to different stages of plant development on yield of soybeans. Crop Sci. 18: 29–34.

Schussler, J.R., M.L. Brenner and W.A. Brun. 1984. Abscisic acid and its relationship to seed filling in soybean. Plant Physiol. 76: 301–306.

Shen, Y. and X. Liu. 2015. Phenological changes of corn and soy-beans over U.S. by Bayesian change-point model. Sustainability 7: 6781–6803.

Shibles, R.M. and C.R. Weber. 1966. Interception of solar radiation and dry matter production by various soybean planting patterns. Crop Sci. 6: 55–59.

Shibles, R.M., I.C. Anderson and A.H. Gibon. 1975. Soybean. pp. 151–190. *In*: Evans, L.T. (ed.). Crop Physiology. Cambridge University Press, London.

Sinclair, T.R. and T. Horie. 1989. Leaf nitrogen photosynthesis, and crop radiation use efficiency: A review. Crop Sci. 29: 90–98.

Sitompul, S.M., D.I. Sari, E. Krishnawati, R.H. Mulia and M. Taufiq. 2015. Pod number and photosynthesis as physiological selection criteria in soybean (Glycine max L. Merrill) breeding for high yield. Agrivita 37: 75–88.

Sofield, I., I.F. Wardlaw, L.T. Evans and S.Y. Zee. 1977. Nitrogen, phosphorus and water contents during grain development and maturation in wheat. Aust. J. Plant Physiol. 4: 799–810.

Southworth, J., R.A. Pfeifer, M. Habeck, J.C. Randolph, O.C. Doering, J.J. Johnston and D.G. Rao. 2002. Changes in soybean yields in the Midwestern United States as a result of future changes in climate, climate variability, and CO_2 fertilization. Clim. Change 53: 447–475. doi:10.1023/A:101526642563.

Specht, J.E. and J.H. Williams. 1984. Contribution of genetic technology to soybean productivity-retrospect and prospect. pp. 49–74. *In*: Fehr, W.R. (ed.). Genetic Contribution to Yield Grain of Five Major Crops Plants. ASA, CSSA, Madison, WI.

Stolf-Moreira R., M.E. Medri, N. Neumaier, N.G. Lemos, J.A. Pimenta, S. Tobita, R.L. Brogin, F.C. Marcelino-Guimarães, M.C.N. Oliveira, J.R. Farias, R.V. Abdelnoor and A.L. Nepomuceno. 2010. Soybean physiology and gene expression during drought. Genetics and Molecular Research 9: 1946–1956.

Summerfield, R.J., F.J. Muehlbauer and R.W. Short. 1986. Flowering in lentil (*Lens culinaris* Medic): The duration of the photoperiodic inductive phase as a function of accumulated daylength above the critical photoperiod. Ann. Bot. 58: 235–248.

Tacarindua, C.R.P., T. Shiraiwa, K. Homma, E. Kumagai and R. Sameshima. 2012. The response of soybean seed growth characteristics to increased temperature under near field conditions in a temperature gradient chamber. Field Crops Res. 131: 26–31.

Tacarindua, C.R.P., T. Shiraiwa, K. Homma, E. Kumagai and R. Sameshima. 2013. The effects of increased temperature on crop growth and yield of soybean grown in a temperature on crop growth and yield of soybean grown in a temperature gradient chamber. Field Crops Res. 154: 74–81.

Thomas, J.F. and C.D. Raper. 1976. Photoperiodic control of seed filling for soybeans. Crop Sci. 16: 667–672.

Thomas, J.F. and C.D. Raper. 1977. Morphological response of soybeans as governed by photoperiod, temperature, and age at treatment. Bot. Gaz. 138: 321–328.

Thomas, J.M.G., K.J. Boote, D. Pan and L.H. Allen Jr. 2010. Elevated temperature delays onset of reproductive growth and reduces seed growth rate of soybean. J. Agrocrop Sci. 1: 19–32.

Thorne, J.H. 1980. Kinetics of 14C photosynthate uptake by developing soybean fruit. Plant Physiol. 65: 975–979.

Thorne, J.H. 1981. Morphology and ultrastructure of maternal seed tissues of soybean in relation to the import of photosynthate. Plant Physiol. 67: 1016–1025.

Tian, L.N. and D.C.W. Brown. 2000. Improvement of soybean somatic embryo development and maturation by abscisic acid treatment. Can. J. Plant Sci. 80: 271–276.

Tzenova, V., Y. Kirkova and G. Stoimenov. 2008. Methods for plant water stress evaluation of soybean canopy. *In*: Balwois 2008—Water Observation and Information System for Decision Support 2008: Ohrid, Republic of Macedonia.

Van Roekel, R.J., L.C. Purcell and M. Salmeron. 2015. Physiological and management factors contributing to soybean potential yield. Field Crops Research 182: 86–97.

Wang, H., L. Zhou, Y. Fu, M.Y. Cheung, F.L. Wong, T.H. Phang, Z. Sun and H.M. Lam. 2012. Expression of an apoplast localized BURP domain protein from soybean (GmRD22) enhances tolerance towards abiotic stress. Plant, Cell & Environment 35: 1932–1947.

Westgate, M.E. 1994. Water status and development of the maize endosperm and embryo during drought. Crop Sci. 34: 76–83.

Wilcox, J.R. 1985. Dry matter partitioning as influenced by competition between soybean isolines. Agron. J. 77: 738–742.

Wu, Y. and D.J. Cosgrove, 2000. Adaptation of roots to low water potentials by changes in cell wall extensibility and cell wall proteins. Journal of Experimental Botany 51(350): 1543–1553.

Xiong, L.M. and J.K. Zhu. 2003. Regulation of abscisic acid biosynthesis. Plant Physiol. 133: 29–36.

Yin, X., M.J. Kropff and M.A. Ynalvez. 1997. Photo periodically sensitive and insensitive phases of preflowering development in rice. Crop Sci. 37: 182–190.

Yoshida, S. 1981. Fundamental of Rice Crop Science. IRRI, Los Banos, Philippines, Pages: 269.

Zheng, S.H., A. Maeda and M. Fukuyama. 2003. Genotypic and environmental variation of lag period of pod growth in soybean. Plant Production Science 6: 243–246.

Zheng, S.H., A. Maeda, Y. Kashiwagi, A. Nakamoto and M. Fukuyama. 2004. Simultaneous growth of pods and seeds set on different racemes in soybean. *In*: 4th International Crop Science Congress.

CHAPTER 6
Plant Nutrition

Nutritional requirements of crops have always been of interest but more so now than ever because of changing economic and environmental dynamics in crop production. Increased demands for limited water supplies for uses other than agriculture, along with the resulting depletion of non-renewable water resources, higher energy and fertilizer costs, adoption of genetically modified crops, and climate change (Allen et al., 2012), alter the landscape of crop production. Concerns over agriculture's contribution to increased water consumption and its possible contamination of water resource sources through current production practices also stimulate a need for more current information about the nutrient requirements of most of our agronomic crops.

Deficiencies in N, P or S all result in a shift in dry matter allocation in favour of root growth (Ericsson, 1995). However, simple measurement of the relative biomass of the shoot and root fails to reveal the many subtleties of the roots' response to changes in nutrient supply. Note, that it seems to be a general rule in plants that primary root growth is much less sensitive to nutritional effects than is the growth of secondary or higher-order roots.

6.1 Nitrogen

The development of N-fixing nodules on legume roots is a highly regulated process. The number of nodules on a root system is controlled by a mechanism called 'autoregulation', in which previously formed or forming nodules suppress the development of further nodules (Schultze and Kondorosi, 1998). Split-root experiments have established that autoregulation acts systemically and that the autoregulatory signal originates in the shoot (Kosslak and Bohlool, 1984).

Nitrate is reported to block or delay the nodulation process at a number of different stages, including both rhizobial infection (by inhibiting root hair curling and infection thread formation) and nodule development, as well as inhibiting nitrogenase activity in established nodules and triggering early nodule senescence (Carroll and Mathews, 1990). Studies on soybean have shown that if the NO_3^- treatment is delayed until 18 h after rhizobial inoculation, its inhibitory effect is greatly diminished (Malik et al., 1987). This indicates that, in this species, the earliest stages of nodulation are the most sensitive.

There appear to be at least two ways by which plants monitor their nutrient supply: Directly through localized changes in nutrient concentration in the external soil solution, or indirectly through changes in the internal nutrient status of the plant itself. The direct pathway has the advantage that it can allow the plant to respond to short-term changes in nutrient availability and can essentially provide the roots with spatial information about the distribution of nutrients within the soil profile, allowing it to concentrate its developmental responses to that region of the soil where they will be of most benefit for nutrient acquisition.

The indirect pathway has the advantage that it enables the plant to integrate its nutritional signals with those coming from range of other physiological processes (such as photosynthesis).

An external supply of NO_3^- is able to trigger the rapid induction of a number of genes, including those for NR, nitrite reductase and NO_3^- transporters. It has been shown that NR activity is not required for the induction to occur, clearly implicating the NO_3^- ion as the signal molecule (Deng et al., 1989;

Pouteau et al., 1989). A number of lines of evidence suggest that the external presence of even very low NO_3^- concentrations ($< 10 \mu M$) can be sensed by plant roots (reviewed in Forde and Clarkson, 1999). Using NR-deficient mutants it has been shown that NO_3^- stimulation of lateral root elongation in *Arabidopsis* (Zhang and Forde, 1998) and localized NO_3^- inhibition of nodulation in legumes (Carroll and Gresshoff, 1986; Jacobsen, 1984) are both independent of NO_3^- assimilation. In the former case, the key role of the NO_3^- ion was confirmed by the finding that localized supplies of other N sources (NH_4^+ and glutamine) could not substitute for NO_3^- (Zhang et al., 1999).

Soybean plants assimilate a large amount of nitrogen during both vegetative and reproductive stages, and the total amount of N assimilated in a plant is highly correlated with the soybean seed yield (Fig. 6.1). One ton of soybean seed requires about 70–90 kg N, which is about four times more than in the case of rice (Hoshi, 1982). Soybean plants assimilate the N from three sources, (i) N derived from symbiotic N_2 fixation by root nodules (Ndfa), (ii) absorbed N from soil mineralized N (Ndfs), and (iii) N derived from fertilizer when applied (Ndff). For the maximum seed yield of soybean, it is necessary to use both N_2 fixation and absorbed N from roots (Harper, 1974; 1987). When only N_2 fixation is available to the plant, vigorous vegetative growth does not occur, resulting in reduced seed yield. On the other hand, a heavy supply of N often depresses nodule development and N_2 fixation activity and induces nodule senescence, which also results in reduced seed yield.

Soybean plants assimilate about 20% of total N until initial flowering stage (R1 stage), and 80% of N during the reproductive stage. On the other hand, rice assimilates about 80% of N until flowering. Therefore, the continuous assimilation of nitrogen after initial flowering stage is essential for good growth and high seed yield in soybean cultivation (Fig. 6.2).

To obtain high seed yield of soybean, good nodulation and high and long lasting nitrogen fixation activity are very important. Nodule formation and nodule growth are influenced by various soil conditions (water content, pH, nutrition) and climatic conditions (solar radiation, temperature, rain fall, etc.). Soybean can fix atmospheric N_2 by their root nodules associated with soil bacteria, bradyrhizobia. In addition, soybean can absorb inorganic nitrogen, such as nitrate and ammonia, from soil or fertilizer. Usually, a high yield of soybean was obtained in a field with high soil fertility. By supplying a constant but low concentration of nitrogen, either from soil or organic manure, soybean growth will occur without depressing nodulation and nitrogen fixation activity. However, it is well known that a high concentration of mineral N depresses nodule formation and nitrogen fixation activity. In particular, nitrate, the most abundant inorganic nitrogen in upland fields, severely inhibits nodulation and nitrogen fixation of soybean when nodulated roots are in direct contact with the soil solution containing nitrate (Fujikake et al., 1994; 2000; Ohyama et al., 2011).

Improving N_2 fixation could even facilitate high productivity (Ronis et al., 1985) and high seed protein content (Leffel et al., 1992). Indeed, rapid N_2 fixation during pod fill (stages R5–R6) was shown to contribute to increased seed yield and seed protein content (Imasande, 1992).

Dinitrogen fixation is, thus, a decisive physiological parameter both for enhanced productivity and higher seed quality (Planchon et al., 1992), traits often reported as negatively correlated (Burton, 1987). In spite of the high energy cost of N_2 fixation, some investigations (Maury et al., 1993) suggested that photosynthetic efficiency in soybean could be adjusted to the photosynthate requirements of the nodules.

However, as a result of the common action of both nitrogen sources, NO_3 assimilation remains a major pathway of N nutrition (Nelson et al., 1984). Assimilation reaches an earlier maximum than fixation, usually at full bloom (stage R2), and declines thereafter (Obaton et al., 1982). NO_3 assimilation is associated with plant biomass gain during vegetative stages and until flowering. In spite of its lower efficiency during pod fill, NO_3 assimilation is also necessary to achieve higher seed yield and seed protein content (Imasandez, 1992; 1998) (Fig. 6.3).

There has recently been concern in Brazil regarding whether biological N_2 fixation (BNF) is capable of meeting the increased N needs of newly released more productive cultivars, as well as doubts about the advantages of annual reinoculation of seeds. Forty experiments were performed over 3 yr in oxisols containing at least 10^3 cells of *Bradyrhizobium* g^{-1} in the State of Paraná, southern Brazil in order to estimate the contributions of BNF and of N fertilizer. The experiments were performed at two sites, Londrina and Ponta Grossa, under conventional (CT) or no-tillage (NT) systems, with two cultivars

[Embrapa 48 (early-maturing) or BRS 134 (medium-maturity group)]. Treatments included non-inoculated controls without or with 200 kg of N ha^{-1}, and inoculation without or with N fertilizer applied at sowing (30 kg of N ha^{-1}), or at the R2 or R4 stage (50 kg of N ha^{-1}). Compared with the non-inoculated control, reinoculation significantly increased the contribution of BNF estimated by the N-ureide technique (on average from 79 to 84%), grain yield (on average 127 kg ha^{-1}, or 4.7%) and total N in grains (on average 6.6%). The application of 200 kg of N fertilizer ha^{-1} drastically decreased nodulation and the contribution of BNF (to 44%), with no further gains in yield.

Application of starter N at sowing decreased nodulation and the contribution of BNF slightly and did not increase yields, while N fertilizer at R2 and R4 stages decreased the contribution of BNF (to 77%) and also yields. Estimates of volatilization of ammonia ranged from 15 to 25% of the N fertilizer applied, and no residual benefits of the N fertilizer in the winter crop were observed.

The results highlight the economic and environmental benefits resulting from replacing N fertilizer with inoculation in Brazil, and reinforce the benefits of reinoculation, even in soils with high populations of *Bradyrhizobium*.

Through extensive management under favorable climatic and field conditions, the plants can produce much higher yields (Japanese record is 7.8 t ha^{-1} grain; Konno, 1976). Several researchers have conducted field studies measuring soybean responses to applied N. There is a significant correlation between the total amount of N accumulated in soybean plants and seed yield (Hoshi, 1982). Under field conditions, nitrogen is generally derived from three sources: N derived from atmospheric dinitrogen (*Ndfa*), N derived from fertilizer (*Ndff*), and N derived from soil (*Ndfs*). Soybean plants fix large amounts of N in low-N soils under optimum environmental conditions (Herridge and Bergersen, 1988). However, sole N$_2$ fixation is often insufficient to achieve the maximum yield level. In contrast, a large supply of N fertilizer severely depresses nodule development and N$_2$ fixation activity, which sometimes results in the reduction of seed yield compared with the level in control plants without N fertilization. However, not all applications of N fertilizer exert an adverse impact on N$_2$ fixation. In soils low in N, a starter dose of as little as 5–10 kg N ha^{-1} can stimulate seedling growth and early nodulation so that both N$_2$ fixation and eventual yield are enhanced (Atkins, 1986). A basal dressing of chemical fertilizers, such as ammonium

Table 6.1 Effects of inoculation and application of N fertilizer on nodulation (nodule number, NN, no.plant^{-1}; and dry weight, NDW, mgplant^{-1}), plant growth (shoot dry weight, SWD, gplant^{-1}) and N accumulation in shoots (total N in shoots, TNS, mgplant^{-1}) of soybean cultivar Embrapa 48 (early-maturing group) at the R2 stage (scale of Fehr and Caviness, 1977). Experiments were performed for 3 consecutive years under conventional or no-tillage systems at Londrina, on an oxisol with established population of bradyrhizobia ($\geq 10^5$ cells g^{-1}) and the results of the 1st and 3rd year are shown here.

Treatment	First year				Third year			
	NN	NDW	SDW	TNS	NN	NDW	SDW	TNS
Conventional tillage								
Non-inoculated (NI)	22a	41a	3.28NS	130a	17a	18a	2.64NS	90a
NI + 200 kgN								
(50% sowing, 50% R2)	09b	07b	2.57	111ab	04b	04b	2.07	74b
Standard inoculation (SI)	17a	30a	2.82	113ab	18a	19a	2.39	98a
SI + 30 kg N at sowing	14a	23a	3.11	120ab	15a	15a	2.21	92a
SI + 50 kg N at R2	16a	23a	2.84	112ab	17a	19a	2.02	90a
SI + 50 kg N at R4	16a	23a	2.76	102b	16a	19a	2.56	95a
No-tillage								
Non-inoculated (NI)	17a	45a	2.75ab	124NS	18a	21a	2.12NS	79NS
NI + 200 kg N								
(50% sowing, 50% R2)	14b	14c	3.72a	157	05b	09b	2.21	87
Standard inoculation (SI)	18a	58a	2.98ab	112	21a	21a	2.39	91
SI + 30 kg N at sowing	20a	35b	3.08ab	128	17a	19a	2.46	87
SI + 50 kg N at R2	16ab	40a	2.47b	110	17a	21a	2.16	85
SI + 50 kg N at R4	18a	52a	2.78ab	121	18a	21a	2.45	90

Means ($n = 6$) for a given tillage system from the same column followed by different letters are significantly different ($P \leq 0.05$, Duncans test); NS, statistically non-significant.

Table 6.2 Effects of inoculation and application of N fertilizer on nodulation (nodule number, NN, no.plant^{-1}; and dry weight, NDW, mgplant^{-1}), plant growth (shoot dry weight, SWD, gplant^{-1}) and N accumulation in shoots (total N in shoots, TNS, mgplant^{-1}) of soybean cultivar BRS 134 (medium-maturing group) at the R2 stage (scale of Fehr and Caviness, 1977). Experiments were performed for 3 consecutive years under the conventional or no-tillage systems at Londrina, on an oxisol with established population of bradyrhizobia ($\geq 10^5$ cells g^{-1}) and the results of the 1st and 3rd year are shown here.

Treatment	First year				Third year			
	NN	NDW	SDW	TNS	NN	NDW	SDW	TNS
Conventional tillage								
Non-inoculated (NI)	12NS	19ab	1.95NS	79NS	17a	30b	1.81NS	73NS
NI + 200 kg N								
(50% sowing, 50% R2)	12	12c	2.56	99	11b	12c	2.00	86
Standard inoculation (SI)	14	22a	2.00	83	19a	47a	1.92	79
SI + 30 kg N at sowing	13	14bc	1.95	84	13b	23c	1.81	71
SI + 50 kg N at R2	14	23a	2.02	89	18a	39a	1.85	73
SI + 50 kg N at R4	15	18ab	1.85	77	19a	42a	1.82	76
No-tillage								
NI	27a	70a	2.16ab	87NS	12a	18a	1.40NS	74a
NI + 200 kg N								
(50% sowing, 50% R2)	20b	21b	2.40a	98	08b	10b	1.57	82a
SI	28a	72a	1.90b	86	14a	18a	1.38	71a
SI + 30 kg N at sowing	20b	35b	1.81b	76	08b	09b	1.38	57b
SI + 50 kg N at R2	24ab	57a	1.71b	80	12a	18a	1.36	77a
SI + 50 kg N at R4	27a	74a	2.19ab	80	12a	17a	1.39	71a

Means ($n = 6$) for a given tillage system from the same column followed by different letters are significantly different ($P \leq 0.05$, Duncans test); NS, statistically non-significant.

sulfate at approximately 15–40 kg N ha^{-1}, is conventionally applied in Japan in order to support seedling growth after the depletion of seed storage nitrogen until the establishment of nitrogen fixation. However, fertilizer efficiency is less than 10% (Takahashi et al., 1991). Similarly, fertilizer N applied during the pod filling stage may enhance seed yield (Afza et al., 1987; Nakano et al., 1987).

Soil acidity is often a limiting factor of the symbiotic nitrogen fixation process. Soils with low pH values lack calcium, and have a surplus of toxic aluminium, so that soybean roots in acidic soils don't have mucous coating on their surface, the purpose of which is to dissolve root pectins, enabling root hair curling and root hair penetration by bacteria. This is very important during the first few days after inoculation, that is, after sowing inoculated seed. Therefore, soils with pH value less than 5.5 (acidic soils) are not suitable for soybean growing, because they lack the necessary conditions for development of useful bacteria whose growth is slowed down or completely halted. Strains found on soybean roots in this type of soil are mostly ineffective, and when cut in half are green in colour. The situation is completely opposite in fertile neutral or mildly alkaline soils, like chernozem. In these types of soil, nitrogen fixing bacteria have not only good conditions for development, but also can survive in large numbers for many years after soybean was grown. In such soils, it is not necessary to perform seed bacterisation if soybean is in rotation every four years. In case of low effects of inoculation on nodule bacteria development, topdressing with 50 kg N/ha in form of calcium ammonium nitrate (27% N) in turn close to flowering or at beginning of flowering is recommended (Vrataric and Sudaric, 2008) the effects and utilization of deep placement ^{15}N-labeled coated urea (CU) and lime nitrogen (LN) were investigated at various growth stages of soybean (*Glycine max* [L.]). Soybean (cv. Enrei) plants were planted in a rotated paddy field in Niigata, Japan. ^{15}N-labeled CU or LN (100 kg N ha^{-1}) was supplied separately at a depth of 20 cm just below the seeding line. Deep placement of LN and CU significantly enhanced the dry weight and N content of the plants compared with the control treatment (Cont) without deep placement. As a result, the seed yield per plant in CU (67.2 g) and LN (70.6 g) was much higher than the Cont (37.1 g). 15N analysis of plants showed that the pattern of labeled N absorption tended to be lower with LN than CU at the R3 and R5 stages, but the recovery rate at R7 was higher in LN (70%) than in CU (61%). Combining the 15N analysis with a relative ureide analysis, the N derived from nitrogen fixation (*Ndfa*), from fertilizer

(*Ndff*) and from soil (*Ndfs*) were evaluated. At the R7 stage, the amount of *Ndfa* was higher in CU (32.1 g m^{-2}) and LN (31.1 g m^{-2}) than in Cont (21.4 g m^{-2}). This positive response for N2 fixation by the deep placement of CU and LN may result from the continuous supply of N from the lower parts of the roots, which promotes shoot growth and extends the photosynthetic activity of the green leaves, resulting in the promotion of nitrogen fixation and seed yield (Fig. 6.4).

Nitrogen deficiency is identified as a yellowing or chlorosis of the leaves lower down in the canopy as nitrogen is remobilized to new growth. N is mobile in plants and it is quickly translocated from old to young organs. For this reason, symptoms of N deficiency (first light green and later green yellow colours of leaves) obtain on the older leaves. In the more over stages it is found falling off the flowers and pods (Vrataric and Sudaric, 2008). Soybean is the most susceptible leguminous to nitrate oversupply. Under these conditions, inhibition of nodule forming and nitrogenise activities were found (Harper and Gipson, 1984). Also, high nitrate in apoplast of soybean had effect on pH increasing, immobilization of iron and developing of iron chlorosis in soybean (Hrustic et al., 1998).

Table 6.3 Seed yield components at maturity (R7 stage) of soybean plants cultivated with control, lime nitrogen and coated urea.

Treatment	Nodes per plant	Pods per plant	Seeds per plant	Seed weight per plant	100 seed weight (g)
Control	59.5b	66.5b	132.1b	37.1b	28.1b
CU	68.3a	81.5a	192.8a	67.2a	34.9a
LN	66.0a	82.3a	204.8a	70.6a	34.5a

Within a column, means followed by the same letter are not significantly different using a least significant difference test ($P < 0.05$; $n = 4$).
Cont, control; CU, coated urea; LN, lime nitrogen.

Recently, Bender et al. (2015) reported on an extensive study of biomass production and nutrient uptake of two Maturity Group (MG) 2.8 and one MG 3.4 soybean cultivars grown on glacial formed soils near 40°N and 42°N latitude. They concluded that supplemental fertility had a modest effect on increasing biomass and yield but none on nutrient portioning. Prior to this study, Ohlrogge's (1960) extensive review of available literature prior to 1960 on the uptake of essential nutrients by soybean was considered the primary source of nutrient uptake information on that crop. He stated, with respect to nitrogen (N), that daily uptake of available soil N plus symbiotic N, acquired via fixation by *Bradyrhizobium japonicum*, consistently increased to a peak of approximately 5.3 kg N ha^{-1} d^{-1} 90 d after seeding and then began rapidly falling off. Lathwell and Evans (1951) stated that soybean yield was closely correlated with the N accumulated throughout the season. They concluded that seed yield was mainly determined by pods per plant which was determined by N availability during the bloom period. Though much of the N uptake by soybean is from symbiosis, it can range from 25% to 75% of the total N required, depending upon available soil moisture (Zapata et al., 1987). Allos and Bartholomew (1959) had reported earlier that only 50%–75% of the total N required for maximum yields could be supplied by symbiotic fixation. Added N fertilizer though has had mixed results with no increase in seed yields being reported in some studies (Long et al., 1965; Maples and Keogh, 1969) and increases in yield in others (Brevedan et al., 1978; Pettiet, 1971).

6.2 Phosphorus

Research on phosphorus (P) in soybean has demonstrated the most common levels of the element found in field grown plants to be between 2.5 and 3.0 mg g^{-1} dry weight (Ohlrogge, 1960). Mederski (1950) found that, in plants grown outdoors in a sand-nutrient culture system, the mean P level of the top growth 30 d after seeding was 6.5 mg g^{-1} of the total dry weight. Later, at 6 d pre-bloom, the mean total P content was 10.5 mg g^{-1}. Hammond et al. (1951) reported shortly thereafter that, at maturity, between 82% and 85% of the P taken up by a soybean plant will be contained in the seed.

P removal by plants is mainly from 10 to 45 kg P, while by soybean is from 15 to 30 kg P/ha/year. The end of growth is the first symptom of P deficiency. Leaves are dark green and in the later stage develop chlorosis and violet color as result of increasing antociane synthesis. Necrotic spots, drying and falling of the leaves is the latest stage of P deficiency. Active nodules (dark pink center) of N-fixing bacteria are absent or few in number under conditions of P deficiencies. Also, decreasing of protein and chlorophyll synthesis was found.

Phosphorus deficiency cause stunted growth, dark green coloration of the leaves, necrotic spots on the leaves, and a purple color to the leaves, occurring first on older leaves.

Excess of P is rare. Plants reducing growth and dark frowning spots in leaves were observed. Intensity of plant development increasing and as results are the earlier flowering, grain forming and senescence. Oversupplies of P could be reason for some nutritional imbalances, for example Zn, Fe, Mn, Cu and B deficiencies.

Phosphorus cycling in soils is a complex phenomenon and depends on several environmental factors, including soil moisture and temperature. The amount of labile or available P available to plants in solution is low and is influenced by soil, plants and microorganisms. Labile P has to be constantly replenished in order to replace plant needs over the course of a plant's life so that P in solution is available to the plant at every growth stage. Contribution and bioavailability of organic P in soil solution is not completely understood. Furthermore, mechanisms controlling the rate of P exchange and availability are not completely understood (Pierzynski et al., 2005). Phosphorus is taken up as an inorganic anion ($H_2PO_4^-$ and HPO_4^{2-}) and, therefore, organic P must be mineralized prior to plant uptake. As apatite is broken down, the P has the potential to bind with metals or salts and become unavailable to plants (Pierzynski et al., 2005). Soils with high clay content will adsorb more P because of the large surface area. Eastern SD soils have higher pH; sometimes alkaline conditions exist with very high levels. Because P transport to the root is predominantly through diffusion, uptake can be reduced (Pierzynski et al., 2005) by soil drying or increased by practices that increase root length (Havlin, 1999). Rate of diffusion is affected by temperature and moisture; therefore, it is not the quantity of P applied that affects availability, but the rate at which P in solution can be replenished. Most P in soils is derived from the weathering of apatite of soluble salts. In such conditions, the P becomes unavailable when it binds with the Ca and forms insoluble compounds. Symbiotic relationships between arbuscular mycorrhizal fungi and soybean creates extension of the root system and allows for more growth and soil exploration for plants. As a result, the roots are able to locate more nutrients for the plant. In no-till/undisturbed soil, this healthy symbiotic mycorrhizal relationship can create a more efficient environment for nutrient uptake. This relationship allows plants to overcome nutrient depletion zones and extend into more soil. If roots do not grow into a new zone of nutrients, rate of uptake may be reduced because the depletion zone in the rhizosphere is replenished very slowly (Taiz et al., 2015). Smith and Read found that P uptake was 5 times higher from soybean roots that were colonized by mychorrhizal fungi than non-infected roots. In no-till soils or soils that have embraced a more diverse rotation, healthy symbiotic mycorrhizal relationships may create effective soil environment for P availability (Smith and Read, 1997).

Phosphorus (P) contents in plants are in wide range, mainly from 0.1 to 0.8% P in dry matter. Reproductive organs, especially of leguminous plants contain high levels of P about 0.6% P. Uptake of P into plants is intensive in the early stages of growth and in period forming of generative organs (Hrustic et al., 1998). Store of P in plants, especially in grain, are mainly in the form of phytic acid. P efficiency

Table 6.4 Yield (kgha⁻¹) and P concentration (gkg⁻¹) under different rates of phosphorus.

Year	Item	P Rate (kg P ha⁻¹)						CD (0.05)
		0	22	45	67	90	Mean	
2013	Yield	2890	2855	2861	2906	2863	2877	NS
	P con	5.1	5.1	5.2	5.3	5.3	5.2	0.1
2014	Yield	4289	4255	4316	4273	4314	4289	NS
	P con	5.3	5.4	5.6	5.7	5.7	5.6	0.1

NS: not significant at $\alpha = 0.05$.

is in close connection with water and temperature regimes in soil. Under optimal soil moisture, P uptake can be up to three-fold higher than in dry soil. Also, oversupplies of water, cold weather and low pH are reducing P uptake in plants.

Across site, analysis of grain yield and seed phosphorus concentration was undertaken for soybean phosphorus experiment. Phosphorus (P) fertilizer rates of 0, 22, 45, 67, and 90 kg P ha^{-1} applied pre-plant broadcast in the spring. Grain yield and seed were collected at maturity (R8) at each site in 2013 and 2014.

Characterization of nodule growth and functioning, phosphorus status of plant tissues and host-plant growth of nodulated soybean (*Glycine max* L. Merr.) plants grown under different phosphorus conditions were studied in order to evaluate the role of phosphorus in symbiotic nitrogen fixation. Phosphorus deficiency treatment decreased the whole plant fresh and dry mass, nodule weight, number and functioning. Under conditions of phosphorus oversupply the decrease in plant growth, nodulation and acetylene reduction was stronger. Phosphorus deficiency significantly affected all phosphorus metabolites. Contents of different phosphorus fractions were decreased under the conditions of phosphorus deficiency.

Table 6.5 Effects of P levels in nutrient solution and cultivars on soybean status.

Parameter	P levels (mM) in solution			Soybean cultivars		
	0.5	1.0	2.0	CKB1	SJ5	CM60
Shoot wt. at R5 (g plant^{-1})	14.1c	20.5a	18.2b	22.2a	14.2c	16.4b
Root wt. at R5 (g plant^{-1})	1.35b	2.36a	1.96a	2.54a	1.42b	1.74b
Total wt. at R5 (g plant^{-1})	15.4c	22.9a	20.2b	24.7a	15.7c	18.1b
Total wt. at mat. (g plant^{-1})	25.8b	36.0a	33.3a	34.2	27.1	33.8
Shoot wt. at mat. (g plant^{-1})	24.6b	34.4a	31.6a	32.5	25.9	32.2
Qa	0.213a	0.139b	0.117b	0.156	0.147	0.167
Eff.b	5.21a	4.77a	3.79b	4.89a	3.72b	5.17a
Shoot P (mg plant^{-1})	118b	258a	292a	236	212	221
Seed P (mg plant^{-1})	47.9b	72.3a	64.4a	59.7a	52.5b	72.4a
Oil (%)	16.9	18.1	17.5	16.9	17.3	18.3
Protein (%)	40.2	38.5	38.3	40.4	37.1	39.4

Duncan's multiple range test (within column, means by the same letter are not significantly at 5%.
a Shoot P-utilization quotient = plant shoot dry weight/mg P in plant shoot of P.
b Shoot P-utilization efficiency (eff.) = (shoot DM)2/shoot P content.

Table 6.6 Content of phosphorus metabolites extracted from soybean leaves and nodules affected by different P nutrition.

P fraction	Leaves		Nodules	
	+P	–P	+P	–P
Total P (µg/gFW)	2531.3	1689.5	3656.3	1968.8
Organic P (µg/gFW)	2377.0	1639.6	3595.6	1930.1
Inorganic P (µg/gFW)	153.1	49.4	60.7	39.1
High energetic (µg/gFW)	3.3	1.9	2.7	2.6
Sugar P (µg/gFW)	996.0	446.3	1331.1	550.7
Nucleotide P (µg/gFW)	229.8	694.1	869.9	786.5
Lipid P (µg/gFW)	1137.0	497.2	1391.9	589.9

6.3 Potassium

Potassium is the second most abundant mineral nutrient in plant tissue after nitrogen and is essential for plant growth. It is not surprising, therefore, that increases in the yield of soybeans (*Glycine max* (L.) Merrill) occurred when potassium fertilizer was added to soils that were low in available potassium. Soybean seed reserves are largely protein and oil. Since potassium deficiency decreases photosynthesis and phloem translocation, assimilate moving to the grain is likely to be reduced and, thus, reduce the oil percentage in the seed. In addition, potassium deficiency adversely affects protein synthesis. Therefore, it was hypothesized that potassium deficiency would limit both oil and protein synthesis to the same degree, thereby maintaining a constant ratio of these major seed reserves.

Soybean is considered to require potassium (K) at a level 30%–50% of the N level required by the plant (Varco, 1999). Over 50 years ago, some of the most striking responses of soybean to added fertilizer were by way of K salts, especially in the South-eastern United States (Ohlrogge, 1960). However, many of those soils were low in available K at the onset and similar consistent increases were not obtained in the Midwest. Fertilizing with K salts is known to result in increases in pods per plant, root nodule number and total mass, improved water stress tolerance, and a decrease in disease incidence (Bharati et al., 1986; Jones et al., 1977). Sale and Campbell (1986) reported that a reduction in K to deficient levels for soybean plants growing in a sand culture reduced seed yields, oil levels, seed [K], and an increase in seed protein concentration. They also reported though that soybean seems to respond to K fertilization as late as anthesis. Jeffers et al. (1982) demonstrated that K nutrition and soybean seed quality were closely correlated. Soybean yields from treatments receiving added fertilizer of 372 kg K ha^{-1} were 2400 kg ha^{-1} compared to 1650 kg ha^{-1} from plots receiving no additional K. A reduction in the incidence of Phomopsis, a seed mold fungus, of 165 g kg^{-1} of seed was also noted. Potassium deficiency occurs on older leaves first, with the visual symptom of chlorosis at the leaf margins and in-between the veins. All, but the newest leaves may show potassium deficiency symptoms in severe cases. Plants absorb potassium as the potassium ion (K$^+$). Potassium is a highly mobile element in the plant and is translocated from the older to younger tissue. Consequently, potassium deficiency symptoms usually occur first on the lower leaves of the plant and progress toward the top as the severity of the deficiency increases. One of the most common signs of potassium deficiency is the yellow scorching or firing (chlorosis) along the leaf margin. In severe cases of potassium deficiency the fired margin of the leaf may fall out. However, with broadleaf crops, such as soybeans and cotton, the entire leaf may shed resulting in premature defoliation of the crop. Potassium deficient crops grow slowly and have poorly developed root systems.

Gill et al. (2008) reported that imbalance and inadequate nutrient supply particularly devoid of K is main reason for low productivity and quality of soybean in India. Yin and Vyn (2004) conducted field experiments at three locations in Ontario, Canada from 1998 through 2000 to estimate the critical leaf K concentrations for conservation-till soybean on K-stratified soils with low to very high soil-test K levels and a 5- to 7-yr history of no-till management. For maximum seed yield, the critical leaf K concentration at the initial flowering stage (R1) of development was 2.43%. This concentration is greater than the traditional critical leaf K values for soybean that are being used in Ontario and in many U.S. Corn Belt states.

Nelson et al. (2005) compared response of soybean to foliar-applied K fertilizer and preplant application. Potassium fertilizer (K$_2$SO$_4$) was either broadcast-applied at 140, 280, and 560 kg K/ha as a preplant application or foliar-applied at 9, 18, and 36 kg K/ha at the V4, R1–R2, and R3–R4 stages of soybean development. Soybean grain yield increased 727 to 834 kg/ha when K was foliar-applied at 36 kg/ha at the V4 and R1–R2 stage of development in 2001 and 2002. Foliar-applied K at the R3–R4 stage of development increased grain yield but not as much as V4 or R1–R2 application timings. Foliar K did not substitute for preplant K in this research. However, foliar K may be a supplemental option when climatic and soil conditions reduce nutrient uptake from the soil.

6.4 Minor Nutrients

With growth, [P], [K], and [Mg] began to decline. Calcium is known to be immobile once taken. Calcium is known to be vital to soybean for cell division root hair development, as a co-factor in several enzymatic functions, and aids in plant disease resistance (Willis, 1989). However, in excessive amounts, it can interfere with the uptake of other cations, especially magnesium (Mg) and Rayar (1981) reported that maximum growth of soybean growing in nutrient solution cultures of varying [Ca] was found to occur at 249.5 μM Ca. Deficiency symptoms were noted at 24.9 μM Ca. At [Ca] > 249.5 μM Ca, tissue concentration is up in the plant and is not translocated from older tissue to developing cells (Varco, 1999).

It is, therefore, essential to have an adequate supply of available Ca throughout the growing season, along with sufficient water, which has been shown to facilitate Ca uptake (Karlen et al., 1982). High Ca may cause necrosis of new growth especially the tips of new leaves and is very rare in Iowa.

Magnesium's most important function in any green plant is the metal co-factor of the chlorophyll porphyrins. It is also utilized in protein construction in plants. Minimal sufficient levels of Mg in soybean have been determined to be approximately 2.6 mg g^{-1} of the dry weight. Deficiencies in Mg usually result in an interveinal chlorosis, a general unhealthy appearance of the plant and are most frequently noticed in plants growing on light sandy soils with low cation exchange capacity (CEC's) and seldom noticed on slits, silty clays or clay soils. As mentioned previously, excesses in Ca can interfere with Mg uptake and Bruns and Abbas (2010) concluded that an excess of K fertilizer on corn (*Zea mays* L.) likely reduced grain yields by interfering with Mg uptake.

Calcium is taken up by roots from the soil. Calcium is an essential plant nutrient. As the divalent cation (Ca^{2+}), it is required for structural roles in the cell wall and membranes, as a counter-cation for inorganic and organic anions in the vacuole, and as an intracellular messenger in the cytosol (Marschner, 1995). Calcium deficiency is rare in nature, but excessive Ca restricts growth of plant communities on calcareous soil solution. It is delivered to the shoot via the xylem. It may traverse the root either through the cytoplasm of cells linked by plasmodesmata (the symplast) or through the spaces between cells (the apoplast). The relative contributions of the apoplastic and symplastic pathways to the delivery of Ca to the xylem are unknown (White, 2001). However, the movement of Ca through these pathways must be finely balanced in order to allow root cells to signal using cytosolic Ca^{2+} concentration ([Ca^{2+}]$_{cyt}$), control the rate of Ca delivery to the xylem, and prevent the accumulation of toxic cations in the shoot.

Reduced seed calcium concentrations have been observed in soybean [*Glycine max* (L.) Merr.] seed lots with reduced seed germination. Soybean plants were grown hydroponically at Ames, IA, 1991, in order to determine if the Ca concentration of the root medium and relative humidity (RH) could influence seed Ca concentration and, subsequently, seed quality. Plants were grown in solution culture in a controlled environment growth chamber, with Ca concentrations of 0, 0.6, 1.2, and 2.5 mM in the root medium from beginning seed (R5) to beginning maturity (R7). In addition, concentrations of 0.6 and 2.5 mM Ca and RH of 55 and 95% were applied from R5 to R7 to plants grown in sand culture in the greenhouse. Treatments were arranged in a randomized compete-block design in both studies, with the blocks replicated in time in the growth-chamber study. Seed Ca concentrations increased with increased Ca supply to the plant. Relative humidity had no effect on either seed Ca concentration or germination. A decrease in the percentage of normal seedlings from 96.7 to 41.8% coincided with a decrease in seed Ca from 2.37 to 0.87 mg g^{-1}. Reduced germination of low-Ca seeds was primarily due to an increase in the percentage of abnormal seedlings. The percentage of normal seedlings was significantly and positively correlated with Ca ($r = 0.83$) and negatively correlated with the B/Ca ratio and B ($r = -0.97$ and -0.83, respectively) in seeds. Results indicate that reduced Ca supply to the plant may reduce seed Ca concentration in addition to altering other seed nutrients.

Differential aluminium tolerance of 'Perry' and 'Chief' soybean varieties, determined previously from growth on acid Bladen soil, was confirmed in nutrient solutions containing Al as the known growth-limiting factor. Differences in Al tolerance between the two varieties were steadily increased as the Ca level of the nutrient solution was reduced from 50 to 8 to 2 ppm.

Aluminium toxicity in soybeans was associated with decreases in concentrations of Ca in the tops and roots of both varieties, but this effect of Al was much more pronounced in the Al-sensitive Chief variety than in the more tolerant Perry. Greater Al sensitivity of the Chief variety was associated with greater susceptibility to a petiole collapse symptom. This symptom was related to a lower Ca concentration in the leaves and petioles, and specifically, related to a lower Ca concentration in the small petiole zone actually showing the collapse. Soil and solution studies indicated that the Ca deficiency observed in acid Bladen soil was Al induced. Aluminium appears to interfere to different degrees in the uptake and use of Ca by these two soybean varieties. The petiole collapse appeared to be a secondary effect of Al injury, and the presence of the symptom was not required for yield reduction by Al. The fact that soybean varieties differ in Al tolerance suggests that plant breeders may be able to develop varieties that can root more effectively in acid, Al-toxic subsoils. Varieties differing in Al tolerance also provide valuable tools for fundamental studies on the physiological nature of Al toxicity in plants. The petiole collapse symptom associated with Al sensitivity may be useful to plant breeders in screening genetic populations of soybeans for Al tolerance.

Greenhouse experiments were conducted in order to study the effects of glyphosate drift on plant growth and concentrations of mineral nutrients in leaves and seeds of non-glyphosate resistant soybean plants (*Glycine max* L.). Glyphosate was sprayed on plant shoots at increasing rates between 0.06 and 1.2% of the recommended application rate for weed control. In an experiment with 3-week-old plants, increasing application of glyphosate on shoots significantly reduced chlorophyll concentration of the young leaves and shoots' dry weight, particularly the young parts of plants. Concentration of shikimate due to increasing glyphosate rates was nearly 2-fold for older leaves and 16-fold for younger leaves in comparison to the control plants without glyphosate spray. Among the mineral nutrients analyzed, the leaf concentrations of potassium (K), phosphorus (P), copper (Cu) and zinc (Zn) were not affected, or even increased significantly in case of P and Cu in young leaves by glyphosate, while the concentrations of calcium (Ca), manganese (Mn) and magnesium (Mg) were reduced, particularly in young leaves. In the case of Fe, leaf concentrations showed a tendency to be reduced by glyphosate. In the second experiment harvested at the grain maturation, glyphosate application did not reduce the seed concentrations of nitrogen (N), K, P, Zn and Cu. Even at the highest application rate of glyphosate, seed concentrations of N, K, Zn and Cu were increased by glyphosate. In contrast, the seed concentrations of Ca, Mg, Fe and Mn were significantly reduced by glyphosate. These results suggested that glyphosate may interfere with uptake and retranslocation of Ca, Mg, Fe and Mn, most probably by binding and, thus, immobilizing them. The decreases in seed concentration of Fe, Mn, Ca and Mg by glyphosate are very specific, and may affect seed quality.

Sulfur is essential in the synthesis of the amino acid cysteine and methionine but has seldom been found deficient in soybean production. In a summarization of research conducted in the South-eastern US, Kamprath and Jones (1986) found that a response to sulfur (S) fertilizer occurred at only two of nine sites. Positive yield responses to S fertility occurred only on soils with ≤ 4.0 mg kg^{-1} S with no responses note for soils with ≥ 8.0 mg kg^{-1} S. Sulfur deficiency is similar to nitrogen deficiency, but the chlorosis occurs on newer leaves because sulfur is not mobile in the plant.

A field experiment on "sulphur nutrition in soybean" was initiated during Kharif 2005–06 at Main Agricultural Research Station, University of Agricultural Sciences, Dharwad (Karnataka state) India. The soils found deficient in sulphur (< 10 mg/kg). To work out the optimum sulphur dose for increased soybean yield and oil content, an experiment was planned and executed in Factorial Randomized Block Design replicated thrice on vertisol during kharif 2006, 2007 and 2008. The treatments comprised of four sulphur levels (0, 10, 20, 30, and 40 kg ha^{-1}). Gypsum as sulphur source was applied along with the basal fertilizer dose (40:80:25 N, P$_2$O$_5$ and K$_2$O kg ha^{-1}). The soybean Cv.JS 335 was sown at 30 × 10 cm spacing. Pooled data (2006–08) revealed that the soybean seed yield was significantly increased with the application of sulphur @ 20 kg ha^{-1} (2534 kg ha^{-1}) compared to sulphur levels; 30 kg ha^{-1} (2494 kg ha^{-1}), 40 kg ha^{-1} (2376 kg ha^{-1}) and 10 kg ha^{-1} (2226 kg ha^{-1}). The oil content, net return and benefit cost ratio followed a similar trend. Based on the findings, the technology "sulphur application @ 20 kg ha^{-1}" was evaluated in farmers field during kharif 2009–10 and 2010–11. The mean results of farm trails over

Table 6.7 Response of soybean to sulphur nutrition (Pooled data, 2006–2008).

Sulphur (kg/ha)	Yield (kg/ha)	Seed yield (g/plant)	Oil content (%)
10	2226	11.10	18.36
20	2534	12.28	18.68
30	2494	12.00	18.35
40	2376	11.44	18.37
Mean	2408	11.68	18.44
Control	2123	10.13	18.33
CD (0.05)	159.83	1.01	0.22

two years (2009–10 and 2010–11) indicated that the soybean seed yield increased to the tune of 12.01% with the application of sulphur @ 20 kg ha^{-1} (through Gypsum @ 100 kg ha^{-1}) along with the basal dose, compared to with no sulphur application (0 kg ha^{-1}).

Field experiments were conducted in order to predict optimum tissue sulphur (S) concentration, nitrogen:sulphur (N:S) ratio in leaf tissue and S levels for maximum grain yields of soybean cultivar PK-327. Various S levels significantly increased the S and N contents in 50-day-old plants, straw and soybean grains. Test weight of soybean grains was also increased. N:S ratio wider than 16.0 in soybean leaves of 50-day-old plants was considered to indicate S starvation. Leaf N:S ratio of 10.83 produced maximum soybean grain yield. Among ammonium sulphate, superphosphate, and elemental S tried as a source of S, ammonium sulphate was found to be the best source of S for soybean. It is recommended that soil testing programmes should include testing for S status so that only optimum levels of S are applied as higher levels of applied S appear to be luxuriantly consumed.

Concentrations and contents of nitrogen, phosphorus, potassium, calcium, magnesium, and sulfur (N, P, K, Ca, Mg, and S) were determined for three irrigated cultivars grown using the early soybean production system (ESPS) on two soils (a sandy loam and a clay) in the Mississippi Delta during 2011 and 2012. Data were collected at growth stages V3, R2, R4, R6, and R8. No change in macro-nutrients due to soil type or years occurred and modern cultivars were similar to data collected > 50 years ago. Mean seed yield of 3328 kg ha^{-1} removed 194.7 kg N ha^{-1}, 16.5 kg P ha^{-1}, 86.0 kg K ha^{-1}, 17.5 kg Ca ha^{-1}, 9.0 kg Mg ha^{-1}, and 10.4 kg S ha^{-1}. Increased yields over the decades are likely due to changed plant architecture and/or pests resistance, improved cultural practices, chemical weed control, and increased levels of atmospheric carbon dioxide (CO_2). The macro-nutrient contents of the whole plant at the observed growth stages are presented in Table 6.8. Two elements (N and P) had significant (P ≤ 0.05) steady increases in content from R2 to R8, while S content significantly increased from R2 to R6 and tended towards an increase from R6 to R8. Nutritionally, soybean is known to be a rich source of protein for human and livestock diets but a comparatively high producer of phythate, a P storage compound in plants that is important in germination but indigestible by humans and monogastric animals. This can lead to deficiencies in iron (Fe) and zinc (Zn) by binding with phythate in the gut, preventing these elements from being absorbed into body (Andrews, 2015).

Sulfur is a component of the essential amino acids cysteine and methionine which are incorporated into the protein contained in the seed. These factors help explain the continued increase in N, P, and S throughout the growing season. Potassium, Ca, and Mg are not as much of a seed component and appear primarily in structural tissue (Ca), water utilization and enzymatic reactions (K), particularly photosynthesis (K and Mg) and the overall seasonal growth of the plant.

Many soybean fertility recommendations are derived from research conducted during the 1930s to 1970s, and may not be adequate in supporting the nutritional needs of the greater biomass accumulation and seed yield associated with current soybean germplasm and production systems. Furthermore, no recent data exist that document the cumulative effects of improved soybean varieties, fertilizer source and placement technologies, and plant health/plant protection advancements on the rate and duration of nutrient accumulation in soybean. A more comprehensive understanding of soybean's nutritional requirements may be realized through this evaluation of the season-long nutrient uptake, partitioning and remobilization patterns in soybean.

Agronomic production practices and soil conditions with a capacity to supply nutrients at the listed quantities in Table 6.9 would be expected to meet soybean nutritional needs for an average yield level of approximately 60 bu/A. The potential for nutrient accumulation in soybean has increased two- to three-fold during the past 80 years as a result of increased DM production and grain yield. Mean grain yield values presented in this study are approximately 30 to 40% greater than the current United States average (USDA-NASS, 2014) and the presented nutrient accumulation information may serve as a resource for anticipated improvements in soybean yield.

Grain nutrient HI values are relative indicators of nutrient partitioning to soybean grain tissues, quantified as the ratio of grain nutrient removal to total nutrient accumulation. Nutrients with high requirements for production (e.g., N, K, Ca) or those that have a high HI value (e.g., N, P, S, Cu) may indicate key nutrients for high yield (Table 6.9). On average, over 80% of accumulated P was removed

Table 6.8 Macro-nutrient concentration (mg g⁻¹) of irrigated soybean at five different growth stages grown in the lower Mississippi River Valley.[†]

Tissue	Element	V3	R2	R4	R6	R8
Leaves	N		47.35	50.72	48.07	34.10
	P		2.94	3.38	2.83	2.21
	K		19.61	18.21	15.33	13.28
	Ca		16.56	16.11	14.85	21.38
	Mg		5.69	5.50	4.00	3.29
	S		2.64	2.80	2.75	2.22
Stems	N		22.04	20.87	15.04	11.10
	P		2.38	2.59	2.35	1.80
	K		30.91	34.36	23.55	14.91
	Ca		16.05	14.87	10.95	10.48
	Mg		5.15	4.96	3.85	3.49
	S		1.72	1.67	1.28	1.09
Pods	N				38.96	40.04
	P				4.47	3.87
	K				28.49	25.94
	Ca				10.98	7.89
	Mg				4.28	3.80
	S				2.20	2.18
Seeds	N					58.51
	P					5.56
	K					21.59
	Ca					5.26
	Mg					2.70
	S					3.13
Residue	N					11.09
	P					1.40
	K					21.27
	Ca					9.54
	Mg					4.26
	S					0.83

[†] Means of three cultivars (AG4303, 94B73, and AG5503), 2 years (2011 and 2012), two sites (Tunica clay and Bosket fine sandy loam), three replications, and three plants.

with harvested soybean grain tissues. Similarly, N, S, and Cu each resulted in nutrient harvest indices greater than 50%. Harvest index values for N, P, and S are similar to previously published values; however, K HI has decreased from nearly 70% (Hammond et al., 1951) to 41% in the current study (Table 6.9). Agronomic production practices that harvest non-grain plant tissues for animal bedding or feed sources, commonplace in key cattle producing regions, may remove, compared to grain harvest, up to an additional 66 lb N, 4 lb P (8 lb P_2O_5), 84 lb K (100 lb K_2O), 7 lb S, 37 lb Mg, 92 lb Ca, 2.78 oz Zn (16 oz = 1 lb), 3.06 oz B, 3.99 oz Mn, and 0.34 oz Cu per acre (Table 6.9).

Grain nutrient HI values are relative indicators of nutrient partitioning to soybean grain tissues, quantified as the ratio of grain nutrient removal to total nutrient accumulation. Nutrients with high requirements for production (e.g., N, K, Ca) or those that have a high HI value (e.g., N, P, S, Cu) may indicate key nutrients for high yield (Table 6.9). On average, over 80% of accumulated P was removed with harvested soybean grain tissues.

Similarly, N, S, and Cu each resulted in nutrient harvest indices greater than 50%. Harvest index values for N, P, and S are similar to previously published values; however, K HI has decreased from nearly 70% (Hammond et al., 1951) to 41% in the current study (Table 6.9). Agronomic production practices that harvest non-grain plant tissues for animal bedding or feed sources, commonplace in key cattle producing regions, may remove, compared to grain harvest, up to an additional 66 lb N, 4 lb P

Table 6.9 Nutrient accumulation associated with producing, on average, 60 bu/A of soybean grain.

Parameter	Maximum total uptake	Removal with grain[†]	Harvest index	Nutrient removal coeff.
Macronutrient	Lbs/A	Lbs/A	%	Lbs/bu
N	245	179	73	2.98
P	19	15	81	0.25
P_2O_5	43	35	81	0.58
K	141	57	41	0.95
K_2O	170	70	41	1.17
S	17	10	59	0.17
Mg	45	08	18	0.13
Ca	101	09	09	0.15
Micronutrient	Oz/A	Oz/A	%	Oz/bu
Zn	4.78	2.00	42	0.033
B	4.64	1.58	34	0.026
Mn	5.30	1.31	25	0.022
Cu	0.90	0.56	62	0.0093

[†] Multiply grain yield by nutrient removal coefficient to obtain the quantity of nutrient removal. Maximum total nutrient uptake, removal with grain, and harvest index (percentage of total nutrient uptake present in the grain) of macro- and micronutrients were averaged over treatments at DeKalb (2012 and 2013) and Champaign (2013).

(8 lb P_2O_5), 84 lb K (100 lb K_2O), 7 lb S, 37 lb Mg, 92 lb Ca, 2.78 oz Zn (16 oz = 1 lb), 3.06 oz B, 3.99 oz Mn, and 0.34 oz Cu per acre (Table 6.9).

The rates and times of acquisition varied among nutrients and were associated with specific vegetative or reproductive growth periods. Nearly 75% of K uptake occurred before the onset of seed filling (Fig. 6.6) compared to the uptake of N, P, S, Mg, Ca, Zn, B, Mn, and Cu, which were more evenly distributed during vegetative and seed-filling growth phases. With the exception of K, maximum rates of nutrient uptake were consistent across macro- and micronutrients and tended to occur during a brief period that bracketed R4. Unlike the rapid uptake of mineral nutrients before tassel emergence in maize (Bender et al., 2013a), nutrient uptake in soybean more closely coincided with DM accumulation, producing a steady, season-long pattern of nutrient assimilation. Soybean nutrient uptake patterns closely resemble those published during the last 80 years (Hammond et al., 1951; Hanway and Weber, 1971), although, in modern cultivars, the proportion of total nutrient accumulation acquired during seed-filling has increased over time.

The differences are especially apparent for N, P, Mg, and Ca, which have increased by as much as 18% during this part of the reproductive period. Collectively, these findings suggest that the improved yield of soybean has concomitantly increased the potential for nutrient accumulation.

Grain acquired nutrients from a combination of (1) direct partitioning to developing grain tissues, and (2) nutrient remobilization from leaf, stem, or flower and pod tissues. Nutrients with relatively high (greater than 50%) HI values included P, N, Cu, and S. Nutrient remobilization from other tissues complemented nutrient acquisition during seed filling to meet grain nutrient demands.

Few studies have investigated changes over time in nutrient uptake and yield, in addition to the study of nutrient stoichiometry as a metric of nutrient limitations in soybean [*Glycine max* (L.) Merr.]. A comprehensive synthesis-analysis was performed by compiling a global historical soybean database of yield, total biomass, and nutrient (N, P, and K) content and concentration in studies published from 1922 to 2015. This period was divided into three eras, based on genetically modified soybean events: Era I (1922–1996), Era II (1997–2006), and Era III (2007–2015). The main findings of this review are: (i) seed yield improved from 1.3 Mg ha^{-1} in the 1930s to 3.2 Mg ha^{-1} in the 2010s; (ii) yield increase was primarily driven by increase in biomass rather than harvest index (HI); (iii) both N and P HIs increased over time; (iv) seed nutrient concentration remained stable for N and declined for both P (18%) and K (13%); (v) stover nutrient concentration remained stable for N, diminished for P, and increased for K; (vi) nutrient ratios portray different trends for N/P (Era I and III > II), N/K (Era I > II and III), and K/P (Era II and III > I); (vii) yield per unit of nutrient uptake (internal efficiency) increased for N

(33%) and P (44%) and decreased for K (11%); and (viii) variations in nutrient internal efficiency were primarily explained by increase in nutrient HI for N and K, but equally explained by both HI for P and seed P concentration. These findings have implications for soybean production and integrated nutrient management to improve yield, nutrient use efficiency, and seed nutrient composition.

Figure 6.7 shows the change in nutrient ratio with era. Nitrogen/phosphorus ratio ranged from 4.9 to 19.0, with greater values for Eras I and III than for Era II. The slopes for Eras I and III are larger (11.5) than for Era II (9.0). This is related to the smaller total nutrient content (N, P, and K) for Era II compared with the other eras. The histogram shows higher frequency of greater N/P ratio for Eras I and III than for Era II. Overall, Pseed and stover P concentration (Pstover) accounted for > 50% of the N/P variation.

Nitrogen/potassium ratio ranged from 1 to 4 (Fig. 6.7B). Eras II and III clustered and have smaller ratios than Era I. Era I showed higher frequency of higher N/K ratio, followed by Eras III and II. For Era I, N/K averaged 3.1 and decreased to 1.9 for Eras II and III. Seed and stover K concentrations accounted for 5 and 10% of the variation in the residuals of the N/K ratio.

Nitrogen/potassium ratio ranged from 1 to 4 (Fig. 6.7B). Eras II and III clustered and have smaller ratios than Era I. Era I showed higher frequency of higher N/K ratio, followed by Eras III and II (Fig. 6.7B1). For Era I, N/K averaged 3.1 and decreased to 1.9 for Eras II and III. Seed and stover K concentrations accounted for 5 and 10% of the variation in the residuals of the N/K ratio.

Potassium/phosphorus ratio ranged from 2 to 11 (Fig. 6.7C). Era I showed greater frequency of smaller K/P ratios (Fig. 6.7C1). Analysis of slopes clustered Eras II and III, with smaller ratios for Era I. The Pstover explained 20% of the variation in the K/P ratio and Pseed only contributed to 5% of the variation (Fig. 6.7C2). Stover P concentration accounted for a larger proportion of the variation in K/P relative to Pseed, in agreement with the larger decrease in Pstover (36%) relative to Pseed (19%) among eras.

For both N and P, IE clustered for Eras II and III and was superior to the IE for Era I. For K, IE clustered for Eras II and III and was inferior to Era I. Average NIE increased 33% (Fig. 6.8A) from Era I (9 kg seed kg^{-1} N) to Eras II and III (12 kg seed kg^{-1} N); average PIE (Fig. 6.8B) increased by 44% from Era I to Eras II and III (90 to 130 kg seed kg^{-1} P ha^{-1}); KIE decreased from 27 to 23 kg seed kg^{-1} K from Era I to Eras II and III (Fig. 6.8C). Variation in seed yield per unit of nutrient uptake was portrayed with two boundaries (dotted lines in Figs. 6.8A–6.8C).

Residuals for the relationship between seed yield and nutrient uptake are presented in Figs. 6.8A1 to 6.8C1. The residuals were regressed against nutrient concentrations in seed and in stover to further dissect the nutrient IEs. The proportion of variance ($R2$) explained by each trait is presented in Figs. 6.8A2 to 6.8C2.

Nitrogen HI accounted for 37% of the variation in NIE, whereas Nseed accounted for 12% of the NIE variation Phosphorous HI accounted for 36% of the variation in PIE, and 28% was explained by Pseed. Potassium HI accounted for 67% of the variation in KIE, with a smaller contribution from Kseed (19%). For all three nutrients, nutrient HI accounted for a large proportion of the variation in the IE, also reflected in the lesser variation relative to NHI. In contrast, Nstover and Pstover decreased 9.5 and 15.5% for Eras II and III relative to Era I, respectively, and Kstover increased by 37% comparing Era I with the average of Eras II and III (Balboa et al., 2018).

6.5 Micronutrients

Together with NPK, calcium (Ca), magnesium (Mg), and sulfur (S) are classified as essential macronutrients. The eight other essential nutrients constitute a distinct group of elements required by plants in very small amounts, described conventionally as micronutrients: copper [Cu], iron [Fe], manganese [Mn], molybdenum [Mo], nickel [Ni], zinc [Zn], boron [B], and chlorine [Cl]. Still, other elements, like selenium (Se), silicon (Si), Cobalt (Co) and sodium (Na), are regarded as nonessential, although they have been found to enhance growth and provide other benefits to plants (Datnoff et al., 2007; Marschner, 2012).

Diagnosing micronutrient deficiencies in the field by assessing crop symptoms is difficult, even for trained agronomists. Look for "multiple evidence" before recommending a micronutrient for a whole field. A combination of crop symptoms in the field, tissue tests, soil tests, test strips, cropping history and other techniques will be used to confirm micronutrient deficiencies and economic yield responses.

Micronutrient deficiencies that do not display symptoms but reduce the yield of a crop are referred to as "hidden hunger". Know the field when assessing for "hidden hunger". If soil tests over a number of years indicate that a micronutrient level is decreasing into the marginal range for that crop, then consider applying the micronutrient in test strips first to see if there is a positive yield response and if that yield response is economical. On the other hand, applying micronutrients when they are not needed may reduce yields and/or economic returns.

Enzymes, superoxide dismutases, catalases, dehydrogenases, oxidases regulation of the activity of urease negatively impacted N metabolism in the plant. There is no doubt that, without micronutrients, many of the processes that drive plant metabolism of N, P, K, Mg, Ca, and S, as well as crop responses to ecological perturbations, would not be optimally functional. Fe, Cu, Mn, and Cl are involved in different aspects of plant photosynthesis, as cofactors for different metabolic processes. Dependent on the enzyme, Fe, Mn, Zn, Cu, Ni, Mo, and Cl all participate in the functioning of different enzymes, including DNA/RNA polymerases, N-metabolizing, ATPases, and numerous other enzymes involved in redox processes (Broadley et al., 2012). This cofactor role of micronutrients is crucial for enzyme and nonenzymic activities in plant metabolism under different environmental conditions. For instance, Zn specifically plays a role in the enzymatic processes involved in the biosynthesis of the plant growth regulator, auxin (Hossain et al., 1997; Fageria, 2002). This is an important function given the role of auxins in enhancing root growth, with agronomic consequences in terms of allowing the plant greater ability to access nutrients and water. Ni is involved with urease enzyme in the N metabolism of plants by converting urea to ammonia. Prior studies with soybean showed that a Ni deficiency induced plant, was leading to urea accumulation and necrosis of shoot (Polaccao et al., 1999; Sirko and Brodzik, 2000). Mo is required for N fixation by both symbiotic and free-living N-fixing bacteria, being a component of the nitrogenase enzyme system (Barron et al., 2009) and, therefore, essential for legume cropping systems, and in efforts to enhance biological N-fixation to supplement mineral N-fertilization.

In terms of abiotic stress mitigation, Zn has also been shown to modulate the activity of enzymes, such as the membrane-bound NADPH oxidase (Cakmak, 2000) involved in the homeostasis of reactive oxygen species that, at relevant levels, mediate important cellular functions, like host defense and signaling during drought or other abiotic stresses (Golldack et al., 2014). In agronomic studies conducted under drought conditions with adequate N and P (70:70 kg ha^{-1}) supply, Bagci et al. (2007) demonstrated that, while inadequate water supply under rain-fed systems could decrease average wheat grain yield by as much as 25%, the addition of Zn (23 kg ha^{-1} as a foliar application) increased wheat yield by 16%. Thus, Zn application lowered loss in yield due to water shortage from 25 to 13%. Cu is specifically essential for carbohydrate metabolism and required for the synthesis of lignin needed for cell wall strengthening (Yruela, 2009; Ryan et al., 2013a, b). The latter role could have implications for the survival of plants under the abiotic stresses of wilting, winds, or rainstorm conditions where plant rigidity is important. As the borate ion, B is involved in cell wall functioning by facilitating the cross-linking of pectic polysaccharides. It also plays a role in the structural integrity of the cytoskeleton (Miwa and Fujiwara, 2010). Chloride plays a role in stomatal regulation; hence, its deficiency results in wilting of the leaf and death of the plant (Broadley et al., 2012).

In most soils, iron is present in large quantities. On average, between 3–5% of soils consists of iron, which makes it the fourth most abundant element in the Earth's crust, after oxygen, silicon and aluminium. However, most iron in soils is unavailable for plant absorption (Meng et al., 2005). For example, iron deficiency is common on calcareous soils (which have a high pH), as iron availability to plants decreases with increasing pH. On the other hand, availability of iron for plants is generally high in acid tropical soils. Copper, zinc, manganese and phosphate are iron antagonists, and high levels of these elements in soils (or in fertilizer) can reduce iron uptake by plants. Thus, information on extractable iron, for example using DTPA, is generally of much more relevance than information on the absolute levels of iron contents in soils.

Iron is an essential nutrient for crops, and plant iron deficiency is long been known to occur in many regions of the world. It can be recognized by the occurrence of intraveinal chlorosis, the appearance of yellow or pale spots on the leaves of plants. Crops suffering from iron deficiency will grow slower than normal and are more susceptible to disease (Chatterjee et al., 2006). In terms of yields, estimated soybean losses in the north and central United States as a result of iron deficiency have been estimated to be in the order of magnitude of 300.000 tons a year (Hansen et al., 2004).

Soybean is susceptible to Fe deficiency. Fe deficiency is a common yield limiting factor for soybean grown on high-pH, calcareous soils, as well as on some seasonally poorly drained soils. Cool and wet periods promote Fe deficiency. Iron may be unavailable for root absorption, not transported after absorption, or may not be utilized by the plant.

In Iowa and Minnesota, over ten million dollars in potential soybean production were lost annually due to iron chlorosis. With the potential increase in alkalinity of Texas soils due to irrigation, reduced soybean production may become a problem. The problem could result from decreased yield per acre or from acreage with decreased productivity due to increased alkalinity. Iron deficiency is not easy or inexpensive to correct in the field. According to Gray et al. (1982), it would take five tons of sulfuric acid per acre to neutralize one per cent calcium carbonate in a 16.5 cm layer of soil.

Fe deficiency results in a characteristic interveinal chlorosis in new leaves and can cause substantial yield loss in soybean. In some years, it developed during early growth stages and disappeared as the plants matured. In more severe cases, chlorosis can persist throughout the entire season. There is wide variation in susceptibility to Fe deficiencies among soybean varieties.

Soybean in chlorotic areas had lower leaf chlorophyll concentrations, stunted growth, and poor nodule development relative to non-chlorotic plants. Also, compared to non-chlorotic areas, soil in chlorotic areas had greater soil moisture contents and concentrations of soluble salts and carbonates (Hansen et al., 2003).

Goos and Johnson (2003) found considerable differences of resistance of soybean varieties to iron chlorosis. Growing of more tolerant varieties is one solution for alleviation of nutritional problems induced by iron deficiency (Table 6.10).

Table 6.10 Chlorosis score in some soybeans.

Soybean low chlorosis score (CS < 2.3)			Soybean high score (CS > 3.2)		
Variety	Originator	CS	Variety	Originator	CS
Trail	N.D.AES	1.7	IA 2042	Iowa AES	3.7
Danatto	N.D.AES	2.0	IA 2041	Iowa AES	3.7
MN 0201	Minn.AES	2.0	MN 1103SP	Minn.AES	3.7
92 M10	Pioneer	2.0	Minnatto	Minn.AES	3.5
IA 1005	Iowa AES	2.0	MN 101SP	Minn AES	3.7
Jim	N.D.AES	2.2	IA2050	Iowa AES	3.3
MN 0203SP	Minn.AES	2.2	IA2033	Iowa AES	3.3
MN 0302	Minn.AES	2.2	MN2101SP	Minn.AES	3.3
Nornatto	N.D.AES	2.2	IA 2050	Iowa AES	3.3
MK 0649	Richland Organics	2.2	Parker	Minn.AES	3.3

CV (Coefficient of variation) = 31.7; LSD 5% = 1.0.

Chaney et al. (1972) reported that dicotyledonous plants might enhance their capacity for iron uptake, in response to a developing deficiency, by increasing their ability to reduce ferric chelates at the root surface. Plants have evolved two separate mechanisms for the acquisition of iron, which can be referred to as Strategy I and Strategy II.

Plant species exhibiting Strategy I show one or more of the following adaptive components:

i) Iron deficiency-induced enhancement of Fe^{3+} reduction to Fe^{2+} at the root surface, with preferential uptake of Fe^{2+} (Chaney et al., 1972);

ii) H$^+$ extrusion (Römheld and Marschner, 1986) that promotes the reduction Fe^{3+} to Fe^{2+}; and

iii) In certain cases, the release of reducing and/or chelating substances by the roots.

Enhanced Fe^{3+} reduction under iron-deficiency, the most typical feature of Strategy I, is characterized by the activation of reductases located on the plasma membrane and presumably in the cell wall (apoplast) of the apical root zones (Bienfait et al., 1983).

Poor seed yield of soybean in Mediterranean-type environments may result from insufficient iron (Fe) uptake and poor biological nitrogen (N) fixation due to high bicarbonate and pH in soils. In a study conducted to evaluate the effects of N and Fe fertilization on growth and yield of double cropped soybean (cv. SA 88, MG III) in a Mediterranean-type environment in Turkey during 2003 and 2004, the soil of the experimental plots was a Vertisol with 176 g CaCO$_3$ kg^{-1} and pH 7.7 and 17 g organic matter kg^{-1} soil. Soybean seeds were inoculated prior to planting with commercial peat inoculants. N fertilizer rates were 0, 40, 80, and 120 kg N ha^{-1}, of which half was applied before planting and the other half at full blooming stage (R2). Fe fertilizer rates were 0, 200 and 400 g Fe EDTA (5.5% Fe and 2% EDTA) ha^{-1}. It was sprayed as two equal portions at two trifoliate (V2) and at five trifoliate stages (V5). Plants were sampled at flower initiation (R1), at full pod (R4) and at full seed (R6) stages. Application of starter N increased biomass and leaf area index at R1 stage, whereas Fe fertilization did not affect early growth parameters. N application continued to have a positive effect on growth parameters at later stages and on seed yield. Fe fertilization increased growth parameters at R4 and R6 stages, and final seed yield in both years. This study demonstrated an interactive effect of N and Fe fertilization on growth and yield of soybean in the soil having high bicarbonate and pH. There was a positive interaction between N and Fe at the N rates up to 80 kg N ha^{-1}. However, further increase in N rate produced a negative interaction. Fertilization of soybean with 80 kg N ha^{-1} and 400 g Fe ha^{-1} resulted in the highest seed yield in both years. It was concluded that the application of starter and top dressed N in combination with two split FeEDTA fertilization can be beneficial to improve early growth and final yield of inoculated soybean in Mediterranean-type soils (Caliskan et al., 2008).

In an experiment with soygreen (3 lb/acre) on soybean, the findings are:

- In-furrow application of Soygreen at planting time showed a positive yield response across all varieties.
- Although yield differences in some locations were minimal, visual differences were noted, as the varieties treated with an in-furrow application of Soygreen were greener and more robust.
- In several plots, varieties treated with Soygreen in-furrow resulted in higher final populations than varieties not treated with Soygreen in high IDC fields (Fig. 6.10).

Zinc (Zn) deficiency in soils has been reported worldwide, particularly in calcareous soils of arid and semiarid regions. In a global soil survey study, Sillanpää (1990) found that about 50% of the soil samples collected in 25 countries were Zn deficient. Zinc deficiency is a particularly widespread micronutrient deficiency in wheat, leading to severe depressions in wheat production and nutritional quality of grains (Graham et al., 1992; Cakmak et al., 1996d; Graham and Welch, 1996).

Zinc as one of the essential micronutrients in plants is necessary for plant growth and development and is involved in many enzymatic activities and IAA formation to increase flower number and fruit set.

However, excessive Zn in plants can profoundly affect normal ionic homeostatic systems by interfering with the uptake, transport, osmotic and regulation of essential ions and result in the disruption of metabolic processes, such as transpiration, photosynthesis and enzyme activities related to metabolism (Sainju et al., 2003). Zinc plays a fundamental role in several critical functions in the cell, such as protein metabolism, gene expression, structural and functional integrity of bio-membranes and photosynthetic carbon metabolism (Cakmak, 2000). In case of zinc deficiency in plants, terminal leaves are small, bud formation is poor and leaves have dead areas.

Zinc (Zn) distribution and transport in plants is affected by the level of Zn supply and plant species. When plants have low to adequate Zn supply, Zn concentrations are usually higher in growing tissue than in mature tissue; this is true for roots, vegetative shoots and reproductive tissues. In plants tolerant of toxic levels of Zn, accumulation has been observed in the root cortex and in leaves. In these tissues, Zn accumulates in cell walls or is sequestered in vacuoles.

Zinc transport in the xylem does not necessarily coincide with that of water. Zinc is a nutrient with variable mobility that is retranslocated to a greater extent when in adequate supply. Zinc movement out of old leaves coincides with their senescence; both can be delayed by Zn deficiency. Species differ widely in their ability to load Zn into seeds; some native plants adapted to nutrient-poor soils have 10 times greater Zn concentrations in their seeds than most cultivated species, but have similar or lower Zn concentrations in their leaves.

The possible involvement of Zinc-Regulated Transporter, Iron-Regulated Transporter (ZRT-IRT)-like proteins (ZIPs) in cellular Zn^{2+} uptake was established by expressing cDNAs from Zn-deficient plants in a yeast *zrt1zrt2* mutant. Some plant members of the so-called Cation Diffusion Facilitator (CDF) family of metal cation/proton antiporters, members of which have also been named ZAT (Zinc Transporter of Arabidopsis thaliana) and MTP (Metal Tolerance Protein or Metal Transport Protein), act in the removal of Zn from the cytoplasm. The first plant MTP was identified serendipitously upon over expression of a cDNA in Arabidopsis and by the similarity of the encoded protein to human ZnTs (van der Zaal et al., 1999). Reverse genetics have been used extensively in the identification of membrane transporters contributing to Zn homeostasis. Plant members of the Natural Resistance-Associated Macrophage Protein (NRAMP) family acting in proton-driven transition metal cation transport were identified by sequence similarity to NRAMPs of other organisms (Belouchi et al., 1995; Thomine et al., 2000). The YSL (Yellow Stripe-Like) family in Arabidopsis was named based on the sequence similarity of these transporters to the *Zea mays* YS1 (Yellow Stripe 1) transporter (Finally, Arabidopsis HMA1 to 4 proteins (Heavy Metal ATPases of the P_{1B}-type ATPases) share considerable similarity with bacterial divalent transition metal cation pumps (Heterologous screening in yeast led to the identification of PCR (Plant Cadmium Resistance) proteins acting in cellular export of various divalent metal cations. One member of this family of membrane transport proteins, PCR2, was later shown to contribute to Zn homeostasis using reverse genetics. A Major Facilitator Superfamily (MFS) transporter, Zinc-Induced Facilitator 1 (ZIF1), was identified to contribute to basal Zn tolerance in Arabidopsis through a classical forward genetic screen.

Soybean, maize and flax are the most susceptible field crops to Zn deficiency. It is often found on sandy soils low in organic matter, on high soil pH and calcarious soils, as well as on soils rich in available P. Cold and wet weather promote Zn deficiency. N improves, while Fe, and especially P, reduce Zn uptake by plants. The first symptom of Zn deficiency in soybean is usually light green colour developing between the veins on the older leaves. New young leaves will be abnormally small. Bronzing of the older leaves may occur.

When the deficiency is severe, leaves may develop necrotic spots. Shortened internodes will give plants a stunted, rosetted appearance (Dahnke et al., 1992).

In order to study the effect of water deficit stress and zinc foliar application on yield, physiological traits and also on seed vigour and seedling emergence percentage in two soybean cultivars, an experiment was conducted as randomized complete block arrangement in split factorial design with three replications. The main factor was drought stress in three levels of optimal irrigation, withholding irrigation from vegetative growth stage and withholding irrigation from flowering stage; subordinate factors were the combination of foliar zinc application in three levels and two cultivars ('L17' and 'Clark 63'). Water deficit stress obviously decreased the yield, soluble protein and chlorophyll content in leaves. Proline and soluble sugars content were significantly increased in response to stress. Water deficit stress increased antioxidant enzymes activity. Also, water deficit stress decreased the germination rate, radicle and plumule dry weight. Foliar application with zinc sulphate increased the yield, germination rate and percentage, radicle and plumule weight. Zinc prevented the harmful effects of stress, which caused decreasing of leaf protein, chlorophyll content and increasing proline and carbohydrate accumulation. In general, foliar application of zinc decreased the harmful effects of oxidative stress due to water deficit stress and improved growth conditions of plants (Karami et al., 2016).

Rose et al. (1981) studied the responses of four soybean varieties (*Lee, Forrest, Bragg* and *Dodds*) to foliar zinc fertilization ($ZnSO_4.7H_2O$ before flowering) at three sites in central and North-west New South Wales. At Narrabri, one spray of 4 kg/ha gave a yield increase of 13%. At Trangie and Breeza, two spray each of 4 kg/ha increased yield by 57% and 208%, respectively. Lee was the least responsive variety at each site and Dodds and Forrest the most responsive to applied zinc. Zinc fertilizer increased plant height, leaf-Zn, oil contents (at two sites) but decreased leaf-P. Leaf-P in untreated plots was indicative of varietal sensitivity to zinc deficiency both within and between sites. Singh et al. (2001) tested twelve nutrient combinations comprising of three levels each of nitrogen (30, 60 and 90 kg N/ha), phosphorus (40, 60 and 80 kg P_2O_5/ha) two levels of potassium (30 and 60 kg K_2O/ha) and a single level of zinc (25 kg Zn/ha) along with control. Zinc fertilization in combination with N, P and K significantly increased the growth attributes and grain yield of soybean. The highest number of pods per plant and grain yield were obtained with the joint application of N, P, K and Zn at the rates of 90, 80, 60, and 25 kg/ha, respectively (Table 6.11).

Soil acidification is also accentuated by abundant pluviometry during winter, causing the main cations to leak from the soil (Mora et al., 2006). On the other hand, lime application is a key factor in decreasing soluble Mn in acid soils with a high Mn content, given that it can increase soil pH (Hue and Mai, 2002). Rhizosphere, which is the narrow zone of soil immediately surrounding the root system, is of great importance for mineral plant nutrition. In this zone, both the mobilization and immobilization of nutrients occur (Marschner, 1995). As shown, a mobilization of Mn^{2+} is produced by the rhizosphere acidification due to the release of H^+ or low molecular weight organic acids (LMWOA) from plants (Rengel and Marschner, 2005). Organic acids released in anion forms from roots can chelate Mn^{2+} released from the MnOx (Mn oxides) (Rayan et al., 2001). Neumann and Römheld (2001) reported that mobilization of micronutrients (including Mn) into the rhizosphere is due mainly to its acidification and complexation with the organic acids (citrate) in various plant species. It has been reported that organic amendments (chip compost and pine bark) applied to melon plants released organic compounds, such as arabinose and malic acid, that can dissolve MnOx (Tsuji et al., 2006). Soil microorganisms can also help Mn mobilization and immobilization, depending on soil conditions (Marschner, 1995). In aerated soils, microorganisms may mobilize Mn through MnOx reduction favored by H^+ root excretion. In contrast, Mn-oxidizing bacteria can decrease Mn availability in aerated and calcareous soils or in poorly aerated and/or submerged soils. Another key factor in the Mn dynamics in soil is organic matter (OM). Given that OM is negatively charged, it has a great Mn adsorption capacity, forming Mn complexes which decrease the amount of exchangeable Mn. However, the Mn adsorbed by OM can be exchanged by the H+ released from the roots (Bradl, 2004).

Manganese (Mn) is an essential micronutrient in most organisms. In plants, it participates in the structure of photosynthetic proteins and enzymes. Its deficit is dangerous for chloroplasts because it affects the water-splitting system of photosystem II (PSII), which provides the necessary electrons for photosynthesis (Buchanan, 2000). However, its excess seems also to be particularly damaging to the

Table 6.11 Response of soybean to fertilization: Pods/plant (P/P), grain/pod (G/P), 100-grain weight and grain yield.

N (kg.ha⁻¹)	P₂O₅ (kg.ha⁻¹)	K₂O (kg.ha⁻¹)	Zinc (g.ha⁻¹)	Pods per plant	Grain per pod	100-grain wt. (g)	Yield (t.ha⁻¹)
60	80	30	0	51.4	2.03	14.1	1.89
90	40	30	0	51.9	2.12	14.4	1.90
90	60	30	0	65.9	2.14	16.6	2.23
90	80	30	0	47.2	2.07	13.5	1.77
90	80	60	0	45.0	1.84	12.3	1.71
90	80	60	25	68.9	2.14	16.1	2.48
LSD 5%				8.51	ns	2.29	0.34

photosynthetic apparatus (Mukhopadhyay and Sharma, 1991). Thus, Mn has two roles in the plant metabolic processes: As an essential micronutrient and as a toxic element when it is in excess (Kochian et al., 2004; Ducic and Polle, 2005). Mn toxicity is favoured in acid soils (Pendias and Pendias, 1992). With decreasing pH, the amount of exchangeable manganese—mainly Mn^{2+} form—increases in the soil solution. This Mn form is available for plants and can be readily transported into the root cells and translocated to the shoots, where it is finally accumulated (Marschner, 1995). In contrast, other forms of Mn predominate at higher pH values, such as Mn (III) and Mn (IV), which are not available and cannot be accumulated in plants (Rengel, 2000). Excessive Mn concentrations in plant tissues can alter various processes, such as enzyme activity, absorption, translocation and utilization of other mineral elements (Ca, Mg, Fe and P), causing oxidative stress (Ducic and Polle, 2005; Lei et al., 2007). With respect to manganese, deficiency has been reported for soybean and for palm trees (Graham et al., 1995; Kee et al., 1995), and both greenhouse and field studies in Brazil have shown that application of this micronutrient may result in yield increases of soybean, wheat and corn (Soni et al., 1996; Fageria, 2002; Mann et al., 2002). Furthermore, for wheat, the application of manganese not only increased yields but also crop manganese content, however, to some extent, at the expense of the crop's copper and iron content.

The threshold of Mn injury as well as tolerance to an excess of this metal is highly dependent on the plant species and cultivars or genotypes within a species (Foy et al., 1988; Horst, 1988).

Manganese moves freely with the transpiration stream in the xylem sap in which its concentration and ionic form may vary widely. Concentrations of 1–3500 µM have been reported for Mn in the xylem sap of a range of plant species. The concentration of Mn present at any time in a particular species would be expected to vary strongly with environmental factors governing Mn and water absorption.

White et al. (1980) calculated that about 40% of the Mn in the xylem exudate of soybean and tomato was present in complexes with citric and malic acids, leaving 60% as free divalent Mn; they obtained experimental evidence for the weak bonding of Mn to ligands in the exudate and showed that procedures of sap collection and analysis could critically affect ligand concentration and pH (White et al., 1980). As would be expected from its chemical properties (Clarkson and Hanson, 1980), Mn formed less stable complexes with components of the xylem sap than did Cu, Fe, Ni, or Zn.

Within the xylem sap, Mn moves freely from roots to shoots in the transpiration stream. For example, 54 Mn moved from roots to shoots of rice at rates comparable with potassium (Ramani and Kannan, 1987). However, while Mn in the xylem sap generally appears to move passively in the transpiration stream, appreciable quantities may, under some conditions, be absorbed from the sap and stored in stem tissues, from which it may later be released for redistribution.

The movement of Mn in the phloem sap of plants has been studied directly by analysis of phloem sap exudates and indirectly by measurements of transport of 54 Mn applied to leaves, changes in the Mn content of leaves, and accretion of Mn and 54 Mn in parts of root systems with no direct Mn supply. The general picture which is emerging is complex. It suggests that Mn accumulated in leaves cannot be remobilised, while that in roots and stems can, and that Mn is highly mobile in the phloem to seeds but immobile to roots.

Transport and distribution patterns of manganese were studied in soybeans (cultivars Lee and Bragg) grown over a wide range of manganese supply levels in solution culture. Increasing the manganese supply from 1.8 µM to 450 µM produced symptoms of manganese toxicity and reduced growth in both cultivars. Symptoms and dry matter yield reductions were more severe in Bragg. Manganese taken up by the plant was preferentially transported to young expanding tissue, while the amount moving to older tissue, including cotyledons, increased with availability. At high manganese supply, the leaves accumulated higher concentrations with age, resulting in a decreasing concentration gradient between old and young leaves. There was no difference between Lee and Bragg in the distribution of manganese to plant parts or in the concentration of manganese in actively growing tissue. Manganese deficiency symptoms developed on both cultivars grown under limiting supply conditions. Partial mobility of the element was indicated: Small amounts were translocated from the old leaves when young expanding leaves were manganese deficient. Nevertheless, considerable amounts remained in the cotyledons, and manganese appeared to have a low mobility when the plant's supply was limited (Henan and Campbell, 1980).

Manganese deficiency symptoms, which often look like those of iron deficiency, appear as interveinal chlorosis (yellow leaves with green veins) on the young leaves, and sometimes tan, sunken spots that appear in the chlorotic areas between the veins (Fig. 6.12). Plant growth may also be reduced and stunted. In whole plant, Bussler (1958) reported that tissues suffering from Mn deficiency have small cell volume, cell walls dominate and inter epidermal tissue is shrunken.

As an essential micronutrient, low Mn levels are absolutely necessary for normal nutrition and development of plants. Normal Mn contents of leaves differ greatly between species (30–500 mg kg^{-1} Mn dry mass, Clarkson, 1988). Nonetheless, when it is present in excessive amounts, it is extremely toxic to plant cells (Migocka and Klobus, 2007). The injury extent of Mn toxicity is approximately proportionate to the concentration of accumulated Mn excess. However, there is considerable inter- and intra-specific variation among Mn levels that induce toxicity as well as the symptoms of this toxicity in plant species (Foy et al., 1988).

In order to study the effects of iron, zinc and manganese on the yield of soybean, an experiment was conducted in Miandoab agriculture research station. Treatments including iron (0, 25 and 50 kg/ha), zinc (0, 25 and 40 kg/ha) and manganese (0 and 40 kg/ha) were arranged as factorial based on randomized complete block design with three replications. All agronomic operations were carried out based on research recommendations. Results showed that 40 kg/ha zinc and manganese led to the highest seed yield (3397 and 3367 kg/ha, respectively). However, zinc and manganese 40 kg/ha produced the highest biological yield (7447 and 7387 kg/ha, respectively). In general, the greatest grain number and seed weight per plant, pod number, and biological and seed yield of soybean were obtained from 40 kg/ha of zinc and manganese (Ghasemian et al., 2010) (Fig. 6.14).

Most crops are not able to mobilize B from vegetative tissues to actively growing, meristematic plant tissues, such as shoots, root tips, flowers, seeds or fruits. Rather, B transport occurs primarily in the xylem channel, resulting from transpiration. Because of this, deficiency symptoms first develop in newly developed plant tissue, such as young leaves and reproductive structures.

Under severe B deficiency, stunted development and death of meristematic growing points are common. Other common reactions include reduced root elongation, failure of flowers to set seeds and fruit abortion. Low B supply may also adversely affect pollination and seed set, without visible leaf deficiency symptoms.

Boron accelerates nitrogenase activity through effective nodule development for nitrogen fixation. Plant reproductive growth is halted with the deficiency of boron. This retarding growth is considered due to the low phloem mobility of boron. In brief, the formation of B complexes with the constituents of cell wall and plasma membrane as well as with the phenolic compounds is a major reason to affect the physiological functions of boron.

Pollard et al. (1977) reported several impairments as a result of B deficiency, such as sugar transport, cell wall synthesis, signification, cell wall structure, carbohydrate metabolism, RNA metabolism, respiration, indole acetic acid (IAA) metabolism, phenol metabolism, and membrane integrity. Symptoms include dying growing tips and bushy stunted growth, extreme cases may prevent fruit set.

Soybean is generally the most tolerant, since it was not affected by boron deficiency at the same level which reduced seed yield of green gram and black gram, and caused a significant percentage of hollow heart in peanut. However, some soybean genotypes, e.g., NW1, are more sensitive to boron deficiency than others (Rerkasem, 1989).

(Cu) is an essential redox-active transition metal involved in many physiological processes in plants because it can exist in multiple oxidation states *in vivo*. Under physiological conditions Cu exists as Cu^{2+} and Cu^+. Cu acts as a structural element in regulatory proteins and participates in photosynthetic electron transport, mitochondrial respiration, oxidative stress responses, cell wall metabolism and hormone signalling (for a review, see Marschner, 1995; Raven et al., 1999). Cu ions act as cofactors in many enzymes, such as Cu/Zn superoxide dismutase (SOD), cytochrome c oxidase, amino oxidase, laccase, plastocyanin and polyphenol oxidase. At the cellular level, Cu also plays an essential role in signalling of transcription and protein trafficking machinery, oxidative phosphorylation and Copper ion mobilization. Thus, plants require Cu as an essential micronutrient for normal growth and development; when this ion is not available, plants develop specific deficiency symptoms, most of which affect young leaves and reproductive organs. The redox properties that make Cu an essential element also contribute to its inherent toxicity. Redox cycling between Cu^{2+} and Cu^+ can catalyze the production of highly toxic hydroxyl radicals, with subsequent damage to DNA, lipids, proteins and other biomolecules (Halliwell and Gutteridge, 1984). Thus, at high concentrations, Cu can become extremely toxic, causing symptoms such as chlorosis and necrosis, stunting, leaf discoloration and inhibition of root growth (van Assche and Clijsters, 1990; Marschner, 1995). At the cellular level, toxicity may result from (i) binding to sulfhydryl groups in proteins, thereby inhibiting enzyme activity or protein function; (ii) induction of a deficiency of other essential ions; (iii) impaired cell transport processes; (iv) oxidative damage (van Assche and Clijsters, 1990; Meharg, 1994).

Copper is important for pollen formation, viability and fertilization. A shortage can, therefore, reduce the number of grains per ear, even when physical deficiency symptoms are not visible (a slight deficiency can reduce yields by as much as 20%).

A field study was conducted in order to determine uptake and distribution of Cu and Zn by soybean [*Glycine max* (L.) Merr.] and corn (*Zea mays* L.) grown on Enon sandy loam (fine, mixed, thermic Ultic Hapludalfs), treated with 0, 25, 50, and 100 Mgha⁻¹ of sewage sludge each year. 'Ransom' soybeans were grown the first year and 'FCX' corn was grown to maturity in the same plots during the second year. In general, sludge significantly increased grain yield of soybeans and corn. Copper concentration in soybean seed was higher than in leaf and stem, but Zn concentration was lower in the seed than in the other tissues, under sludge treatment. Copper in corn leaf increased more than in stem and grain as the rate of sludge addition. Zinc increased in corn leaf and stem as the rate of sludge increased, but Zn in grain was not affected.

Copper (Cu) is seldom deficient in soil. Only on soils high in organic matter and under conditions pH above 6.0 is Cu likely be deficient. The color of legume and forage plants deficient in Cu tends to be grayish-green, blue-green or olive green. The internodes become shortened and produce a bushy type of plant (Sauchelli, 1969). Soybean has low requirements for Cu.

Introduction of high-yielding varieties and higher use of nitrogen (N), phosphorus (P), and potassium (K), however, increased crop production several fold after the green revolution but this has led to micronutrient deficiency in most of the Indian soils (Singh, 2001; Sahrawat et al., 2010). Copper and Mo are likely to become critical in the future for sustaining high productivity in certain areas of India (Singh, 2004).

Indian soils are low in total Mo content, i.e., traces to 12 mgkg⁻¹ (Sakal, 2001), and about 11% of soils in India are deficient in available Mo (Singh, 2001). Mo is one of the important micronutrients that helps in biological N fixation. Mo is a component of the enzymes nitrogenase and nitrate reductase, which is required in N fixation and also plays an important role in P utilization and protein synthesis (Jones, 1987). The critical limit is the threshold level of a given nutrient in the soil which helps separate deficient soils from non-deficient ones.

The critical limit of Mo for different crops has been established by many studies. Available Mo concentration in soils of Gujarat range between 0.06 to 0.23 mg kg⁻¹ and 1–26% of land area is found deficient; available Mo was 0.05 mg kg⁻¹ for Jhargram soils in West Bengal.

Molybdenum is involved in several enzyme systems including nitrate reductase, xanthine oxidase, aldehyde oxidase and sulphate oxidase (Nicholas, 1975). Molybdenum is also required in the synthesis of ascorbic acid and is implicated in making iron (Fe) physiologically available in plants.

Since Mo is closely involved in the N metabolism of plants, its deficiency resembles N deficiency. Molybdenum-deficient plants grow poorly, i.e., their leaves become pale and wither.

Several research studies confirm that the presence of large quantities of Mo in plants does not produce harmful effects or cause significant reductions in the crops. Animals, however, are sensitive to a high-Mo forage ration, which could lead to Mo-induced Cu deficiency.

Conversely, plants are more sensitive to molybdenum deficiency than to toxicity. Deficiency symptoms for most micronutrients appear on the young leaves at the top of the plant because most micronutrients are not easily translocated. Mo is an exception that is easily translocated and its deficiency symptoms appear on the entire plant, but the symptoms normally show themselves as a nitrogen shortage, especially seen in legumes. These symptoms are connected to the function of molybdenum in nitrogen metabolism, for example, its part on N_2 fixation and nitrate reduction. However, plants suffering from great deficiency often show symptoms that are related only to molybdenum (Anke and Seifert, 2007). For instance, molybdenum deficient leaf edges tend to roll up. Marginal chlorosis of older leaves may happen. In many occasions, necrosis follows and the whole plant is stunted. Young leaves become mottled and their leaf edges are very narrow. In addition, molybdenum deficiency reduces photosynthesis, lower sugar and vitamin C contents. Vitamin C content of leaves can decrease by 25% compare to the normal level. Affected plants gather nitrate in their leaves. Protein content of the plants also decrease (Marschner, 1995). Mo deficiency also has great effects on the growth of plants (Gupta and Lipsett, 1981; Kaiser et al., 2005). Agarwala et al. (1978) saw that, in Mo-deficient plants, low growth and a reduction in leaf size are very common and, in dicotyledonous species, much more reduction in size can be observed. Some literature established that Mo application can affect growth stimulation on acidic mineral soils (Ndakidemi, 2005; Bambara and Ndakidemi, 2010).

Prior to 1990, little emphasis was placed on the importance of Mo in soybean production in South Africa. During the 1990's, however, yield responses of up to 170 per cent, due to Mo seed treatment, were recorded with soybean on acid, heavy Hutton soils in KwaZulu-Natal, highlighting the crucial role of this element in ensuring maximum soybean yields.

Rhodes and Nangju (1979) conducted two field experiments in order to evaluate the effectiveness of several pelleting materials in increasing the yields of cowpea and soybean on an acid soil in Sierra Leone. Applied either alone or in combination with phosphate rock, Mo increased the growth and yield of cowpeas but had no effect on soybean growth and yield, although it significantly increased the number of nodules per plant.

Haque and Bundu (1980) observed that inoculation and Mo application (0.4 g sodium molybdate/100 g seed) increased soybean grain yield and protein content by 154% and 14.6%, respectively, over the control on an upland soil in Sierra Leone.

Rhodes and Kpaka (1982) found that applying Mo increased DM, pod weight and seed yield of cowpeas. Pelleting seed with nitromolybdenum at 0.4 g/100 g seed increased seed yield by 1.39 t/ha, or 21%, over the control. This method of applying Mo should prove attractive to small farmers because it is simple, cheap, does not require spraying equipment and is less subject to the vagaries of rain and wind.

Nickel (Ni) is an essential micronutrient required for plants' metabolism due to its role as a structural component of urease and hydrogenase, which, in turn, perform nitrogen (N) metabolism in many legume species. Seed treatment with cobalt, molybdenum and *Bradyrhizobium* strains has been widely practiced in order to improve crops. Additionally, seed treatment together with Ni fertilization of soybean might improve the efficiency of biological nitrogen fixation (BNF), boosting grain dry matter yield, and N content. So, this study was to evaluate the effect of soybean seed treatment with Ni rates (0, 45, 90, 135, 180, 360, and 540 mg kg^{-1}) on BNF, directly by the ^{15}N natural abundance method ($\delta^{15}N‰$) and by measurement of urease [E.C. 3.5.1.5] activity, as well as indirectly by nitrogenase (N-ase) activity [E.C. 1.18.6.1]. Soybean plants (cultivar *BMX Potência RR*) were grown in a sandy soil up to the R7 developmental stage (grain maturity), at which point, the nutrient content in the leaves, chlorophyll content, urease, and N-ase activities, Ni and N content in the grains, nodulation (at R1—flowering stage), as well as the contribution of biological nitrogen fixation (BNF; $\delta^{15}N‰$) were evaluated. The proportion of N derived from N_2 fixation varied from 77 to 99% using the natural ^{15}N abundance method and non-nodulating *Panicum miliaceum* and *Phalaris canariensis* as references. A Ni rate of 45 mg kg^{-1} increased BNF by 12% compared to the control. The increased N uptake in the grains was closely correlated with chlorophyll content in the leaves, urease, and N-ase activities, as well as with nodulation. Grain dry matter yield and aerial part dry matter yield increased, respectively, by 84 and 51% in relation to the control plants at 45 mg kg^{-1} Ni via seed treatment. Despite Ni concentration being increased with Ni-seed treatment, Ni rates higher than 135 mg kg^{-1} promoted negative effects on plant growth and yield. In these experimental conditions, seed treatment with low Ni rates caused higher dry matter yield of plants and grains, N content in the grains, and in the aerial part by increasing of BNF (Lavres et al., 2016).

6.6 Beneficial Elements

Two field experiments were carried out in order to study the effect of cobalt on soybean growth and productivity. Experiments were conducted at Research and production Station, National Research Centre, El-Nobaria, Beheara Governorate, Delta Egypt under drip irrigation system during two the successive seasons of 2010 and 2011. Cobalt added once, i.e., 0, 4, 8, 12, 16 and 20 ppm.

The obtained results indicate that: Cobalt concentrations significantly improve all growth, seeds and oil yield quantity and quality compared with control. Cobalt at 12 ppm gave the highest figures of all nodulation, growth and yield parameters as well as minerals composition and chemical constituents. Increasing cobalt in plant media reduced iron content; there are antagonistic relationships between two elements (Fe and Co). Increasing cobalt concentration in plant media more than 12 ppm gave the adverse promotive effect.

Cobalt concentrations significantly improve all growth, seeds and oil yield quantity and quality compared with control.

The beneficial effect of Co for legumes carrying out dinitrogen fixation was established in the early 1950's (Ahmed and Evans, 1960; Delwiche et al., 1961). Shortly thereafter, it was determined that this effect of Co was related to an absolute requirement for this element by the symbiotic microorganism, whether or not it was engaged in symbiotic dinitrogen fixation (Dilworth et al., 1979). However, the demand for Co may be greater during dinitrogen fixation than when nitrogen is supplied to the microorganism Asu*11Jniu1n (Kliewer and Evans, 1963). The rationale for the Co requirement by dinitrogen-fixing organisms is based upon its function in the vitamin B, coenzyme'cobalairin, which is involved in several important biochemical Processes in the dinitrogen-fixing microorganisms (Dilworth et al., 1979). In soils deficient in Co, the extent of Rhizobium infection of legumes is greatly reduced and the development of symbiotic dinitrogen-fixing capacity may be delayed several weeks (Marschner, 1986 and references therein). Enhancing the extent of injection under these conditions can occur if Co is exogenously supplied. It is still unclear whether Co has any direct functions in higher plants. Clearly the effects of this nutrient in higher plants which do not engage in symbiotic dinitrogen fixation are minor (Marschner, 1986 and references therein). For legumes, the proposal that Co is only required by the dinitrogen-fixing microorganism is supported by the early observation that Co increased growth and nitrogen content of legume shoots, but was without effect when mineral nitrogen was supplied (Ozanne et al., 1963; Powrie, 1964).

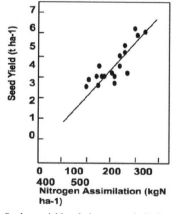

Fig. 6.1 Soybean yield and nitrogen assimilation relation.

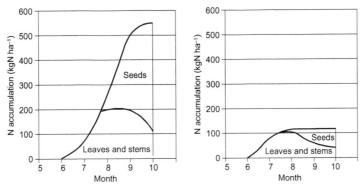

Fig. 6.2 Comparison of the nitrogen assimilation and distribution pattern of soybean and rice.

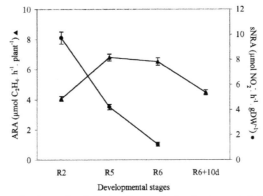

Fig. 6.3 Nitrogen fixation and assimilation changes during developmental stages in soybean.

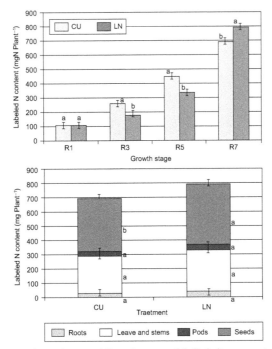

Fig. 6.4 Influence of coated urea and lime nitrogen on labelled nitrogen content in soybean.

Fig. 6.5 Potassium deficiency in soybean.

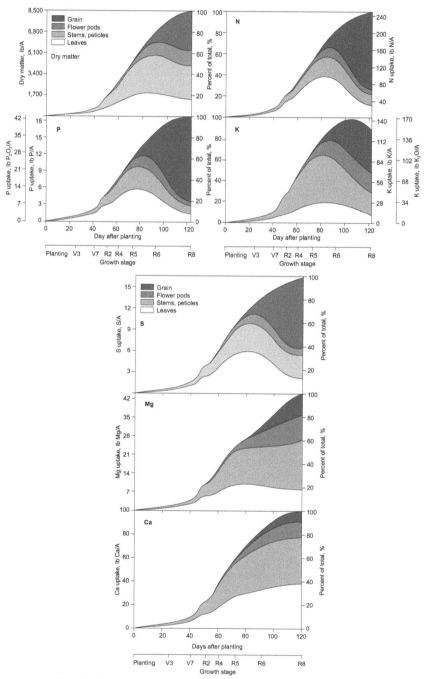

Fig. 6.6 Dynamics of macronutrient uptake and harvest index in soybean.

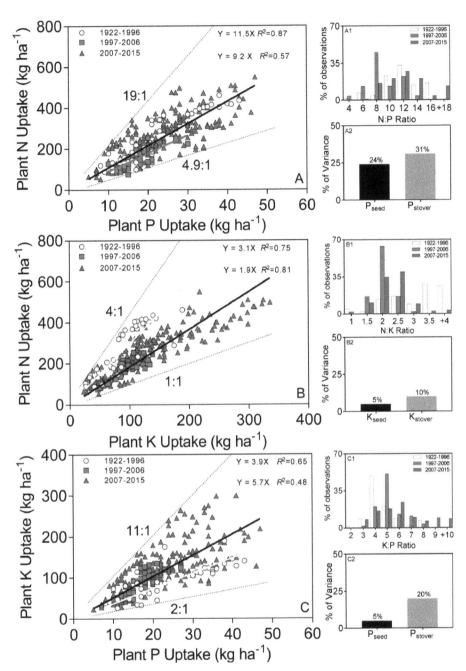

Fig. 6.7 (A) Plant N uptake as a function of plant P uptake, (B) plant N uptake as a function of K uptake, and (C) plant K uptake as a function of plant P uptake. N/P, N/K, and P/K ratio by era (A1, B1, and C1, respectively). Percentage of variance ($R2$) explained by the linear regression between (A2) residuals of Fig. A as a function of seed and stover P concentration, (B2) residuals of Fig. B as a function of seed and stover K concentration, and (C2) residuals of Fig. C as a function of seed and stover P concentration. Dotted lines indicate boundaries for maximum and minimum ratio for each dataset. Due to lack of significant difference ($p > 0.05$) in slopes, Eras I and II where pooled in Fig. A (black line); Eras II and III were pooled in Fig. B and C (black line). Linear function for pooled dataset: (A) $Y = 11.5X$ ($R2 = 0.87$), (B) $Y = 1.9X$ ($R2 = 0.81$), (C) $Y = 5.7X$ ($R2 = 0.48$).

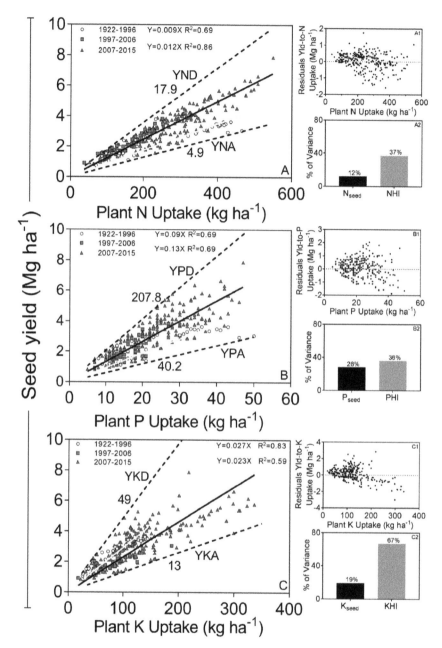

Fig. 6.8 Seed yield as a function of (A) plant N uptake, (B) plant P uptake, and (C) plant K uptake. (A1), (B1), and (C1) show residuals for the fitted functions in (A), (B), and (C), respectively. (A2) Proportion of variation ($R2$) provided by the linear regression between residuals of Fig. 6.8A as a function of seed N concentration and N harvest index (NHI); (B2) Residuals of Fig. 6.8B as a function of seed P concentration and P harvest index (PHI); and (A3) residuals of Fig. 6.8C as a function of seed K concentration and K harvest index (KHI). YNA, yield N accumulation; YND, yield N dilution; YKA, yield K accumulation; YKD, yield K dilution; YPA, yield P accumulation; YPD, yield P dilution. Due to lack of significant difference ($p > 0.05$) in slopes, Eras I and II were pooled in Figs. 6.7A to 6.7C (black line) Linear function for pooled dataset (A) $Y = 0.012X$ ($R2 = 0.86$), (B) $Y = 0.13X$ ($R2 = 0.69$), (C) $Y = 0.023X$ ($R2 = 0.59$).

Fig. 6.9 Iron deficiency in soybean.

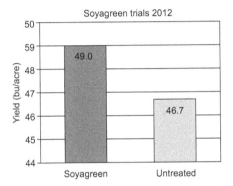

Fig. 6.10 Effects of soygreen on soybean yield.

Fig. 6.11 Zn deficiency in soybean.

Fig. 6.12 Manganese deficiency in soybean.

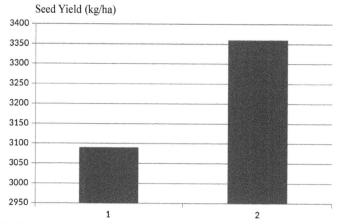

Fig. 6.13 Mn application (1 = 0 g/ha, 2 = 40 g/ha) effects on seed yield of soybean.

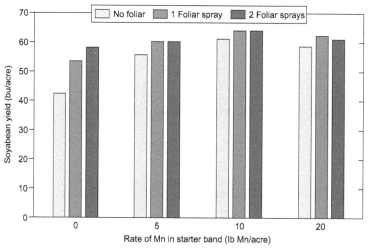

Fig. 6.14 Effects of Mn starter band on yield of soybean.

Fig. 6.15 Beneficial effect of Ni on soybean growth.

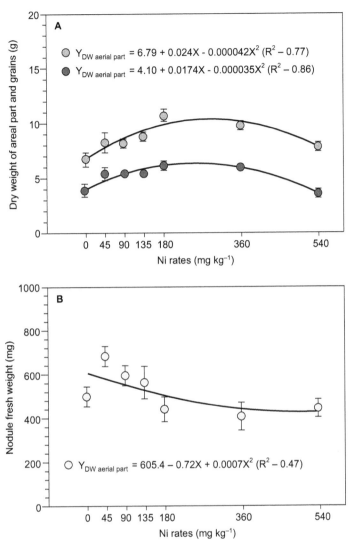

Fig. 6.16 Effect of Ni on yield (A) and nodule fresh weight (B).

References

Afza, R., G. Hardarson, F. Zapata and S.K.A. Danso. 1987. Effects of delayed soil and foliar N fertilization on yield and N₂ fixation of soybean. Plant Soil 97: 361–368.

Agarwala, S.C., C.P. Sharma, S. Farooq and C. Chatterjee. 1978. Effects of molybdenum deficiency on the growth and metabolism of corn plants raised in sand culture. Can. J. Bot. 56: 1905–1909.

Allen, L.H., K.J. Boote, J.W. Jones, P.H. Jones, R.R. Valle, B. Acock, H.H. Rogers and R.C. Dahlman. 2012. Responses of vegetation to rising carbon dioxide: Photosynthesis, biomass, and seed yield of soybean. Global Biogeochemical Cycles. [Online]. doi:10.1029/GB001i001p00001.

Allos, H.F. and W.V. Bartholomew. 1959. Replacement of symbiotic fixation by available nitrogen. Soil Science 87: 61–66.

Andrews, R. 2015. Phytates and phytic acid. [Online]. http://www.precisionnutrition.com/all-about-phytates-phytic-acid.

Anke, M. and M. Seifert. 2007. The biological and toxicological importance of molybdenum in the environment and in the nutrition of plants, animals and man. Acta Biologia Hungarica 58: 311–324.

Atkins, C.A. 1986. The legume: Rhizobium symbiosis. Limitations to maximizing nitrogen fixation. Outlook Agric. 15: 128–134.

Bagci, S.A., H. Ekiz, A. Yilmaz and I. Cakmak. 2007. Effects of zinc deficiency and drought on grain yield of field-grown wheat cultivars in Central Anatolia. J. Agron. Crop Sci. 193: 198–206.

Balboa, G.R., V.O. Sadras and I.A. Ciampitti. 2018. Shifts in soybean yield, nutrient uptake, and nutrient stoichiometry: A historical synthesis–analysis. Crop Sci. 58: 43–54.

Bambara, S. and P.A. Ndakidemi. 2010. The potential roles of lime and molybdenum on the growth, nitrogen fixation and assimilation of metabolites in nodulated legume: A special reference to Phaseolus vulgaris L. Afr. J. Biotechnol. 8: 2482–2489.

Barron, A.R., N. Wurzburger, J.P. Bellenger, S.J. Wright, A.M.L. Kraepiel and L.O. Hedin. 2009. Molybdenum limitation of asymbiotic nitrogen fixation in tropical forest soils. Nat Geosci. 2: 42–45. doi:10.1038/ngeo366.

Belouchi, A., M. Cellier, T. Kwan, H.S. Saini, G. Leroux and P. Gros. 1995. The macrophage-specific membrane protein NRAMP controlling natural resistance to infections in mice has homologues expressed in the root system of plants. Plant Mol. Biol. 29: 1181–1196.

Bender, R.R., J.W. Haegele, M.L. Ruffo and F.E. Below. 2013. Nutrient uptake, partitioning, and remobilization in modern, transgenic insect-protected maize hybrids. Agron. J. 105: 161–170.

Bender, R.R., J.W. Haegele and F.E. Below. 2015. Nutrient uptake, partitioning, and remobilization in modern soybean varieties. Agronomy Journal 107: 563–573.

Bergersen, F.J., G.L. Tuner, R.R. Gault, D.L. Chase and J. Brockwell. 1985. The natural abundance of 15N in an irrigated soybean crop and its use for calculation of nitrogen fixation. Aust. J. Agric. Res. 36: 411–423.

Bharati, M.P., D.K. Whigham and R.D. Voss. 1986. Soybean Response to Tillage and Nitrogen, Phosphorus, and Potassium Fertilization1. Agron. J. 78: 947–950. doi:10.2134/agronj1986.00021962007800060002x.

Bienfait, H.F., R.J. Bino, A.M. van der Bliek, J.F. Duivenvoorden and J.F. Fontaine. 1983. Characterization of ferric reducing activity in roots of Fe-deficient *Phaseolus vulgaris*. Physiol. Plant. 59: 196–202.

Bradl, H. 2004. Adsorption of heavy metal ions on soils and soils constituents. J. Colloid. Interf. Sci. 277: 1–18.

Brevedan, R.E., D.B. Egli and J.E. Leggett. 1978. Influence of N nutrition on flower and pod abortion and yield of soybeans. Agronomy Journal 70: 81–84. doi:10.2134/agronj1978.00021962007000010009x.

Broadley, M., P. Brown, I. Cakmak, Z. Rengel and F. Zhao. 2012. Functions of nutrient: Micronutrients. pp. 243–248. *In*: Marschner, P. (ed.). Marschner's mineral nutrition of higher plants, 3rd edn. Elsevier, Oxford.

Bruns, H.A. and H.K. Abbas. 2010. Additional potassium did not decrease aflatoxin or fumonisin nor increase corn yields. Crop Management, 9 [Online]. doi:10.1094/CM-2010-0216-01-RS.

Buchanan, B.B., W. Gruissem and R.L. Jones. 2000. Biochemistry and Molecular Biology of Plants. Rockville, MD: American Society of Plant physiologists.

Burton, J.W. 1987. Quantitative genetics: Results relevant to soybean breeding. pp. 211–247. *In*: Wilcox, J.R. (ed.). Soybeans: Improvement, Production and Uses. 2nd edition, Agron. Monogr., 16. ASA, CSSA and SSSA, Madison, Wiscon-sin, USA.

Bussler, W. 1958. Manganvergiftung bei höheren Pflanzen. Z. Pflanzenernaehr. Bodenkd. 85: 256–265.

Cakmak, I., A. Yilmaz, M. Kalayci, H. Ekiz, B. Torun, B. Erenoglu and H.J. Braun. 1996d. Zinc deficiency as a critical problem in wheat production in Central Anatolia. Plant Soil. 180: 165–172.

Cakmak, I. 2000. Tansley review no. 111—possible roles of zinc in protecting plant cells from damage by reactive oxygen species. New Phytol. 146: 185–205. doi:10.1046/j.1469-8137.2000.00630.

Cakmak, I. 2002. Plant nutrition research priorities to meet human needs for food in sustainable ways. Plant and Soil 247: 3–24.

Caliskan, S., I. Ozkaya, M.E. Caliskan and M. Arslan. 2008. The effects of nitrogen and iron fertilization on growth, yield and fertilizer use efficiency of soybean in a Mediterranean-type soil. Field Crops Res. 108: 126–132.

Carroll, B.J. and P.M. Gresshoff. 1986. Isolation and initial characterization of constitutive nitrate reductase-deficient mutants nr328 and nr345 of soybean (*Glycine max*). Plant Physiol. 81: 572–576.

Carroll, B.J. and A. Mathews. 1990. Nitrate inhibition of nodulation in legumes. pp. 159–180. *In*: Gresshoff, P.M. (ed.). Molecular Biology of Symbiotic Nitrogen Fixation. CRC Press, Boca Raton, FL.

Chaney, R.L., J.C. Brown and L. Tiffin. 1972. Obligatory reduction of ferric chelates in iron uptake by soybeans. Plant Physiol. 50: 208–213.

Chatterjee, C., R. Gopal and B.K. Dube. 2006. Impact of iron stress on biomass, yield, metabolism and quality of potato (*Solanum tuberosum* L.). Sci. Hort. 108: 1–6.

Clarkson, D.T. and J.B. Hanson. 1980. The mineral nutrition of higher plants. Ann. Rev. Plant. Physiol. 31: 239–298.

Curie, C., G. Cassin, D. Couch, F. Divol, K. Higuchi, M. Le Jean, J. Misson, A. Schikora, P. Czernic and S. Mari. 2009. Metal movement within the plant: Contribution of nicotianamine and yellow stripe 1-like transporters. Ann. Bot. 103: 1–11.

Dahnke, W.C., C. Fanning and A. Cattanach. 1992. Fertilizing soybean, Available from: www.ag.ndsu.edu/pubs/plantsci/soilfert/sf719w.htm (10. 03. 2011).

Datnoff, L.E., W.H. Elmer and D.M. Huber. 2007. Mineral nutrition and plant disease. The American Phytopathological Society, St. Paul.

Delwiche, C.C., C.M. Johnson and H.M. Reisenauer. 1961. Influence of cobalt on nitrogen fixation by Medicago. Plant Physiol. 36: 73–78.

Deng, M.D., T. Moureaux and M. Caboche. 1989. Tungstate, a molybdate analog inactivating nitrate reductase, deregulates the expression of the nitrate reductase structural gene. Plant Physiol. 91: 304–309.

Dilworth, M.J., A.D. Robson and D.L. Chatel. 1979. Cobalt and nitrogen fixation in Lupinus angustifolius L. II. Nodule formation and functions. New Phytologist 83: 63–79.

Drew, M.C. 1975. Comparison of the effects of a localized supply of phosphate, nitrate, ammonium and potassium on the growth of the seminal root system, and the shoot, in barley. New Phytol. 75: 479–490.

Ducic, T. and A. Polle. 2005. Transport and detoxification of manganese and copper in plants. Braz. J. Plant Physiol. 17: 103–112.

Ericsson, T. 1995. Growth and shoot:root ratio of seedlings in relation to nutrient availability. Plant Soil 168–169: 205–214.

Fageria, N.K. 2002. Influence of micronutrients on dry matter yield and interaction with other nutrients in annual crops. Pesq. Agropec. Bras. 37: 1765–1772.

Forde, B.G. and D.T. Clarkson. 1999. Nitrate and ammonium nutrition of plants: Physiological and molecular perspectives. Advances in Botanical Research 30: 1–90.

Foy, C., B. Scott and J. Fisher. 1988. Genetic differences in plant tolerance to manganese toxicity. pp. 293–307. *In*: Graham, R.D., R.J. Hannam and N.J. Uren (eds.). Manganese in Soil and Plants. Kluwer Academic Publishers, Dordrecht, The Netherlands.

Ghasemian, V., A. Ghalavand, A. Soroosh Zadeh and A. Pirzad. 2010. The effect of iron, zinc and manganese on quality and quantity of soybean seed. J. Phytol. 2(11): 73–79.

Gill, M.S., V.K. Singh and A.A. Shukla, 2008. Potassium Management for Enhancing Productivity in Soybean-based Cropping System. Indian J. of Fertilizers 4: 13–22.

Golldack, D., C. Li, H. Mohan and N. Probst. 2014. Tolerance to drought and salt stress in plants: Unraveling the signaling networks. Front. Plant Sci. 5: 151.

Goos, R.J. and B.E. Johnson. 2000. A comparison of three methods for reducing iron-deficiency chlorosis in soybean. Agron. J. 92: 1135–1139.

Graham, R.D., J.S. Ascher and S.C. Hynes. 1992. Selecting zinc efficient cereal genotypes for soils of low zinc status. Plant Soil 146: 241–250.

Graham, M.J., C.D. Nickell and R.G. Hoeft. 1995. Inheritance of tolerance to manganese deficiency in soybean. Crop Science 35: 1007–1010.

Graham, R.D. and R.M. Welch. 1996. Breeding for staple food crops with high micronutrient density. Working Papers on Agricultural Strategies for Micronutrients, No. 3. International Food Policy Research Institute, Washington, D.C.

Gray, C., H.D. Pennington and J. Matocha. 1982. Identifying and correcting iron deficiency in field crops. Texas Agric. Ext. Ser. L-723.

Gupta, U.C. and J. Lipsett. 1981. Molybdenum in soils, plants, and animals. Advances in Agronomy 34: 73–115.

Halliwell, B. and J.M.C. Gutteridge. 1984. Oxygen toxicity, oxygen radicals, transition metals and disease. Biochem. J. 219: 1–14.

Hammond, L.C., C.A. Black and A.G. Norman. 1951. Nutrient uptake by soybeans on two Iowa soils. Iowa Agricultural Experiment Station Research Bulletin 384: 463–512.

Hansen, N.C., M.A. Schmitt, I.E. Anderson and J.S. Strock. 2003. Iron deficiency of soy bean in the Upper Midwest and associated soil properties. Agron. J. 95: 1595–1601.

Hansen, N.C., V.D. Jolley, S.L. Naeve and R.J. Goos. 2004. Iron deficiency of soybean in the North central US and associated soil properties. Soil Sci. and Plant Nutr. 50: 983–987.

Hanway, J.J. and C.R. Weber. 1971. Dry matter accumulation in soybean (*Glycine max* (L) Merrill) plants as influenced by N, P, and K fertilization. Agron. J. 63: 263–266.

Haque, I. and H.S. Bundu. 1980. Effects of inoculation, N, Mo and mulch on soybean in Sierra Leone. Communications in Soil Science and Plant Analysis 11: 477–483.

Harper, J.E. 1974. Soil and symbiotic nitrogen requirements for optimum soybean production. Crop Sci. 14: 255–260.

Harper, J.E. and A.H. Gipson. 1984. Differential nodulation tolerance to nitrate among legume species. Crop Science 24: 797–801.

Harper, J.E. 1987. Nitrogen metabolism. pp. 497–533. *In*: Soybeans: Improvement, Production and Uses. 2nd ed. Agronomy Monograph no. 16. ASA-CSSA-SSSA.

Havlin, J. 1999. Soil fertility and fertilizers: An introduction to nutrient management. 6th ed. Upper Saddle River, N.J., Prentice Hall.

Henan, D.P. and L.G. Campbell. 1980. Transport and distribution of manganese in two cultivars of soybean (*Glycine max* Merr.). Aust. J. Agric. Cul. Res. 3: 943.

Herridge, D.F. and F.J. Bergersen. 1988. Symbiotic nitrogen fixation. pp. 46–65. *In*: Wilson, J.R. (ed.). Advances in Nitrogen Cycling in Agricultural Ecosystems. Wallingford: CAB International.

Horst, W.J. 1988. The physiology of Mn toxicity. pp. 175–188. *In*: Webb, M.J., R.O. Nable, R.D. Graham and R.J. Hannam (eds.). Manganese in Soil and Plants. Kluwer Academic Publishers, Dodrecht, The Netherlands.

Hoshi, S. 1982. Nitrogen fixation, growth and yield of soybean. pp. 5–33. *In*: Nitrogen Fixation in Root, Edited by: Japanese Society of Soil Science and Plant Nutrition. Tokyo, Japan, Hakuyusha Publishers.

Hossain, B., N. Hirata, Y. Nagatomo, R. Akashi and H. Takaki. 1997. Internal zinc accumulation is correlated with increased growth in rice suspension culture. J. Plant Growth Reg. 16: 239–243. doi: 10.1007/PL00007003.

Hrustic, M., M. Vidic and D.J. Jockovic. 1998. Soja/Soybean. Institut za ratarstvo I povrtarstvo Novi Sad, Soja protein d.d. Becej, Novi Sad – Becej, Serbia.

Hue, N. and Y. Mai. 2002. Manganese toxicity in watermelon as affected by lime and compost amended to a Hawaiian acid. Oxisol. Hort. Sci. 37: 656–661.

Imsande, J. 1992. Agronomic characteristics that identify high yield, high protein soybean genotypes. Agron. J. 84: 409–414.

Imsande, J. 1998. Nitrogen deficit during soybean pod filling and increased plant biomass by vigorous N2 fixation. Eur. J. Agron. 8: 1–11.

Jacobsen, E. 1984. A new pea mutant nodulating effectively in the presence of nitrate. *In*: Veeger, C. and W.E. Newton (eds.). Advances in Nitrogen Fixation Research. p 597. Martinus Nijhoff/Dr. W. Junk Publ., The Hague.

Jeffers, D.L., A.F. Schmitthenner and M.E. Kroetz. 1982. Potassium fertilization effects on phomopsis seed infection, seed quality, and yield of soybeans1. Agron. J. 74: 886–890.

Jones, V.S. 1987. Fertilizers and Soil Fertility, New Delhi: Prentice Hall of India.

Jones, G.D., J.A. Lutz Jr. and T.J. Smith. 1977. Effects of phosphorus and potassium on soybean nodules and seed yield. Agronomy Journal 69: 1003–1006.

Kaiser, B.N., K.L. Gridley, J.N. Brady, T. Phillips and S.D. Tyerman. 2005. The role of molybdenum in agricultural plant production. Annals of Botany 96: 745–754.

Kamprath, E.J. and U.S. Jones. 1986. Plant response to sulfur in the south-eastern United States. pp. 323–43. *In*: Tabatabia, M.A. (ed.). Sulfur in Agriculture. Madison, WI: Agronomy Monograph. 27, ASA, CSSA, and SSSA.

Karami, S., S.A.M. Modarres Sanavy, S. Ghanehpoor and H. Keshavarz. 2016. Effect of foliar zinc application on yield, physiological traits and seed vigor of two soybean cultivars under water deficit. Not. Sci. Biol. 8: 181–191.

Karlen, D.L., P.G. Hunt and T.A. Matheny. 1982. Accumulation and distribution of K, Ca, and Mg by selected determinate soybean cultivars grown with and without irrigation. Agronomy Journal 74: 347–354.

Kee, K.K., K.J. Goh and P.S. Chew. 1995. Investigation into manganese deficiency in mature oil palms (E-Guinensis) in Malaysia. Fertilizer Research 40: 1–6.

Kliewer, M. and H.J. Evans. 1963. Cobamide coenzyme contents of soybean nodules and nitrogen-fixing bacteria in relation to physiological conditions. Plant Physiol. 38: 99–104.

Kochian, L.V., O.A. Hoekenga and M.A. Piñeros. 2004. How do crop plants tolerate acid soils? Mechanisms of aluminum tolerance and Phosphorous efficiency. Annu. Rev. Plant Biol. 55: 459–93.

Konno, S. 1976. Physiological study on the mechanism of seed production of soybean. Bul. Nat. Inst. Agric. Sci. D27: 139–295.

Kosslak, R.M. and B.B. Bohlool. 1984. Suppression of nodule development of one side of a split-root system of soybeans caused by prior inoculation of the other side. Plant Physiol. 75: 125–130.

Lathwell, D.J. and C.E. Evans. 1951. Nitrogen uptake from solution by soybeans at successive stages of growth. Agronomy Journal 43: 264–270.

Lavres, J., G.C. Franco and G.M.S. Câmara. 2016. Soybean seed treatment with nickel improves biological nitrogen fixation and urease activity. Front. Environ. Sci. 4: 37.

Lei, X.G., J.M. Porres, E.J. Mullaney and H. Brinch-Pedersen. 2007. Phytase source, structure and applications. pp. 505–529. *In*: Polaina, J. and A.P. MacCabe (eds.). Industrial Enzymes. Structure, Function and Applications. Springer, ISBN-78-1-4020-5376-4, Dordrecht, The Netherlands.

Long, O.H., J.R. Overton, E.W. Counce and T. McCutchen. 1965. Lime and fertilizer experiments on soybeans, 391. Knoxville, TN: University of Tennessee Agriculture Experiment Station Bulletin.

Leffel, R.C., P.B. Cregan, A.P. Bolgiano and D.J. Thibeau. 1992. Nitrogen metabolism of normal and high-seed-protein soybean. Crop Sci. 32: 747–750.

Malik, N.S.A., H.E. Calvert and W.D. Bauer. 1987. Nitrate-induced regulation of nodule formation in soybean. Plant Physiol. 84: 266–271.

Mann, E.N., P.M. de Resende, R.S. Mann, J.G. de Carvalho and E.V.D. Von Pinho. 2002. Effect of manganese application on yield and seed quality of soybean. Pesquisa Agropecuaria Brasileira 37: 1757–1764.

Maples, R. and J.L. Keogh. 1969. Soybean fertilization experiments. Arkansas Agriculture Experiment Station Report Series 178. Fayette, AR: University of Arkansas.

Marschner, H. 1995. Mineral nutrition of higher plants. 2nd edn. London: Academic Press.

Marschner, P. 2012. Marschner's mineral nutrition of higher plants. 3rd edn. Elsevier, Oxford.

Maury, P., S. Suc, M. Berger and C. Planchon. 1993. Response of photochemical processes of photosynthesis to dinitrogen fixation in soybean. Plant Physiol. 101: 493–497.

Mederski, H.J. 1950. Relation of varying phosphorus supply to dry matter production, and to N and P partition during the development of the soybean plant. Abstracts of Doctoral Dissertations, Ohio State University. Vol. 6.

Meharg, A.A. 1994. A critical review of labelling techniques used to quantify rhizosphere carbon-flow. Plant Soil 166: 55–62.

Meng, F., Y. Wei and X. Yang. 2005. Iron content and bioavailability in rice. J. Trace Elements in Medicine and Biology 18: 333–338.

Migocka, M. and G. Klobus. 2007. The properties of the Mn, Ni and Pb transport operating at plasma membranes of cucumber roots. Physiol. Plant. 129: 578–587.

Miwa, K. and T. Fujiwara. 2010. Boron transport in plants: Co-ordinated regulation of transporters. Ann. Bot. 105: 1103–1108. doi: 10.1093/aob/mcq044 PubMedPubMedCen.

Mora, M., M. Alfaro, S. Jarvis, R. Demanet and P. Cartes. 2006. Soil aluminium availability in Andisols of southern Chile and its effect on forage production and animal metabolism. Soil Use Manage. 22: 95–101.

Mukhopadhyay, M. and A. Sharma. 1991. Manganese in cell metabolism of higher plants. Bot. Rev. 57: 117–149.

Nakano, H., M. Kuwahara and I. Watanabe. 1987. Supplemental nitrogen fertilizer to soybeans. II. Effect of application rate and placement on seed yield and protein yield. Jpn. Crop Sci. 56: 329–336.

Ndakidemi, P.A. 2005. Nutritional characterization of the rhizosphere of symbiotic cowpea and maize plants in different cropping systems. Doctoral degree Thesis. Cape Peninsula University of Technology, Cape Town, South Africa. p. 150.

Nelson, K.A., P.P. Motavally and M. Nathan. 2005. Response of no-till soybean [*Glycine max* (L.) Merr.] to timing of preplant and foliar potassium applications in a claypan soil. Agronomy Journal 97: 832–838.

Nelson, D.R., R.J. Bellville and C.A. Porter. 1984. Role of nitrogen assimilation in seed development of soybean. Plant Physiol. 74: 128–133.

Neumann, G. and V. Romheld. 2001. The release of root exudates as affected by the plants physiological status. pp. 41–93. *In*: Pinto, R., Z. Varanini and P. Nannipieri (eds.). The Rhizosphere: Biochemistry and Organic Substances at the Soil-Plant Interface. Marcel Dekker. New York.

Nicholas, D.J.D. 1975. The function of trace elements in plants. pp. 181–198. *In*: Nicholas, D.J.D. and A.R. Edan (eds.). Trace Elements in Soil–Plant–Animal Systems. Academic Press, New York, USA.

Obaton, M., M. Miquel, P. Robin, G. Conejero, A.M. Domenach and R. Bardin. 1982. Influence du déficit hydrique sur l'activité nitrate-réductase et nitrogénase chez le soja. C.R. Acad. Sci. Paris 294: 1007–1012.

Ohlrogge, A.J. 1960. Mineral nutrition of soybeans. Advances in Agronomy 12: 229–263.

Ozanne, P.G., E.A.N. Greenwood and T.C. Shaw. 1963. The cobalt requirement of subterranean clover in the field. Aus. J. Biol. Sci. 20: 809–818.

Parvej, M.R., N.A. Slaton, L.C. Purcell and T.L. Roberts. 2015. Potassium fertility effects yield components and seed potassium concentration of determinate and indeterminate soybean. Agron. J. 107: 943–950.

Planchon, C., N. Burias and M. Berger. 1992. Nitrogenase activity, protein content and yield in soybean. Eur. J. Agron. 1: 195–200.

Pendias, K. and H. Pendias. 1992. Trace Elements in Soils and Plants. USA, CRC Press, p. 365.

Pettiet, J.V. 1971. Are nitrogen fertilizers needed for soybean? Mississippi Agricultural and Forestry Experiment. Station Information. Sheet 1147. Mississippi: Mississippi State University.

Pettigrew, W.T. 2008. Potassium influences on yield and quality production for maize, wheat, soybean and cotton. Physiol. Plant. 133: 670–681.

Pierzynski, G.M., R.W. McDowell and J.T. Sims. 2005. Chemistry, cycling, and potential movement of inorganic phosphorus in soils. pp. 53–86. *In*: Sims and Sharpley (eds.). Phosphorus: Agriculture and the Environment. ASA, CSSA, and SSSA, Madison, WI.

Polaccao, J., S. Freyermuth, J. Gerendas and S. Cianzio. 1999. Soybean genes involved in nickel insertion into urease. J. Exp. Bot. 50: 1149–1156.

Pollard, A.S., A.J. Parr and B.C. Loughman. 1977. Boron in relation to membrane function in higher plants. J. Exp. Bot. 28: 831–84.

Pouteau, S., I. Cherel, H. Vaucheret and M. Caboche. 1989. Nitrate reductase mRNA regulation in *Nicotiana plumbaginifolia* nitrate reductase-deficient mutants. Plant Cell 1: 1111–1120.

Powrie, J.K. 1964. The effect of cobalt on the growth of young lucerne on a siliceous sand. Plant Soil 21: 81–93.

Raven, J.A., M.C.W. Evans and R.E. Korb. 1999. The role of trace metals in photosynthetic electron transport in O_2-evolving organisms. Photosynth. Res. 60: 111–149.

Rayar, A. 1981. Effect of calcium concentration on growth and ion uptake in soybean plants in solution culture. Z. Pflanzenphysiologie 105: 59–64. doi:10.1016/S0044-328X(81)80008-6.

Rengel, Z. 2000. Manganese uptake and transport in plants. Met. Ions Biol. Syst. 37: 57–87.

Rengel, Z. and P. Marschner. 2005. Nutrient availability and management in the rhizosphere: Exploiting genotypic differences. New Phytol. 168: 305–312.

Rerkasem, B., R.W. Bell and J.F. Loneragan. 1989. Effects of seed and soil boron on early seedling growth of black and green gram (*Vigna munga* and *V. radiata*). pp. 281–285. *In*: van Beusichem, M.L. (ed.). Plant Nutrition-Physiology and Applications. Dordrecht, The Netherlands: Kluwer Academic Publishers.

Rhodes, E.R. and D. Nangju. 1979. Effects of pelleting cowpea and soybean seed with fertilizer dusts. Exp. Agric. 15: 27–32.

Rhodes, E.R. and M. Kpaka. 1982. Effects of nitrogen, molybdenum and cultivar on cowpea growth and yield on an Oxisol. Commun. Soil Sci. Plant Anal. 13(4): 279–283.

Römheld, V. and H. Marschner. 1986. Mobilization of iron in the rhizosphere of different plant species. Adv. Plant Nutr. 2: 155–204.

Ronis, D.H., D.J. Sammons, W.J. Kenworthy and J.J. Meisinger. 1985. Heritability of total and fixed N content of the seed in two soybean populations. Crop Sci. 25: 1–4.

Rose, I.A., W.L. Felton and L.W. Banks. 1981. Response of four soybean varieties to foliar zinc fertilizer. Aust. J. Exp. Agri. and Animal Husb. 21(109): 236–24.

Ryan, B.M., J.K. Kirby, F. Degryse, H. Harris, M.J. McLaughlin and K. Scheiderich. 2013a. Copper speciation and isotope fractionation in plants, uptake and translocation mechanisms. New Phytol. 199: 367–368.

Ryan, J., A. Rashid, J. Torrent, S.K. Yau, H. Ibrikci, R. Sommer, E.B. Erenoglu and D.L. Sparks. 2013b. Micronutrient constraints to crop production in the Middle East-West Asia region: Significance, research and management. Adv. Agron. 122: 1–84.

Ryan, P., E. Delhaize and D. Jones. 2001. Function and mechanism of organic anion exudation from plant roots. Annu. Rev. Plant Physiol. Plant Mol. Biol. 52: 527–560.

Sadras, V.O. 2006. The N:P stoichiometry of cereal, grain legume and oilseed crops. Field Crops Res. 95: 13–29.

Sale, P.W.G. and L.C. Campbell. 1986. Yield and composition of soybean seed as a function of potassium supply. Plant and Soil 96: 317.

Sainju, U.M., R. Dris and B. Singh. 2003. Mineral nutrition of tomato. Food, Agriculture and Environment 1(2): 176–183.

Sauchelli, V. 1969. Trace Elements in Agriculture. Van Nostrand, New York.

Schultze, M. and A. Kondorosi. 1998. Regulation of symbiotic root nodule development. Ann. Rev. Genet. 32: 33–57.

Shaukat, A. and H.J. Evans. 1960. Cobalt: A micronutrient element for the growth of soybean plants under symbiotic conditions. Soil Sci. 90: 205–210.

Sillanpää, M. 1990. Micronutrients assessment at the country level: An international study. FAO Soils Bulletin 63. Food and Agriculture Organization of the United Nations, Rome.

Singh, M.V. 2001. Evaluation of micronutrient status in different agro-ecological zones of India. Fert. News 46: 25–42.

Singh, M.V. 2004. Transfer of scientific knowledge to useful field practices. IFA International Symposium on Micronutrients. 23–25 Feb 2004, New Delhi.

Sirko, A. and R. Brodzik. 2000. Plant ureases: Roles and regulation. Acta Biochim. Polon. 47: 1189–1195.

Smith, S.E. and D.J. Read. 1997. Mycorrhizal symbiosis. 2nd ed. Academic Press, London.

Soni, M.L., A. Swarup and M. Singh. 1996. Influence of rates and methods of manganese application on yield and nutrition of wheat in a reclaimed sodic soil. J. Agri. Sci. 127: 433–439.

Taiz, L., E. Zeiger, I.M. Mollar and A. Murphy. 2015. Assimilation of inorganic nutrients. pp. 353–376. *In*: Tiaz, L. (ed.). Plant Physiology and Development. Sinauer Associates.

Takahashi, Y., T. Chinushi, Y. Nagumo, T. Nakano and T. Ohyama. 1991. Effect of deep placement of controlled release nitrogen fertilizer (coated urea) on growth, yield and nitrogen fixation of soybean plants. Soil Sci. Plant Nutr. 37: 223–231.

Thomine, S., R.C. Wang, J.M. Ward, N.C. Crawford and J.I. Schroeder. 2000. Cadmium and iron transport by members of a plant metal transporter family in Arabidopsis with homology to Nramp genes. PANAS, USA 97: 4991–4996.

Tsuji, M., H. Mori, T. Yamamoto, S. Tanaka, Y. Kang, K. Sakurai, and K. Iwasaki. 2006. Manganese toxicity of melon plants growing on an isolated soil bed after steam sterilization. J. Soil Sci. Plant Nutr. 77: 257–263.

Van Assche, E. and H. Clijsters. 1990. Effects of metals on enzyme activity in plants. Plant Cell Environ. 13: 195–206.

Van der Zaal, B., L. Neuteboom, J. Pinas, A. Chardonnens, H. Schat, V. Jac and P. Hooykaas. 1999. Overexpression of a novel Arabidopsis gene related to putative zinc-transporter genes from animals can lead to enhanced zinc resistance and accumulation. Plant Physiol. 119: 11047–11055.

Varco, J.J. 1999. Nutrition and fertility requirements. pp. 53–70. *In*: Heatherly, G. and H.F. Hodges (eds.). Soybean Production in the Midsouth. Boca Raton, FL: CRC Press.

Vrataric, M. and A. Sudaric. 2008. Soja/Soybean. Sveuciliste J.J. Strossmayera u Osijeku I Poljoprivredni institut Osijek, Hrvatska.

White, P.J. 2001. The pathways of calcium movement to the xylem. Journal of Experimental Botany 52: 891–899.

White, M.C., F.D. Baker, R.L. Chaney and A.M. Decker. 1980. Metal complexation in xylem fluid. II. Theoretical equilibrium model and computational computer program. Plant Physiol. 67: 301–310.

White, M.C., R.L. Chaney and A.M. Decker. 1980. Metal complexation in xylem fluid. III. Electrophoretic evidence. Plant Physiol. 67: 311–316.

Willis, H. 1989. Growing great soybeans, 1. Austin, TX: Acres USA, March, 6–8.

Yin, X.Y. and T.J. Vyn. 2004. Critical leaf potassium concentrations for yield and seed quality of conservation-till soybean. Soil Science Society of America Journal 68: 1626–1635.

Yruela, I. 2009. Copper in plants: Acquisition, transport and interactions. Funct. Plant Biol. 26: 409–430.

Zapata, F., S.K.A. Danso, G. Hardarson and M. Fried. 1987. Time course of nitrogen fixation in field-grown soybean using nitrogen-15 methodology. Agron. J. 79: 172–176.

Zhang, H., A.J. Jennings, P.W. Barlow and B.G. Forde. 1999. Dual pathways for regulation of root branching by nitrate. Proc. Natl. Acad. Sci. USA 96: 6529–6534.

CHAPTER 7
Photosynthesis

Synthesis of the photosynthetic apparatus requires large amounts of N, the proportion of leaf N allocated to the chloroplast amounts to approximately 75% (Hák et al., 1993). When grown in high irradiance, leaves generally have a ratio of 1.0:1.4 mol N in ribulose-1,5-bisphosphate carboxylase/oxygenase (RuBPCO) vs. thylakoid N, while under low irradiance, more N is partitioned into the thylakoid components (Evans, 1989). Significant correlations between photosynthesis and leaf N content have been documented for a large number of species, including soybean (Evans, 1989). A positive correlation between leaf N or N fertilization rate and chlorophyll (Chl) content is well documented for a large number of plant species and has been investigated for rapid N status determination using Chl meters in most major crops, including corn (*Zea mays* L.), rice (*Oryza sativa* L.), cotton (*Gossypium hirsutum* L.), wheat (*Triticum aestivum* L.) and numerous other plant species (Mauromicale et al., 2006). Taking Chl meter readings is easy and quick, does not necessitate destructive sampling, and has been employed to predict Chl content in a large number of plant species, including soybean (Yadava, 1986; Marquard and Tipton, 1987). Thompson et al. (1996) observed strong correlations of Chl content with SPAD readings and areal leaf mass (ALM) in soybean, and suggested that a portable SPAD Chl meter may be used to select for genotypes differing in ALM. Generally, plants modulate leaf anatomy and physiology to irradiance, developing thicker leaves with a greater mesophyll to surface-area ratio. Common adjustments to high irradiance also include a reduction in the total Chl per unit leaf area and an increased Chl a/b ratio (Anderson, 1986). Due to its predictable response to irradiance, the Chl a/b ratio has been proposed as a bioassay to assess the light environment of a plant (Dale and Causton, 1992). Other modulations in response to irradiance include effects on chloroplast ultra-structure and Chl associations with photosystems (PS) 1 and 2 (Evans, 1989; Green and Durnford, 1996). Under crop production conditions, the irradiance of most leaves is influenced by plant density and row distance and modulates various aspects of the photosynthetic apparatus. Thus, genotype specific Chl composition and content characteristics could have implications for crop management.

Reducing chlorophyll (Chl) content may improve the conversion efficiency of absorbed photosynthetically active radiation into biomass and, therefore, yield in dense monoculture crops by improving light penetration and distribution within the canopy. The effects of reduced Chl on leaf and canopy photosynthesis and photosynthetic efficiency were studied in two reportedly robust reduced-Chl soybean mutants, *Y11y11* and *y9y9*, in comparison to the wild-type (WT) "Clark" cultivar. Both mutants were characterized during the 2012 growing season, whereas only the *Y11y11* mutant was characterized during the 2013 growing season. Chl deficiency led to greater rates of leaf-level photosynthesis per absorbed photon early in the growing season when mutant Chl content was ~ 35% of the WT, but there was no effect on photosynthesis later in the season when mutant leaf Chl approached 50% of the WT. Transient benefits of reduced Chl at the leaf level did not translate to improvements in canopy-level processes. Reduced pigmentation in these mutants was linked to lower water use efficiency, which may have dampened any photosynthetic benefits of reduced Chl, especially since both growing seasons experienced significant drought conditions. These results, while not confirming the hypothesis or an earlier published study in which the *Y11y11* mutant significantly outyielded the WT, do demonstrate that soybean significantly overinvests in Chl. Despite a > 50% Chl reduction, there was little negative impact on biomass accumulation or yield, and the small negative effects present were likely due to pleiotropic effects of the mutation. This outcome points to an opportunity to reinvest nitrogen and energy resources, that would otherwise be used in pigment-proteins, into increasing biochemical photosynthetic capacity, thereby improving canopy photosynthesis and biomass production.

A physiologically-based steady-state model of whole leaf photosynthesis (WHOLEPHOT) is used to analyze observed net photosynthesis daily time courses of soybean, *Glycine max* (L.) Merr., leaves observations during two time periods of the 1978 growing season are analyzed and compared. After adjustment of model of soybean, net photosynthesis rates are calculated with the model in response to measured incident light intensity, leaf temperature, air carbon dioxide concentration and leaf diffusion resistance. The steady-state calculations closely approximate observed, net photosynthesis. Results of the comparison reveal a decrease in photosynthetic capacity in leaves sampled during the second time period, which is associated with decreasing ability of leaves to respond to light intensity and internal air space carbon dioxide concentration, increasing mesophyll and stomatal resistance (Tenhunen et al., 1980) (Fig. 7.1).

A comparative analysis of photosynthetic rate of corn (C4 plant) and soybean (C3), given below, indicated the difference and ontogenic pattern of changes (Fig. 7.2).

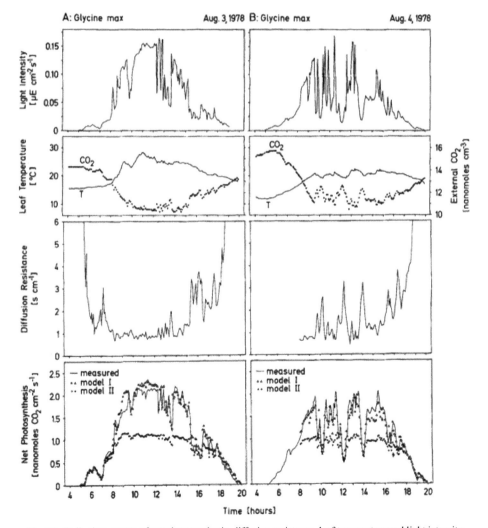

Fig. 7.1 Daily time course of net photosynthesis, diffusion resistance, leaf temperature and light intensity.

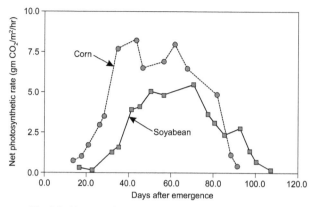

Fig. 7.2 Changes of net photosynthetic rate during life span.

7.1 Carbon Dioxide

Despite the wide uses of potted plants, information on how pot size affects plant photosynthetic matter production is still considerably limited. This study investigated with soybean plants how transplantation into larger pots affects various characteristics related to photosynthetic matter production. The transplantation was analyzed to increase leaf photosynthetic rate, transpiration rate, and stomatal conductance without affecting significantly leaf intercellular CO_2 concentration, implicating that the transplantation induced equally increases in the rate of CO_2 diffusion via leaf stomata and the rate of CO_2 fixation in leaf photosynthetic cells. Analyses of Rubisco activity and contents of a substrate (ribulose-1,5-bisphosphate (RuBP)) for Rubisco and total protein in leaf suggested that an increase in leaf Rubisco activity, which is likely to result from an increase in leaf Rubisco content, could contribute to the transplantation-induced increase in leaf photosynthetic rate. Analyses of leaf major photosynthetic carbohydrates and dry weights of source and sink organs revealed that transplantation increased plant sink capacity that uses leaf starch, inducing a decrease in leaf starch content and an increase in whole plant growth, particularly, growth of sink organs. Previously, in the same soybean species, it was demonstrated that negative correlation exists between leaf starch content and photosynthetic rate and that accumulation of starch in leaf decreases the rate of CO_2 diffusion within leaf. Thus, it was suggested that the transplantation-induced increase in plant sink capacity decreasing leaf starch content could cause the transplantation-induced increase in leaf photosynthetic rate by inducing an increase in the rate of CO_2 diffusion within leaf and thereby substantiating an increase in leaf Rubisco activity *in vivo*.

The expected doubling of present-day atmospheric CO_2 levels in the next century has been calculated to approximately halve photorespiration relative to net photosynthesis *(A)* in C3 plants, excluding any acclimation or temperature effects (Sharkey, 1988; Long, 1991). Thus, for most C3 plants at current temperatures, the ratio of carbon fixed by photoassimilation will increase compared with the carbon lost through photorespiration (Long, 1991). This increase in *A* should contribute to the productivity increase expected from atmospheric CO_2 enrichment.

The air pollutant, tropospheric O_3, is a potentially interacting factor with the effect of CO_2 enrichment on productivity (Allen, 1990). Although damaging concentrations of O_3 during the growing season tend to be regionalized, tropospheric O_3 is steadily increasing on a global scale (Fishman, 1991). Net photosynthesis, growth and yield are suppressed in many plants by the levels of O_3 currently found in industrialized countries, but the effects of O_3 on photorespiration are less well documented (Miller, 1988).

The effects of CO_2 enrichment and O_3 on photosynthesis and photorespiration are mediated in part through a common enzyme, ribulose-1,5-bisphosphate carboxylase/oxygenase (Rubisco) (Long, 1994). Carbon dioxide and O_2 are competitive substrates for Rubisco, and their partial pressures affect the rates of ribulose-1,5-bisphosphate (RuBP) carboxylation and oxygenation (Farquhar et al., 1980). Normally, the most significant regulator of flux through these pathways is the relative concentration of CO_2 and O_2 at the active site of Rubisco (Wallsgrove et al., 1992). This is why atmospheric CO_2 enrichment is expected to promote photosynthetic carbon reduction over photorespiratory carbon oxidation. Ozone, on the other hand, suppresses photosynthesis in part by decreasing Rubisco activity and content (Pell et al., 1994). Photorespiration would be expected to decline if Rubisco activity decreased, although its relative sensitivity to O_2 is unknown.

Photorespiration is the light-dependent release of CO_2 that is sensitive to the O_2 concentration and that originates mainly from the metabolism of compounds through the glycolate pathway (Canvin, 1979). Photorespiration begins with the oxygenation of RuBP by Rubisco to form phosphoglycolate, which is hydrolysed to glycolate by a phosphatase in the chloroplast and then excreted (Ogren, 1984). In the peroxisome, glycolate is oxidized to glyoxylate by glycolate oxidase and then transaminated to glycine using either glutamate or serine (Ogren, 1984). Catalase decomposes the H_2O_2 formed during glycolate oxidation. Glycine enters a mitochondrion where almost all the CO_2 evolved in photorespiration comes from the oxidation of glycine to form serine and ammonia (Ogren, 1984). Serine then passes back to the peroxisome where it is transaminated to hydroxypyruvate and reduced by hydroxypyruvate reductase to glyceric acid. Glyceric acid may be returned to the chloroplast and phosphorylated to form 3-phosphoglyceric acid, thus completing the photorespiratory carbon oxidation cycle (Ogren, 1984).

It has been suggested that decreased photorespiration due to CO_2 enrichment might be accompanied by decreases in the activities of photorespiratory pathway enzymes (Webber et al., 1994). Nitrogen resources utilized in this pathway would, therefore, be conserved. However, decreased levels of catalase, an enzyme that putatively helps protect plants against injury from O_2 (Willekens et al., 1994), might be detrimental to plants exposed to elevated O_3 in combination with elevated CO_2 (Polle et al., 1993). The effects of O_3 on the photorespiratory enzymes, glycolate oxidase and hydroxypyruvate reductase, have not been reported. Glycine and serine, the products of glycolate metabolism, accumulate in the cell, and their levels can be influenced by CO_2 concentration (Servaites and Ogren, 1977). Elevated CO_2 decreased the concentration of glycine in leaf tissues of an *Arabidopsis thaliana* mutant unable to convert glycine to serine (Somerville and Somerville, 1983) and lowered glycine-serine levels in excised wheat (*Triticum aestivum* L.) leaves (Sen Gupta, 1988), presumably due to suppressed photorespiration. In contrast, glycine and serine levels increased in leaves of kidney bean (*Phaseolus vulgaris* L.) after exposure to 200 nmolmol^{-1} O_3 in controlled environments, and it was concluded that photorespiration was enhanced by exposure to O_3 (Ito et al., 1985).

The effects of elevated carbon dioxide (CO_2) and ozone (O_3) on soybean [*Glycine max* (L.) Merr.] photosynthesis and photorespiration-related parameters were determined periodically during the growing season by measurements of gas exchange, photorespiratory enzyme activities and amino acid levels. Plants were treated in open-top field chambers from emergence to harvest maturity with seasonal mean concentrations of either 364 or 726 μmol mol^{-1} CO_2 in combination with either 19 or 73 nmol mol l^{-1} O_3 (12 h daily averages). On average, at growth CO_2 concentrations, net photosynthesis *(A)* increased 56% and photorespiration decreased 36% in terminal mainstem leaves with CO_2-enrichment. Net photosynthesis and photorespiration were suppressed 30% and 41%, respectively, by elevated O_3 during late reproductive growth in the ambient CO_2 treatment, but not in the elevated CO_2 treatment. The ratio of photorespiration to *A* at growth CO_2 was decreased 61% by elevated CO_2. There was no statistically significant effect of elevated O_3 on the ratio of photorespiration to *A*. Activities of glycolate oxidase, hydroxypyruvate reductase and catalase were decreased 10–25% by elevated CO_2, and by 46–66% by elevated O_3 at late reproductive growth. The treatments had no significant effect on total amino acid or glycine levels, although serine concentration was lower in the elevated CO_2 and O_3 treatments at several sampling dates. The inhibitory effects of elevated O_3 on photorespiration-related parameters were generally commensurate with the O_3-induced decline in *A*.

The responses of soybean to elevated [CO_2] have been investigated extensively using enclosure studies. A meta-analysis of 111 of these studies showed that elevated [CO_2] (from 450 to 1,250 μmol mol^{-1}) increased the soybean leaf CO_2 assimilation rate by 39%, but decreased Rubisco activity by 11%

(Ainsworth et al., 2002). However, these studies were all conducted in protected environments, either in laboratory or using field enclosures or open-top chambers (OTC). Above conditions could modify the microclimate and amplify the elevated [CO_2] effects on plant growth and development (McLeod and Long, 1999; Long et al., 2004). The Soybean Free-Air CO_2 Enrichment (SoyFACE) facility (Rogers et al., 2004) was the first experimental setup designed to investigate the response of soybean to elevated [CO_2] under open-air conditions. FACE systems enable plants to be grown under natural conditions for long periods and have been used internationally at more than 30 sites (Ainsworth et al., 2008). Trees, crops or pastures are grown within rings with a regulated supply of CO_2 fed into the experimental area, and the amount and position of CO_2 release controlled by sensors to maintain targeted CO_2 levels. The leaf ultrastructure of plants could be improved under elevated [CO_2], and the amount of starch in chloroplasts and cytoplasmic lipids increased (Zuo et al., 2002), which is another way to evaluate the changes of leaf carbohydrate under elevated [CO_2]. Within the photosynthetically active radiation (PAR) region of the electromagnetic spectrum (400–700 nm), the reflectance of vegetation is controlled by the absorption characteristics of Chl *a* and *b*. These pigments provide energy for the reactions of photosynthesis and for the carotenoids, which protect the reaction centres from excess light and help them to intercept PAR as auxiliary pigments of Chl *a*. Therefore, changes in pigment concentrations relate strongly to the physiological status and productivity of a plant (Blackburn, 1998). Jiang et al. (2006) and Zhao et al. (2003) investigated the photosynthetic pigments basis for changes in soybeans leaf development at elevated [CO_2] in OTC, and showed that Chl *a*, Chl *b*, total Chl contents, and carotenoid contents in leaves were increased at elevated [CO_2].

Photosynthetic acclimation is frequently associated with an accumulation of leaf non-structural carbohydrates and a decrease in N concentration in the leaf and plant (Stitt and Krapp, 1999; Nowak et al., 2004). In the SoyFACE experiment, non-nodulating soybeans showed down regulation of Vc,max under elevated [CO_2], while nodulating varieties maintained the same photosynthetic capacity under ambient and elevated [CO_2] (Ainsworth et al., 2004).

A study was undertaken to investigate the effect of elevated (550 ± 17 mmol mol^{-1}) CO_2 concentration ([CO_2]) on leaf ultrastructure, leaf photosynthesis and seed yield of two soybean cultivars [*Glycine max* (L.) Merr. cv. Zhonghuang13 and cv. Zhonghuang 35] at the Free-Air Carbon dioxide Enrichment (FACE) experimental facility in North China.

Photosynthetic acclimation occurred in soybean plants exposed to long-term elevated [CO_2] and varied with cultivars and developmental stages. Photosynthetic acclimation occurred at the beginning bloom (R1) stage for both cultivars, but at the beginning seed (R5) stage only for Zhonghuang 13. No photosynthetic acclimation occurred at the beginning pod (R3) stage for either cultivar. Elevated [CO_2] increased the number and size of starch grains in chloroplasts of the two cultivars. Soybean leaf senescence was accelerated under elevated [CO_2], determined by unclear chloroplast membrane and blurred grana layer at the beginning of the bloom (R1) stage. The different photosynthesis responses to elevated [CO_2] between cultivars at the beginning seed (R5) contributed to the yield difference under elevated [CO_2]. Elevated [CO_2] significantly increased the yield of Zhonghuang 35 by 26% with the increased pod number of 31%, but not for Zhonghuang 13, which had no changes of pod number. It was concluded that the occurrence of photosynthetic acclimation at the beginning seed (R5) stage for Zhonghuang 13 restricted the development of extra C sink under elevated [CO_2], thereby limiting the response to elevated [CO_2] for the seed yield of this cultivar (Hao et al., 2012) (Fig. 7.3; Table 7.1 and Fig. 7.4).

While photosynthesis of C_3 plants is stimulated by an increase in the atmospheric CO_2 concentration, photosynthetic capacity is often reduced after long-term exposure to elevated CO_2. This reduction appears to be brought about by end product inhibition, resulting from an imbalance in the supply and demand of carbohydrates. A review of the literature revealed that the reduction of photosynthetic capacity in elevated CO_2 was most pronounced when the increased supply of carbohydrates was combined with small sink size. The volume of pots in which plants were grown affected the sink size by restricting root growth. While plants grown in small pots had a reduced photosynthetic capacity, plants grown in

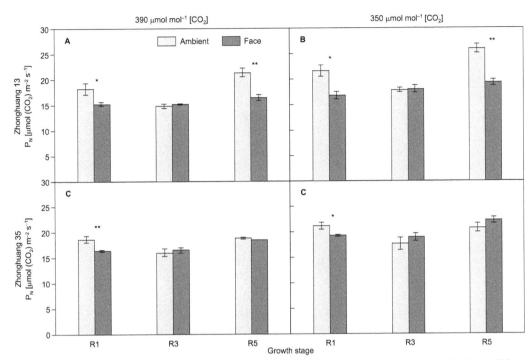

Fig. 7.3 Net photosynthesis in soybean at elevated CO_2 in different growth stages grown under ambient and FACE conditions.

Table 7.1 Photosynthetic pigment contents of soybean leaf grown in ambient and FACE plots. Values are means ± SE of variables across the three replicates, and the statistical significance level *P* values for the effects of [CO_2] treatment, cultivar, growth stage and their interaction. Chl – chlorophyll; Z13 – Zhonghuang 13; Z35 – Zhonghuang 35; R1, R3 – beginning bloom stage and beginning pod stage, respectively. FM – fresh mass.

Growth stage	Cultivar	Growth (CO₂)	Chl a (g g⁻¹FM)	Chl b (g g⁻¹FM)	Carot. (g g⁻¹FM)	Chl a+b (g g⁻¹FM)	Chl a/b
R1	Z13	Ambient	1.01	0.32	0.35	1.34	3.16
	Z35	FACE	1.15	0.40	0.41	1.55	2.88
		Ambient	0.92	0.32	0.33	1.24	2.93
		FACE	0.95	0.30	0.33	1.25	3.15
R3	Z13	Ambient	1.79	0.67	0.62	2.46	2.75
	Z35	FACE	1.63	0.60	0.56	2.22	2.72
		Ambient	2.15	0.84	0.77	2.99	2.58
		FACE	2.14	0.84	0.78	2.99	2.56
P values Cultivar Growth stage			0.98 0.07 0.00	0.99 0.04 0.00	0.93 0.04 0.00	0.99 0.06 0.00	0.91 0.41 0.00
CO₂ × Cv. CO₂ × G.S.			0.89 0.28	0.88 0.37	0.99 0.35	0.96 0.30	0.13 0.81

the field showed no reduction or an increase in this capacity. Pot volume also determined the effect of elevated CO_2 on the root/shoot ratio: The root/shoot ratio increased when root growth was not restricted and decreased in plants grown in small pots. It is suggested that plants growing in the field will maintain a high photosynthetic capacity as the atmospheric CO_2 level continues to rise.

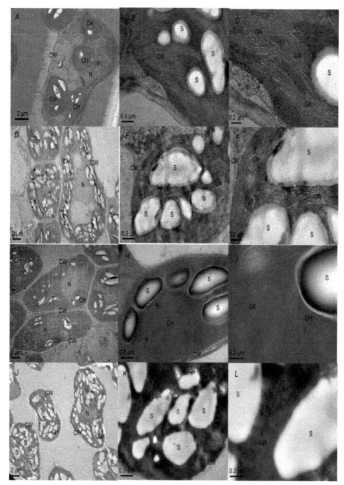

Fig. 7.4 Chloroplast ultrastructure in mesophyll cells of soybean leaf grown in ambient and FACE plots at R1 stage. A–F: Zhonghuang13. G–L: Zhonghuang 35. A–C, G–I: chloroplast ultrastructure of soybean leaf grown in ambient (× 5,000; × 30,000; × 60,000). D–F, J–L: Chloroplast ultrastructure of soybean leaf grown under elevated [CO_2] (× 5,000; × 30,000; × 80,000). S – starch grain; GR – grana layer; O – osmophore; CM – chloroplast membrane; CH – chloroplast; CW – cell wall; N – nucleus. R1 – beginning bloom stage.

7.2 Water Stress

The association between photosynthesis and yield is likely to be strong during the seed filling period (R5–R7). Soybean leaf photosynthesis begins to decline around this period (Sinclair, 1980; Lugg and Sinclair, 1981), and this decline includes two types: Slow and rapid (Mathew et al., 2001). The conflicting reports regarding the relationship between leaf photosynthesis and yield may be due to the phenological stage at which the measurements were taken (Kumudimi, 2002). Leaf photosynthesis is composed of mesophyll photosynthetic capacity and stomatal opening; the latter factor controls leaf photosynthesis through the CO_2 influx into the leaves. Stomatal opening also has another role in controlling H_2O efflux from the leaves; this varies in response to various environmental conditions to avoid excess transpiration and, as a result, leaf water shortage. Therefore, leaf photosynthesis is sensitive to leaf water potential (indicator of leaf water status); leaf photosynthesis of crop plants usually begins to drops in response to the decline in leaf water potential through the decline in stomatal conductance (indicator of stomatal opening) (Berkowitz, 1998). However, the sensitivity of soybean leaves to soil drought is lower;

stomatal conductance (indicator of leaf water status) (Nielsen, 1990) and, as a result, leaf photosynthesis (Boyer, 1970) is relatively unaffected until leaf water potential drops below –1.1 MPa and then drops dramatically (Nielsen, 1990). These factors strongly affect the recovery from severe soil drought stress: Leaf photosynthesis generally decreases with the decline in stomatal conductance even under mild soil drought condition in order to avoid excessive water loss from the leaves and severe damage to mesophyll photosynthetic capacity (Berkowitz, 1998). So, improvement of soil water condition (i.e., precipitation) can cause the recovery of leaf photosynthesis quickly thanks to the maintenance of mesophyll photosynthetic capacity. However, soybean leaves keep high stomatal conductance under relatively severe soil drought conditions (Nielsen, 1990), and this can easily result in severe damage to mesophyll photosynthetic capacity. In addition, soybean leaves can also keep leaf greenness under severe soil drought condition (Nagasuga et al., 2016). So, the symptom of drought stress in soybean leaves is not clear, and the recovery of leaf photosynthesis is impossible if soybean leaves are drooped by severe soil drought. Similarly, soybean leaf senescence can occur without a concomitant loss of leaf greenness (Kumudini, 2002). It is too difficult to measure the net leaf photosynthetic capacity and to evaluate the relationship between leaf photosynthesis and yield accurately.

7.3 High Temperature

Effects of temperature on canopy photosynthesis and CGR are characterized by an optimal temperature response range falling between minimal and maximal optimal temperatures, and suboptimal and supra optimal temperatures falling below and above the optimal range, respectively (Hollinger and Angel, 2009). The part of the photosynthetic apparatus most sensitive to heat stress is photosystem II. Specifically, the splitting of water to provide electrons to the light reactions is inhibited (Paulsen, 1994). Temperatures falling below the minimal optimal level reduce canopy photosynthesis and CGR through reduced reaction rates and/or enzyme inactivation. Studies conducted under constant day-time temperatures (12–16 hours per day) across an extended period generally have reported an optimal temperature range for photosynthesis of 25–35°C (Gesch et al., 2001; Vu et al., 1997). However, under natural growing conditions, maximal daily temperature usually occurs for only 1–2 hours (Louisiana Agric. Exp. Stn., 2010). When heat stress studies are conducted under more realistic conditions of short-term stress, temperature had to be raised to 42–43°C in order to have a deleterious effect on soybean photosynthesis (Ferris et al., 1998). These results are corroborated by Fitter and Hay (1987) who stated that, for plants from most climatic regions, temperatures of 45–55°C for 30 minutes were sufficient to cause irreversible damage to the photosynthetic apparatus. In conclusion, under typical growing conditions, the optimal temperature range for soybean canopy photosynthetic rate appears to be 25–40°C. A similar optimal temperature range of 26 to slightly above 36°C for crop growth rate has also been reported (Baker et al., 1989; Hofstra, 1972). Adverse effects on yield were entirely due to high day-time temperatures rather than night-time temperatures (Gibson and Mullen, 1996). At the crop level, heat-stress induced reductions in canopy photosynthesis affect yield components being formed at the time of the stress. Stresses occurring during flowering and pod formation (R1–R5) affect seed number, whereas stress during seed filling (R5–R7) reduces seed size (Gibson and Mullen, 1996). Both reductions were linked with lower photosynthetic rates. Concomitant with these reductions in canopy photosynthesis and yield components are decreased TDM and plant size. Soybean yield was as sensitive to heat stress during flowering/pod formation (R1–R5) as during seed filling (R5–R7).

Soybean (*Glycine max* L. Merr. cv. Bragg) was grown season-long in eight sunlit, controlled-environment chambers at two day-time $[CO_2]$ levels of 350 (ambient) and 700 (elevated) μmol mol^{-1}. Dry bulb day/night maximum/minimum air temperatures, which followed a continuously and diurnally varying, near sine-wave control set point that operated between maximum (day-time, at 1500 EST) and minimum (night-time, at 0700 EST) values, were controlled at 28/18 and 40/30°C for the ambient-CO_2 plants, and at 28/18, 32/22, 36/26, 40/30, 44/34 and 48/38°C for the elevated-CO_2 plants. Hence, it was to assess the upper threshold tolerance of photosynthesis and carbohydrate metabolism with increasing temperatures at elevated $[CO_2]$, as it is predicted that air temperatures could rise as much as 4–6°C within the 21st century with a doubling of atmospheric [CO_2]. Leaf photosynthesis measured at growth [CO_2] and temperature was greater for elevated-CO_2 plants and was highest at 32/22°C, but markedly declined at temperatures above 40/30°C. Growth temperatures from 28/18 to 40/30°C had little effect on midday total activity and protein content of Rubisco, while higher temperatures substantially reduced them. Conversely, midday Rubisco *rbcS* transcript abundance declined with increasing temperatures from 28/18 to 48/38°C. Elevated-CO_2 plants exceeded the ambient-CO_2 plants in most aspects of carbohydrate metabolism. Under elevated [CO_2], midday activities of ADPG pyrophosphorylase and sucrose-P synthase and invertase paralleled net increases in starch and sucrose contents, respectively. They were highest at 36/26–40/30°C, but declined at higher or lower growth temperatures. Thus, in the absence of other climatic stresses, soybean photosynthesis and carbohydrate metabolism would perform well in the rising atmospheric [CO_2] and temperature conditions predicted for the 21st century.

Variability, along with occurrences of short episodes of high temperature (HT), will be one of the important characteristics of future climates. Climate models predict that global surface temperatures will continue to rise (IPCC, 2007). Lobell and Asner (2003) reported that each 1°C increase in the average growing season temperature leads to a 17% decrease in soybean yield. Photosynthesis was the main physiological process that was affected by HT, often before other cell functions were impaired. The photosynthetic rate in C3 plants is generally maximum at about 30°C, with significant reductions for each additional degree increase in temperature (Wise et al., 2004). Leaf photosynthesis was usually inhibited when leaf temperatures exceeded approximately 38°C (Berry and Bjorkman, 1980). In general, chlorophyll fluorescence is also a good measure of the photosynthetic activity in leaves (Krause and Weis, 1991; Govindjee, 1995). Photosystem II (PS II) is temperature sensitive and down regulates photosynthesis under HT conditions. High-temperature stress can also cause loss of grana stacking and denaturation of the PS-II complexes and light-harvesting complex II (Shutilova et al., 1995). However, the exact mechanisms causing lower photosynthesis under HT stress in soybeans are still not clearly understood and need attention. Earlier research in soybean showed that under HT stress the stomatal conductance and photosynthetic rate decreased (Djanaguiraman and Prasad, 2010). The lower photosynthetic rate can be caused by decreased stomatal conductance or by nonstomatal factors. The carbon isotope ratio can provide information on stomatal and nonstomatal factors influencing photosynthetic rate. The ratio of 13C to 12C in plant tissue is less than the ratio of 13C to 12C in the atmosphere, indicating that plants discriminate against 13C during photosynthesis. The ratio of 13C to 12C in C3 plants (δ13C) varies mainly because of the discrimination that occurs during diffusion and enzymatic processes. The rate of diffusion of 13CO_2 across the stomatal pore is lower than that of 12CO_2 by a factor of 4.4‰ (Farquhar et al., 1989). Hence, measuring the carbon isotope ratio in leaves can help determine if the decreased photosynthetic rate is an effect of stomatal conductance or of the activation state and/or content of the enzyme Rubisco.

High-temperature stress in soybean causes premature leaf senescence and lower leaf photosynthesis (Djanaguiraman and Prasad, 2010). In addition, HT stress also decreased the activity of the antioxidant defence system. Premature leaf senescence under HT stress was caused by the increased production of reactive oxygen species (ROS) (Djanaguiraman and Prasad, 2010). There is limited anatomical evidence (ultrastructural changes in the chloroplast, thylakoid membranes, mitochondria, and plasma membrane) to explain the mechanisms of decreased activity of PS-II photochemistry (Fv/Fm, the ratio of variable fluorescence to maximum fluorescence) and decreased photosynthetic rate under HT stress. Anatomical changes under HT stress showed smaller cell size, closure of stomata, and greater stomatal density (Anon et al., 2004) in other species. Zhang et al. (2005) showed that HT stress severely damaged the mesophyll cells and increased the permeability of the plasma membrane in grapes (*Vitis vinifera*; Linn.). At the subcellular level, major structural modifications occurred in the chloroplasts and mitochondria, leading to significant changes in photosynthesis (Zhang et al., 2005). Such studies on crop species, and particularly on soybean, are limited in number.

Earlier studies showed that photosynthetic rates in plants were related to the ratio F_o/F_m (Ristic et al., 2008; Djanaguiraman and Prasad, 2010). An increased F_o/F_m ratio (which indicates thylakoid membrane damage) is associated with a lower chlorophyll content in wheat (Ristic et al., 2008) and soybean (Djanaguiraman and Prasad, 2010).

Soybean (*Glycine max* L. Merr.) genotype K 03-2897 was grown in a controlled environment in order to study the effects of high-temperature (HT) stress on ultrastructural changes in leaves and its relationship with photosynthetic rate. The aims of the study were to (i) quantify the effects of HT stress during flowering stage on photosynthetic rate and (ii) observe the anatomical and ultrastructural changes in leaves of soybean grown under HT. Plants were exposed to HT (38/28°C) or optimum temperature (OT; 28/18°C) for 14 d at flowering stage R2. High-temperature stress significantly decreased the leaf photosynthetic rate and stomatal conductance by 20.2 and 12.8%, respectively, compared with those at OT. However, HT stress significantly increased the thicknesses of the palisade and spongy layers and the lower epidermis above those at OT. In addition, HT stress damaged the plasma membrane, chloroplast membrane, and thylakoid membranes. The mitochondrial membranes, cristae, and matrix were distorted under HT stress. High-temperature stress increased the content of leaf reducing sugars by 82.6% above that at OT. Leaves under HT stress had a higher carbon isotope ratio compared with that at OT. Decreases in photosynthesis at HT stress were mediated through anatomical and structural changes in the cell and cell organelles, particularly the chloroplast and mitochondria (Djanaguiraman et al., 2011) (Figs. 7.5 and 7.6).

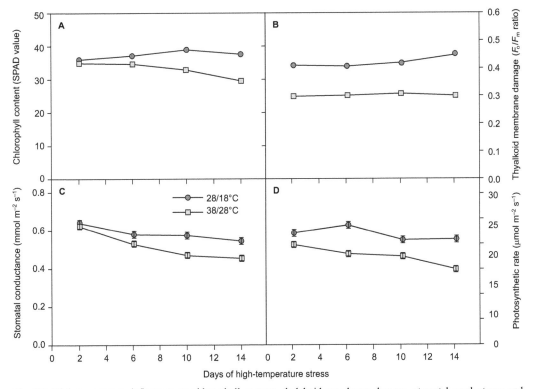

Fig. 7.5 High temperature influences on chlorophyll content, thylakoid membrane damage, stomatal conductance and photosynthetic rate in soybean.

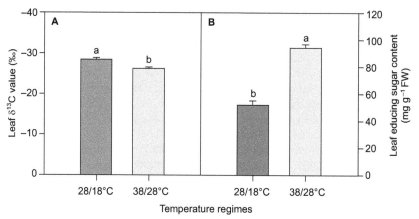

Fig. 7.6 Leaf ^{13}C value and reducing sugar content as affected by temperature in soybean.

7.4 Cold Temperature

Soybean seedlings often experience short-term cold temperature during the growing season that may affect subsequent growth and production. The effects of short-term cold temperature on soybean (*Glycine max* [L.] Merr. 'Hutcheson') growth and development, biomass partitioning, photosynthesis, and carbohydrate metabolism showed that short-term cold temperature delays reproductive stages by reducing photosynthesis and altering biomass allocation. Soybean plants were grown in controlled environmental conditions at a range of day/night temperatures (23/18°C, and 33/28°C) and exposed to a cold treatment of 8°C for 24 h at the V5 and R1 stages. The cold treatments delayed R1 for plants grown at 28/23°C, delayed R2 for plants grown at all three temperatures by up to 7 d, and prolonged the time periods between R1 and R2 stages. Leaf photosynthesis in the treated plants was 81%, 75%, and 79% of controls 5 h after the treatment for plants grown at 23/18°C, 28/23°C, and 33/28°C, respectively. A greater reduction in photosynthesis was obtained after the second cold treatment was applied at R1. Total soluble carbohydrate in leaves was reduced by the short-term cold temperature 5 h after the cold treatment. Two cycles of cold temperatures also increased the partitioning of total biomass to vegetative shoots, but decreased the partitioning to flowers and pods for all three temperatures. Results indicate that cold temperature injury delayed soybean reproductive stages and that the delays, at least in part, resulted from decreased leaf photosynthesis, reduced photosynthate availability, and altered biomass partitioning, favoring vegetative over reproductive growth. The magnitude and sensitivity of soybean to short-term cold temperatures varied and was dependent on growth temperature and developmental stage.

CO_2-gas exchange behaviour and direct chlorophyll fluorescence kinetics were investigated in order to obtain a better insight into the stress-induced changes that occur in soybean (*Glycine max* L. Merr.) during cold stress. CO_2 assimilation during photosynthesis was measured by infrared gas analysis, while the function of photosystem II (PSII) was assessed by fast-phase chlorophyll fluorescence at different periods of exposure to low night temperature (8 Celsius). The rate of CO_2 assimilation decreased by 87% after a single night of cold stress in the cultivar Fiskeby V. Analysis of A:C$_i$ response curves and the polyphasic fast chlorophyll fluorescence rise, using the JIP test, showed that the observed inhibition of photosynthesis was due to mesophyll limitation. Cold stress markedly decreased the carboxylation efficiency of Fiskeby V and also resulted in severe impairment of PSII function, mainly through reaction centre deactivation and reduced electron transport capacity. It was predicted that CO_2-gas exchange analysis and the JIP test will prove invaluable in future studies of the basis of cold injury and tolerance in soybean.

7.5 Salinity

An experiment was conducted to determine if salinity stress alters the response and tolerance of soybean to defoliation. Four soybean [*Glycine max* (L.) Merr.] cultivars ('Tachiutaka,' 'Tousan 69,' 'Dare' and 'Enrei') in a growth chamber were exposed to two salinity treatments (0 and 40 mM NaCl) and two defoliation treatments (with and without defoliation). The interactive effects of salinity stress and defoliation on growth rate, leaf expansion, photosynthetic gas exchange, and sodium (Na^+) accumulation were determined. The decrease in growth rate resulting from defoliation was more pronounced in plants grown under salinity stress than in those grown without the stress. Without salinity stress, defoliated plants of all four cultivars had leaf-expansion similar rates to those of the undefoliated ones, but the photosynthetic rates of their remaining leaves were higher than those of undefoliated plants. However, with salinity stress, defoliated 'Tachiutaka' and 'Tousa 69' had lower leaf expansion and photosynthetic rates than undefoliated plants. For cultivars 'Dare' and 'Enrei,' the defoliated plants had leaf-expansion rates similar to undefoliated ones, but the photosynthetic rate of the remaining leaves did not increase. Except for cultivar 'Dare,' defoliated plants grown under salinity stress had higher Na^+ accumulation in leaves than undefoliated ones, and this result may be related to slow leaf expansion and photosynthesis. Salinity stress negatively affects soybean response and tolerance of defoliation, and the effects varied according to the salt tolerance of the cultivar.

7.6 Nutrition

The impact of starter nitrogen fertilizer on soybean root activity, leaf photosynthesis, grain yield and their relationship was undertaken in field experiments conducted in 2013 and 2014, using a randomized complete block design, with three replications. Nitrogen was applied at planting at rates of 0, 25, 50, and 75 kg N ha^{-1}. In both years, starter nitrogen fertilizer benefited root activity, leaf photosynthesis and, consequently, yield. Statistically significant correlation was found among root activity, leaf photosynthetic rate, and grain yield at the developmental stage. The application of N25, N50, and N75 increased grain yield by 1.28%, 2.47%, and 1.58% in 2013 and by 0.62%, 2.77%, and 2.06% in 2014 compared to the N0 treatment. Maximum grain yield of 3238.91 kg ha^{-1} in 2013 and 3086.87 kg ha^{-1} in 2014 were recorded for N50 treatment. Grain yield was greater for 2013 than 2014, possibly due to more favorable environmental conditions. This research indicated that applying nitrogen as starter is necessary to increase soybean yield in Sangjiang River Plain in China.

Photosynthetic pigments, such as chlorophyll (Chl) a, Chl b and carotenoids concentration, and chlorophyll fluorescence (CF) have widely been used as indicators of stress and photosynthetic performance in plants, although photosynthetic pigments and CF are partly interdependent due to absorption and re-remittance of solar energy in green leaves. To investigate the relationship between photosynthetic pigments and CF, soybean plants were grown in controlled environments at three levels of phosphorus (P) treatments (0.50, 0.10, 0.01 mM) under ambient and elevated CO_2 (aCO_2, 400 and eCO_2, 800 µmol mol^{-1}, respectively). The significant effect of treatments (P and CO_2) on soybean were confirmed by the decrease in plant height, mainstem node number, and leaf area associated with reductions in the rates of stem elongation, node addition, and leaf area expansion under P deficiency. However, eCO_2 stimulated these growth parameters. Under P deficiency, the estimated CF parameters, total Chl,

Table 7.2 Effects of starter nitrogen fertilizer on leaf photosynthetic rate at different stages (µmol CO_2m^{-2} s^{-1}).

Nitrogen (kg/ha)	V4			R2			R4		
	2013	2014	Mean	2013	2014	Mean	2013	2014	Mean
0	18.79	15.78	15.90	24.87	22.20	23.54	20.23	17.35	18.79
25	19.91	16.84	17.98	25.06	22.11	23.59	20.54	17.53	19.04
50	21.13	19.07	19.29	25.91	26.24	26.08	20.51	18.11	19.31
75	17.93	18.14	16.35	22.54	24.27	23.41	20.06	17.84	18.95

Chl a, and Chl b concentrations decreased but carotenoids concentration was fairly stable, indicating its possible role in photo protection in soybean. The CF parameters showed a good relationship with chlorophyll concentration but a relatively poor correlation with Chl a/b ratio or carotenoids. However, TChl/Carotenoids ratio showed the strongest linear relationship with CF parameters, such as efficiency of energy harvesting by photosystem II reaction centers ($r^2 = 0.70$) and photochemical quantum yield ($r^2 = 0.60$). This relationship was not affected by growth CO_2 conditions. The high correlation between TChl/Carotenoids emphasized the importance of the quantification of both chlorophyll and carotenoids concentrations to understand the photochemistry and underlying processes of minimizing photo protection mechanisms in the given environmental conditions.

The effects of phosphorus nutrition on various aspects of photosynthetic metabolism have been examined for soybean plants (*Glycine max*) grown in growth chambers. Orthophosphate was supplied at two levels in 0.5-strength Hoagland's solution. At the end of the 19-d growth period, plants grown at 10 μM KH_2PO_4 (low-P plants) had undergone a 40% drop in net CO_2 exchange (averaged over a 16-h light period), as compared with control plants grown with 200 μM KH_2PO_4. Low-P resulted in reductions in the initial activities of five, and in the total activities of seven, Calvin-cycle enzymes. Notable exceptions were the initial and total activities of chloroplastic fructose-1,6-bisphosphatase (EC 3.1.3.11) which were increased by 85 and 53%, respectively, by low-P. Low-P decreased leaf 3-phosphoglycerate (PGA) levels most (by 80%), ribulose-1,5-bis-phosphate (RuBP) less (by 47%) while triose-phosphate (TP) was not significantly changed. The results indicate that photosynthetic CO_2-fixation in low-P plants was limited more by RuBP regeneration than by ribulose-1,5-bisphosphate carboxylase/oxygenase (EC 4.1.1.39) activity. Ribulose-1,5-bisphosphate regeneration in low-P plants did not appear to be limited by ATP and-or NADPH supply because ATP/ADP and NADPH/NADP$^+$ ratios were increased by 60 and 37%, respectively, by low-P, and because TP/PGA ratios were higher in low-P plants. Low-P may diminish RuBP regeneration and, hence, photosynthesis, by reducing Calvin-cycle enzyme activity, in particular, the initial activity of ribulose-5-phosphate kinase (EC 2.7.1.19) (44% reduction), and by enhancing the flux of carbon into starch biosynthesis (Fredeen et al., 1990) (Fig. 7.7; Table 7.3).

Low-P had a lesser effect on photosynthesis; light saturated photosynthetic rates at ambient and saturated CO_2 levels were lowered by 55 and 45%, respectively, after 19 days of low-P treatment. Low-P treatment increased starch concentrations in mature leaves, expanding leaves and fibrous roots; sucrose concentrations, however, were reduced by low-P in leaves and increased in roots. Foliar F-2,6-BP levels were not affected by P treatment in the light but in darkness they increased with high-P and decreased with low-P. The increase in the starch/sucrose ratio in low-P leaves was correlated primarily with changes in the total activities of enzymes of starch and sucrose metabolism.

Potassium is an important nutrient element requiring high concentration for photosynthetic metabolism. The potassium deficiency in soil could inhibit soybean (*Glycine max* (L.) Merr.) photosynthesis and result in yield reduction. Two representative soybean varieties Tiefeng 40 (tolerance to K$^+$ deficiency) and GD8521 (sensitive to K$^+$ deficiency) were hydroponically grown to measure the

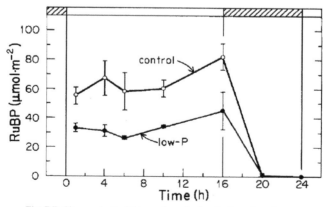

Fig. 7.7 Changes in RuBP in soybean leaf under low phosphorus.

Table 7.3 Effect of low-P and Control on initial and total activities of selected Calvin cycle enzymes in soybean leaves after 1.5 h of illumination at 500 μmole.m^{-2}.s^{-1} PFD (n = 3, ±SD).

Enzyme Activity	Low-P (μmol.m^{-2}s^{-1})	Control (μmol.m^{-2}s^{-1})	Low-P as % of Control
Ribulose-1,5-bisphosphate carboxylase/oxygenase			
Initial	90 ± 9	119 ± 14	76
Total	98 ± 12	138 ± 6	71
3-phosphoglyceric acid kinase			
Initial	47 ± 2	86 ± 3	55
Total	50 ± 9	64 ± 4	78
NADP-glyceraldehyde-3-phosphate dehydrogenase			
Initial	50 ± 11	74 ± 2	68
Total	72 ± 16	86 ± 13	84
Triose-phosphate isomerise			
Total	365 ± 29	488 ± 112	75
Fructose-1,6-bisphosphate aldolase			
Total	19.8 ± 2.2	24.4 ± 0.95	81
Fructose-1,6-bisphosphate phosphatase			
Initial	12.2 ± 0.4	6.6 ± 0.8	185
Total	15.3 ± 1.4	10.0 ± 0.9	153
Transketolase			
Total	5.06 ± 0.3	8.75 ± 0.5	58
Sedoheptulose-1,7-bisphosphate phosphatise			
Initial	1.88 ± 0.11	2.94 ± 0.17	64
Total	5.26 ± 0.5	4.37 ± 0.55	120
Ribulose-5-phosphate kinase			
Initial	39 ± 4	70 ± 1	56
Total	89 ± 14	95 ± 6	94

photosynthesis, chlorophyll fluorescence parameters and Rubisco activity under different potassium conditions. With the K-deficiency stress time extending, the net photosynthetic rate (P_n), transpiration rate (T_r) and stomatal conductance (G_s) of GD8521 were significantly decreased under K-deficiency condition, whereas the intercellular CO_2 concentration (C_i) was significantly increased. In contrast, the variations of Tiefeng 40 were almost little under K-deficiency condition, which indicated tolerance to K$^+$ deficient variety could maintain higher efficient photosynthesis. On the 25th d after treatment, the minimal fluorescence (F_0) of GD8521 was significantly increased and the maximal fluorescence (F_m), the maximum quantum efficiency of PSII photochemistry (F_v/F_m), actual photochemical efficiency of PSII (ϕ_{PSII}), photochemical quenching (q_P), and electron transport rate of PSII (*ETR*) were significantly decreased under K$^+$ deficiency condition. In addition, the Rubisco content of GD8521 was significantly decreased in leaves. It is particularly noteworthy that the chlorophyll fluorescence parameters and Rubisco content of Tiefeng 40 were unaffected under K$^+$ deficiency condition. On the other hand, the non-photochemical quenching (q_N) of Tiefeng 40 was significantly increased. The dry matter weight of Tiefeng 40 was little affected under K$^+$ deficiency condition. Results indicated that Tiefeng 40 could avoid or relieve the destruction of PSII caused by exceeded absorbed solar energy under K-deficiency condition and maintain natural photosynthesis and plant growth. It was an essential physiological mechanism for low-K-tolerant soybean under K-deficiency stress.

Aluminium is a light metal that makes up 7% of the earth's crust, occurring in the form of harmless oxides and aluminosilicates. If the soil becomes acidic, Al is solubilized into toxic forms like

[Al $(H_2O)_6]^{3+}$, generally referred to Al^{3+}, which is now present in 40% of the arable lands in the world. Excess Al^{3+} in soil enters roots, resulting in reduced plant vigor and yield (Ciamporová, 2002). The initial symptom of Al toxicity is the inhibition of root elongation, which has been proposed to be caused by a number of different mechanisms, including Al interactions within the cell wall (Massot et al., 1999), the plasma membrane (Piñeros and Kochian, 2001), or the symplast (Kochian, 1995). Al stress was found to be accompanied by decreasing chlorophyll concentrations and photosynthetic rates in *Picea abies* (Schlegel and Godbold, 1991), rice (Shi, 2004), and *Dimocarpus longana* (Xiao, 2002). Due to reduction in absorbing surfaces (Stienen, 1986), root-water permeability (Zhao et al., 1987) and stomata aperture (Hampp and Schnabl, 1975), Al has generally been found to decrease transpiration rates. In contrast, Schlegel and Godbold (1991) observed enhanced transpiration rates of spruce needles due to Al. Further investigations on aerial tissues are needed to seek better stress elimination solutions or strategies for growing soybean plants in Al contaminated soil.

In a study, the photosynthetic and related morphological and physiological characteristics of two soybean (*Glycine max* Merrill.) varieties were evaluated in response to aluminium (Al) stress in soil. The pot-grown soybean plants were cultured with different supplemental aluminium, and measurements was conducted during the 5-foliate period. Results indicate that Al at low concentrations in the soil is helpful to growth, and Al is toxic to plants only when the concentration exceeds a certain threshold. Increased leaf area, root surface area, specific leaf weight (SLW), and lower malondialdehyde levels were found in soybean plants under a 200 mg/kg Al^{3+} treatment. However, higher aluminium concentrations (800 mg/kg) caused declining chlorophyll contents, depressed photosynthesis rates (PN), enhanced transpiration rates, and decreased PAR utilization efficiency (PUE) and water utilization efficiency. No significant difference in stomatal conductance or leaf water potential was observed among soybean plants under the various aluminium treatments. Moreover, higher aluminium concentration significantly increased lipid peroxidation, decreased cell membrane stability, and changed the activities of superoxide dismutase (SOD) in the leaves of both plants. It is concluded that soybean plants maintain relatively higher SLW, PN, PUE, WUE, SOD activity in order to cope with high aluminium stress.

In a study, Tiefeng 31 seedling leaves were sprayed with 1500 mg L^{-1} $KHCO_3$ and $NaHCO_3$, respectively, and the photosynthetic rate, soluble sugar, photosynthetic pigment and other physiological indices were determined in order to explore the effect of K^+, Na^+ and HCO_3^- on soybean seedlings photosynthesis. The results showed that, compared with water spraying control, K^+, Na^+ and HCO_3^- improved soybean seedlings' photosynthetic rate, the content of soluble sugar, chlorophyll a, chlorophyll b, and the ratio of chlorophyll b in total chlorophyll, promoted soybean seedlings photosynthesis by providing more assimilatory power to dark reactions carbon fixation, and enhanced the activities of ATP enzyme, photophosphorylation, and PEPC. In addition, HCO_3^- played a more important role in promoting chlorophyll b synthesis and improving photosynthetic rate. $NaHCO_3$ significantly increased the content of Rubisco. For other physiological indexes but the content of Rubisco, the promotion of K^+ is more than Na^+ (Hao et al., 2013) (Tables 7.4 and 7.5).

Table 7.4 Effects of Na^+, K^+ and HCO_3^- on content of Rubisco of soybean seedlings leaves.

Treatment	Rubisco content	Δ%	Rubisco occupies soluble protein (%)	Δ%
$KHCO_3$	5.9270 ± 1.1030	−5.17	35.29	−18.06
$NaHCO_3$	8.1260 ± 2.2016	30.02	49.16	14.14
CK	6.2500 ± 1.1450	–	43.07	–

Table 7.5 Effects of Na K and HCO_3^- on photosynthetic rate of the leaf of soybean seedlings.

Treatment	Photosynthetic rate ($\mu mol.m^{-2}s^{-1}$)	Δ%
$KHCO_3$	9.5200 ± 0.3233	13.20
$NaHCO_3$	8.9100 ± 0.1386	5.95
CK	8.4100 ± 0.1848	–

7.7 Shade Effect

Shade influences the interception of light with both difference in quality and quantity and with varied energy. Thus, photosynthetic changes are also different in plants. Obviously, shade reduces the net photosynthetic rate of soybean. It was also observed that, when normal light intercepts after the stress, there is a lag phase in recovery. Depending on the plant population interception of light vary, in high population at reproductive phase shade there is a decline in photosynthetic rate.

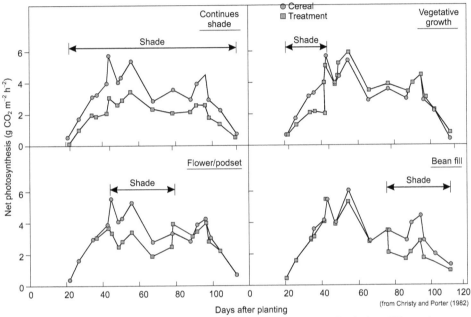

Fig. 7.8 Net photosynthesis changes in soybean after planting to harvest under shade at different phases.

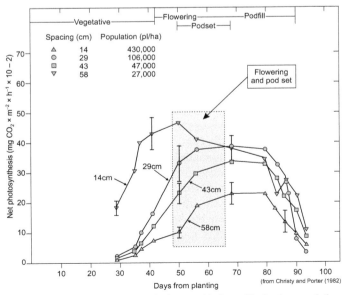

Fig. 7.9 Variations in net photosynthetic rate with time with planting population.

Photosynthesis (CER (μmol CO_2 m^{-2} leaf area s^{-1})), stomatal conductance (*gsw*), and intercellular [CO_2] (*Ci*) of soybean (*Glycine max* L. Merr.) grown using the early soybean production system (ESPS) of the midsouth were determined. Three irrigated cultivars were grown using ESPS on Bosket (Mollic Hapludalfs) and Dundee (Typic Endoaqualf) soils in 2011 and 2012 at Stoneville, MS.

Single leaf CER, *gsw*, and *Ci* were determined at growth stages R3, R4, and R5 using decreasing photosynthetic photon flux densities (PPFD, μmolm^{-2} s^{-1}) beginning at 2000 PPFD and decreasing by 250 PPFD increments to 250 PPFD. Photosynthesis changes fit a quadratic polynomial for all fixed variables and range from 6.0 and 9.0 CER at 250 PPFD and 22.0 to 28.0 CER at 2000 PPFD. No cultivar differences in CER, *gsw*, or *Ci* were noted at any growth stage or site either year. In 2012, CER, *gsw*, and *Ci* were lower when measured at R5 than the two previous growth stages, which was not observed in 2011. The R5 sampling in 2012 had accumulated 39 to 70 more growing degree units at 10°C base temperature (GDU 10's) than in 2011 and were likely more mature. Increased soybean yields from ESPS appear not to result from higher leaf CER.

The use of hydroponic systems for cultivation in controlled climatic conditions and the selection of suitable genotypes for the specific environment help in improving crop growth and yield. It was hypothesized that plant performance in hydroponics could be further maximized by exploiting the action of plant growth-promoting organisms (PGPMs). However, the effects of PGPMs on plant physiology have scarcely been investigated in hydroponics. Within a series of experiments aimed to identify the best protocol for hydroponic cultivation of soybean [*Glycine max* (L.) Merr.], the effects of a PGPMs mix, containing bacteria, yeasts, mycorrhiza and trichoderma beneficial species on leaf anatomy, photosynthetic activity and plant growth were evaluated in soybean cv.Pr91m10 in closed nutrient film technique (NFT). Plants were grown in a growth chamber under semi-aseptic conditions, inoculated at seed, seedling and plant stages, and compared to non-inoculated (control) plants. Light and epi-fluorescence microscopy analyses showed that leaves of inoculated plants had higher density of smaller stomata (297 vs. 247 n/mm^2), thicker palisade parenchyma (95.0 vs. 85.8 mm), and larger intercellular spaces in the mesophyll (57.5% vs. 52.2%), compared to non-inoculated plants. The modifications in leaf functional anatomical traits affected gas exchanges; in fact, starting from the reproductive phase, the rate of leaf net photosynthesis (NP) was higher in inoculated than in control plants (8.69 vs. 6.13 mmol CO_2 m^{-2}s^{-1} at the beginning of flowering). These data are consistent with the better maximal PSII photochemical efficiency observed in inoculated plants (0.807 vs. 0.784 in control); conversely, no difference in leaf chlorophyll content was found. The PGPM-induced changes in leaf structure and photosynthesis lead to an improvement of plant growth (C29.9% in plant leaf area) and seed yield (C36.9%) compared to control. Our results confirm that PGPMs may confer benefits in photosynthetic traits of soybean plants even in hydroponics (i.e., NFT), with positive effects on growth and seed production, prefiguring the potential application of beneficial microorganisms in plant cultivation in hydroponics (Paradiso et al., 2017).

7.8 Glyphosate

Photosynthesis is the major biochemical process occurring in photoautotrophic organisms and is known to be affected by various anthropogenic factors. Some herbicides were found to directly interrupt photosynthetic electron transport. For example, 3-(3,4-dichlorophenyl)-1,1-dimethylurea (DCMU) is known to block the electron flow between QA and QB by competing for QB binding sites (Tóth et al., 2005). Other herbicides, such as glyphosate, will affect photosynthesis indirectly by inhibiting the biosynthesis of carotenoids, chlorophylls, fatty acids, or amino acids (Fedtke and Duke, 2005). As an EPSPS competitive inhibitor, glyphosate blocks the shikimate pathway, inhibiting the biosynthesis of secondary metabolites in plants, including compounds related to photosynthesis, such as quinones (Dewick, 1998). However, it is unclear how glyphosate leads to plant death, and hypotheses, such as depletion of protein stocks and drainage of C from other vital pathways, have been put forward (Duke and Powles, 2008). A closer look at glyphosate's effects on photosynthetic processes may shed light on this hypothesis. Indeed, numerous field and greenhouse studies have indicated a decreased photosynthetic rate in plants following glyphosate exposure (Zobiole et al., 2012).

Global production of glyphosate-resistant (GR) soybean [*Glycine max* (L.) Merr.] continues to increase annually; however, there are no particular specific fertilizer recommendations for the transgenic varieties used in this system, largely because reports of glyphosate effects on mineral nutrition of *GR* soybeans are lacking. Several metabolites or degradation products of glyphosate have been identified or postulated to cause undesirable effects on GR soybeans. In a work, increasing glyphosate rates in different application on cv. '*BRS 242 GR*' in order to evaluate photosynthetic parameters, macro- and micronutrient uptake and accumulation and shoot and root dry biomass production was undertaken. Increasing glyphosate rates revealed a significant decrease in photosynthesis, macro and micronutrients accumulation in leaf tissues and also decreases in nutrient uptake. The reduced biomass in *GR* soybeans represents additive effects from the decreased photosynthetic parameters as well as lower availability of nutrients in tissues of the glyphosate treated plants.

Previous greenhouse studies have demonstrated that photosynthesis in some cultivars of first-(GR1) and second-generation (GR2) glyphosate-resistant soybean was reduced by glyphosate. The reduction in photosynthesis that resulted from glyphosate might affect nutrient uptake and lead to lower plant biomass production and, ultimately, reduced grain yield. Therefore, a field study was conducted in order to determine if glyphosate-induced damage to soybean (*Glycine max* L. Merr. cv. Asgrow AG3539) plants observed under controlled greenhouse conditions might occur in the field. The study evaluated photosynthetic rate, nutrient accumulation, nodulation, and biomass production of GR2 soybean receiving different rates of glyphosate (0, 800, 1200, 2400 g a.e. ha^{-1}) applied at V2, V4, and V6 growth stages. In general, plant damage observed in the field study was similar to that in previous greenhouse studies. Increasing glyphosate rates and applications at later growth stages decreased nutrient accumulation, nodulation, leaf area, and shoot biomass production. Thus, to reduce potential undesirable effects of glyphosate on plant growth, application of the lowest glyphosate rate for weed-control efficacy at early growth stages (V2 to V4) is suggested as an advantageous practice within current weed control in GR soybean for optimal crop productivity.

It is generally claimed that glyphosate kills undesired plants by affecting the 5-enolpyruvylshikimate-3-phosphate synthase (EPSPS) enzyme, disturbing the shikimate pathway. However, the mechanisms leading to plant death may also be related to secondary or indirect effects of glyphosate on plant physiology. Moreover, some plants can metabolize glyphosate to aminomethylphosphonic acid (AMPA) or be exposed to AMPA from different environmental matrices.

AMPA is a recognized phytotoxin, and its co-occurrence with glyphosate could modify the effects of glyphosate on plant physiology. An overall picture of alterations caused by environmental exposure to glyphosate and its metabolite AMPA and their effects on several physiological processes are summarized. It particularly focuses on photosynthesis, from photochemical events to C assimilation and translocation, as well as oxidative stress.

7.9 Photosynthesis in Relation to Yield

Improvement of soybean [*Glycine max* (L.) Merr.] yield and photosynthesis physiology have been achieved over decades of cultivar breeding. Identification of the mechanisms involved in shoot-root interactions would be beneficial for the development of yield improvement breeding strategies. The objectives of this study were to investigate soybean shoot-root interactions with different-year released soybean cultivars and to evaluate their effects on grain yield and yield components. Soybean grafts used in this study were constructed with two record-yield cultivars, Liaodou 14 (L14) and Zhonghuang 35 (Z35), and eleven cultivars released in 1966–2006 from the United States of America and China. The grafting experiments were conducted as pot-culture experiments and repeated in 2014 and 2015. Results showed that net photosynthesis rate (PN) was positively correlated to both root activity and root bleeding sap mass (RBSM) during the R6 reproductive stage. Moreover, different year-released soybean shoots had all exhibited capabilities of changing the root activity and architecture of L14 and Z35 rootstocks to "generation"-specific patterns during all reproductive stages. However, these influences were independent of the photosynthetic strength. Yield analysis had demonstrated that high-yielding root systems (L14 and Z35 rootstocks) could cause more than 15% of yield increase in seven out of eleven common scions in a scion genotype-dependent manner. For Williams-descendant cultivar scions, L14 and Z35 rootstocks promoted yields mainly by increasing the seed number (SN), but those scions of Amsoy-descendent cultivars showed mainly seed weight (SW) increases when grafted onto L14 and Z35 rootstocks. On the other hand, although most tested common rootstocks did not show significant influence over the final yields in record-yield L14 and Z35 scions, they were obviously capable of shifting the formation of yield components when compared to L14 and Z35 self-grafting controls. Taken together, soybean shoots could influence the root physiology and played a crucial role in the determination of yield potentials. Synergistically with shoots, soybean roots played a more supportive role during the realization of yield potentials through root activities.

Soybeans [*Glycine max* (L.) Men. cv. Bragg] were grown throughout their life cycle in six controlled-environment chambers that maintained CO_2 concentrations at 330, 450, 600, and 800 µmol CO_2 mol^{-1} air. Air temperatures were controlled, and solar radiation and dewpoint temperatures were measured continually for each chamber. Net photosynthesis and transpiration were computed on 5-min intervals. The CO_2 enrichment had no apparent effect on the rate of mainstem node development, but leaf area increased at a faster rate proportional to the CO_2 concentration. Day-time CO_2 exchange rate (CER) increased with increasing CO_2 and followed the same trend as final yield. Midday CER was typically about 90 µmol CO_2 m^{-1} of ground area s^{-1} for 800 µmol mol^{-1} and 60 µmol CO_2 m^{-2} ground area per second for 330 µmol mol^{-1} chambers. Canopy transpiration rates, leaf area index and vapor pressure deficits were used to compare bulk canopy resistances among CO_2 treatments. Bulk canopy resistance (stomatal and boundary-layer) of the 800 µmol mol^{-1} treatment was about 1.6 times that of the 330 µmol mol^{-1} CO_2 treatment. These results imply that the expected continued increase in global atmospheric CO_2 will increase both agricultural productivity and water use efficiency.

Experimental work was undertaken to investigate development of some photosynthetic characters and their relationship with seed yield during the genetic improvement of soybean. 38 soybean cultivars representing genetic improvement progress in northeast China, released from the period of 1923 to 2005 (82 years), were grown. The experiment was performed in 2005 and 2006. Some photosynthetic characteristics, such as leaf area, chlorophyll content and specific leaf weight (SLW), were measured at R2 stage (flowering stage). Days to maturity and yield were measured in each plot. The results of two years study showed that number of days to maturity decreased from 140.3 d to 125.6 d depending on the year of release, but soybean seed yield increased significantly from 1282.3 Kg to 2310.7 Kg during the 82 years of breeding. The net photosynthetic rate (Pn), stomatal conductance (Gs), apparent mesophyll conductance (Pn/Ci), transpiration (Tr), chlorophyll content and SLW all increased in bred cultivars during the improvement, however, the leaf area and ratio of Ci/Ca (intercellular CO_2 concentration, Ci; ambient CO_2 concentration, Ca) decreased in the same time. Yield was positively correlated with Pn, Tr, Gs, Pn/Ci, chlorophyll content and SLW, whereas the leaf area, Ci/Ca and WUE were negatively correlated. Results revealed that Pn and Pn/Ci are the effective selection indexes for seed yield and can be used in future soybean breeding programs. We also found that the increase of Tr is higher than Pn. As a result, WUE is decreased as the increase of yield with year of release. Therefore, the main cost of high yield in new cultivars is water expense. A work was to investigate development of some photosynthetic characters and their relationship with seed yield during the genetic improvement of soybean. For this objective, we grew 38 soybean cultivars, representing genetic improvement progress in northeast China, released from the period of 1923 to 2005 (82 years). The experiment was performed in 2005 and 2006. Some photosynthetic characteristics, such as leaf area, chlorophyll content and specific leaf weight (SLW), were measured at R2 stage (flowering stage). Days to maturity and yield were measured in each plot. The results of two years study showed that number of days to maturity decreased from 140.3 d to 125.6 d, depending on the year of release, but soybean seed yield increased significantly from 1282.3 Kg to 2310.7 Kg during the 82 years of breeding. The net photosynthetic rate (Pn), stomatal conductance (Gs), apparent mesophyll conductance (Pn/Ci), transpiration (Tr), chlorophyll content and SLW all increased in bred cultivars during the improvement, however, the leaf area and ratio of Ci/Ca (intercellular CO_2 concentration, Ci; ambient CO_2 concentration, Ca) decreased in the same time. Yield was positively correlated with Pn, Tr, Gs, Pn/Ci, chlorophyll content and SLW, whereas the leaf area, Ci/Ca and WUE were negatively correlated. Our results revealed that Pn and Pn/Ci are the effective selection indexes for seed yield and can be used in future soybean breeding programs. We also found that the increase of Tr is higher than Pn. As a result, WUE decreases as the yield increases with year of release. Hence, the main cost of high yield in new cultivars is water expense (Liu et al., 2012) (Fig. 7.10 and Table 7.6).

Intrinsic water use efficiency (WUEintr), the ratio of photosynthesis to stomatal conductance to water, is often used as an index for crop water use in breeding projects. However, WUEintr conflates variation in these two processes and, thus, may be less useful as a selection trait than knowledge of

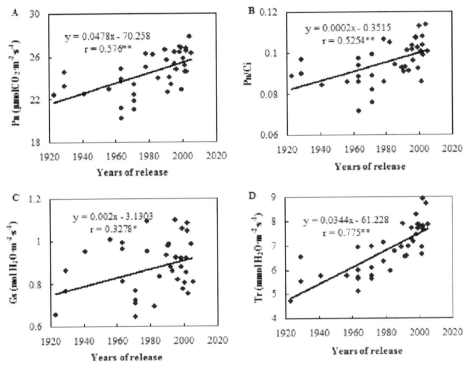

Fig. 7.10 Changes in Pn, Pn/Ci, Gs and Tr with year of release.

Table 7.6 Correlation coefficients between yield and Pn, Gs, Ci/Ca, Pn/Ci, Tr, chlorophyll content and SLW of 38 soybean cultivars with means of 2005 and 2006. 38 soybean cultivars representing genetic improvement progress in northeast China during the period of 1923 to 2005 (82 years). The experiment was performed in 2005 and 2006.

	Yield	Pn	SLW	Chlo	Gs	Ci	Ci/Ca	Pn/Ci	Tr
Pn	0.6102								
SLW	0.4464	0.446							
Chlo	0.3075	0.5127	0.4939						
Gs	0.1613	0.5644	0.2981	0.4444					
Ci	−0.282	−0.011	−0.200	−0.117	0.312				
Ci/Ca	−0.349	−0.205	−0.244	−0.327	0.1206	0.915			
Pn/Ci	0.5211	0.7919	0.4996	0.5985	0.4035	−0.36	−0.55		
Tr	0.6277	0.6715	0.4836	0.3716	0.4739	−0.04	−0.23	0.6587	
WUE	−0.516	−0.21	−0.164	−0.003	−0.125	0.092	0.075	0.0606	−0.5

Analysis of correlation were performed according to the means of 2005 and 2006. Ci, intercellular CO_2 concentration; Ca, ambient CO_2 concentration; Chl, chlorophyll content; Pn, net photosynthetic rate (Pn); Pn/Ci, apparent mesophyll conductance; Gs, stomatal conductance; Tr, transpiration; WUE, water usage efficiency; SLW, specific leaf weight.

both components. Photosynthetic capacity was defined as the variation in WUEintr that would occur if genotypes of interest had the same stomatal conductance as a reference genotype and only differed in photosynthesis; similarly, the contribution of stomatal conductance to WUEintr was calculated assuming a constant photosynthetic capacity across genotypes. Genotypic differences in stomatal conductance had the greatest effect on WUEintr (26% variation when well-watered), and was uncorrelated with the effect of photosynthetic capacity on WUEintr. Thus, photosynthetic advantages of 8.3% were maintained under drought. The maximal rate of Rubisco carboxylation, generally the limiting photosynthetic process for soybeans, was correlated with photosynthetic capacity. As this trait was not interactive with leaf temperature, and photosynthetic capacity differences were maintained under mild drought, the observed patterns of photosynthetic advantage for particular genotypes are likely to be consistent across a range of environmental conditions. This suggests that it is possible to employ a selection strategy of breeding water-saving soybeans with high photosynthetic capacities in order to compensate for otherwise reduced photosynthesis in genotypes with lower stomatal conductance (Gilbert et al., 2011) (Fig. 7.11).

Two pot experiments were conducted to assess the effect of various degree of defoliation on photosynthesis, dry matter production and yield in soybean. Defoliation significantly increased rate of photosynthesis and transpiration and leaf conductance of the soybean genotypes however, it decreased dry matter production and yield.

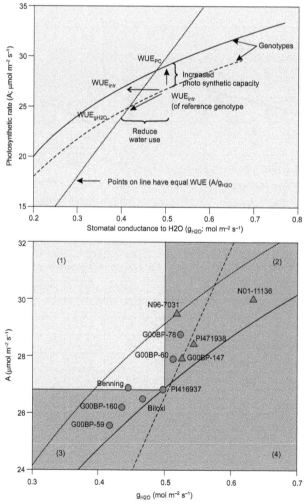

Fig. 7.11 Photosynthetic rate and stomatal conductance to WUEintr relationships.

Photosynthesis can be limited by sink capacity (i.e., ability to use photosynthate). After flowering, the major sink in grain crops is the number and potential size of the seed formed. Decreased sink capacity, as may be induced by removing filling grains, can feedback to decrease photosynthetic capacity (Peet and Kramer, 1980). It may be expected, however, that breeding selects for the cultivars that are able to make maximum use of photosynthetic capacity. For example, if weather favours increased photosynthesis, an effectively selected cultivar should have sufficient capacity for formation of grain to use the additional photosynthate. However, a recent detailed analysis that reviewed the magnitude of seed dry weight changes in response to manipulations in assimilate availability during seed filling for wheat, maize and soybean has concluded that, in all three crops, yield is usually more limited by sink than by source (i.e., photosynthesis) (Borrás et al., 2004).

The rate of photosynthesis (CER) varied considerably between plants and was comparable to those observed previously (Campbell et al., 1988). The ratio of TPN/CER at Pmax, called PIR (photosynthate interest rate) in pod formation, was found to be an important factor in the relationship between Pmax and TPN or TSW. After taking this into consideration, close relationships were then found between Pmax and TPN (R^2 = 0.803 and 0.894) that led to the development of PIR model. The estimated TSW as a function of N, Pmax and PIR with the PIR model was highly correlated to the observed TSW (R^2 = 0.944) (Fig. 7.12). This supports the inclusion of photosynthetic rate and PIR as selection parameters in addition to pod number. Sedghi and Amanpour-Balaneji (2010) concluded pod number and pre-flowering net photosynthesis as the best selection criteria in soybean for grain yield. Harrison et al. (1981) succeeded to use CAP (canopy apparent photosynthesis) as a selection parameter to increase yield of soybean lines, and to select the best lines (F3-derived lines). The use of PIR as a selection criteria has never been reported previously, but in line with a conclusion that source/sink ratio is an important trait for further genetic improvement of soybean yield (Kumudini, 2002).

A field study was conducted in order to assess the growth parameters controlling the dry matter and seed yield of soybean. The result shows that growth rate was slow during vegetative phase in all genotypes. A relatively smaller portion of total dry mass (TDM) was produced before flower initiation and the bulk of it after anthesis. Maximum absolute growth rate (AGR) was observed during pod filling stage in all genotypes due to maximum leaf area (LA) development and leaf area index (LAI) at this stage. Plant characters, like LAI and AGR, contributed to higher TDM production. Results indicate that a high yielding soybean genotype should possess larger LAI, higher TDM production ability and higher AGR at all growth stages (Malek et al., 2012) (Table 7.7).

The historical positive correlation between photosynthesis and yield in soybean suggests that improving photosynthetic efficiency might be a promising target for further yield gains. As defined by Monteith (1977), yield of a crop at any given location is the product of the incident photosynthetically active radiation, and the efficiencies with which it intercepted (ei), the intercepted PAR is converted into biomass (ec), and the efficiency with which the biomass is partitioned into seed (ep), also termed harvest

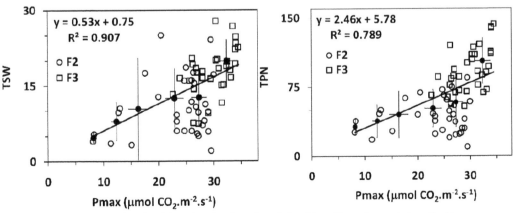

Fig. 7.12 Total seed weight and total pod number relationships with maximum photosynthesis.

Table 7.7 Some physiological parameters, yield components and yield in five soybean genotypes.

Genotype	BAU-21	BAU-70	BAU-80	Shohag	BARI soybean5	F test
Chlorophyll (mg g^{-1} FW)	2.16	2.17	2.21	2.12	2.09	NS
Photosynthesis (μmol CO$_2$ dm^{-2}s^{-1})	20.66	24.24	19.22	21.45	19.89	*
Pods plant^{-1}	24.10	31.23	21.40	17.90	20.8	**
Seeds pod^{-1}	2.12	2.19	2.17	2.19	2.13	NS
100-seed weight (g)	15.11	14.55	16.08	15.51	15.88	*
Seed wt. plant^{-1} (g)	7.71	9.95	7.47	6.08	7.03	**
Seed yield (t ha^{-1})	2.57	3.31	2.49	2.02	2.34	**
Harvest index (%)	33.22	37.73	33.24	28.28	34.16	*

Same letter(s) in a column does not differ significantly at P ≤ 0.05 by DMRT; NS = Not significant; ** and * indicate significance at 1 and 5% level of probability, respectively.

index (Zhu et al., 2010). It has been argued that ei and ep have been maximized for modern crops (Zhu et al., 2010), including soybean. The canopy of a modern cultivar of soybean growing in central Illinois was shown to intercept ~ 90% (ei = 0.9), of the incident PAR integrated over the growing season, and to partition ~ 60% (eP = 0.6) of the biomass energy into seed (Zhu et al., 2010). These achievements in ei and ep seem to leave little room for further improvement in soybean. As ei represents the interception efficiency over the growing season, greater yield may be obtainable if, at a given location, the growing season could be extended. Growing seasons in the corn belt of the United States are generally limited by temperature, and in the western United States by moisture. Identification of germplasm capable of development and maintenance of leaves at lower temperature or lower water potentials could allow breeding of more productive lines, as could an improved understanding of the gene networks affecting these characters. In the absence of growing season extension, ec remains an important, and mathematically perhaps the only remaining target for improvement of yield potential. Recent estimates of maximum theoretical ec for soybean and other C3 plants range from 4.1 to 4.6%, at current [CO$_2$] and 30°C (Zhu et al., 2008; Amthor, 2010). Soybean grown in productive soils in central Illinois achieved ec of 1.6% at an atmospheric CO$_2$ concentration of 380 ppm (Zhu et al., 2010), falling well short of the theoretical maximum. The observation that neither ei nor ep increased while leaf photosynthesis, ec and seed yield increased in soybean exposed to season-long elevation of CO$_2$ concentration suggests that attempts to increase ec by altering photosynthetic metabolism could have similar beneficial effects on seed yield (Zhu et al., 2010). Targets for improving ec and enhancing C3 photosynthesis have been the subject of a number of recent papers (Ort et al., 2011; Parry et al., 2011), and are collectively hypothesized to boost yield potential by up to 50% (Long et al., 2006b). In the following section are the potential targets for improving ec, highlighting those with the greatest potential to be realized in soybean in the next 20 years. A natural starting point for improving ec in C3 plants is ribulose 1,5-bisphosphate carboxylase/oxygenase (Rubisco), the primary enzyme of CO$_2$ fixation, which is competitively inhibited by O$_2$ (Spreitzer and Salvucci, 2002). One strategy for improving the performance of Rubisco is to alter its specificity for CO$_2$ relative to O$_2$ (von Caemmerer and Evans, 2010). Although increasing the specificity of Rubisco would increase photosynthesis when ribulose-1,5-bisphosphate (RuBP) is limiting (von Caemmerer and Evans, 2010), there is a trade-off between specificity and catalytic rate (Zhu and Spreitzer, 1996). The average specificity factor for C3 crop canopies was modelled to exceed the optimal level for today's atmospheric carbon dioxide concentration (CO$_2$) (Zhu et al., 2004a). Therefore, an 'optimal canopy' might have a Rubisco with low specificity and high catalytic rate in the upper canopy, and high specificity, lower catalytic rate Rubisco in the lower canopy (Zhu et al., 2010). Rubisco enzymes with a range of specificities and catalytic capacities are found in naturally occurring photosynthetic organisms (Galmes et al., 2005). However, *Limonium gibertii*, a plant adapted to a hot, arid environment, has significantly higher catalytic rates than average C3 species, and also maintains a higher specificity (Galmes et al., 2005). This provides a potential model for avoiding the observed trade-off between specificity and catalytic capacity, and a tool for improving both properties in crops (Parry et al., 2007). It also suggests that more efficient Rubiscos are likely to be discovered. However, there are significant technical barriers to overcome before soybean can be efficiently transformed with foreign Rubisco or engineered by mutagenesis (Parry et al., 2007; Peterhansel et al., 2008). A major complexity is the need to replace both the plastid-encoded large subunit and the nuclear-encoded small subunit in order to ensure that an effective holoenzyme is expressed in the plastid. Foreign small and large subunits have been success-fully engineered into tobacco (Whitney and Andrews, 2001; Dhingra et al., 2004) and key steps controlling the assembly of the Rubisco holoenzyme were recently reported (Liu et al., 2010). Still, a 'better' Rubisco is yet to be engineered in higher plants (Whitney et al., 2011). Recently, the genes encoding two forms of soybean Rubisco activase, which is key to the activation and stability of Rubisco, were cloned and characterized (Yin et al., 2010). Expression of these genes was positively correlated with Rubisco activity, photosynthetic rate and seed yield. Thus, altering Rubisco activase may provide another approach for enhancing soybean photosynthesis and productivity (Spreitzer and Salvucci, 2002; Yin et al., 2010). Although engineering improved Rubisco might be technically challenging in the short term, altering plant investment in other enzymes of primary metabolism is currently feasible. In particular, over-expression of sedoheptulose-1,7-bisphosphatase (SBPase) in tobacco increased photosynthesis and biomass production (Raines, 2006). This remains to be tested in soybean in a field setting, but it is a promising target for enhancing photosynthesis. Using a dynamic metabolic model of C metabolism, Zhu et al. (2007) found that the current partitioning of nitrogen among the enzymes of C3 carbon metabolism was not optimized to today's atmospheric [CO_2]. The model predicted that SBPase should be increased, consistent with the experimental data, and also predicted that ADP glucose pyrophosphorylase could be increased, while photorespiratory enzymes could be decreased. In addition to optimizing concentration of specific Calvin cycle enzymes, increasing overall investment in photosynthetic proteins may also be beneficial. In *Arabidopsis*, investment in proteins involved in primary metabolism was positively correlated with biomass accumulation in 129 accessions (Sulpice et al., 2010). Therefore, it appears that increased investment in photosynthetic capacity is a potential strategy for at least increasing biomass and potentially improving seed yield in C3 crops. In soybean and other C3 plants, the oxygenation reaction and subsequent photorespiration account for a significant loss of energy and, as mentioned previously, C3 crops appear to over-invest in photorespiratory enzymes (Zhu et al., 2007). Therefore, engineering plants to reduce photorespiration would improve conversion efficiency. One strategy for reducing photorespiration is to engineer a CO_2 concentrating mechanism into plants (Hibberd et al., 2008). CO_2 is a competitive inhibitor of the oxygenase reaction of Rubisco. The dicarboxylate cycle of C4 photosynthesis serves as a light driven pump, concentrating CO_2 at Rubisco, to a sufficient level to largely eliminate photorespiration. One solution would, therefore, be to engineer Kranz anatomy and the C4 pathway into soybean. However, this appears to require many changes, not only the expression of two photosynthetic tissue types in the place of one, but also expression of the C4 and C3 enzymes and transporters in the correct tissues and organelles. Given that the gene networks underlying the development of C4 structure and function remain incompletely understood, such transformations were considered long-term goals (Zhu et al., 2010). However, an alternative viewpoint is that, as C4 photosynthesis is broadly similar across flowering plants, yet has evolved multiple times (Sage, 2004), there may be relatively simple and conserved pathways that would facilitate a rapid conversion of C3 to C4 plants (Hibberd et al., 2008). Considerable effort is being invested in work towards engineering the C4 syndrome into rice, and clearly, if successful, would indicate the path for converting other C3 crops (Hibberd et al., 2008). In the shorter term, a strategy may be to engineer the CO_2 concentrating mechanism of cyanobacteria into soybean chloroplasts. For example, *Synechococcus* has membrane proteins that actively pump both bicarbonate and CO_2 into the photosynthetic cell. A further sophistication in some species is the presence of carboxysomes, an ordered structure that encloses Rubisco and carbonic anhydrase within a coat protein. This creates a local high concentration of CO_2 at the site of Rubisco where the conversion of bicarbonate to CO_2 is accelerated (Badger et al., 2002). Chloroplasts are considered to have evolved from ancestral cyanobacterial symbionts, which may have lost these concentrating mechanisms as plants evolved from carbon-limited aquatic systems to the assumed high CO_2 world of the first terrestrial plants. Given the relationship of plastids to cyanobacteria, reintroducing these prokaryotic genes may be feasible. Indeed, Lieman-Hurwitz et al. (2003) produced transgenic *Arabidopsis* and *Nicotiana tabacum* plants that expressed the ictB gene involved in bicarbonate accumulation in *Synechococcus*. These plants had significantly lower CO_2 compensation points of photosynthesis, showing decreased photorespiration and significantly increased rates of leaf CO_2 uptake when CO_2 availability was limiting, but not when it was saturating. Given this success with two other dicotyledonous species, this would appear a promising target for improving soybean photosynthesis. Another strategy for reducing photorespiration, and in particular the energy lost in the current photorespiratory pathway of soybean C3 crops, is to express the key genes of one of the *Escherichia coli* pathways for the metabolism of glycolate to phosphoglycerate, which has been successfully done in *Arabidopsis* (Kebeish et al., 2007). Here, a three reaction pathway of conversion of two molecules of glycolate to one of phosphoglycerate was engineered into the chloroplast. The pathway bypasses the photorespiratory reactions normally involving the cytosol, peroxisomes and mitochondria, resulting in reduced metabolite flow through photorespiration, enhanced carbon assimilation and improved growth in transgenic plants (Kebeish et al., 2007; Peterhansel et al., 2008). Although this pathway still releases one molecule of CO_2 for every two molecules of glycolate formed, it has two advantages. Firstly, the CO_2 is released within the chloroplast, which more effectively increases CO_2 concentration around Rubisco. Secondly, no ammonia is released, which in normal higher plant C2 metabolism requires a large amount of reductive power to reassimilate. The pathway does produce NADH, which would lead to a nucleotide imbalance; however, based on parallel directed evolution of metabolic pathways in *E. coli*, modification to NADPH utilization is unlikely to represent a major barrier. To avoid parallel use of the native C2 metabolic pathway, the plastid glycolate transporter would also need to be knocked-out. It has been suggested that photorespiration has a photoprotective role, particularly in young, expanding soybean leaves exposed to high light at the top of the canopy (Jiang et al., 2006). However, reduction of photorespiration by open-air elevation of CO_2 in the field was not found to cause any loss of photosystem II (PSII) operating efficiency, as an indicator of photoprotection or photoinhibition, at any stage in the plant life cycle (Rogers et al., 2004; Bernacchi et al., 2006). Further, as noted by Jiang et al. (2006), leaf movement and xanthophyll de-epoxidation also play key parts in protection. The xanthophyll cycle can be up-regulated, with apparently little additional investment, and as de-epoxidation is inducible, photosynthetic efficiency is only lowered under conditions of excess light to improve soybean yield.

References

Ainsworth, E.A., P.A. Davey, C.J. Bernacchi, O.C. Dermody, E.A. Heaton, D.J. Moore, P.B. Morgan, S.L. Naidu, H.S.Y. Ra, X.G. Zhu et al. 2002. A meta-analysis of elevated [CO_2] effects on soybean (*Glycine max*) physiology, growth and yield. Glob. Change Biol. 8: 695–709.

Ainsworth, E.A., A. Rogers, R. Nelson and S.P. Long. 2004. Testing the 'source–sink' hypothesis of down regulation of photosynthesis in elevated [CO_2] in the field with single gene substitutions in *Glycine max*. Agricultural and Forest Meteorology 122: 85–94.

Ainsworth, E.A., A.D.B. Leakey, D.R. Ort and S.P. Long. 2008. FACE-ing the facts: Inconsistencies and interdependence among field, chamber and modeling studies of elevated [CO_2] impacts on crop yield and food supply. New Phytologist 179: 5–9.

Allen Jr, L.H. 1990. Plant responses to rising carbon dioxide and potential interactions with air pollutants. Journal of Environmental Quality 19: 15–34.

Amthor, J.S. 2010. From sunlight to phytomass: on the potential efficiency of converting solar radiation to phyto-energy. New Phytologist 188: 939–959.

Anderson, J.M. 1986. Photoregulation of the composition, function, and structure of thylakoid membranes. Annu. Rev. Plant Physiol. 37: 93–136.

Anon, S., J.A. Fernandez, J.A. Franco, A. Torrecillas, J.J. Alarcon and M.J. Sanchez-Blanco. 2004. Effects of water stress and night temperature preconditioning on water relations and morphological and anatomical changes of *Lotus creticus* plants. Sci. Hortic. (Amsterdam, Neth.) 101: 333–342.

Badger, M.R., D. Hanson and G.D. Price. 2002. Evolution and diversity of CO_2 concentrating mechanisms in cyanobacteria. Functional Plant Biology 28: 161–173.

Baker, J.T., L.H.J. Allen and K.J. Boote. 1989. Response of soybean to air temperature and carbon dioxide concentration. Crop Sci. 29: 98–105.

Berkowitz, G.A. 1998. Water and salt stress. pp. 226–237. *In*: Photosynthesis: A Comprehensive Treatise. Cambridge, U.K.: Cambridge University Press.

Bernacchi, C.J., A.D.B. Leakey, L.E. Heady et al. 2006. Hourly and seasonal variation in photosynthesis and stomatal conductance of soybean grown at future CO_2 and ozone concentrations for 3 years under fully open-air field conditions. Plant Cell and Environment 29: 2077–2090.

Berry, J.A. and O. Bjorkman. 1980. Photosynthetic response and adaptation to temperature in higher plants. Annu. Rev. Plant Physiol. 31: 491–543.

Björkman, O., N.K. Boardman, J.M. Anderson, S.W. Thorne, D.J. Goodchild and N.A. Pyliotis. 1972. Effect of light intensity during growth of *Atriplex patula* on the capacity of photosynthetic reactions, chloroplast components and structure. Carnegie Inst. Year Book 71: 115–135.

Blackburn, G.A. 1998. Spectral indices for estimating photosynthetic pigment concentrations: A test using senescent tree leaves. International Journal of Remote Sensing 19: 657–675.

Borrás, L., G.A. Slafer and M.E. Otegui. 2004. Seed dry weight response to source-sink manipulations in wheat, maize and soybean: A quantitative reappraisal. Field Crops Research 86: 131–146.

Boyer, J.S. 1970. Leaf enlargement and metabolic rates in corn, soybean, and sunflower at various leaf water potentials. Plant Physiology 46: 233–235.

Canvin, D.T. 1979. Photorespiration: Comparison between C3 and C4 plants. pp. 368–96. *In*: Gibbs, M. and E. Latzko (eds.). Photosynthesis II. Encyclopedia of Plant Physiology. Berlin: Springer-Verlag.

Ciamporová, M. 2002. Morphological and structural responses of plant roots to aluminium at organ. Biol. Plant. 45: 161–171.

Dale, M.P. and D.R. Causton. 1992. Use of the chlorophyll *a/b* ratio as a bioassay for the light environment of a plant. Funct. Ecol. 6: 190–196.

Dewick, P.M. 1998. The biosynthesis of shikimate metabolites. Natural Product Reports 15: 17–58.

Dhingra, A., A.R. Portis and H. Daniell. 2004. Enhanced translation of a chloroplast-expressed RbcS gene restores the subunit levels and photosynthesis in nuclear RbcS antisense plant. Proc. Natl. Acad. Sci. 101: 6315–6320.

Djanaguiraman, M. and P.V.V. Prasad. 2010. Ethylene production under high temperature stress causes premature leaf senescence in soybean. Funct. Plant Biol. 37: 1071–1084.

Djanaguiraman, M., P.V.V. Prasad, D. Boyle and W.T. Schapaugh. 2011. High-temperature stress and soybean leaves: Leaf anatomy and photosynthesis. Crop Science 51: 2125–2131.

Duke, S.O. and S.B. Powles. 2008. Glyphosate: A once-in-a-century herbicide. Pest Management Science 64: 319–325.

Evans, J.R. 1989. Photosynthesis and nitrogen relationships in leaves of C_3 plants. Oecologia 78: 9–19.

Farquhar, G.D., J.R. Ehleringer and K.T. Hubick. 1989. Carbon isotope discrimination and photosynthesis. Annu. Rev. Plant Physiol. Plant Mol. Biol. 40: 503–537.

Fedtke, K. and S. Duke. 2005. Herbicides. pp. 247–330. *In*: Hock, B. and E. Elstner (eds.). Plant Toxicology. New York: Marcel Dekker.

Ferris, R., T.R. Wheeler, P. Hadley and R.H. Ellis. 1998. Recovery of photosynthesis after environmental stress in soybean grown under elevated CO_2. Crop Sci. 38: 948–955.

Fredeen, A.L., T.K. Raab, I.M. Rao and N. Terry. 1990. Effects of phosphorus nutrition on photosynthesis in *Glycine max* (L.) Merr. Planta 18: 399–405.

Galmes, J., J. Flexas, A.J. Keys, J. Cifre, R.A.C. Mitchell, P.J. Madgwick, R.P. Haslam, H. Medrano and M.A.J. Parry. 2005. Rubisco specificity factor tends to be larger in plant species from drier habitats and with persistent leaves. Plant, Cell & Environment 28: 571–579.

Gesch, R.W., J.C.V. Vu, L.H.J. Allen and K.J. Boote. 2001. Photosynthetic responses of rice and soybean to elevated CO_2 and temperature. Recent Res. Devel. Plant Physiol. 2: 125–137.

Gibson, L.R. and R.E. Mullen. 1996. Soybean seed quality reductions by high day and night temperature. Crop Sci. 36: 1615–1619.

Gilbert, M.E., M.A. Zwieniecki and N.M. Holbrook. 2011. Independent variation in photosynthetic capacity and stomatal conductance leads to differences in intrinsic water use efficiency in 11 soybean genotypes before and during mild drought. Journal of Experimental Botany 62: 2875–2887.

Govindjee, R. 1995. Sixty-three years since Kautsky: Chlorophyll a fluorescence. Aust. J. Plant Physiol. 22: 131–160.

Green, B.R. and D.G. Durnford. 1996. The chlorophyll-carotenoid proteins of oxygenic photosynthesis. Annu. Rev. Plant Physiol. Plant Mol. Biol. 47: 685–714.

Hák, R., U. Rinderle-Zimmer, H.K. Lichtenthaler and L. Nátr. 1993. Chlorophyll *a* fluorescence signatures of nitrogen deficient barley leaves. Photosynthetica 28: 151–159.

Hampp, R. and H. Schnabl. 1975. Effect of aluminium ions on 14CO_2-fixation and membrane system of isolated spinach chloroplasts. Z. Pflanzenphysiol. 76: 300–306.

Hao, X.Y., X. Han, S.K. Lam, T. Wheeler, H. Ju, P. Li and E.D. Lin. 2012. Effects of fully open-air [CO_2] elevation on leaf ultrastructure, photosynthesis, and yield of two soybean cultivars. Photosynthetica 50: 362–370.

Hao, J., C. Huang, H. Lu and Y. Yu. 2013. Influence of K^+, Na^+ and HCO_3^- on photosynthesis of soybean seedlings. Global Journal of Science Frontier Research (Biological science) 13: 820405.

Harrison, S.A., H.R. Boerma and D.A. Ashley. 1981. Heritability of canopy-apparent photosynthesis and its relationship to seed yield in soybeans. Crop Sci. 21: 222–226.

Hibberd, J.M., J.E. Sheehy and J.A. Langdale. 2008. Using C4 photosynthesis to increase the yield of rice—rationale and feasibility. Current Opinion in Plant Biology 11: 228–231.

Hofstra, G. 1972. Response of soybeans to temperature under high light intensities. Can. J. Plant Sci. 52: 535–543.

Hollinger, S.E. and J.R. Angel. 2009. Weather and crops. *In*: Nafziger, E. (ed.). Illinois Agronomy Handbook 24th edition, Illinois Agr. Ext. Pub. C1394, Champaign, IL.

IPCC. 2007. Intergovernmental Panel on Climate Change fourth assessment report. Climate change 2007. Cambridge University Press, Cambridge, UK.

Ito, O., F. Mitsumori and T. Totsuka. 1985. Effects of NO_2 and O_3 alone or in combination on kidney bean plants (*Phaseolus vulgaris* L.): Products of $13CO_2$ assimilation detected by 13C nuclear magnetic resonance. Journal of Experimental Botany 36: 281–289.

Jiang, Y.L., Y.G. Yao, Q.G. Zhang et al. 2006. Changes of photosynthetic pigment contents in soybean under elevated atmospheric CO_2 concentrations. Crop Res. 2: 144–146 [In Chin.].

Jiang, C.D., H.Y. Gao, Q. Zou, G.M. Jiang and L.H. Li. 2006. Leaf orientation, photorespiration and xanthophyll cycle protect young soybean leaves against high irradiance in field. Environmental and Experimental Botany 55: 87–96.

Kebeish, R., M. Niessen, K. Thiruveedhi, R. Bari, H.-J. Hirsch, R. Rosenkranz, N. Stabler, B. Schonfeld, F. Kreuzaler and C. Peter-hansel. 2007. Chloroplastic photorespiratory bypass increases photosynthesis and biomass production in *Arabidopsis thaliana*. Nature Biotechnology 25: 593–599.

Kochian, L.V. 1995. Cellular mechanisms of aluminium toxicity and resistance in plants. Annu. Rev. Plant Physiol. Plant Mol. Biol. 46: 237–260.

Krause, G.H. and E. Weis. 1991. Chlorophyll fluorescence and photosynthesis: The basics. Annu. Rev. Plant Physiol. Plant Mol. Biol. 42: 313–349.

Kumudini, S. 2002. Trial and tribulations: A review of the role of assimilate supply in soybean genetic yield improvement. Field Crops Research 75: 211–222.

Lieman-Hurwitz, J., S. Rachmilevitch, R. Mittler, Y. Marcus and A. Kaplan. 2003. Enhanced photosynthesis and growth of transgenic plants that express ictB, a gene involved in HCO_3^- accumulation in cyanobacteria. Plant Biotechnology Journal 1: 43–50.

Liu, C.M., A.L. Young, A. Starling-Windhof et al. 2010. Coupled chaperone action in folding and assembly of hexadecameric Rubisco. Nature 463: 197–202.

Liu, G., C. Yang, K. Xu, Z. Zhang, D. Li and Z. Chen. 2012. Development of yield and some photosynthetic characteristics during 82 years of genetic improvement of soybean genotypes in northeast China. Aus. J. Crop Science 6: 1416–1422.

Lobell, D.B. and G.P. Asner. 2003. Climate and management contributions to recent trends in U.S. agricultural yields. Science 299: 1032.

Long, S.P. 1991. Modification of the response of photosynthetic productivity to rising temperature by atmospheric CO_2 concentrations: Has its importance been underestimated? Plant, Cell and Environment 14: 729–739.

Long, S.P. 1994. The potential effects of concurrent increases in temperature, CO_2 and O_3 on net photosynthesis, as mediated by RubisCO. pp. 21–38. *In*: Alscher, R.G. and A.R. Wellburn (eds.). Plant Responses to the Gaseous Environment. London: Chapman and Hall.

Long, S.P., E.A. Ainsworth, A. Rogers and D.R. Ort. 2004. Rising atmospheric carbon dioxide: Plants FACE the future. Annual Review of Plant Biology 55: 591–628.

Long, S.P., X.-G. Zhu, S.L. Naidu and D.R. Ort. 2006b. Can improvement in photosynthesis increase crop yield? Plant, Cell and Environment 29: 315–330.

Lugg, D.G. and T.R. Sinclair. 1981. Seasonal changes in photosynthesis of field-grown soybean leaflets. 2. Relation to nitrogen content. Photosynthetica 15: 138–144.

Malek, M.A., M.M.A. Mondal, M.R. Ismail, M.Y. Rafii and Z. Berahim. 2012. Physiology of seed yield in soybean: Growth and dry matter production. African J. Biotech. 11: 7643–7649.

Marquard, R.D. and J.L. Tipton. 1987. Relationship between extractable chlorophyll and an *in situ* method to estimate leaf greenness. Hort. Science 22: 1327.

Massot, N., M. Llugany, C. Poschenrieder and J. Barcelo. 1999. Callose production as indicator of aluminium toxicity in bean cultivars. Journal of Plant Nutrition 22: 1–10.

Mathew, J.P., S.J. Herbert, S. Zhang, A.F. Rautenkranz and G.V. Litchfield. 2001. Differential response of soybean yield components to the timing of light enrichment. Agronomy Journal 92: 1156–1161.

Mauromicale, G., A. Ierna and M. Marchese. 2006. Chlorophyll fluorescence and chlorophyll content in field-grown potato as affected by nitrogen supply, genotype, and plant age. Photosynthetica 44: 76–82.

McLeod, A. and S.P. Long. 1999. Free air carbon dioxide enrichment (FACE) in global change research: A review. Advances in Ecological Research 28: 1–55.

Miller, J.E. 1988. Effects on photosynthesis, carbon allocation, and plant growth associated with air pollutant stress. pp. 287–314. *In*: Heck, W.W., O.C. Taylor and D.T. Tingey (eds.). Assessment of Crop Loss from Air Pollutants. London: Elsevier Applied Science.

Monteith, J.L. 1977. Climate and the efficiency of crop production in Britain. Philosophical Transactions of the Royal Society B 281: 277–294.

Nagasuga, K., A. Fukunaga, C. Higashi and T. Umezaki. 2016. The response of vegetative growth to soil water condition in native soybean cultivar 'Misato-zairai'. Japanese Journal of Crop Science 85: 138–143. In Japanese with English abstract.

Nielsen, D.C. 1990. Scheduling irrigations for soybeans with the crop water stress index (CWSI). Field Crops Research 23: 103–116.

Nowak, R.S., D.S. Ellsworth and S.D. Smith. 2004. Functional responses of plants to elevated atmospheric CO_2—do photosynthetic and productivity data from FACE experiments support early predictions? New Phytol. 162: 253–280.

Ogren, W.L. 1984. Photorespiration: Pathways, regulation, and modification. Annual Review of Plant Physiology 35: 415–42.

Ort, D.R., X.-G. Zhu and A. Melis. 2011. Optimizing antenna size to maximize photosynthetic efficiency. Plant Physiology 155: 79–85.

Paradiso, R., C. Arena, V. De Micco, M. Giordano, G. Aronne and S. De Pascale. 2017. Changes in leaf anatomical traits enhanced photosynthetic activity of soybean grown in hydroponics with plant growth promoting microorganisms. Front. Plant Sci., https://doi.org/10.3389/fpls.2017.00674.

Parry, M.A.J., P.J. Madgwick, J.F.C. Carvalho and P.J. Andralojc. 2007. Prospects for increasing photosynthesis by overcoming the limitations of Rubisco. Journal of Agricultural Science 145: 31–43.

Parry, M.A.J., M. Reynolds, M.E. Salvucci, C. Raines, P.J. Andralojc, X.-G. Zhu, G.D. Price, A.G. Condon and R.T. Furbank. 2011. Raising yield potential of wheat. II. Increasing photosynthetic capacity and efficiency. Journal of Experimental Botany 62: 453–567.

Paulsen, G.M. 1994. High temperature responses of crop plants. pp. 365–389. *In*: Boote, K.J., T.R. Sinclair and G.M. Paulsen (eds.). Physiology and Determination of Crop Yield. American Society of Agronomy, Madison, WI.

Peet, M.M. and P.J. Kramer. 1980. Effects of decreasing source-sink ratio in soybeans on photosynthesis, photo-respiration, transpiration and yield. Plant, Cell and Environment 3: 201–206.

Pell, E.J., N.A. Eckardt and R.E. Click. 1994. Biochemical and molecular basis for impairment of photosynthetic potential. Photosynthesis Research 39: 453–62.

Pineros, M.A. and L.V. Kochian. 2003. Differences in whole-cell and single-channel ion currents across the plasma membrane of mesophyll cells from two closely related Thlaspi species. Plant Physiology 131: 583–594.

Polle, A., T. Pfirrmann, S. Chakrabarti and H. Rennenberg. 1993. The effects of enhanced ozone and enhanced carbon dioxide concentrations on biomass, pigment and antioxidative enzymes in spruce needles (*Picea abies* L.). Plant, Cell and Environment 16: 311–16.

Ristic, Z., U. Bukovnik, I. Momcilovic, J. Fu and P.V.V. Prasad. 2008. Heat-induced accumulation of chloroplast protein synthesis elongation factor, EF-Tu, in winter wheat. J. Plant Physiol. 165: 192–202.

Rogers, A., D.J. Allen, P.A. Davey et al. 2004. Leaf photosynthesis and carbohydrate dynamics of soybeans grown throughout their life-cycle under free air carbon dioxide enrichment. Plant, Cell and Environment 27: 449–458.

Schlegel, H. and D.L. Godbold. 1991. The influence of Al on the metabolism of spruce needles. Water Air Soil Poll. 57-58: 131–138.

Sedghi, M. and B. Amanpour-Balaneji. 2010. Sequential Path model for grain yield in soybean. Not. Sci. Biol. 2(3): 104–109.

Sen Gupta, U.K. 1988. Effect of increasing CO_2 concentration on photosynthesis and photorespiration in wheat leaf. Current Science 57: 145–146.

Servaites, J.C. and W.L Ogren. 1977. Chemical inhibition of the glycolate pathway in soybean leaf cells. Plant Physiology 60: 461–466.

Sharkey, T.D. 1988. Estimating the rate of photorespiration in leaves. Physiologia Plantarum 73: 147–152.

Shi, G.Y. 2004. Effect of aluminium on growth and some physiological function of rice seedlings. Guihaia 24(1): 77–80.

Shutilova, N., G. Semenova, V. Klimov and V. Shnyrov. 1995. Temperature-induced functional and structural transformations of the photosystem II oxygen-evolving complex in spinach sub chloroplast preparations. Biochem. Mol. Biol. Int. 35: 1233–1243.

Sinclair, T.R. 1980. Leaf CER from post flowering to senescence of field-grown soybean cultivars. Crop Science 20: 196–200.

Somerville, S.C. and C.R. Somerville. 1983. Effect of oxygen and carbon dioxide on photorespiratory flux determined from glycine accumulation in a mutant of *Arabidopsis thaliana*. Journal of Experimental Botany 34: 415–424.

Spreitzer, R.J. and M.E. Salvucci. 2002. Rubisco: Interactions, associations and the possibilities of a better enzyme. Annu. Rev. Plant Biol. 53: 449–475.

Stienen, H. 1986. Veränderungen in Wasserhaushalt junger Koniferen in säurer and Al3+ haltiger Nähr und Bodenlösung. Forstarchiv 57: 227–231.

Stitt, M. and A. Krapp. 1999. The interaction between elevated carbon dioxide and nitrogen nutrition: The physiological and molecular background. Plant Cell Environ. 22: 583–621.

Tenhunen, J.D., O.L. Lange, P.C. Harley, A. Meyer and D.M. Gates. 1980b. The diurnal time course of net photosynthesis of soybean leaves: Analysis with a physiologically based steady-state photosynthesis model. Oecologia 46: 314–321.

Thompson, J.A., L.E. Schweitzer and R.L. Nelson. 1996. Association of specific leaf weight, an estimate of chlorophyll, and chlorophyll content with apparent photosynthesis in soybean. Photosynth. Res. 49: 1–10.

Tóth, S.Z., G. Schansker and R.J. Strasser. 2005. In intact leaves, the maximum fluorescence level (FM) is independent of the redox state of the plastoquinone pool: A DCMU-inhibition study. Biochimica et Biophysica Acta–Bioenergetics 1708: 275–282.

von Caemmerer, S. and J.R. Evans. 2010. Enhancing C3 photosynthesis. Plant Physiology 154: 589–592.

Vu, J.C.V., L.H. Allen Jr, K.J. Boote and G. Bowes. 1997. Effects of elevated CO_2 and temperature on photosynthesis and Rubisco in rice and soybean. Plant Cell. Environ. 20: 68–76.

Wallsgrove, R.M., A. Baron and A.K. Tobin. 1992. Carbon and nitrogen cycling between organelles during photorespiration. pp. 79–96. *In*: Tobin, A.K. (ed.). Plant Organelles. Cambridge University Press.

Webber, A., G.-Y. Nie and S.P. Long. 1994. Acclimation of photosynthetic proteins to rising atmospheric CO_2. Photosynthesis Research 39: 413–25.

Whitney, S.M. and T.J. Andrews. 2001b. The gene for the ribulose-1, 5-bisphosphate carboxylase/oxygenase (Rubisco) small subunit relocated to the plastid genome of tobacco directs the synthesis of small subunits that assemble into Rubisco. Plant Cell 13: 193–205.

Whitney, S.M., R.L. Houtz and H. Alonso. 2011. Advancing our understanding and capacity to engineer nature's CO_2-sequestering enzyme, Rubisco. Plant Physiology 155: 27–35.

Willekens, H., W. Van Camp, M. van Montagu, D. Inze', C. Langebartels and H.J. Sandermann. 1994. Ozone, sulphur dioxide, and ultraviolet B have similar effects on mRNA accumulation of antioxidant genes in *Nicotiana plumbagimfolia* L. Plant Physiology 106: 1007–1014.

Wise, R.R., A.J. Olson, S.M. Schrader and T.D. Sharkey. 2004. Electron transport is the functional limitation of photosynthesis in field-grown Pima cotton plants at high temperature. Plant Cell Environ. 27: 717–724.

Xiao, X.X. 2002. The physiological and biochemical response of Longan (*Dimocarpus Longana* Lour.) to aluminium stress and rectification of aluminium toxicity. Fujian J. Agr. Sci. 17: 182–185.

Yadava, U.L. 1986. A rapid and non-destructive method to determine chlorophyll in intact leaves. Hort. Science 21: 1449–1450.

Yin, Z., F. Meng, H. Song, X. Wang, X. Xu and D. Yu. 2010. Expression quantitative trait loci analysis of two genes encoding Rubisco activase in soybean. Plant Physiology 152: 1625–1637.

Zhang, J.H., W.D. Huang, Y.P. Liu and Q.H. Pan. 2005. Effects of temperature acclimation pre-treatment on the ultrastructure of mesophyll cells in young grape plants (*Vitis vinifera* L. cv. Jingxiu) under cross-temperature stresses. J. Integr. Plant Biol. 47: 959–970.

Zhao, X.J., E. Sucoff and E.J. Stadelmann. 1987. Al^{3+} and Ca^{2+} alteration of membrane permeability of *Quercus rubra* cortex cells. Plant Physiol. 83: 159–162.

Zhao, T.H., Y. Shi and G.H. Huang. 2003.Effect of doubled CO_2 and O_3 concentration and their interactions on ultrastructure of soybean chloroplast. Chin. J. Appl. Ecol. 14: 2229–2232 [In Chin.].

Zhu, G. and R.J. Spreitzer. 1996. Directed mutagenesis of chloroplast ribulose-1,5-bisphosphate carboxylase/oxygenase. Loop 6 substitutions complement for structural stability but decrease catalytic efficiency. The Journal of Biological Chemistry 271: 18494–18498.

Zhu, X.-G., D.R. Ort, J. Whitmarsh and S.P. Long. 2004a. The slow reversibility of photosystem II thermal energy dissipation on transfer from high to low light may cause large losses in carbon gain by crop canopies: A theoretical analysis. Journal of Experimental Botany 55: 1167–1175.

Zhu, X.-G., E. de Sturler and S.P. Long. 2007. Optimizing the distribution of resources between enzymes of carbon metabolism can dramatically increase photosynthetic rate: A numerical simulation using an evolutionary algorithm. Plant Physiology 145: 513–526.

Zhu, X.-G., S.P. Long and D.R. Ort. 2010. Improving photosynthetic efficiency for greater yield. Annual Review of Plant Biology 61: 235–26.

Zobiole, L.H.S., R.J. Kremer, R.S. de Oliveira Jr. and J. Constantin. 2012. Glyphosate effects on photosynthesis, nutrient accumulation, and nodulation in glyphosate-resistant soybean. Journal of Plant Nutrition and Soil Science 175: 319–330.

Zuo, B.Y., Q. Zhang, G.Z. Jiang, K.Z. Bai and T.Y. Kuang. 2002. Effects of doubled CO_2 concentration on ultrastructure, supramolecular architecture and spectral characteristics of chloroplasts from wheat. Acta Bot. Sin. 44: 908–912.

CHAPTER 8
Respiration

A grain is a living organism that breathes. During respiration, starch and oxygen are converted to carbon dioxide, water and heat. The major factors affecting the respiration of stored grain are grain moisture content, temperature, and composition of the inter granular air.

8.1 Seeds

The dynamic of oxygen (O_2) and carbon dioxide (CO_2) concentration was characterized in soybean (*Glycine max* (L.)) samples hermetically stored in glass jars at 15, 25 and 35°C and 13, 15 and 17% moisture content (m.c., wet base). Two correlations were used for smoothing gas concentration in time: Linear and exponential. Then, the respiration rate at each temperature and m.c. combination was calculated as storage time progressed and oxygen was consumed and two predictive models for respiration were proposed: Model I (temperature and m.c. dependent) and Model II (temperature, m.c. and oxygen dependent). It was observed that respiration rate increased with storage m.c. and temperature. However, respiration rate was not mainly affected by O_2 until a critical concentration limit of about 2% was reached. Respiration rates were from 0.341 to 22.684 mg O_2/(kgDM d) and from 0.130 to 20.272 mg CO_2/(kgDM d) for a range of storage conditions of 13–17% m.c. and 15–35°C temperature. The respiration rate of soybean seeds obtained in this study resulted significantly lower than the rates reported in the literature for other grains at similar temperature and a_w (water activity) storage condition. For hermetic storage simulations in which O_2 concentration is not expected to drop below 2%, the simplest model (Model I) could be used, but if the O_2 concentration of the hermetic system is expected to be depleted, Model I would underestimate the time at which O_2 is consumed, therefore, a model with O_2 dependency is recommended instead (Model II) (Ochandio et al., 2012) (Fig. 8.1).

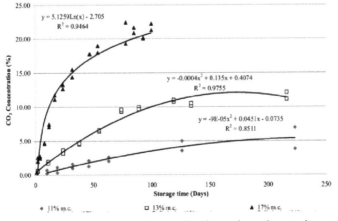

Fig. 8.1 Carbon dioxide evolution by soybean seeds under varying moisture and temperature.

Currently, the highest interest with respect to the assessment of seed physiological quality is to obtain reliable results in a relatively short period of time. This initiative allows for prompt decisions during different phases of seed production, primarily after physiological maturity. Research was performed in order to verify the efficiency and rapidity of the method of Pettenkofer to determine the respiratory activity and to differentiate vigor levels of soybean seed lots. Three lots of soybean seeds cv. 8000 were used. Seed performance was determined by respiratory activity, compared to the following tests: Standard germination, germination first count, electrical conductivity, seedling emergence, seedling shoot and root length and total dry mass. Results ranked seed lots according to differences in physiological quality. Seed imbibition and conditioning period in Pettenkofer's equipment were enough to detect differences in vigor among seed lots, showing that the determination of the respiratory activity is a promising procedure to identify differences in vigor levels among soybean seed lots.

Carbon dioxide evolution was used to determine the respiration rate and carbon dioxide production per unit mass of soybean at different storage temperature and moisture content. Treatments included soybean moisture content of about 14, 18, 22 and 26% (wet basis), stored in temperature of about 15, 20, 25, 30, cycling from 15 to 25 and cycling from 20 to 30°C on a 24 h basis. The results indicated that respiration of soybean increased by increasing storage temperature and moisture content. However, the respiration of soybean seems to be constant with the storage time at low storage temperature. Relations of carbon dioxide production from soybean seeds versus storage time were fitted. This relation seems to be a polynomial equation. Storage temperature was the most important effecting factor on the respiration of soybean. An increase in the storage temperature leads to an increase in the respiration rate of soybean per unit mass of dry matter (Sorour and Uchino, 2004) (Fig. 8.2).

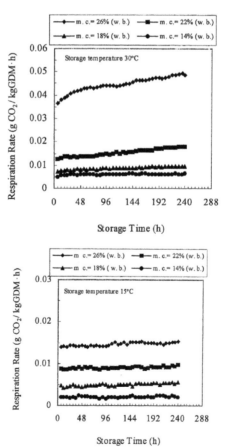

Fig. 8.2 Effect of storage time on respiration rate of soybean seeds.

8.2 Leaf and Plants

The rate of plant respiration is linked to the rate of metabolism and growth due to requirements for ATP, reductant, and carbon skeletons during cell maintenance, division, and expansion (Hunt and Loomis, 1979; Lambers et al., 1983). For example, respiration rates are often lower in species with intrinsically slower growth rates (Poorter et al., 1991). Moreover, respiration is rapid in tissues with high energy demands, such as thermogenic floral spadices (Meeuse, 1975), and in rapidly growing tissues, such as the elongation zone of roots (Lambers et al., 1996). Plant respiration can also increase rapidly in response to both biotic and abiotic stress (for review, see Lambers et al., 1996). Conversely, decreases in respiratory rate often occur as plant tissues age (Winkler et al., 1994). Various factors may be responsible for these changes, including substrate availability, enzyme activation, specific protein degradation or *de novo* protein synthesis, and alterations in mitochondrial numbers.

The extent to which such changes in respiration rate alter the rate of oxidative phosphorylation also depends on the partitioning of electron flux between the Cyt and the alternative pathways of electron transport. The Cyt pathway (terminating at COX) couples the reduction of O_2 to water with the translocation of protons across the inner mitochondrial membrane, thereby building a proton-motive force that drives ATP synthesis. The alternative pathway branches directly from Q and reduces O_2 to water without further proton translocation. This pathway appears to consist of a single-subunit cyanide-resistant quinol oxidase, AOX. Electron flow via AOX in plants can allow carbon flux through the TCA cycle when ADP is limiting, thereby providing carbon skeletons for other cellular processes (Lambers and Steingröver, 1978). This pathway may also protect against harmful reactive O_2 generation when the Q pool is highly reduced (Purvis and Shewfelt, 1993; Wagner and Krab, 1995), allowing respiration to proceed in the presence of nitric oxide (Millar and Day, 1996), and help avoid the production of fermentation products when pyruvate accumulates (Vanlerberghe et al., 1995).

The effects of light on plant respiration have been studied for many years (Azcón-Bieto et al., 1983; Hill and Bryce, 1992; Kromer, 1995). Light is known to regulate gene expression of several key respiratory enzymes through the action of phytochrome, including cytochrome *c* oxidase (Hilton and Owen, 1985) and phosphoenolpyruvate carboxylase (Sims and Hague, 1981). It has also been suggested that blue light can cause an increase in total respiration (Kowallik, 1982). Furthermore, there is a differential expression of alternative oxidase genes between light-grown and dark-grown tissues (Obenland et al., 1990; Finnegan et al., 1997). Other workers have reported an indirect effect of light and photosynthesis on respiration, with the cellular concentration of sugars suggested to play an important role in the regulation of respiration in leaves (Azcón-Bieto et al., 1983) and roots (Bingham and Farrar, 1988).

Light effects on electron flow through the cyanide-resistant respiratory pathway, oxygen isotope fractionation and total respiration were studied in soybean (*Glycine max* L.) cotyledons. During the first 12 h of illumination, there was an increase in both electron partitioning through the alternative pathway and oxygen isotope fractionation by the alternative oxidase. The latter probably indicates a change in the properties of the alternative oxidase. There was no engagement of the alternative oxidase in darkness and its fractionation was 27%. In green cotyledons, 60% of the respiration flux was through the alternative pathway and the alternative oxidase fractionation was 32‰. Exposing previously illuminated tissue to continuous darkness induced a decrease in the electron partitioning through the alternative pathway. However, this decrease was not directly linked with the low cellular sugar concentration resulting from the lack of light, because 5 min of light every 12 h was sufficient to keep the alternative pathway engaged to the same extent as plants grown under control conditions.

Measurements of respiration were made on leaf discs from glasshouse-grown soybean (*Glycine max* [L.] Merr. cv 'Corsoy') plants in the presence and absence of cyanide (KCN) and salicylhydroxamic acid (SHAM). O_2 uptake by mature leaves measured at 25°C was stimulated by 1 millimolar KCN (63%) and also by 5 millimolar azide (79%). SHAM, an inhibitor of the alternative oxidase and a selection of other enzymes, also stimulated O_2 uptake by itself at a concentration of 10 millimolar. However, in combination, KCN and SHAM were inhibitory. The rate of O_2 uptake declined consistently with leaf age. The stimulation of O_2 uptake by KCN and by SHAM occurred only after a certain stage of leaf development had been reached and was more pronounced in fully expanded leaves. In young leaves,

O_2 uptake was inhibited by both KCN and SHAM individually. The uncoupler, p-trifluoromethoxy carbonyl-cyanide phenylhydrazone, stimulated leaf respiration at all ages studied, the stimulation being more pronounced in fully expanded leaves. The uncoupled rate was inhibited by KCN and SHAM individually. The capacity of the cytochrome path declined with leaf age, paralleling the decline in total respiration. However, the capacity of the alternative path peaked at about full leaf expansion, exceeding the cytochrome capacity and remaining relatively constant. These results are consistent with the presence in soybean leaves of an alternative path capacity that seems to increase with age, and they suggest that the stimulation of O_2 uptake by KCN and NaN_3 in mature leaves was mainly by the SHAM-sensitive alternative path. The stimulation of O_2 uptake by SHAM was not expected, and the reason for it is not clear.

Respiration is an important component of plant carbon balance, with rapidly growing plants daily losing about one third of the carbon fixed in photosynthesis (McCree and Amthor, 1982). Predicting how plant carbon balance may be affected by the rising concentration of carbon dioxide in the atmosphere ($[C_a]$) is of importance both from a global carbon budget perspective and in predicting plant growth. While there are many measurements of responses of photosynthesis of plants to simulated increases in $[C_a]$, and biochemical models that often provide a reasonable approximation of the observed photosynthetic responses, how respiration will respond to increases in $[C_a]$ remains uncertain (Drake et al., 1999; Gonzalez-Meler et al., 2004).

There have been numerous measurements of respiration of plants grown at ambient and elevated $[CO_2]$ in controlled environment chambers (e.g., Grimmer and Komor, 1999; Sakai et al., 2001), including soybean (e.g., Griffin et al., 2001b), but diverse responses have been found. In most studies in controlled environment chambers, light and temperature were kept constant from day to day and often during days; the relevance of these observations to respiration rates under field conditions is uncertain. It was to determine if the responses of respiration of soybean leaves to elevated $[CO_2]$ in the field were similar to those found under controlled environment conditions.

Plants draw CO_2 from the atmosphere and make sugars through the process of photosynthesis. However, they also release some CO_2 during respiration as they use the sugars to generate energy for self-maintenance and growth. How elevated CO_2 affects plant respiration will, therefore, influence future food supplies and the extent to which plants can capture CO_2 from the air and store it as carbon in their tissues.

While there is broad agreement that higher atmospheric CO_2 levels stimulate photosynthesis in C3 plants, such as soybean, no such consensus exists on how rising CO_2 levels will affect plant respiration.

There's been a great deal of controversy about how plant respiration responds to elevated CO_2. Some summary studies suggest it will go down by 18 percent, some suggest it won't change, and some suggest it will increase by as much as 11 percent.

Understanding how the respiratory pathway responds when plants are grown at elevated CO_2 is key to reducing this uncertainty, Leakey said. His team used microarrays, a genomic tool that can detect changes in the activity of thousands of genes at a time, to learn which genes in the high CO_2 plants were being switched on at higher or lower levels than those of the soybeans grown at current CO_2 levels.

Rather than assessing plants grown in chambers in a greenhouse, as most studies have done, Leakey's team made use of the Soybean Free Air Concentration Enrichment (Soy FACE) facility at Illinois. This open-air research lab can expose a soybean field to a variety of atmospheric CO_2 levels—without isolating the plants from other environmental influences, such as rainfall, sunlight and insects.

Some of the plants were exposed to atmospheric CO_2 levels of 550 parts per million (ppm), the level predicted for the year 2050 if current trends continue. These were compared to plants grown at ambient CO_2 levels (380 ppm).

The results were striking. At least 90 different genes coding the majority of enzymes in the cascade of chemical reactions that govern respiration were switched on (expressed) at higher levels in the soybeans grown at high CO_2 levels. This explained how the plants were able to use the increased supply of sugars from stimulated photosynthesis under high CO_2 conditions to produce energy, Leakey noted, the rate of respiration increased 37 percent at the elevated CO_2 levels.

The enhanced respiration is likely to support greater transport of sugars from leaves to other growing parts of the plant, including the seeds.

The expression of over 600 genes was altered by elevated CO_2 in total, which will help us to understand how the response is regulated and also hopefully produce crops that will perform better in the future.

While measurements of whole-plant respiration rates would be desirable from a modelling or productivity perspective, experimental separation of plant from soil respiration remains problematic, and with regard to temperature acclimation, roots and shoots do not experience the same temperatures in the field. Therefore, the more specific hypotheses concerning responses of respiration to $[CO_2]$ and temperature were addressed at the single-leaf rather than at the whole-plant level. Both Atkin et al. (2001) and Griffin et al. (2002) found that leaf respiration rates differed depending on whether diurnal changes in temperature occurred for the whole shoot or just for the measured leaf, and emphasized that accurate estimates of leaf respiration required that natural temperature patterns occurred for the whole shoot.

Elevated $[CO_2]$ increased daytime net carbon dioxide fixation rates per unit of leaf area by an average of 48%, but had no effect on night-time respiration expressed per unit of area, which averaged 53 mmol m^{-2} d^{-1} (1.4 µmol m^{-2} s^{-1}) for both the ambient and elevated $[CO_2]$ treatments. Leaf dry mass per unit of area was increased on average by 23% by elevated $[CO_2]$, and respiration per unit of mass was significantly lower at elevated $[CO_2]$. Respiration increased by a factor of 2.5 between 18 and 26°C average night temperature, for both $[CO_2]$ treatments.

These results do not support predictions that elevated $[CO_2]$ would increase respiration per unit of area by increasing photosynthesis or by increasing leaf mass per unit of area, nor the idea that acclimation of respiration to temperature would be rapid enough to make dark respiration insensitive to variation in temperature between nights.

Photosynthetic and respiratory exchanges of CO_2 by plants with the atmosphere are significantly larger than anthropogenic CO_2 emissions, and these fluxes will change as growing conditions are altered by climate change. Understanding feedbacks in CO_2 exchange is important in predicting future atmospheric $[CO_2]$ and climate change. At the tissue and plant scale, respiration is a key determinant of growth and yield. Although the stimulation of C_3 photosynthesis by growth at elevated $[CO_2]$ can be predicted with confidence, the nature of changes in respiration is less certain. This is largely because the mechanism of the respiratory response is insufficiently understood. Molecular, biochemical and physiological changes in the carbon metabolism of soybean in a free-air CO_2 enrichment experiment were investigated over 2 growing seasons. Growth of soybean at elevated $[CO_2]$ (550 µmol·mol^{-1}) under field conditions stimulated the rate of night time respiration by 37%. Greater respiratory capacity was driven by greater abundance of transcripts encoding enzymes throughout the respiratory pathway, which would be needed for the greater number of mitochondria that have been observed in the leaves of plants grown at elevated $[CO_2]$. Greater respiratory quotient and leaf carbohydrate content at elevated $[CO_2]$ indicate that stimulated respiration was supported by the additional carbohydrate available from enhanced photosynthesis at elevated $[CO_2]$. If this response is consistent across many species, the future stimulation of net primary productivity could be reduced significantly. Greater foliar respiration at elevated $[CO_2]$ will reduce plant carbon balance, but could facilitate greater yields through enhanced photoassimilate export to sink tissues.

Physiological processes of plants are largely affected by the alteration of the surrounding environmental temperature. The ability of plants to cope with extreme temperature is a complex process and is determined by environmental factors and also by the genetic capability of the plant. In general, stability of life processes in most plants is comparatively wide, ranging from several degrees above zero to around 35°C (Zrobek-sokolnik, 2012). The increase of temperature up to a certain level increases plant growth, photosynthesis, respiration and enzyme activity, but after that these parameters tend to decline (Fig. 8.3). Respiration rapidly increases with temperature and drops drastically after an extreme tolerable temperature. Photosynthesis is a comparatively less sensitive process than respiration but its declining pattern is similar. The average rate of enzymatic reactions increases twofold with every 10°C increase in temperature within the range. The optimal temperatures for structural integrity and activity of most enzymes are within the range of 30–45°C; enzymes are irreversibly denatured and inactivated at temperatures higher than 60°C, with the exception of thermophilous organisms. Thus, each life process has its own referred critical or lethal temperature, after that it cannot proceed and causes permanent damage to cell structures and ultimately the cell, then plant death as well (Zrobek-sokolnik, 2012).

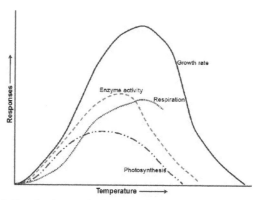

Fig. 8.3 Growth, photosynthesis and respiration changes with temperature.

It has long been known that respiration is not simply a constant fraction of photosynthesis (McCree, 1970), and respiration responds to carbon dioxide concentration and temperature quite differently than the response of photosynthesis. The rate of respiration by plants increases substantially with short-term increases in temperature, often doubling with a 10°C increase in temperature. This has led some to speculate that higher global temperatures resulting from elevated atmospheric carbon dioxide concentrations would stimulate respiration (e.g., Woodwell et al., 1983). However, this suggestion does not take the acclimation of respiration to temperature or the response of respiration to carbon dioxide concentration into account. Despite a large short-term response of respiration rate to temperature, respiration often acclimates to growth temperature so that respiration rates measured in the growth environment remain constant across a range of growth temperatures (e.g., Billings et al., 1971). Temperature acclimation of respiration has been primarily examined in mature leaves, and its relationship to the growth and maintenance model of whole-plant respiration (Amthor, 1989) remains unclear. Analysis of whole plant respiration using the growth and maintenance model has indicated that growth respiration varies with temperature only as relative growth rate varies, and that maintenance respiration increases with increasing growth temperature (McCree and Silsbury, 1978; McCree and Amthor, 1982). Increasing carbon dioxide concentration can affect respiration rates both directly and indirectly. Direct effects of the carbon dioxide concentration during the measurement on the rate of respiration have been found in several species (e.g., Ziska and Bunce, 1994), including soybean (Bunce, 1990; 1995; Thomas and Griffin, 1994). Respiration rates are generally lower at elevated carbon dioxide concentrations. Indirect effects of growth carbon dioxide concentration on respiration rate may result from changes in relative growth rate or from changes in tissue composition (Ryan, 1991; Ziska and Bunce, 1994), but they have seldom been separated from direct effects (Bunce, 1995).

Diurnal changes of respiration have been examined in various plant species (Azcon-Bieto and Osmond, 1983; Breeze and Elston, 1978; Challa, 1976; Farrar, 1980; Ludwig et al., 1975; McCree, 1970; 1974; Mousseau, 1977). The initial rate of respiration at the start of darkness usually depends on the photosynthetic rate in the preceding light period, and, as a general trend, the respiration rate continuously decreases with time during dark period. This behavior of respiration could be the result of the change in the level of photosynthetically produced carbohydrate. It could, therefore, be said that the rate of respiration is determined by the level of its substrate. However, one can find several papers suggesting that the rate of respiration is not necessarily controlled only by the level of substrate. For an example, a temporary increase of respiration in the course of continuous decrease during the dark period has been observed in several species (Challa, 1976; Ludwig et al., 1975; Mousseau, 1977). From this, one could speculate that respiration can be accelerated in accordance with the plant requirement of energy supply for the fulfillment of metabolic performance, as suggested by Beevers (1970). Furthermore, it has also been reported that high nitrogen plants have a high respiration rates, suggesting that high nitrogen levels in a plant necessitates more respiratory products (Wilson, 1982).

In the report, the diurnal change in the respiration in soybean plants under different environmental or physiological conditions was examined, aiming to elucidate the mechanism of the determination of respiration rate (Fig. 8.4).

The experiment was to determine how respiration of soybeans may respond to potential increases in atmospheric carbon dioxide concentration and growth temperature. Three cultivars of soybeans (*Glycine max* L. Merr.), from maturity groups 00, IV, and VIII, were grown at 370, 555 and 740 ppm carbon dioxide concentrations at 20/15, 25/20, and 31/26°C day/night temperatures. Rates of carbon dioxide efflux in the dark were measured for whole plants several times during exponential growth. These measurements were made at the night temperature and the carbon dioxide concentration at which the plants were grown. For the lowest and highest temperature treatments, the short term response of respiration rate to measurement at the three growth carbon dioxide concentrations was also determined. Elemental analysis of the tissue was used to estimate the growth conversion efficiency. This was combined with the observed relative growth rates in order to estimate growth respiration. Maintenance respiration was estimated as the difference between growth respiration and total respiration. Respiration rates were generally sensitive to short term changes in the measurement carbon dioxide concentration for plants grown at the lowest, but not the highest carbon dioxide concentration. At all temperatures, growth at elevated carbon dioxide concentrations decreased total respiration measured at the growth concentration, with no significant differences among cultivars. Total respiration increased very little with increasing growth temperature, despite an increase in relative growth rate. Growth respiration was not affected by carbon dioxide treatment at any temperature, but increased with temperature because of the increase in relative growth rate. Values calculated for maintenance respiration decreased with increasing carbon dioxide concentration and also decreased with increasing temperature. Calculated values of maintenance respiration were sometimes zero or negative at the warmer temperatures. This suggests that respiration

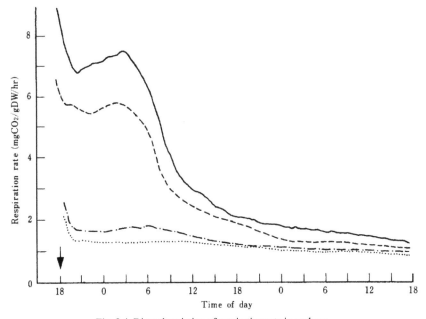

Fig. 8.4 Diurnal variation of respiration rate in soybean.

Arrow shows the onset of the dark. Air temperature, humidity and light intensity were 25°C, 21°C as the dew point and 610 μeinstein/cm²/s (PAR).

———— : 5th foliage leaf fully expanded.

– – – – : 6th foliage leaf fully expanded and flowering started.

—·—·— : First pod, the length of which reached 2 cm, appeared on the top four internodes on the main stem.

.... : Yellow pod appeared on the main stem.

rates measured in the dark may not have reflected average 24-h rates of energy use. The results indicate that increasing atmospheric carbon dioxide concentration may reduce respiration in soybeans, and respiration may be insensitive to climate warming (Bunce and Ziska, 1996) (Table 8.1).

Water stress is considered one of the most important factors limiting plant performance and yield worldwide (Boyer, 1982). Effects of water stress on a plant's physiology, including growth (McDonald and Davies, 1996), signaling pathways (Chaves et al., 2003), gene expression (Bray, 1997; 2002), and leaf photosynthesis (Lawlor and Cornic, 2002; Flexas et al., 2004), have been studied extensively. Surprisingly, in comparison with other physiological processes, studies examining the effects of water stress on respiration are few (Hsiao, 1973; Amthor, 1989), despite the importance of respiration in ecosystem annual net productivity (Valentini et al., 1999) and the fact that ecosystem respiration is strongly affected by water availability (Bowling et al., 2002). Another important point to consider is the effect of water stress on the electron partitioning between the cytochrome and the cyanide-resistant, alternative pathway and its consequences for ATP synthesis. Unfortunately, the few studies that have focused on alternative respiration as affected by water stress (González-Meler et al., 1997) used specific inhibitors for the cytochrome (KCN) and alternative (salicylhydroxamic acid [SHAM]) respiratory pathways; this methodology is now known to be invalid (Lambers et al., 2005). Currently, the only available system to

Table 8.1 Respiration rates of three soybean cultivars grown at three carbon dioxide concentrations and three temperature regimes, and analysis of variance. Respiration rates were measured at the night temperatures and carbon dioxide concentrations. Rates are averages of single measurements on each of three plants at each of four measurement times per treatment for each cultivar.

Carbon dioxide (ppm)	Temperature Day/night (°C)	Maple Glen	Clark	CNS
370	20/15	24 ± 3	24 ± 5	23 ± 7
370	25/20	25 ± 6	24 ± 3	26 ± 5
370	31/26	25 ± 0	19 ± 7	21 ± 6
555	20/15	17 ± 2	19 ± 1	19 ± 1
555	25/20	20 ± 0	21 ± 2	21 ± 9
555	31/26	16 ± 5	19 ± 5	28 ± 2
740	20/15	14 ± 5	15 ± 0	15 ± 4
740	25/20	19 ± 1	16 ± 9	18 ± 9
740	31/26	18 ± 1	20 ± 5	22 ± 9

Main effect means

CO_2	Temperature	Cultivar
370 (23±9)	20/15 (19±2)	Maple Glen (20±1)
555 (20±3)	25/20 (21±6)	Clark (20±1)
740 (17±9)	31/26 (21±3)	CNS (22±0)

ANOVA

Source	df	Mean square	Fvalue	Pvalue
CO2	9	77±9	10±87	0±0001
Temperature	2	187±5	2±08	0±1262
Cultivar	2	137±7	1±53	0±2181
CO2-Temperature	4	141±7	1±58	0±1810
CO2-Cultivar	4	91±7	1±02	0±3975
Temperature-Cultivar	4	58±6	0±65	0±6266
CO2-Temperature-Cultivar	8	55±3	0±62	0±7651
Residual	297	90±0		

measure electron partitioning between the two respiratory pathways and their actual activities is the use of the oxygen-isotope-fractionation technique (Ribas-Carbo et al., 2005). This technique is based on the fact that the two terminal oxidases fractionate ^{18}O differently, with the cytochrome oxidase discriminating less than the alternative oxidase (AOX; Guy et al., 1989). This differential fractionation is the basis of a methodology that allows the assessment of the electron partitioning between the cytochrome and alternative respiratory pathways in the absence of added inhibitors (Ribas-Carbo et al., 1995, 1997; Gastón et al., 2003). Using this technique, it has been observed that the electron transport through the alternative pathway increases under phosphate limitation (González-Meler et al., 2001), during chilling recovery (Ribas-Carbo et al., 2000), and by the application of allelochemicals (Peñuelas et al., 1996) or herbicides inhibiting branched-chain amino acid synthesis (Gastón et al., 2003).

The biochemical regulation of the electron partitioning between the cytochrome and alternative pathways is quite complex because the two pathways that compete for electrons from the ubiquinone pool have their own regulation (Lambers et al., 2005). The cytochrome pathway, including Complex III and Complex IV, is strongly regulated by the proton gradient between the matrix and the intermembrane space, which in turn is directly affected by the mitochondrial "respiratory control" or ATP/ADP ratio (Millenaar and Lambers, 2003). Changes in the ATP/ADP ratio could result from a change in the kinetics of the ATPase or from a change in the balance between ATP synthesis and demand. Furthermore, the cytochrome c oxidase can be inhibited by natural inhibitors, such as carbon monoxide, cyanide, sulfide, or nitric oxide, among others (Moore and Siedow, 1991; Millar and Day, 1996). On the other hand, the alternative pathway has several mechanisms of biochemical regulation. AOX can be activated by pyruvate and other α-ketoacids (Millar et al., 1993). Furthermore, the AOX in plants is present as a dimer and is regulated by a disulfide/sulfhydryl system (Umbach and Siedow, 1993; Vanlerberghe et al., 1999), with the two subunits linked by disulfide bridges and with the reduced form being the active form and the oxidized form being almost inactive (Vanlerberghe et al., 1999). It has been suggested that a mitochondrial thioredoxin might be involved in the regulation of the redox status of this disulfide bridge (Umbach and Siedow, 1993). Finally, the maximum capacity of the AOX is determined by the total amount of AOX protein present (Vanlerberghe et al., 1994). The expression of the AOX has previously been shown to increase under several stress situations (Millenaar and Lambers, 2003; Lambers et al., 2005), such as low temperature, low phosphate availability (Parsons et al., 1999; Juszczuk et al., 2001), inhibition of the cytochrome pathway (Wagner et al., 1992), the application of inhibitors of mitochondrial protein synthesis (Day et al., 1996), and herbicides inhibiting branched-chain amino acid synthesis (Aubert et al., 1997). This increase is remarkably strong under conditions where the production of reactive oxygen species (ROS), especially superoxide, occurs (Wagner and Krab, 1995). Water stress, particularly combined with light, increases the risk of oxidative stress by increasing the presence of ROS in different cell compartments (Bartoli et al., 2004). Under these conditions, the alternative respiratory path could play a role in preventing the formation of damaging ROS. A possible role of the alternative pathway is in protecting plants from ROS or in sustaining respiration under situations where the cytochrome pathway is restricted (Lambers et al., 2005). It has also been proposed that the cyanide-resistant respiratory pathway could be involved in the prevention of the formation of ROS (Purvis, 1997; Maxwell et al., 1999) and programmed cell death (Vanlerberghe et al., 2002). Finally, it has been suggested that ROS could be part of a signal or communication pathway indicating that there is a restriction of the activity of the cytochrome pathway (Foyer and Noctor, 2003).

The effect of water stress on respiration and mitochondrial electron transport has been studied in soybean (*Glycine max*) leaves, using the oxygen-isotope-fractionation technique. Treatments with three levels of water stress were applied by irrigation to replace 100%, 50%, and 0% of daily water use by transpiration. The levels of water stress were characterized in terms of light-saturated stomatal conductance (g_s): Well irrigated ($g_s > 0.2$ mol H_2O m^{-2} s^{-1}), mildly water stressed (g_s between 0.1 and 0.2 mol H_2O m^{-2} s^{-1}), and severely water stressed ($g_s < 0.1$ mol H_2O m^{-2} s^{-1}). Although net photosynthesis decreased by 40% and 70% under mild and severe water stress, respectively, the total respiratory oxygen uptake (V_t) was not significantly different at any water-stress level. However, severe water stress caused a significant shift of electrons from the cytochrome to the alternative pathway. The electron partitioning through the alternative pathway increased from 10% to 12% under well-watered or mild water-stress

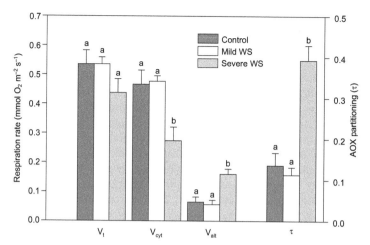

Fig. 8.5 Respiration rate and AOX partitioning under water stress.

conditions to near 40% under severe water stress. Consequently, the calculated rate of mitochondrial ATP synthesis decreased by 32% under severe water stress. Unlike many other stresses, water stress did not affect the levels of mitochondrial alternative oxidase protein. This suggests a biochemical regulation (other than protein synthesis) that causes this mitochondrial electron shift (Ribas-Carbo et al., 2005) (Fig. 8.5).

The leakage of solutes from cotyledons of soybeans (cv. Chppewa 64) was markedly stimulated by a chilling treatment (1 to 4°C) during the 1st minute of imbibitions, but chilling after even 1 minute of water uptake resulted in little or no leakage increase. The respiratory rate of soybean particles was reduced more than 60%. Chilling treatment (15 minutes at 1 to 4°C) was given during the first minutes of imbibitions, and little or no reduction was obtained if the chilling treatment was begun at 5 to 15 minutes after the start of imbibition. Using KCN as an inhibitor of cytochrome oxidase pathway of respiration and salicylhydroxamic acid as an inhibitor of the alternative pathway, it was found that the chilling injury involved a major reduction in the cytochrome pathway in whole axes and cotyledons and an engagement of the alternative pathway of respiration in cotyledon tissue. The suggestion is made that the chilling injury involves lesion resulting from temperature stress during the reorganization of membranes with water entry, and that both the leakage and the respiratory effects are consequences of these membrane lesions.

The influence of low temperature on soybean (*Glycine max* (L.) Merr. cv. Wells) energy transduction via mitochondrial respiration and dehydrogenase was investigated in the study during imbibition and germination. Mitochondria were isolated from embryonic axes of seeds treated at 10 and 23°C (control) by submergence in H_2O for 6 hours and maintenance for an additional 42 hours in a moist environment. Arrhenius plots of initial respiration rates revealed that those from cold treated axes had respiratory control (RC) ratios of near 1.0 above an inflection in the plot at 8°C. Arrhenius plot of control axes mitochondrial respiration showed RC ratios of 2.8 above and 5.0 below an inflection temperature of 12.5°C. Energies of activation for mitochondrial respiration between 20 and 30°C for the cold and control treatments were 7.8 and 15.6 kcal/mole, respectively. These data indicate possible differences in mitochondria membranes, degree of mitochondrial integrity, and mitochondrial enzyme complement between the two treatments. Glutamate dehydrogenase (GDH), malate dehydrogenase (MDH), alcohol dehydrogenase (ADH), Glucose-6-phosphate dehydrogenase (G-6-PDH), and NADP-isocitrate dehydrogenase (NADP-ICDH) were assayed from whole seeds and axes (after germination) during the 48 hours of temperature treatments. Activity of these dehydrogenases decreased during the first 6 hours, with the exception of MDH. After germination at 23°C (48 hours), all five dehydrogenases increased in activity. Arrhenius plots of cotyledon dehydrogenases activities indicated that one inflection temperature between 6 and 18°C was present for each enzyme assayed. Differences were seen in Arrhenius plots of axes dehydrogenases activities with the two temperature treatments in the cases of GDH and MDH from mitochondrial pellets and with differences in enzyme extraction media. These data suggest that the temperature treatments yield differences in mitochondrial enzyme complement. There were no detectable inflection temperatures for the activities of G6P-DH and ADH extracted from axes. Arrhenius plots of NADP-ICDH activity indicated extreme cold sensitivity. The slopes of the plots axes NADP-ICDH were very similar to those for mitochondrial respiration (23°C), suggesting that this enzyme may limit mitochondrial respiration at low temperatures in soybean tissues grown at moderate temperatures.

High night-time temperatures can result in wasteful respiration and a lower net amount of dry matter accumulation in plants. The rate of respiration of plants increases rapidly as the temperature increases, approximately doubling for each 10.6°C increase. During the day, soybean plants accumulate starch in their leaves. At night, the starch is broken down and exported from their leaves. When nights are cool, the amount of starch exported is reduced, resulting in high leaf starch the following day, which can disrupt photosynthesis. Night-time temperatures have to exceed 29.4°C before any noticeable reduction in soybean yield is experienced (10%) (Anonymous, 2012a).

The respiration rate increases at higher temperatures, causing a greater loss in seed weight. Gulluoglu et al. (2006) reported that soybean cannot synthesize some of the necessary hormones in a sufficient level to control its growth and development under high temperature and low humidity, consequently, the plant cannot show its real yield potential. Marking (1986) indicated that high temperature pollen viability and accelerates ethylene production that increase flower abortion and plant senescence. Yield decrease due to high temperature and (CO_2) could be due to the effect of on reproductive at both organ and process levels. High temperature inhibits pollen germination and pollen tube growth and genotypes differ in their sensitivity (Huan et al., 2000; Kakani et al., 2002).

8.3 Root

Carbon (C) expended in root respiration can amount to 10 ± 30% of net photosynthesis under favourable conditions (Lambers et al., 1996) and has been found to increase under unfavourable conditions, such as temperature extremes (Sisson, 1983; Bouma et al., 1997a), drought (Sisson, 1989), low light intensity (Hansen and Jensen, 1977), low nitrate supply (van der Werf et al., 1992), and low phosphorus P availability (Lynch and Beebe, 1995; Nielsen et al., 1998). Root respiration can be partitioned into costs for growth, maintenance, and ion uptake. Poorter et al. speculated about fast-growing species having lower specific respiratory costs for ion uptake than slow-growing species (Poorter et al., 1991). However, Scheurwater et al. determined specific costs for growth, maintenance of biomass and ion transport in fast- and slow-growing grass species and concluded that respiratory costs associated with ion uptake are clearly higher in the slow-growing species (Scheurwater et al., 1998). This is mostly due to the relatively high nitrate efflux (inefficient net nitrate uptake) in slow-growing grasses (Scheurwater et al., 1999). When plants are deprived of phosphate for extended periods, vacuolar P stores become exhausted, and this is followed by reductions in cytoplasmic P concentration (Lauer et al., 1989).

Other work suggests that alternative pathways of glycolytic carbon flow and mitochondrial electron transport allows respiration to proceed in plants during severe P deficiency (Theodorou and Plaxton, 1993). Respiration of pea roots was not found to be affected by P starvation (Rychter et al., 1992). When plants are P stressed to an extent that growth is reduced, respiratory costs of growth and nutrient uptake will be affected.

The R_D varied little across temperatures under control P treatment (Fig. 8.6C). However, compared to MLT and OT, R_D was almost doubled at warmer temperatures (i.e., MHT and HT) under P deficiency, especially, between 10 and 12 weeks after planting (Fig. 8.6D). The total seasonal net C-gain followed an almost similar pattern as total dry matter harvested for the entire season across treatments (Fig. 8.6E, F). Relative to the OT, total net C-gain and dry matter at HT was lower or almost similar under sufficient and deficient P levels, respectively. In respect to P effects, P deficiency had smaller values of total net C-gain (13–33%) and total dry matter (14–26%) at and below OT. However, at warmer than OT both parameters were almost similar to the observations made under sufficient P condition (Fig. 8.6E, F).

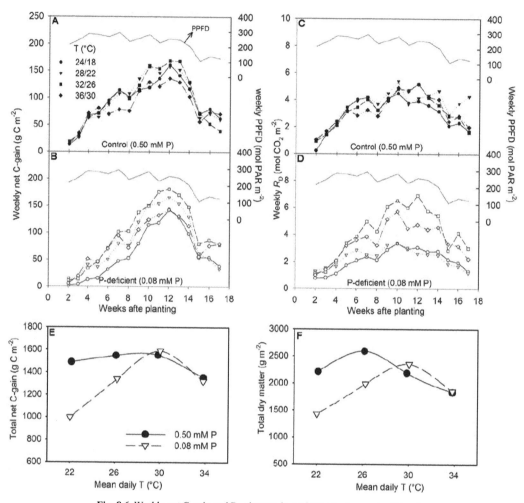

Fig. 8.6 Weekly net C-gain and R_D changes in soybeans under varying P level.

Plant species vary in their response to flooding, but soybeans are generally sensitive to it (Scott et al., 1989). Flooding soybeans has been reported to significantly reduce seed yield compared with non-flooded soybeans (Scott et al., 1989). Flood damage in soybean may be the result of death of root cortical tissue, reduced vigor, and wilting as a result of hypoxic conditions (Mozafar et al., 1992). Lack of oxygen diffusion and increased CO_2 around the root system also can inhibit nitrogen fixation (Bennett and Albrecht, 1984). Flooding has been shown to decrease respiration, causing an increase in membrane permeability, which can result in plant cell death (Barta, 1985; Chirkova, 1978). This increase in membrane permeability results in "leakiness" that increases loss of nutrients from the tissue (Gronewald and Hanson, 1982).

Flooding is quite a common phenomenon and can occur after a short period of heavy rain, especially in poorly-drained soils. Waterlogged soils lead to a deficient supply of oxygen to the roots, in view of the diffusion rate of oxygen in water being some 10000 times slower than in air (Armstrong et al., 1994). Once the dissolved oxygen concentration falls below that necessary to maintain a normal rate of tissue respiration, by definition, the state of hypoxia is reached. The reduced rate of respiration of waterlogged roots leads to adjustments in cell metabolism that help the plant tolerate the stress. The main metabolic change involves pyruvate which switches from oxidative to fermentation metabolism, whose principal products are lactate, ethanol and alanine (Ricard et al., 1994).

Despite these metabolic changes contributing to flooding tolerance, different species show large variations in their degree of tolerance, depending not only on metabolic adaptability but also morpho-anatomic adaptations, such as formation of aerenchyma and pneumatophors. Other factors also appear to be important. For example, it has been observed that the presence of nitrate in the flooding medium enhances tolerance (Magalhães et al., 2002), although the mechanism involved is not understood (see Sousa and Sodek, 2002a). In a previous study (Thomas and Sodek, 2005), it was shown that tolerance of soybean to flooding was enhanced by NO_3^- but not by other N sources. Little is known of the uptake and metabolism of nitrate under O_2 deficiency. Moreover, the data are conflicting. The earlier work of Lee (1978) suggests that uptake of nitrate is restricted under hypoxia of the root system and almost ceased under anoxia. Trought and Drew (1981) and Buwalda and Greenway (1989) also found reduced uptake of nitrate for wheat seedlings under anaerobic conditions. In contrast, Morard and collaborators (2004) report that nitrate loss from the medium with excised tomato roots was strongly stimulated by anoxia. As to the metabolism of nitrate, Lee (1978) reported diminished reduction and assimilation. A hypothesis that nitrate respiration may explain the beneficial effect on flooding tolerance was proposed, since the recycling of NAD by nitrate reduction (nitrate respiration) would substitute for ethanol formation in this role (Roberts et al., 1985). Support for the hypothesis may be found in the work of Fan and collaborators (1997) where nitrate reduced the production of ethanol in rice coleoptiles. However, in excised roots of barley (Lee, 1978) and rice (Reggiani et al., 1985b) more, not less, ethanol was produced in the presence of nitrate, thus throwing doubt upon the hypothesis. More recent evidence suggests that the beneficial effect of nitrate may involve its metabolic product nitrite (Libourel et al., 2006; Stoimenova et al., 2007).

Soybean plants are very tolerant of excess water and anaerobiosis, but are injured under anaerobic conditions when root-zone CO_2 concentrations increase to those found in flooded soybean fields (30%). Rice, a flooding-tolerant species, is much more tolerant of elevated root-zone CO_2 levels than is soybean. Results suggest that CO_2 toxicity is a factor affecting soybean tolerance to flooded soils.

Early exposure of soybean plants to flooding stress causes severe damage due to rapid imbibition of water by the cotyledons and destruction to the root systems (Nakayama et al., 2004). Its yield was estimated to be reduced to 25% due to flooding injuries in Asia, North America, and other regions of the world where soybean is rotated with rice in paddy fields. Oosterhuis et al. (1990) reported a reduction in soybean yield of 17–43% at the vegetative stage and 50–56% at the reproductive stage due to flooding stress. This stress leads to a shift to alternative pathways of energy generation. The shortage of oxygen under flooding stress results in a shift from aerobic to anaerobic respiration. A low diffusion rate of oxygen under flooding stress is a limiting factor for plant survival, and most plants die under limited oxygen supply (Voesenek et al., 2006).

Aerenchyma provides a low-resistance O_2 transport pathway that enhances plant survival during soil flooding. When in flooded soil, soybean produces aerenchyma and hypertrophic stem lenticels. The aims of this study were to investigate O_2 dynamics in stem aerenchyma and evaluate O_2 supply via stem lenticels to the roots of soybean during soil flooding. Oxygen dynamics in aerenchymatous stems were investigated using Clark-type O_2 microelectrodes, and O_2 transport to roots was evaluated using stable-isotope (18)O_2 as a tracer, for plants with shoots in air and roots in flooded sand or soil. Short-term experiments also assessed venting of CO_2 via the stem lenticels. The radial distribution of the O_2 partial pressure (pO_2) was stable at 17 kPa in the stem aerenchyma 15 mm below the water level, but rapidly declined to 8 kPa at 200–300 microm inside the stele. Complete submergence of the hypertrophic lenticels at the stem base, with the remainder of the shoot still in air, resulted in gradual declines in pO_2 in stem aerenchyma from 17.5 to 7.6 kPa at 13 mm below the water level, and from 14.7 to 6.1 kPa at 51 mm below the water level. Subsequently, re-exposure of the lenticels to air caused pO_2 to increase again to 14–17 kPa at both positions within 10 min. After introducing (18)O_2 gas via the stem lenticels, significant (18)O_2 enrichment in water extracted from roots after 3 h was confirmed, suggesting that transported O_2 sustained root respiration. In contrast, slight (18)O_2 enrichment was detected 3 h after treatment of stems that lacked aerenchyma and lenticels. Moreover, aerenchyma accelerated venting of CO_2 from submerged tissues to the atmosphere. Hypertrophic lenticels on the stem of soybean, just above the water surface, are entry points for O_2, and these connect to aerenchyma and enable O_2 transport into roots in flooded soil. Stems that develop aerenchyma, thus, serve as a 'snorkel' that enables O_2 movement from air to the submerged roots.

Little information is available about oxygen transport in the nodules of soybeans and of other legumes. This topic is of substantial interest since oxygen partial pressure strongly affects the rate of nitrogen fixation and acetelene reduction in soybean nodules (Mague and Burris, 1972). Also, examination of oxygen transport in soybean nodules might give information about the possible role of leghemoglobin in oxygen transport.

Bergersen (1962a) measured the respiration rate of soybean nodules as a function of pO_2 and found a unique two-step curve. Also, such measurements have obtained somewhat different results. From these results and from calculations based on nodule anatomy, a theory of nodule respiration was developed which differs from that of Bergersen and which helps to explain the role that leghemoglobin may play in oxygen transport in the nodule (Tjepkema, 1971).

The respiration rate of individual soybean nodules was measured as a function of pO_2 and temperature. At 23°C, as the pO_2 was increased from 0.1 to 0.9 atm, there was a linear increase in respiration rate. At 13°C, similar results were obtained, except that there was an abrupt saturation of respiration at approximately 0.5 atm pO_2. When measurements were made on the same nodule, the rate of increase in respiration with pO_2 was the same at 13°C and 23°C. Additional results were that 5% CO in gas phase had no effect on respiration, except for a small decrease in the pO_2 at which respiration became saturated (Fig. 8.7). Also, nodules still attached to the soybean root displayed the same respiratory behavior as detached nodules. A model for oxygen transport in the nodule is presented, which explained these results quantitatively. The essence of the model is that the respiration rate of the central tissue of the nodule is almost entirely determined by the rate of oxygen diffusion to the respiratory enzymes. Evidence is given that the nodule cortex is the site of almost all the resistance to oxygen diffusion within the nodule (Tjepkema and Yocum, 1973).

Soil waterlogging is a common abiotic stress worldwide in cultivated areas and influences the composition and productivity of soybean (*Glycine max* L. Merril) and most other crop species (Kokubun, 2013). In waterlogged soils, gas exchanges between root systems and soil porous spaces are limited due to oxygen diffusion resistance that is around 10,000 times higher in water than in the air (Bailey-Serres et al., 2012).

Decrease in the oxygen level is the main factor that causes stress, leading to a chain signaling that unleashes a series of metabolic changes (Horchani et al., 2009), such as N metabolism and inter conversion of amino acids (Puiatti and Sodek, 1999; Oliveira et al., 2013), changes in carbohydrates levels and energetic metabolism (Sousa and Sodek, 2002a), in an effort to secure plant survival and growth when exposed to hypoxic stress (Geigenberger, 2003). Anaerobic metabolism is activated under low oxygen concentration and, as a consequence, there is a significant decrease in energy production that is derived mainly from glycolysis, in contrast to oxidative phosphorylation (Sairam et al., 2009; Zabalza

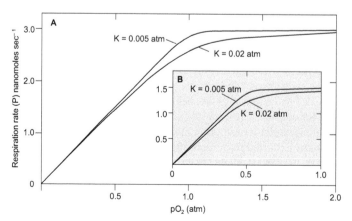

Fig. 8.7 Respiration rate and partial pressure of oxygen relationship.

et al., 2009). Plant survival under these conditions depends almost exclusively on anaerobic metabolism (Sousa and Sodek, 2002a).

Due to the lack of O_2 as the final electron acceptor, there is an accumulation of intermediates of Krebs cycle, $NAD(P)^+$ levels decrease, pyruvate accumulate and ATP levels decrease. All these modifications act as a signal for further adaptive responses (Horchani et al., 2009). Enzymes of two important pathways are induced under anaerobic conditions: Lactic fermentation is a one-step reaction from pyruvate to lactate, catalyzed by lactate dehydrogenase (LDH) with the regeneration of NAD^+. Ethanolic fermentation is a two-step process regenerating NAD^+, in which pyruvate is first decarboxylated to acetaldehyde by pyruvate decarboxylase (PDC), and acetaldehyde is subsequently converted to ethanol by alcohol dehydrogenase (ADH) (Tadege et al., 1999; Zabalza et al., 2009) (Figs. 8.8 and 8.9).

Waterlogging blocks the oxygen supply to the root system, inhibiting respiration and greatly reducing the energy status of cells that affect important metabolic processes. This study evaluated fermentative metabolism and carbohydrate contents in the root systems of two soybean (*Glycine max* L. Merril) genotypes under hypoxic and post-hypoxic conditions. Nodulated plants (genotypes Fundacep 53 RR and

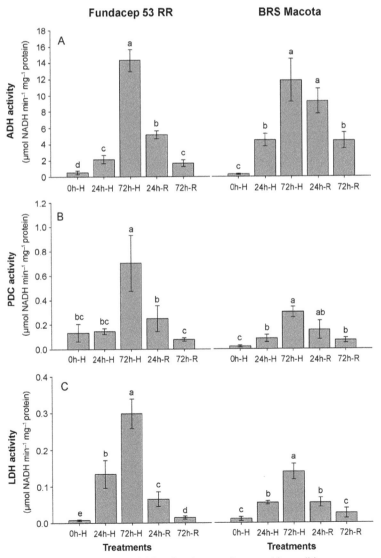

Fig. 8.8 Enzymic activities of soybeans under anaerobic conditions.

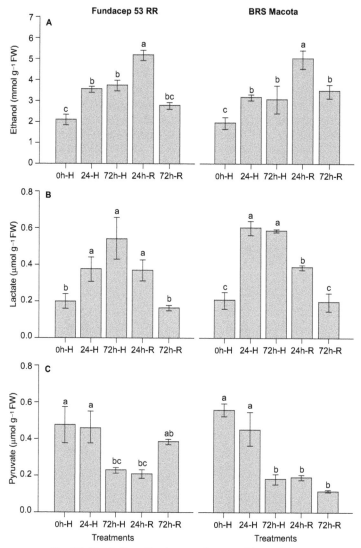

Fig. 8.9 Metabolites of soybeans under anaerobic conditions.

BRS Macota) were grown in vermiculite and transferred to a hydroponic system at the reproductive stage. The root system was submitted to hypoxia by flowing N_2 (nitrogen) gas in a solution for 24 and 72 h. For recovery, plants returned to normoxia condition by transfer to vermiculite for 24 and 72 h. Fermentative enzyme activity, levels of anaerobic metabolites and carbohydrate content were all quantified in roots and nodules. The activity of alcohol dehydrogenase, pyruvate decarboxylase and lactate dehydrogenase enzymes, as well as the content of ethanol and lactate, increased with hypoxia in roots and nodules, and subsequently returned to pre-hypoxic levels in the recovery phase in both genotypes. Pyruvate content increased in nodules and decreased in roots. Sugar and sucrose levels increased in roots and decreased in nodules under hypoxia in both genotypes. Fundacep RR 53 was more responsive to the metabolic effects caused by hypoxia and post-hypoxia than BRS Macota, and it is likely that these characteristics contribute positively to improving adaptation to oxygen deficiency.

Changes in the respiratory rate and the contribution of the cytochrome (Cyt) *c* oxidase and alternative oxidase (COX and AOX, respectively) were investigated in soybean (*Glycine max* L. cv Stevens) root seedlings using the ^{18}O-discrimination method. In 4-d-old roots respiration proceeded almost entirely

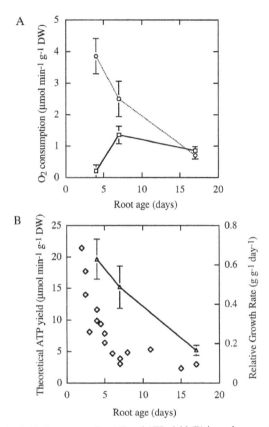

Fig. 8.10 O_2 consumption (A) and ATP yield (B) in soybean root.

via COX, but by d 17, more than 50% of the flux occurred via AOX. During this period, the capacity of COX, the theoretical yield of ATP synthesis, and the root relative growth rate all decreased substantially (Fig. 8.10). In extracts from whole roots of different ages, the ubiquinone pool was maintained at 50% to 60% reduction, whereas pyruvate content fluctuated without a consistent trend. In whole-root immunoblots, AOX protein was largely in the reduced, active form at 7 and 17 d, but was partially oxidized at 4 d. In isolated mitochondria, Cyt pathway and succinate dehydrogenase capacities and COX I protein abundance decreased with root age, whereas both AOX capacity and protein abundance remained unchanged. The amount of mitochondrial protein on a dry-mass basis did not vary significantly with root age. It is concluded that decreases in whole-root respiration during growth of soybean seedlings can be largely explained by decreases in maximal rates of electron transport via COX. Flux via AOX is increased so that the ubiquinone pool is maintained in a moderately reduced state.

8.4 Nodules

The respiratory cost of symbiotic N fixation in soybean was estimated by 4 methods. When nitrogenase activity and root respiration were decreased by treating roots briefly with 1.0 atm oxygen, the respiration associated with nitrogenase was estimated as 2.1 micro mol carbon dioxide/micro mol acetylene. This value was 2.9 and 4.08 micro mol carbon dioxide/micro mol acetylene when nitrogenase activity and respiration were decreased by addition of nitrate and nodule removal, respectively.

Most rhizobia are obligate aerobes. The C_4-dicarboxylic acids supplied by the host plant must be metabolized through the tricarboxylic acid (TCA) cycle in the bacteroid. Malate, the primary carbon source of bacteroids, is converted to pyruvate and CO_2 via the NAD^+-dependent malic enzyme (DME). Pyruvate is subsequently decarboxylated by pyruvate dehydrogenase to form acetyl-coenzyme A (acetyl-CoA) and enters the TCA cycle. DME is required to provide pyruvate to the TCA cycle in *S. meliloti* and *Azorhizobium caulinodans* (Zhang et al., 2012). *S. meliloti* possesses two distinct malic enzymes, an NAD(P)$^+$-dependent enzyme (DME) (EC 1.1.1.39) and a strictly NADP$^+$-dependent enzyme (TME) (EC 1.1.1.40) (Voegele et al., 1999). In addition, acetyl-CoA can be produced alternatively by phosphoenolpyruvate carboxykinase (PCK) which catalyzes the decarboxylation of OAA to PEP. Symbiotic phenotypes of *pck* mutants vary depending on the host plant. *pckA* mutants of *R. leguminosarum* MNF3085 fix nitrogen at rates comparable to wild type, while *pckA* mutants of *S. meliloti* show a reduced N$_2$ fixation capacity (McKay et al., 1985; Finan et al., 1991). In *R. leguminosarum* and *S. fredii* strain NGR234, DME or a pathway involving PCK and pyruvate kinase (PYK) can synthesize the precursors required for SNF (Zhang et al., 2012). Enzymes participating in the TCA cycle have been identified in *B. japonicum* strain USDA110 (Sarma and Emerich, 2006), *R. leguminosarum* (McKay et al., 1985), *S. meliloti* (Djordjevic, 2004), and *R. tropici* (Romanov et al., 1994). However, TCA cycle involvement probably varies among rhizobia. Evidence shows that *S. meliloti, R. tropici*, and *R. leguminosarum* utilize the full oxidative TCA cycle to provide ATP, precursors of amino acid synthesis, as well as reducing equivalents for N$_2$ fixation. Enzymes of the TCA cycle appear to be essential for nitrogen fixation in *S. meliloti*, since mutations in succinate dehydrogenase (*sdh*), malate dehydrogenase (*mdh*), isocitrase dehydrogenase (*icd*), 2-oxoglutarate dehydrogenase, aconitase (*acnA*), and citrate synthase (*gltA*) abolish N$_2$ fixation, despite the fact that nodules were formed (Koziol et al., 2009). In contrast, *B. japonicum* shows a higher metabolic plasticity. Mutations in fumarase (*fumC*), ICDH (*idhA*), alpha-ketoglutarate dehydrogenase (*agdA*), and acotinase (*acnA*) in *B. japonicum* USDA110 still show phenotypes capable of fixing nitrogen (Shah and Emerich, 2006). Transcriptome analyses of *R. etli* bacteroids have suggested that the TCA cycle is inactive in these bacteroids (Vercruysse et al., 2011), but additional experiments are required in order to explain how the bacteroids obtain the ATP necessary for nitrogen fixation in these cases. A comparative metabolic profiling study between free-living *R. leguminosarum* and pea bacteroids showed that the TCA cycle is not the only path to oxidize dicarboxylic acids derived from the host plant in the bacteroids (Terpolilli et al., 2016). Metabolic profiling and flux analysis revealed that pea bacteroids divert acetyl-CoA into TCA and the production of lipid or polyhydroxybutyrate (PHB). These findings suggest new pathways for electron allocation in nitrogen-fixing bacteroids where lipogenesis may be a requirement in legume nodules.

In legume nodules, treatments, such as detopping or nitrate fertilization, inhibit nodule metabolism and N fixation by decreasing the nodule's permeability to O diffusion, thereby decreasing the infected cell O concentration O$_i$ and increasing the degree to which nodule metabolism is limited by O availability. In the present study, nodule oximetry was used to asses and compare the role of O limitation in soybean (*Glycine max* L. Merr.) nodules inhibited by either drought or detopping. Compared to detopping, drought caused only minor decreases in *Oi,* and when the external O concentration was increased to raise *Oi,* the infected cell respiration rate in the drought-stressed plants was not stimulated as much as it was in the nodules of the detopped plants. Unlike those in detopped plants, nodules exposed to moderate drought stress displayed an O-sufficient respiration rate that was significantly lower than that in control nodules. Despite possible side effects of oximetry in altering nodule metabolism, these results provided direct evidence that, compared to detopping, O, limitation plays a minor role in the inhibition of nodule metabolism during drought stress and changes in nodule permeability are the effect, not the cause, of a drought-induced inhibition of nodule metabolism and the O-sufficient rate of respiration (del Castillo and Layzell, 1995) (Fig. 8.11).

There is much information on the negative effect of nitrate on the nodulation induced by rhizobia (Alcantar-Gonzales et al., 1988). Until the end of the 80's, it was believed that the presence of nitrate in medium completely inhibited or drastically reduced the efficiency of infection in strains that possessed any NR activity (Alcantar-Gonzales et al., 1988). However, it was shown that, at least in some Rhizobium strains, the presence of nitrate did not affect this process negatively. Serrano and Chamber (1990) investigated three Bradyrhizobium sp. (Lupinus) strains differing in NR activity: The first strain possessed constitutive NR (cNR), the second inducible NR (iNR), and the third did not express NR activity (NR–). The results demonstrated that the strains with cNR and iNR activity were capable of infecting roots sufficiently even at relatively high (12 mM) nitrate concentration, whereas the NR–strain completely lost

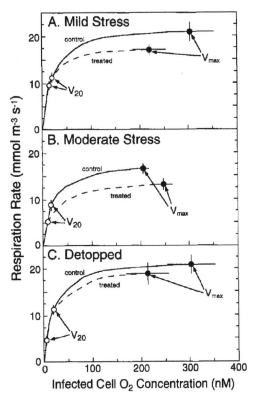

Fig. 8.11 Respiration rate changes with infected cell O_2 concentration in water stressed and detopped plants of soybean.

the ability to form nodules. Moreover, at nitrate concentration of 1–2 mM, the strains with NR activity (either constitutive or inducible) infected roots slightly more efficiently than in the absence of nitrate. The described "patterns" of infection ability resulted in production of durable nodules, which was observed three weeks after. These results show that there is a possibility of formation of symbiotic associations, in which the presence of nitrate does not inhibit nodulation. The effect of NR activity on the efficiency of root infection might be explained by the removal of nitrate ions from the bacterial environment (O'Hara and Daniel, 1985).

One of possible ways in which nitrate can affect the gas diffusion resistance is its direct influence (Kuzma et al., 1995). About 90% of the nitrate applied onto the nodule surface accumulated in the nodule cortex (Sprent et al., 1987). Vessey and Waterer (1992) suggested that high nitrate concentration could cause formation of a local pH gradient or electrochemical gradient due to differences in the concentration of nitrate or in their reduction rate between cell layers. This was believed to affect gas permeability of cortex cell layer, and, consequently, to cause conditions in which nitrogenase activity was limited by deficiency of $NAD(P)H_2$ and ATP (Minchin et al., 1989; Sprent et al., 1987). The intensity of lupin bacteroid respiration corresponded to as much as 83% of total respiration activity in the infected zone (Ratajczak et al., 1996). In such a situation, lowered bacteroid respiration activity, caused by an increase in the resistance to O_2 diffusion, could be the reason for significant ATP deprivation.

The infected cells of soybean (*Glycine max*) root nodules require ATP production for ammonia assimilation and purine synthesis under microaerobic conditions. It is likely that the bulk of this demand is supplied through mitochondrial oxidative phosphorylation. Mitochondria purified from root nodules respired and synthesized ATP in sub-micromolar oxygen concentrations, as measured by leghaemoglobin spectroscopy and luciferase luminescence. Both oxygen uptake and the apparent ATP/O ratio declined significantly as the oxygen concentration fell below 100 μmol mol^{-1}. Cytochrome-pathway respiration by root nodule mitochondria had a higher apparent affinity for oxygen ($K^\wedge$ 50 μmol mol^{-1}) than did mitochondria isolated from roots ($K^\wedge$ 125 μmol mol⁻¹). Electron micrographs showed that mitochondria predominated at the periphery of infected cells adjacent to gas-filled intercellular spaces, where the oxygen concentration is predicted to be highest. Calculations of oxygen concentration and nitrogen fixation rates on an infected cell basis suggest that the measured rates of ATP production by isolated mitochondria are sufficient for the quantifiable *in vivo* requirements of ammonia assimilation and purine synthesis (Millar et al., 1995) (Fig. 8.12).

Nodule mitochondria are highly sensitive to the respiratory inhibitor antimycin A. The antimycin-resistant oxygen uptake is 5–10% of the rate of control mitochondrial respiration. The high sensitivity to this inhibitor means that non-phosphorylating pathways are absent from the nodules and the energetic effectiveness of mitochondria is very high. The latter is proved by the good respiratory control, observed in mitochondria from soybean plant nodules. Mitochondria, isolated from nodules of plants that are treated with succinate and α-ketoglutarate during inoculation showed highest respiratory control. Significant differences in the rate of oxygen uptake by mitochondria isolated from soybean plant nodules were observed in all treatments investigated. The rates of oxygen consumption by mitochondria isolated from nodules of citrate treated plants were close to these of the control plants. However, treatment with succinate, malate and α-ketoglutarate resulted in a significant increase in mitochondrial oxygen uptake. Results support the relation between plant photosynthesis and bacteroid respiration. Photosynthetic intensity and the oxygen uptake of bacteroids were the lowest in control soybean plants and the ones treated with citrate during inoculation. In the cases when the intensity of photosynthesis was high (treatments with succinate, malate and α-ketoglutarate during inoculation), the rate of oxygen uptake of the bacteroids was the highest. Results obtained indicate that the oxidative capacity of mitochondria and bacteroids from root nodules of soybean treated with organic acids succinate and α-ketoglutarate was stimulated, possibly leading to more effective nodule nitrogen fixation.

The end products of SNF in soybean (*Glycine max* (L.) Merr.) root nodules, namely ureides, are transported to the upper parts of the plant to supply nitrogen. Symbiotic nitrogen fixation provides a vital advantage for the production of soybean, compared with most grain crops, in that soybean fixes the nitrogen required for its growth and for the production of the high-protein content in seed and oil. The process of SNF is dramatically affected by drought, salt, cold and heavy metal stresses. Since SNF is such an important yield-determining factor, a lack in understanding these complexes inevitably delays progress

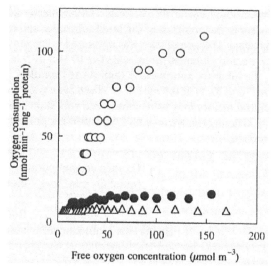

Fig. 8.12 Microaerobic respiration utilizing NADH as substrate by isolated nodule bacteroids (A) and nodule mitochondria in the presence (•) or absence (O) of antimycin A and n-propylgallate over a range of free oxygen concentrations. Oxygen consumption and free oxygen concentration were calculated by monitoring the concentration of LbO₂ in a cuvette in the presence of ADP (600 mmol mol⁻¹). The substrate NADH was added at 1 mol mol⁻¹, and the inhibitors antimycin A at 1 mol mol⁻¹, and nPG at 0.2 mol mol⁻¹.

towards the genetic improvement of soybean genotypes and also complicates decisions with regard to the suitability of certain genotypes for the various soybean producing areas in South Africa. The largest soybean producing areas in South Africa are situated at high altitudes, with minimum daily temperatures which can be critically low and impede the production of soybean. Soybean is chilling sensitive, with growth, development and yield being negatively affected at temperatures below 15°C. Dark chilling (low night temperature) stress has proved to be one of the most important restraints to soybean production in South Africa. Among the symptoms documented in dark chilling sensitive soybean genotypes are reduced growth rates, loss of photosynthetic capacity and pigment content, as well as premature leaf senescence and severely inhibited SNF. Existing knowledge about stress-induced nodule senescence is based on fragmented information in the literature obtained in numerous, and often diverse, legume species. The precise nature and sequence of events participating in nodule senescence has not yet been fully explained. The main objectives of this investigation were to characterise the natural senescence process in soybean nodules under optimal growth conditions and to characterise the alteration of the key processes of SNF in a chilling sensitive soybean genotype during dark chilling. Moreover, the experiment was undertaken in order to establish whether recovery in nodule functionality following a long term dark chilling period occurred, or whether nodule senescence was triggered, and if sensitive biochemical markers of premature nodule senescence could be identified. A known chilling sensitive soybean genotype, PAN809, was grown under controlled growth conditions in a glasshouse. To determine the baseline and change over time for key parameters involved in SNF, a study was conducted under optimal growing conditions over a period of 6 weeks commencing 4 weeks after sowing. The cluster of crown nodules were monitored weekly and analysis included nitrogenase activity, ureide content, respiration rate, leghemoglobin content, sucrose synthase (SS) activity and sucrose content. Further investigations focused on induced dark chilling effects on nodule function to determine the alterations in key parameters of SNF. Plants were subjected to dark chilling (6°C) for 12 consecutive nights and kept at normal day temperatures (26°C). The induced dark chilling was either only shoot (SC) exposure or whole plant chilling (WPC). These treatments were selected since, in some areas of South Africa, cold nights result not only in shoot chilling (SC) but also in low soil temperatures, causing direct chilling of both roots and shoots. To determine if premature nodule senescence was triggered, the recovery following 12 consecutive nights of chilling treatment was monitored for another 4 weeks. It was established that the phase of optimum nitrogenase activity under optimal growing conditions occurred during 4 to 6 weeks after sowing, whereafter a gradual decline commenced. This decline was associated with a decline in nitrogenase protein content and an increase in ureide content. The stability of SS activity and nodule respiration showed that carbon-dependent metabolic processes were stable for a longer period than previously mentioned parameters. The negative correlation that was observed between nitrogenase activity and nodule ureide content pointed towards the possible presence of a feedback inhibition trigger on nitrogenase activity. A direct effect of dark chilling on nitrogenase activity and nodule respiration rate led to a decline in nodule ureide content that occurred without any limitations on the carbon flux of the nodules (i.e., stable sucrose synthase activity and nodule sucrose content). The effect on SC plants was much less evident but did indicate that currently unknown shoot-derived factors could be involved in the minor inhibition of SNF. It was concluded that the repressed rates of respiration might have led to increased O_2 concentrations in the nodule, thereby inhibiting the nitrogenase protein and so the production of ureides. It was found that long term chilling severely disrupted nitrogenase activity and ureide synthesis in nodules. Full recovery in all treatments occurred after 2 weeks of suspension of dark chilling, however, this only occurred when control nodules already commenced senescence. This points toward reversible activation of the nitrogenase protein with no evidence in support of premature nodule senescence. An increase in intercellular air space area was induced by long term dark chilling in nodules, specifically by the direct chilling of nodules (WPC treatment). The delayed diminishment of intercellular air space area back to control levels following dark chilling may be an important factor involved in the recovery of nitrogenase activity because enlarged air spaces would have favoured gaseous diffusion, and hence deactivation of nitrogenase, in an elevated O_2 environment (due to supressed nodule respiration rates). These findings revealed that dark chilling did not close the diffusion barrier, as in the case of drought and other stress factors, but instead opened it due to an increase in air space areas in all regions of the nodule. In conclusion, this study established that dark chilling did not initiate premature nodule senescence and that SNF demonstrated resilience, with full recovery possible following even an extended dark chilling period involving low soil temperatures.

References

Alcantar-Gonzales, G.M., A. Migianac-Maslow and M.L. Champigny. 1988. Effect of nitrate supply on energy balance and acetylene reduction and nitrate reductase activities of soybean root nodules infected with Bradyrhizobium japonicum. CR Acad. Sci. Paris 307: 145–152.

Amthor, J.S. 1989. Respiration and Crop Productivity. Berlin: Springer-Verlag.

Anonymous. 2012a. http://www.cornandsoybeandigest.com/corn/high-temperature-effects-corn-soybeans.

Armstrong, W., M.E. Strange, S. Cringl and P.M. Beckett. 1994. Microelectrode and modeling study of oxygen distribution in roots. Annals of Botany 74: 287–299.

Atkin, O.K. and W.R. Cummins. 1994. The effect of nitrogen source on growth, nitrogen economy and respiration of two high Arctic plant species differing in relative growth rate. Funct. Ecol. 8: 389–399.

Atkin, O.A., C. Holly and M.C. Ball. 2001. Acclimation of snow gum (*Eucalyptus pauciflora*) leaf respiration to seasonal and diurnal variations in temperature: The importance of changes in the capacity and temperature sensitivity of respiration. Plant, Cell and Environment 23: 15–26.

Aubert, S., R. Bligny, D.A. Day, J. Whelan and R. Douce. 1997. Induction of alternative oxidase synthesis by herbicides inhibiting branched-chain amino acid synthesis. Plant J. 11: 649–657.

Azcon-Bieto, J., H. Lambers and D.A. Day. 1983. Respiratory properties of developing bean and pea leaves. Aust. J. Plant Physiol. 10: 237–245.

Azcón-Bieto, J. and C.B. Osmond. 1983. Relationship between photosynthesis and respiration. The effect of carbohydrate status on the rate of CO_2 production by respiration in darkened and illuminated wheat leaves. Plant Physiology 71: 574–581.

Bailey-Serres, J., T. Fukao, D.J. Gibbs, M.J. Holdsworth, S.C. Lee, F. Licausi, P. Perata, L.A.C.J. Voesenek and J.T. van Dongen. 2012. Making sense of low oxygen sensing. Trends in Plant Science 17: 129–138.

Barta, A.L. 1985. Metabolic response of Medicago sativa L. and Lotus corniculatus L. roots to anoxia. Plant Cell Environ. 9: 127–131.

Bartoli, C.G., F. Gomez, D.E. Martinez and J.J. Guiamet. 2004. Mitochondria are the main target for oxidative damage in leaves of wheat (*Triticum aestivum* L.). J. Exp. Bot. 55: 1663–1669.

Beevers, H. 1970. Respiration in plants and its regulation. pp. 209–214. *In*: Setlik, I. (ed.). Prediction and Measurement of Photosynthetic Productivity. PUDOC, Wageningen.

Bennett, J.M. and S.L. Albrecht. 1984. Drought and flooding effects on N_2 fixation, water relations, and diffusive resistance of soybean. Agron. J. 76: 735–740.

Bergersen, F.J. 1962a. The effects of partial pressure of oxygen upon respiration and nitrogen fixation by soybean root nodules. J. Gen. Microbiol. 29: 113–125.

Billings, W.D., P.J. Godfrey, B.F. Chabot and D.P. Borque. 1971. Metabolic acclimation to temperature in arctic and alpine ecotypes of *Oxyriadigyna*. Arctic and Alpine Research 3: 277–290.

Bingham, I.J. and J.F. Farrar. 1988. Regulation of respiration in roots of barley. Physiologia Plantarum 73: 278–285.

Bouma, T.J., K.L. Nielsen, D.M. Eissenstat and J.P. Lynch. 1997a. Estimating respiration of roots in soil: Interactions with soil CO_2, soil temperature and soil water content. Plant and Soil 195: 221–232.

Bowling, D.R., N.G. McDowell, B.J. Bond, B.E. Law and J.R. Ehleringer. 2002. C-13 content of ecosystem respiration is linked to precipitation and vapor pressure deficit. Oecologia 131: 113–124.

Boyer, J.S. 1982. Plant productivity and environment. Science 218: 443–448.

Bray, E. 1997. Plant responses to water deficit. Trends Plant Sci. 2: 48–54.

Bray, E. 2002. Abscisic acid regulation of gene expression during water-deficit stress in the era of the *Arabidopsis* genome. Plant Cell Environ. 25: 153–161.

Breeze, V. and J. Elston. 1978. Some effects of temperature and substrate content upon respiration and the carbon balance of field beans (*Vicia faba* L.). Ann. Bot. 42: 863–876.

Bunce, J.A. 1990. Short- and long-term inhibition of respiratory carbon dioxide efflux by elevated carbon dioxide. Annals of Botany 65: 637–642.

Bunce, J.A. 1995. The effect of carbon dioxide concentration on respiration of growing and mature soybean leaves. Plant, Cell and Environment 18: 575–581.

Bunce, J.A. and L.H. Ziska. 1996. Responses of respiration to increases in carbon dioxide concentration and temperature in three soybean cultivars. Annals of Botany 77: 507–514.

Buwalda, F. and H. Greenway. 1989. Nitrogen uptake and growth of wheat during O_2 deficiency in root media containing NO_3^- only, or NO_3^- plus NH_4^+. New Phytol. 111: 161–166.

Challa, H. 1976. An analysis of the diurnal course of growth carbon dioxide exchange and carbohydrate reserve content of cucumbers. PUDOC, Wageningen, pp. 1–88.

Chaves, M.M., J.P. Maroco and J. Pereira. 2003. Understanding plant responses to drought—from genes to the whole plant. Funct. Plant Biol. 30: 239–264.

Chirkova, T.V. 1978. Some regulatory mechanisms of plant adaptation to temporal an-aerobiosis. pp. 137–154. *In*: Plant Life in An-aerobic Environments. Ann. Arbor Science, Ann Arbor, MI.

Day, D.A., K. Krab, H. Lambers, A.L. Moore, J.N. Siedow, A.M. Wagner and J.T. Wiskich. 1996. The cyanide-resistant oxidase: To inhibit or not to inhibit, that is the question. Plant Physiol. 110: 1–2.

Del Castillo, L.D. and D.B. Layzell. 1995. Drought stress, permeability to O_2 diffusion and the respiratory kinetics of soybean root nodules. Plant Physiol. 107: 1187–1194.

Djordjevic, M.A. 2004. *Sinorhizobium meliloti* metabolism in the root nodule: A proteomic perspective. Proteomics 4: 1859–1872.

Drake, B.G., J. Azcon-Bieto, J. Berry, J. Bunce, J. Dijkstra, J. Farrar, R.M. Gifford, M.A. Gonzalez-Meler, G. Kock and H. Lambers. 1999. Does elevated atmospheric CO_2 concentration inhibit mitochondrial respiration in green plants? Plant, Cell and Environment 22: 649–657.

Farrar, J.F. 1980. The pattern of respiration rate in the barley plant. Ann. Bot. 46: 71–76.

Finan, T.M., E. McWhinnie, B. Driscoll and R.Watson. 1991. Complex symbiotic phenotypes result from gluconeogenic mutations in *Rhizobium meliloti*. Mol. Plant Microbe Interact. 4: 386–392.

Finnegan, P.M., J. Whelan, A.H. Millar, Q. Zhang, M.K. Smith, J.T. Wiskich and D.A. Day. 1997. Differential expression of the multigene family encoding the soybean mitochondrial alternative oxidase. Plant Physiology 114: 455–466.

Flexas, J., J. Bota, F. Loreto, G. Cornic and T.D. Sharkey. 2004. Diffusive and metabolic limitations to photosynthesis under drought and salinity in C_3 plants. Plant Biol. 6: 269–279.

Foyer, C.H. and G. Noctor. 2003. Redox sensing and signalling associated with reactive oxygen in chloroplasts, peroxisomes and mitochondria. Physiol. Plant. 199: 355–364.

Gastón, S., M. Ribas-Carbo, S. Busquets, J.A. Berry, A. Zabalza and M. Royuela. 2003. Changes in mitochondrial electron partitioning in response to herbicides inhibiting branched-chain amino acid biosynthesis in soybean. Plant Physiol. 133: 1351–1359.

Geigenberger, P. 2003. Response of plant metabolism to too little oxygen. Current Opinions in Plant Biology 6: 247–256.

González-Meler, M.A., R. Matamala and J. Peñuelas. 1997. Effects of prolonged drought stress and nitrogen deficiency on the respiratory O_2 uptake of bean and pepper leaves. Photosynthetica 34: 505–512.

Gonzalez-Meler, M.A., L. Giles, R.B. Thomas and J.N. Siedow. 2001. Metabolic regulation of leaf respiration and alternative pathway activity in response to phosphate supply. Plant Cell Environ. 24: 205–215.

Gonzalez-Meler, M.A., L. Taneva and R.J. Trueman. 2004. Plant respiration and elevated atmospheric CO_2 concentration: Cellular responses and global significance. Annals of Botany 94: 647–656.

Green, L.S. and D.W. Emerich. 1997. The formation of nitrogen fixing bacteroids is delayed but not abolished in soybean infected by an a-ketoglutarate dehydrogenase deficient mutant of *Bradyrhizobium japonicum*. Plant Physiol. 114: 1359–1368.

Griffin, K.L., O.R. Anderson, M.D. Gastrich, J.D. Lewis, G. Lin, W. Schuster, J.R. Seemann, D.T. Tissue, M.H. Turnbull and D. Whitehead. 2001. Plant growth in elevated CO_2 alters mitochondrial number and chloroplast fine structure. Proceedings of the National Academy of Science 98: 2473–2478.

Griffin, K.L., M. Turnbull, R. Murthy, G. Lin, J. Adams, B. Farnsworth, T. Madato, G. Bazin, M. Potasnak and J.A. Berry. 2002. Leaf respiration is differentially affected by leaf vs. stand-level night-time warming. Global Change Biology 8: 479–485.

Grimmer, C. and E. Komor. 1999. Assimilate export by leaves of *Ricinus communis* L. growing under normal and elevated carbon dioxide concentrations: The same rate during the day, a different rate at night. Planta 209: 275–281.

Gronewald, J.W. and J.B. Hanson. 1982. Adenine nucleotide content of corn roots as affected by injury and subsequent washing. Plant Physiol. 69: 1252–1256.

Gulluoglu, L., H. Arioglu and M. Arslan. 2006. Effect of some plant growth regulators and nutrient complexes on above-ground biomass and seed yield of some soybean grown under heat-stressed environment. Journal of Agronomy 5(2): 126–130.

Guy, R.D., J.A. Berry, M.L. Fogel, D.H. Turpin and H.G. Weger. 1992. Fractionation of the stable isotopes of oxygen during respiration by plants: The basis for a new technique. pp. 442–453. *In*: Lambers, H. and L.H.W. van der Plas (eds.). Molecular, Biochemical and Physiological Aspects of Plant Respiration. Academic Publishing, The Hague, The Netherlands.

Hansen, G.K. and C.R. Jensen. 1977. Growth and maintenance respiration in whole plants, tops and roots of Lolium multiforum. Physiologia Plantarum 39: 155–164.

Hill, S.A. and J.H. Bryce. 1992. Malate metabolism and light-enhanced respiration in barley mesophyll protoplasts. pp. 221–230. *In*: Lambers, H. and L.H.W. van der Plas (eds.). Molecular, Biochemical and Physiological Aspects of Plant Respiration. SPB Academic Publishing. The Hague.

Hilton, J.R. and P.W. Owen. 1985. Phytochrome regulation of extractable cytochrome oxidase activity during early germination of *Bromus sterilis* and *Lactuca sativa* L. cv. Grand Rapids seeds. New Phytologist 100: 163–171.

Horchani, F.H. Khayati, P. Raymond, R. Brouquisse and S. Aschi-Smiti. 2009. Contrasted effects of prolonged root hypoxia on tomato root and fruit (*Solanum lycopersicum*) metabolism. Journal of Agronomy and Crop Science 195: 313–318.

Hsiao, T. 1973. Plant responses to water stress. Annu. Rev. Plant Physiol. 24: 519–570.

Huan, F., A. Lizhe, T. Ling Ling, H. Zonf Dong and W. Xunling. 2000. Effect of enhanced ultraviolet-B radiation on pollen germination and tube growth of 19 taxa *in vitro*. Environment and Experimental Botany 43: 45–53.

Hunt, W.F. and R.S. Loomis. 1979. Respiration modelling and hypothesis testing with a dynamic model of sugar beet growth. Ann. Bot. 44: 5–17.

Juszczuk, I.M., E. Malusa and A.M. Rychter. 2001. Oxidative stress during phosphate deficiency in roots of bean plants (*Phaseolus vulgaris*). J. Plant Physiol. 158: 1299–1305.

Kakani, V.G., P.V.V. Parasad, P.Q. Craufurd and T.R. Wheeler. 2002. Response of *in vitro* pollen germination and pollen tube growth of Groundnut (*Arachis hypogaea* L.) genotype to temperature. Plant Cell and Environment 25: 1651–1661.

Kokubun, M. 2013. Genetic and cultural improvement of soybean for waterlogged conditions in Asia. Field Crops Research 152: 3–7.

Kowallik, W. 1982. Blue light effects on respiration. Annual Review of Plant Physiology 33: 51–72.

Koziol, U., L. Hannibal, M.C. Rodríguez, E. Fabiano, M.L. Kahn and F. Noya. 2009. Deletion of citrate synthase restores growth of *Sinorhizobium meliloti* 1021 aconitase mutants. J. Bacteriol. 191: 7581–7586.

Kromer, S. 1995. Respiration during photosynthesis. Annual Review of Plant Physiology and Plant Molecular Biology 46: 45–70.

Kuzma, M.M., A.F. Topunow and D.B. Layzell. 1995. Effects of temperature on infected cell O_2 concentration and adenylate levels in attached soybean nodules. Plant Physiol. 107: 1103–13.

Lambers, H. and E. SteingroÈver. 1978. Efficiency of root respiration of a cold tolerant and cold-intolerant Senecio species as affected by low oxygen tension. Physiologia Plantarum 42: 179–184.

Lambers, H., R.K. Szaniawski and R. de Visser. 1983. Respiration for growth, maintenance and ion uptake: An evaluation of concepts, methods, values and their significance. Physiol. Plant. 58: 55–61.

Lambers, H., I. Scheurwater and O.K. Atkin. 1996. Respiratory patterns in roots in relation to their functioning. pp. 323–362. *In*: Waisel, Y., A. Eshel and V. Kafakki (eds.). Plant Roots: The Hidden Half. Ed 2. Marcel Dekker, New York.

Lambers, H., S.A. Robinson and M. Ribas-Carbo. 2005. Regulation of respiration *in vivo*. Chapter 1. pp. 1–15. *In*: Lambers, H. and M. Ribas-Carbo (eds.). Plant Respiration: From Cell to Ecosystem, Vol 18. Advances in Photosynthesis and Respiration Series. Springer, Dordrecht, The Netherlands.

Lauer, M.J., D.G. Blevins and H. Sierputowska-Gracz. 1989. 31P-nuclearmagnetic resonance determination of phosphate compartmentation in leaves of reproductive soybeans as affected by phosphate nutrition. Plant Physiology 89: 1331–1336.

Lawlor, D.W. and G. Cornic. 2002. Photosynthetic carbon assimilation and associated metabolism in relation to water deficits in higher plants. Plant Cell Environ. 25: 275–294.

Lee, R.B. 1978. Inorganic nitrogen metabolism in barley roots under poorly aerated conditions. J. Exp. Bot. 29: 692–708.

Libourel, I.G.L., P.M. van Bodegom, M.D. Fricker and R.G. Ratcliffe. 2006. Nitrite reduces cytoplasmic acidosis under anoxia. Plant Physiol. 142: 1710–1717.

Ludwig, L.J., D.A. Charles-Edwards and A.C. Withers. 1975. Tomato leaf photosynthesis and respiration in various light and carbon dioxide environments. pp. 29–36. *In*: Marcelle Junk, R. (ed.). Environmental and Biological Control of Photosynthesis. The Hague.

Lynch, J.P. and S.E. Beebe. 1995. Adaptation of bean (*Phaseolus vulgaris* L.) to low phosphorus availability. Hort. Science 30: 1165–1171.

Magalhães, J.R., F.I.L.M. Silva, I. Salgado, O. Ferrarese-Filho, P. Rockel and W.M. Kaiser. 2002. Nitric oxide and nitrate reductase in higher plants. Physiol. Mol. Biol. Plants 8: 11–17.

Maque, T.H. and R.H. Burris. 1972. Reduction of acetylene and nitrogen by field grown soybeans. New Phytologist 71: 275–286.

Marking, S. 1986. Heat stress shrivels bean bushels. Soybean Digest; May–June 44–45 pp, USA.

Maxwell, D.P., Y. Wang and L. McIntosh. 1999. The alternative oxidase lowers mitochondrial reactive oxygen production in plant cells. Proc. Natl. Acad. Sci. USA 96: 8271–8276.

McCree, K.J. 1970. An equation for the rate of dark respiration of white clover plants grown under controlled conditions. pp. 221–229. *In*: Setlik, I. (ed.). Prediction and Measurement of Photosynthetic Productivity. Wageningen: Center for Agricultural Publishing and Documentation.

McCree, K.J. 1974. Equations for the rate of dark respiration of white clover and grain sorghum as functions of dry weight, photosynthetic rate and temperature. Crop Sci. 14: 509–514.

McCree, K.J. and J.H. Silsbury. 1978. Growth and maintenance requirements of subterranean clover. Crop Science 18: 13–18.

McCree, K.J. and M.E. Amthor. 1982. Effects of diurnal variation in temperature on the carbon balances of white clover plants. Crop Science 22: 822–827.

McDonald, A.J.S. and W.J. Davies. 1996. Keeping in touch: Responses of the whole plant to deficits in water and nitrogen supply. Adv. Bot. Res. 22: 229–300.

McKay, I., A. Glenn and M. Dilworth. 1985. Gluconeogenesis in Rhizobium leguminosarum MNF3841. J. Gen. Microbiol. 131: 2067–2073.

Meeuse, B.J.D. 1975. Thermogenic respiration in aroids. Annu. Rev. Plant Physiol. 26: 117–126.

Millar, A.H., J.T. Wiskich, J. Whelan and D.A. Day. 1993. Organic acid activation of the alternative oxidase of plant mitochondria. FEBS Lett. 329: 259–262.

Millar, A.H., D.A. Day and F.J. Bergersen. 1995. Microaerobic respiration and oxidative phosphorylation by soybean nodule mitochondria: Implications for nitrogen fixation. Plant Cell Environ. 18: 715–726.

Millar, A.H. and D.A. Day. 1996. Nitric oxide inhibits the cytochrome oxidase but not the alternative oxidase of plant mitochondria. FEBS Lett. 398: 155–158.

Millenaar, F.F. and H. Lambers. 2003. The alternative oxidase: *In vivo* regulation and function. Plant Biol. 5: 2–15.

Minchin, F.R., M. Becana and J.I. Sprent. 1989. Short-term inhibition of legume N_2 fixation by nitrate. II. Nitrate effect on nodule oxygen diffusion. Planta 180: 46–52.

Moore, A.L. and J.N. Siedow. 1991. The regulation and nature of the cyanide-resistant alternative oxidase of plant mitochondria. Biochim. Biophys. Acta 1059: 121–140.

Mousseau, M. 1977. Night respiration in relation to growth, photosynthesis and development of *Chenopodium polyspermum* in long and short days. Plant Sci Lett. 9: 339–346.

Mozafar, A., R. Gamperle and J. Loch. 1992. Root aeration inhibits the recovery of soybean from flooding-induced chlorosis under non calcareous conditions. J. Plant Nut. 15: 1927–1933.

Nakayama, N., S. Hashimoto, S. Shimada, M. Takahashi, Y.H. Kim, T. Oya and J. Arihara. 2004. The effect of flooding stress at the germination stage on the growth of soybean in relation to initial seed moisture content. Jpn. J. Crop Sci. 73: 323–329.

Nielsen, K.L., T.J. Bouma, J. Lynch and D.M. Eissenstat. 1998. Effects of phosphorus availability and vesicular-arbuscular mycorrhizas on the carbon budget of common bean (*Phaseolus vulgaris*). New Phytologist 139: 647±656.

Obenland, D., R. Dielthelm, R. Shibles and C. Stewart. 1990. Relationship of alternative respiratory capacity and alternative oxidase amount during soybean seedling growth. Plant Cell Physiol. 31: 897–901.

Ochandio, D., R. Bartosik, A. Yommi and L. Cardoso. 2012. Carbon dioxide concentration in hermetic storage of soybean (*Glycine max*) in small glass jars. pp. 495–500. *In*: Navarro, S., H.J. Banks, D.S. Jayas, C.H. Bell, R.T. Noyes, A.G. Ferizli, M. Emekci, A.A. Isikber and K. Alagusundaram (eds.). Proc. 9th. Int. Conf. on Controlled Atmosphere and Fumigation in Stored Products, Antalya, Turkey. 15–19 October 2012, ARBER Professional Congress Services, Turkey.

O'Hara, G.W. and R.M. Daniel. 1985. Rhizobial denitrification: A review. Soil Biol. Biochem. 17: 1–9.

Oosterhuis, D.M., H.D. Scott, R.E. Hampton and S.D. Wullschleger. 1990. Physiological responses of two soybean cultivars to short-term flooding. Environ. Exp. Bot. 30: 85–92.

Parsons, H.L., J.Y.H. Yip and G.C. Vanlerberghe. 1999. Increased respiratory restriction during phosphate-limited growth in transgenic tobacco lacking alternative oxidase. Plant Physiol. 121: 1309–1320.

Peñuelas, J., M. Ribas-Carbo and L. Giles. 1996. Effects of allelochemicals on plant respiration and oxygen isotope fractionation by the alternative oxidase. J. Chem. Ecol. 22: 801–805.

Poorter, H., A. van der Werf, O.K. Atkin and H. Lamber. 1991. Respiratory energy requirements of roots vary with the potential growth rate of the plant species. Physiol Plant. 83: 469–475.

Puiatti, M. and L. Sodek. 1999. Waterlogging affects nitrogen transport in the xylem of soybean. Plant Physiology and Biochemistry 37: 767–773.

Purvis, A.C. and R.L. Shewfelt. 1993. Does the alternative pathway ameliorate chilling injury in sensitive plant tissues? Physiol. Plant. 88: 712–718.

Purvis, A.C. 1997. Role of the alternative oxidase in limiting superoxide production by plant mitochondria. Physiol. Plant. 100: 165–170.

Reggiani, R., I. Brambilla and A. Bertani. 1985b. Effect of exogenous nitrate on anaerobic metabolism in excised rice roots. II. Fermentation activity and adenylic energy charge. J. Exp. Bot. 36: 1698–1704.

Ribas-Carbo, M., J.A. Berry, D. Yakir, L. Giles, S.A. Robinson, A.L. Lennon and J.N. Siedow. 1995a. Electron partitioning between the cytochrome and alternative pathways in plant mitochondria. Plant Physiol. 109: 829–837.

Ribas-Carbo, M., R. Aroca, M.A. Gonzalez-Meler, J.J. Irigoyen and M. Sanchez-Diaz. 2000. The electron partitioning between the cytochrome and alternative respiratory pathways during chilling recovery in two cultivars differing in chilling sensitivity. Plant Physiol. 122: 199–204.

Ribas-Carbo, M., S.A. Robinson and L. Giles. 2005. The application of the oxygen-isotope technique to assess respiratory pathway partitioning. Chapter 3. pp. 31–42. *In*: Lambers, H. and M. Ribas-Carbo (eds.). Plant Respiration: From Cell to Ecosystem, Vol 18. Advances in Photosynthesis and Respiration Series. Springer, Dordrecht, The Netherlands.

Ribas-Carbo, M., N.L. Taylor, L. Giles, S. Busquets, P.M. Finnegan, D.A. Day, H. Lambers, H. Medrano, J.A. Berry and J. Flexas. 2005. Effects of water stress on respiration in soybean leaves. Plant Physiol. 139: 466–473.

Ricard, B., I. Couée, P. Raymond, P.H. Saglio, V. Saint-Ges and A. Pradet. 1994. Plant metabolism under hypoxia and anoxia. Plant Physiol. Biochem. 32: 1–10.

Roberts, J.K.M., F. Andrade and I.C. Anderson. 1985. Further evidence that cytoplasmic acidosis is a determinant of flooding intolerance in plants. Plant Physiol. 77: 492–494.

Romanov, V.I., I. Hernandezlucas, E. Martinezromero, I. Hernandez-Lucas and E. Martinez-Romero. 1994. Carbon metabolism enzymes of *Rhizobium tropici* cultures and bacteroids. Appl. Environ. Microbiol. 60: 2339–2342.

Ryan, M.G. 1991. Effects of climate change on plant respiration. Ecological Applications 1: 157–167.

Rychter, A.M., M. Chauveau, J.-L. Bomsel and C. Lance. 1992. The effect of phosphate deficiency on mitochondrial activity and adenylate levels in bean roots. Physiologia Plantarum 84: 80–86.

Sairam, R.K., D. Kumutha, C. Viswanathan and C.M. Ramesh. 2009. Waterlogging-induced increase in sugar mobilization, fermentation, and related gene expression in the roots of mung bean (*Vigna radiata*). Journal of Plant Physiology 166: 602–616.

Sakai, H., K. Yagi, K. Kobayashi and S. Kawashima. 2001. Rice carbon balance under elevated CO_2. New Phytologist 150: 241–249.

Sarma, A.D. and D.W. Emerich. 2006. A comparative proteomic evaluation of culture grown vs nodule isolated *Bradyrhizobium japonicum*. Proteomics 6: 3008–3028.

Scheurwater, I., C. Cornelissen, F. Dictus, R. Welschen and H. Lambers. 1998. Why do fast- and slow-growing grass species differ so little in their rate of root respiration, considering the large differences in rate of growth and ion uptake? Plant, Cell and Environment 21: 995–1005.

Scheurwater, I., D.T. Clarkson, J.V. Purves, G. van Rijt, L.R. Saker, R. Welschen and H. Lambers. 1999. Relatively large nitrate efflux can account for the high specific respiratory costs for nitrate transport in slow-growing grass species. Plant Soil 215: 123–134.

Scott, H.D., J. DeAngulo, M.B. Daniels and L.S. Wood. 1989. Flood duration effects on soybean growth and yield. Agron. J. 81: 631–636.

Serrano, A. and M. Chamber. 1990. Nitrate reduction in *Bradyrhizobium* sp. (Lupinus) strains and its effects on their symbiosis with *Lupinus luteus*. J. Plant Physiol. 136: 240–246.

Shah, R. and D.W. Emerich. 2006. Isocitrate dehydrogenase of *Bradyrhizobium japonicum* is not required for symbiotic nitrogen fixation with soybean. J. Bacteriol. 188: 7600–7608.

Sims, T.L. and R.D. Hague. 1981. Light-stimulated increase of translatable mRNA for phosphoenolpyruvate carboxylase in leaves of maize. Journal of Biological Chemistry 256: 8252–8256.

Sisson, W.B. 1989. Carbon balance of Panicum coloratum during drought and non-drought in the northern Chihuahuan desert. Journal of Ecology 77: 799–810.

Sorour, H. and T. Uchino. 2004. The effect of storage condition on the respiration of soybean. J. JSAM 66: 66–74.

Sousa, C.A.F. and L. Sodek. 2002a. The metabolic response of plants to oxygen deficiency. Braz. J. Plant Physiol. 14: 83–94.

Sousa, C.A.F. and L. Sodek. 2002b. Metabolic changes in soybean plants in response to waterlogging in the presence of nitrate. Physiology Molecular Biology of Plants 8: 97–104.

Sprent, J.I., C. Giannakis and W. Wallace. 1987. Transport of nitrate and calcium into legume root nodules. J. Exp. Bot. 38: 1121–1128.

Stoimenova, M., A.U. Igamberdiev, K.J. Gupta and R.D. Hill. 2007. Nitrite-driven anaerobic ATP synthesis in barley and rice root mitochondria. Planta 226: 465–474.

Tadege, M., I.I. Dupuis and C. Kuhlemeier. 1999. Ethanolic fermentation: New functions for an old pathway. Trends in Plant Science 4: 320–325.

Theodorou, M.E. and W.C. Plaxton. 1993. Metabolic adaptations of plant respiration to nutritional phosphate deprivation. Plant Physiology 101: 339–344.

Thomas, R.B. and K.L. Griffin. 1994. Direct and indirect effects of atmospheric carbon dioxide enrichment of leaf respiration of *Glycine max* (L.) Merr. Plant Physiology 104: 355–361.

Thomas, A.L. and L. Sodek. 2005. Development of the nodulated soybean plant after flooding of the root system with different sources of nitrogen. Braz. J. Plant Physiol. 17: 29.

Trought, M.C.T. and M.C. Drew. 1981. Alleviation of injury to young wheat plants in anaerobic solution cultures in relation to the supply of nitrate and other inorganic nutrients. J. Exp. Bot. 32: 509–522.

Tjepkema, J.D. 1971. Oxygen transport in the soybean nodule and the function of leghemoglobin. Doct. dissert. Univ. of Michigan. Ann Arbor, Michigan, USA.

Tjepkema, J.D. and C.S. Yocum. 1973. Respiration and oxygen transport in soybean nodules. Planta 115: 59–72.

Umbach, A.L. and J.N. Siedow. 1993. Covalent and non-covalent dimers of the cyanide-resistant alternative oxidase protein in higher plant mitochondria and their relationship to enzyme activity. Plant Physiol. 103: 845–854.

Valentini, R., G. Matteucci, A.J. Dolman, E.-D. Schulze, C. Rebmann, E.J. Moore, A. Granier, P. Gross, N.O. Jensen, K. Pilegaard et al. 1999. Respiration as the main determinant of carbon balance in European forests. Nature 404: 861–865.

Van der Werf, A., R. Welschen and H. Lambers. 1992. Respiratory losses increase with decreasing inherent growth rate of a species and with decreasing nitrate supply: A search for explanations for these observations. pp. 421–432. *In*: Lambers, H. and L.H.W. Van der Plas (eds.). Molecular, Biochemical and Physiological Aspects of Plant Respiration. The Hague: SPB Academic Publishing.

Vanlerberghe, G.C., A.E. Vanlerberghe and L. McIntosh. 1994. Molecular genetic alteration of plant respiration. Silencing and overexpression of alternative oxidase in transgenic tobacco. Plant Physiol. 106: 1503–1510.

Vanlerberghe, G.C., D.A. Day, J.T. Wiskich, A.E. Vanlerberghe and L. McIntosh. 1995. Alternative oxidase activity in tobacco leaf mitochondria: Dependence upon tricarboxylic acid cycle-mediated redox regulation and pyruvate activation. Plant Physiol. 109: 353–361.

Vanlerberghe, G.C., J.Y.H. Yip and H.L. Parsons. 1999. *In organello* and *in vivo* evidence of the importance of the regulatory sulfhydryl/disulfide system and pyruvate for alternative oxidase activity in tobacco. Plant Physiol. 121: 793–803.

Vanlerberghe, G.C., C.A. Robson and J.Y.H. Yip. 2002. Induction of mitochondrial alternative oxidase in response to a cell signal pathway down-regulating the cytochrome pathway prevents programmed cell death. Plant Physiol. 129: 1829–1842.

Vercruysse, M., B.T. Fauvart, S. Beullens et al. 2011. A comparative transcriptome analysis of Rhizobium etli bacteroids: Specific gene expression during symbiotic nongrowth. Mol. Plant Microbe In. 24: 1553–1561.

Vessey, J.K. and J. Waterer. 1992. In search of the mechanism of nitrate inhibition of nitrogenase activity in legume nodules: Recent developments. Physiol Plant. 84: 171–176.

Voesenek, L.A.C.J., T.D. Colmer, R. Pierik, F.F. Millenaar and A.J.M. Peeters. 2006. How plants cope with complete submergence. New Phytol. 170: 213–226.

Voegele, R.T., M.J. Mitsch and T.M. Finan. 1999. Characterization of two members of a novel malic enzyme class. Biochim. Biophys. Acta 1432: 275–285.

Wagner, A.M. and K. Krab. 1995. The alternative respiration pathway in plants: Role and regulation. Physiol. Plant. 95: 318–325.

Wagner, A.M., W.A.M. Van Emmerik, J.H. Zwiers and H.M.C.M. Kaagman. 1992. Energy metabolism of *Petunia hybrida* cell suspensions growing in the presence of antimycin A. pp. 609–614. *In*: Lambers, H. and L.H.W. Van der Plas (eds.). Molecular, Biochemical and Physiological Aspects of Plant Respiration. SPB Academic Publishing, The Hague, The Netherlands.

Wilson, D. 1982. Response to selection for dark respiration rate of mature leaves in *Lolium perenne* and its effects on growth of young plants and simulated swards. Ann. Bot. 49: 303–312.

Winkler, M.N., B.T. Mawson and T.A. Thorpe. 1994. Alternative and cytochrome pathway respiration during shoot bud formation in cultured *Pinus radiata* cotyledons. Physiol. Plant. 90: 144–151.

Woodwell, G.M., J.E. Hobbie, R.A. Houghton, J.M. Mellilo, B. Moore, B.J. Peterson and G.R. Shaver. 1983. Global deforestation: Contribution to atmospheric carbon dioxide. Science 222: 1081–1086.

Zabalza, A., J.T. van Dongen, A. Froehlich, S.N. Oliver, B. Faix, K.J. Gupta, E. Schmäzlin, M. Igal, L. Orcaray, M. Royuela and P. Geigenberger. 2009. Regulation of respiration and fermentation to control the plant internal oxygen concentration. Plant Physiology 149: 1087–1098.

Zhang, Y., T. Aono, P. Poole and T.M. Finan. 2012. NAD(P)+-malic enzyme mutants of *Sinorhizobium* sp. strain ngr234, but not *Azorhizobium caulinodans* ors571, maintain symbiotic N2 fixation capabilities. Appl. Environ. Microbiol. 78: 2803–2812.

Ziska, L.H. and J.A. Bunce. 1994. Direct and indirect inhibition of single leaf respiration by elevated CO_2 concentrations: Interaction with temperature. Physiologia Plantarum 90: 130–138.

Zróbek-sokolnik, A., 2012. Temperature stress and responses of plants. pp. 113–134. *In*: Ahmad, P. and M.N.V. Prasad (eds.). Environmental Adaptations and Stress Tolerance of Plants in the Era of Climate Change. New York: Springer.

CHAPTER 9
Nitrogen Metabolism and Biological Nitrogen Fixation

Soybean seeds contain a large amount of protein N and the total amount of N assimilated in a plant is highly correlated with the soybean seed yield. One ton of soybean seed requires about 70–90 kg N, which is about four times more than in the case of rice (Hoshi, 1982). Soybean plants assimilate the N from three sources, N derived from atmospheric nitrogen by symbiotic N_2 fixation in root nodules (Ndfa), absorbed N derived from soil mineralized N (Ndfs), and N derived from fertilizer when applied (Ndff) (Fig. 9.1). For the maximum seed yield of soybean, it is necessary to use both N_2 fixation and absorbed N from roots (Harper, 1974; 1987). Sole N_2 fixation is often insufficient to support vigorous vegetative growth, which results in the reduction of seed yield. On the other hand, a heavy supply of N fertilizer often depresses nodule development and N_2 fixation activity, inducing nodule senescence, which sometimes results in the reduction of seed yield. In addition, N fertilizer often causes over luxuriant growth, which results in lodging or poor pod formation. Therefore, no nitrogen fertilizer is applied for soybean cultivation, or only a small amount of N fertilizer is applied as a starter N to promote initial growth.

Biological N fixation requires plant's reduced carbon (C) and energy, as reviewed by Kaschuk et al. (2009). For soybean, BNF requires 6–7 g C g^{-1} N in comparison to 4 g C g^{-1} N for assimilation of mineral N; integrated over the growing season, the difference in cost is substantial, with potential implications for seed yield and seed protein or oil concentrations. The cost of BNF can be partially compensated by increase in photosynthesis of plants associated with rhizobia (Kaschuk et al., 2009) or shifts in allocation of biomass. For instance, nodulated roots accumulated less biomass compared with plants growing with

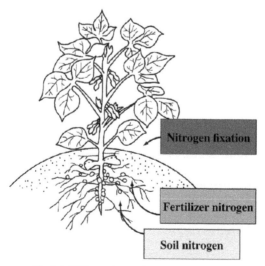

Fig. 9.1 Different forms of nitrogen in soybean.

high soil N supply (Salon et al., 2001) and lower biomass partitioning to seeds associated with increasing BNF. Thus, the crop can accommodate the cost of BNF by five non-mutually exclusive mechanisms, whereby N fixation (a) reduces shoot growth and seed yield, or maintains shoot growth and seed yield by (b) enhanced photosynthesis (Kaschuk et al., 2009), or (c) reduced root:shoot ratio (Lambers, 1983), or maintains shoot growth but reduces seed yield by (d) reducing seed oil and protein concentration in seed, or (e) the fraction of shoot biomass allocated to seed (i.e., harvest index; HI).

Furthermore, there is an agronomic interest on the role of mineral N to support high seed yield (Menza et al., 2017; Tamango and Ciampitti, 2017) and avoid protein dilution (Rotundo et al., 2016; Medic et al., 2014). A recent review of Mourtzinis et al. (2018) concluded that N fertilization has a small and inconsistent effect on soybean seed yield. This conclusion is, however, largely based on generic trials where coarse fertilization regimes were established to shift the contribution of mineral N and BNF. In contrast, a full-N treatment devised with a careful experimental protocol to ensure an ample N supply during the entire crop season increased soybean seed yield by 11% in relation to unfertilized controls, with a range from no effect for stressful environments (ca. 2500 kg ha^{-1}) but increases of 900 kg ha^{-1} in high potential environments (ca. 6000 kg ha^{-1}) (Menza et al., 2017).

Legumes rely on soil mineral nitrogen (N) and biological N fixation (BNF). The interplay between these two sources is biologically interesting and agronomically relevant as the crop can accommodate the cost of BNF by five non-mutually exclusive mechanisms, whereby BNF reduces shoot growth and seed yield, or maintains shoot growth and seed yield by enhanced photosynthesis, or reduced root:shoot ratio, or maintains shoot growth but reduces seed yield by reducing the fraction of shoot biomass allocated to seed (harvest index), or reducing concentration of oil and protein in seed. Researchers explore the impact of N application on the seasonal dynamics of BNF, and its consequences for seed yield, with emphasis on growth and shoot allocation mechanisms. Trials were established in 23 locations across the US Midwest under four N conditions. Fertilizer reduced the peak of BNF up to 16% in applications at the full flowering stage. Seed yield declined 13 kg ha^{-1} per % increase in RAUR. Harvest index accounted for the decline in seed yield with increasing BNF. This indicates the cost of BNF was met by a relative change in dry matter allocation against the energetically rich seed, and in favor of energetically cheaper vegetative tissue (Fig. 9.2).

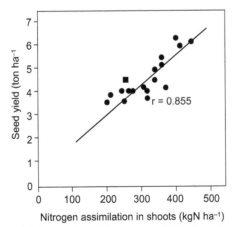

Fig. 9.2 Seed yield and nitrogen assimilation in soybean.

9.1 Nitrogen Assimilation

Soybean (*Glycine max* (L.) Merr.) is a symbiotic nitrogen-fixing crop. In order to increase grain yield, it is important to know how soybean plants respond to nitrogen topdressing for the improvement of nitrogen utilisation. In two soybean cultivars with different grain yield potentials and applied 13 nitrogen topdressing treatments were undertaken to determine optimal topdressing time and nitrogen metabolism.

Nitrogen treatments included a base fertiliser and single topdressings at different times, in 10-day intervals from 10 to 120 days after emergence (DAE). Among the nitrogen treatments, the optimal times for topdressing were at 40 DAE or 90 DAE to increase grain yield, and both soybean cultivars also had higher nitrate reductase (NR) and glutamine synthetase (GS) activities with topdressing at these times. Higher expression of the *NR2* gene was associated with upregulated NR activity in leaves of both cultivars at the early-mature stage. With topdressing at 90 DAE, higher *GS1* expression and GS activity were found in the leaves of the higher yielding cultivar at the full-seed stage and the early-mature stage. With topdressing at 90 DAE, the higher yielding cultivar had a higher nitrate metabolism capacity at the late reproductive stages than the lower (common) yielding cultivar.

Leguminous plants form root nodules with rhizobia that fix atmospheric dinitrogen (N_2) for the nitrogen (N) nutrient. Combined nitrogen sources, particular nitrate, severely repress nodule growth and nitrogen fixation activity in soybeans (*Glycine max* [L.] Merr.). A microarray-based transcriptome analysis and the metabolome analysis were carried out for the roots and nodules of hydroponically grown soybean plants treated with 5 mM of nitrate for 24 h and compared with control without nitrate. Gene expression ratios of nitrate vs. the control were highly enhanced for those probe sets related to nitrate transport and assimilation and carbon metabolism in the roots, but much less so in the nodules, except for the nitrate transport and asparagine synthetase. From the metabolome analysis, the concentration ratios of metabolites for the nitrate treatment vs. the control indicated that most of the amino acids, phosphorous-compounds and organic acids in roots were increased about twofold in the roots, whereas in the nodules most of the concentrations of the amino acids, P-compounds and organic acids were reduced, while asparagine increased exceptionally. These results may support the hypothesis that nitrate primarily promotes nitrogen and carbon metabolism in the roots, but mainly represses this metabolism in the nodules.

The outlines of absorption and metabolism of ammonium and nitrate in plant cells are well known. Ammonium, which is the most reduced form of nitrogen, and nitrate, which is the most oxidized form of nitrogen, are two major inorganic nitrogen compounds in soil. The NH_4^+ ion is absorbed through the membrane-bound protein, ammonium transporter. The NO_3^- ion is absorbed through the nitrate transporter with $2H^+$ co-transport. There are two types of nitrate transporter, a high affinity nitrate transporter system (HATS) and a low affinity nitrate transporter system (LATS) (Crawford and Glass, 1998).

The changes in NO_3^- concentration in solution were monitored by a nutrient solution circulation system, in which the UV absorbance was detected by a UV detector. The relationship between nitrate absorption rate and nitrate concentration shows that the kinetics of this pattern indicates the presence of only one HATS which Km value is 19 µ mole in soybean roots. The value is comparable to that in potato cultivars from 11.2–17.3 µ mole (Sharifi and Zebarth, 2006).

Using the same system, the characteristics of NO_3^- absorption was investigated. The effect of pH of the solution from 5 to 8 on NO_3^- absorption rate was investigated. The NO_3^- absorption rate was highest at pH 5, and decreased with increasing pH. At pH 8, NO_3^- absorption was stopped or some in roots was excreted to the solution. This result was in accordance with H^+ co-transport of NO_3^- absorption, because the higher the pH the lower the $[H^+]$ concentration (electrochemical potential). It shows the NO_3^- absorption rate under medium temperature from 15 to 45°C. At low temperature 15°C, NO_3^- absorption rate was low only 20% of that at 25°C. The NO_3^- absorption rate increased at 35°C, but it completely stopped at 45°C. Concerning to temperature effect, this experiment was conducted just after changing the temperature from room temperature around 25°C. After several hours of incubation at 45°C, the NO_3^- absorption recovered by adaptation to high temperature.

The diurnal rhythm in NO_3^- absorption by intact soybean plants was investigated by sampling the culture solution every 15 min and analyzed by ion chromatography (Ohyama et al., 1989b). When 10 mgN of nitrate was initially supplied in the solution, the soybean absorbed NO_3^- almost linearly, irrespective of the time during the day and night period or NO_3^- concentration in the medium. The NO_3^- absorption rate was different between day (1.10 mgN L^{-1} h^{-1}) and night period (0.77 mgN L^{-1} h^{-1}), and the level of the absorption rate in the night was about 60–75% of that in the daytime. The temporary interruption of NO_3 absorption was observed twice a day at dawn and dusk. The same trends were observed when the solution with 25 mgN L^{-1} of nitrate was supplied. The absorption rate during day time was 0.93 mgN L^{-1}

h⁻¹, and that during night time was 0.68 mgN L⁻¹ h⁻¹. It was suggested that this rhythmic pattern of NO_3^- absorption was not directly controlled by the shoots, because the rhythm continued under the extended dark period (Fig. 9.3) or by cutting the shoots (Ohyama et al., 1989b). When the roots were put in the water bath under constant temperature at 30°C, the rhythm of NO_3^- absorption disappeared, suggesting that the nitrate absorption rate of soybean roots is controlled by monitoring the root temperature changes. Different results were reported by Delhon et al. (1995a, b). They reported diurnal regulation of NO_3^- uptake in soybean plants, and the NO_3^- absorption was monitored in the non-nodulated soybean plant during 14/10 h light/dark period at a constant temperature of 26°C. During the night, NO_3^- uptake rate and nitrate reduction was reduced. The accumulation of NO_3^- and asparagine were observed in the roots in dark period.

Some of the absorbed NO_3^- is known to excrete from cell to apoplast, which is called nitrate efflux. Efflux from soybean roots occurred only in the presence of NO_3^- in the medium (Ohyama et al., 1989b). The precise mechanism of nitrate efflux has not been understood yet, but this process may play a role in fine tuning of net NO_3^- absorption rate, which means influx rate minus efflux rate. The efflux of NO_3^- from the roots to the medium was confirmed when the soybean plants were transferred from the 15NO_3^- medium to the non-labeled NO_3^- medium (Ohyama et al., 1989b). The results suggested the presence of three steps of efflux: First, a rapid efflux (5 mgN hr⁻¹) for the initial few min, and the second efflux (0.9 mgN hr⁻¹) for about 10 min, and the third efflux (0.4 mgN hr⁻¹) continued for a several hrs. When plants were transferred from 15NO_3^- solution to N-free solution, the efflux was very low compared with non-labeled NO_3^- solution. This suggests that nitrate efflux system in soybean roots operate only in the presence of external NO_3^- (Fig. 9.4).

Nitrate is reported to improve tolerance of plants towards oxygen deficiency caused by waterlogging of the root system, but the mechanism underlying the phenomenon remains poorly understood. Researchers studied the metabolism of nitrate in roots exposed to hypoxia, using soybean plants growing in a hydroponic system after suspending aeration and covering the surface of the nutrient solution with mineral oil. Nitrate depletion from the medium was more intense under hypoxia than normoxia, but in the presence of chloramphenicol, consumption under hypoxia was significantly reduced. Nitrite accumulated in the medium in the state of hypoxia and this effect was partially eliminated by chloramphenicol. Nitrate consumption sensitive to chloramphenicol was attributed to bacterial activity. Endogenous root nitrate was strongly reduced under hypoxia, indicating mobilization. Although the transport of nitrate to the shoot via the xylem was also reduced under hypoxia, the severity of this reduction was dependent on the concentration of nitrate in the medium, suggesting that at least some of the nitrate in the xylem came from the medium. Root nitrate reductase was also strongly reduced under hypoxia, but recovered rapidly on return to normoxia. Overall, the data are consistent with two main metabolic fates for chloramphenicol-

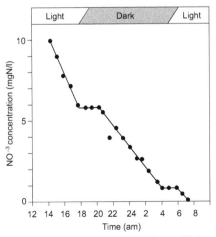

Fig. 9.3 Nitrate concentration in soybean with time.

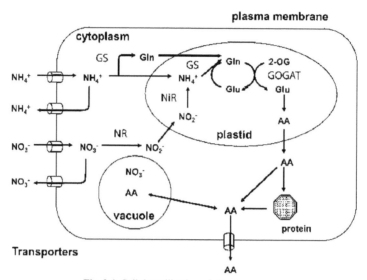

Fig. 9.4 Cellular utilization of nitrogen source.

insensitive nitrate depletion under hypoxia: The reduction of some nitrate to nitrite (despite the reduced nitrate reductase activity) followed by its release to the medium (at least one-third of the nitrate consumed followed this route), and the transport of nitrate to the shoot. Nevertheless, it is highly unlikely that these metabolic routes account for all the nitrate consumed.

Although nitrate (NO_3^-) but not ammonium (NH_4^+) improves plant tolerance to oxygen deficiency, the mechanisms involved in this phenomenon are just beginning to be understood. By using gas chromatography–mass spectrometry, the metabolic fates of $^{15}NO_3^-$ and $^{15}NH_4^+$ in soybean plants (*Glycine max* L. Merril cv. IAC-23) subjected to root hypoxia were investigated. This stress reduced the uptake of $^{15}NO_3^-$ and $^{15}NH_4^+$ from the medium and decreased the overall assimilation of these nitrogen sources into amino acids in roots and leaves. Root $^{15}NO_3^-$ assimilation was more affected by hypoxia than that of $^{15}NH_4^+$, resulting in enhanced nitrite and nitric oxide release in the solution. However, $^{15}NO_3^-$ was translocated in substantial amounts by xylem sap and considerable $^{15}NO_3^-$ assimilation into amino acids also occurred in the leaves, both under hypoxia and normoxia. In contrast, $^{15}NH_4^+$ assimilation occurred predominantly in roots, resulting in the accumulation of mainly ^{15}N–alanine in this tissue during hypoxia. Analysis of lactate levels suggested higher fermentation in roots from NH_4^+-treated plants compared to

Table 9.1 Nitrate reductase activity of soybean roots during a 4 d period of hypoxia followed by return to normoxia.[a]

Root nitrate reductase activity[b] (μmol NO_2^{-1} gfw^{-1}h^{-1})		
Days	normoxia	hypoxia
0	0.91(±0.16)	0.91(±0.16)
1		0.56(±0.03)
2		0.35(±0.02)
3		0.28(±0.04)
4	1.56(±0.17)	0.28(±0.02)
5		0.92(±0.22)
6	1.24(±0.13)	0.98(±0.14)

[a] *In vitro* assay in the presence of EDTA.
[b] Mean (±SE).

the NO_3^- treatment. Thus, foliar NO_3^- assimilation may be relevant to plant tolerance to oxygen deficiency, since it would economize energy expenditure by hypoxic roots. Additionally, the involvement of nitric oxide synthesis from nitrite in the beneficial effect of NO_3^- is known.

Some part of the NO_3^- absorbed in the root cell is reduced to nitrite (NO_2^-) by nitrate reductase (NR) in cytosol, then reduced to ammonia by nitrite reductase (NiR) in plastids followed by assimilation via GS/GOGAT pathway to amino acids. When a high concentration of NO_3^- is supplied, a part of NO_3^- is temporarily stored in vacuoles. Some part of NO_3^- is transported cell to cell via symplast pathway and effluxed in the stele and transported via xylem with transpiration stream in the form of NO_3^-.

NR catalyzes the reaction as follows:

$$NO_3^- + NAD(P)H + H^+ \rightarrow NO_2^- + NAD(P)^+ + H_2O$$

Plant NR requires NADH or NAD(P)H, which means both NADH and NADPH as electron donors. In soybean, there are two types of NAD(P)H-NR and one type of NADH-NR (Harper, 1987).

NiR catalyzes the reaction as follows:

$$NO_2^- + 6\ Fdred \rightarrow NH_4^+ + 6Fdox + 2H_2O$$

where Fdred is a reduced type of ferredoxin and Fdox is an oxidized type of ferredoxin.

Assimilation and transport of nitrate was compared with nitrite and ammonium using ^{15}N labelled compounds (Ohyama et al., 1989a). The nodulated soybean plants were treated with a culture solution containing 10 mgN L^{-1} (0.7 mM) $^{15}NO_3^-$, $^{15}NO_2^-$, or $^{15}NH_4^+$, and the assimilation and transport of N originated from these compounds was compared for 24 hr.

The absorption rate of N and their partitioning patterns among roots, nodules, stems and leaves were very similar for $^{15}NO_3^-$ and $^{15}NH_4^+$. The N from both N sources was rapidly transported to the stems and leaves and readily assimilated into the protein (80% ethanol insoluble fraction). During 24 hr of ^{15}N feedings, approximately 70% of N originated from $^{15}NO_3^-$ and $^{15}NH_4^+$ was partitioned in the leaves and stems. However, the absorption of $^{15}NO_2^-$ was about half as much as those of $^{15}NO_3^-$ and $^{15}NH_4^+$. In addition, the N originating from $^{15}NO_2^-$ was accumulated in the roots, and not readily transported to the shoots (only about 20% partitioning in leaves and stems after 24 hr). Different from soybean, ammonium uptake was twice as fast as nitrate uptake in lupin plants which were supplied with 2.8 mM NH_4NO_3 (Atwell, 1992).

After the addition of $^{15}NO_3^-$ in the solution, the asparagine concentration increased markedly, indicating that asparagine is a major assimilatory compound of NO_3^- in soybean roots. When $^{15}NH_4^+$ was supplied in the solution, the concentration of glutamine in roots increased very rapidly in 4 hr, then asparagines concentration increased linearly in 24 hr. On the other hand, when $^{15}NO_2^-$ was supplied, glutamine levels increased rapidly in the roots, but the concentration of asparagine did not (Ohyama et al., 2017) (Fig. 9.5).

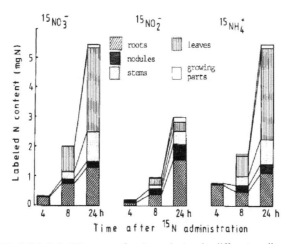

Fig. 9.5 Labeled N content of soybean plant under different medium.

Nitrogen assimilation and transport of the plants supplied with $^{15}NO_3^-$ was investigated by analyzing xylem sap collected from decapitated soybean plants (Ohyama et al., 1989c). The ^{15}N abundance of xylem sap increased very rapidly, about 8% of N in xylem sap collected during first 15 min originated form the N absorbed $^{15}NO_3^-$. The time course of the changes in the ^{15}N abundance of xylem sap indicated that some part of absorbed NO_3^- is very rapidly transported through xylem, but the other part may be transported slowly once stored in the roots. Xylem sap collected at flowering stage contained about the same levels of NO_3^- and asparagine as primary compounds of NO_3^- transport in soybean.

Concerning nitrate assimilation and transport in the roots, it was proposed a hypothetic scheme (Ohyama et al., 1989a). Some part of NO_3^- absorbed in the roots is immediately exported to the shoots, whereas another part of NO_3^- is temporarily stored in the vacuoles of root cells then gradually released to the xylem. On the other hand, some other part of NO_3^- is reduced and assimilated in the roots and synthesized in asparagine. Some part of the asparagine is transported immediately after assimilation in the root cytoplasm, while another part is stored in vacuoles and released gradually. The degradation product of root protein may be exported as in the form of asparagine.

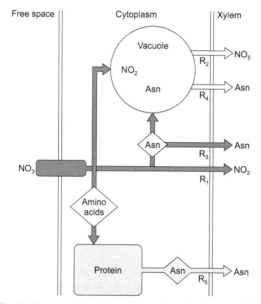

Fig. 9.6 Transport of metabolites in soybean in nitrate nutrition.

9.2 Abiotic Stress

Ozone can interfere with nitrogen metabolism of soybean [*Glycine max* (L.) Merr.], but the effect of O_3 on nitrogen metabolism of field-grown soybean has received only limited attention. A two-year field study was conducted in order to determine the effects of soil moisture deficit and O_3 on nitrogen metabolism of 'Davis' soybean. Two soil moisture regimes, well-watered and water-stressed (WS), were established using different irrigation frequencies. Ozone treatments ranged from 0.025 to 0.107 µL L^{-1} (7-h day^{-1} seasonal means) and were charcoal filtered, nonfiltered, and two (1983) or four (1984) levels of O_3 addition to nonfiltered ambient air entering open-top field chambers. Growing conditions were quite different between the two years. The 1983 season was dry and hot throughout vegetative growth. In contrast, 1984 was wetter and cooler, with no requirement for irrigation during vegetative growth. Effects due to WS were more pronounced than those due to O_3. Nitrogen fixation, estimated by the acetylene reduction (AR) technique, was decreased by both O_3 and WS treatments in both years. As O_3 concentration increased, there was a decrease in total AR activity and a more rapid decline of activity during reproductive growth. Water deficit decreased AR activity significantly both years. The activity of nitrate reductase (NR) was also reduced by WS treatments. Ozone effects on NR activity varied between years. Rates of N accumulation were generally lower in higher O_3 and WS treatments. The partitioning of N among various plant parts and concentrations of N within plant parts were altered in WS treatments during all stages of growth and, to a far lesser extent, by O_3 treatments during reproductive growth. While there were no significant O_3—water regime interactions, the results demonstrate that soil water status has a pronounced effect on plant response to O_3.

Nitrogen requirement of legumes can be met by inorganic N assimilation and symbiotic N_2 fixation; in practice, they obtain N through both processes. A reduction of nitrate to nitrite (NO_2^-) is catalyzed by nitrate reductase (NR; EC 1.6.6.4), an inducible enzyme whose activity depends on the availability of nitrate and light (Campbell, 1999). At high concentrations, nitrate inhibits both nodulation and N_2 fixation in almost all legume species (Becana and Sprent, 1987). Like most legumes, soybean has the potential to fix atmospheric nitrogen through symbiotic relationships with soil organisms (Stefan and Christian, 2002).

A field experiment was conducted in order to study the impact of the exclusion of the solar UV components on growth, photosynthesis and nitrogen metabolism in soybean (*Glycine max*) varieties PK-472, Pusa-24, JS 71-05, JS-335, NRC-7 and Kalitur. The plants were grown in specially designed UV exclusion chambers wrapped with filters to exclude UV-B or UV-A/B and transmitted all UV. Exclusion of UV significantly enhanced the growth of the aerial parts as well as the growth of the below ground parts in all of the six soybean varieties. Nitrate reductase activity (NRA) was significantly reduced, whereas leghemoglobin (Lb) content, total soluble protein, net photosynthesis (P_n) and α-tocopherol content were enhanced after UV exclusion. The exclusion of solar UV-A/B enhanced all parameters to a larger extent than the exclusion of solar UV-B in four of the six varieties of soybean except for NRC-7 and Kalitur. These two varieties responded more to UV-B exclusion compared to UV-A/B exclusion. A significant inverse correlation between the NRA and the number of nodules per plant was observed. The extent of response in all parameters was greater in PK-472 and JS71-05 than that in Kalitur and JS-335 after UV exclusion. The exclusion of UV augmented the growth of nodules, Lb content and α-tocopherol levels and conferred higher rates of P_n to support better growth of nodules. Control plants (+UV-A/B) seemed to fulfill their N demand through the assimilation of NO_3^- resulting in lower symbiotic nitrogen fixation and higher NR activity (Fig. 9.7) (Baroniya et al., 2014).

Plant GS, like other eukaryotic GS, is made up of eight subunits (Meister, 1973) and is probably assembled as two tetramers stacked one upon the other. The native enzyme in plants has a molecular mass ranging from 320 to 380 kD, each subunit having a molecular mass of between 38 and 45 kD. The GS_1 genes of several plants, especially legumes, have been cloned and sequenced (Temple et al., 1995). The GS_1 genes in all plants studied are members of small gene families, and the different members

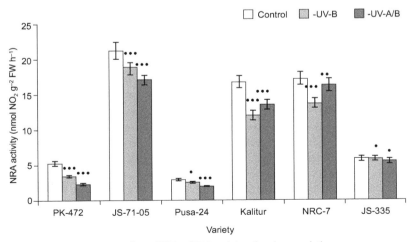

Fig. 9.7 Effect of UV on NRA activity of soybean varieties.

show a unique expression pattern suggesting that the gene members are differentially regulated (Temple et al., 1995).

Based on the characterization of GS_1 cDNAs and their expression pattern, two classes of GS_1 genes have so far been identified in soybean (*Glycine max*) (Roche et al., 1993; Marsolier et al., 1995; Temple et al., 1996); however, the presence of multiple GS_1 polypeptides in the roots and nodules of soybean, based on two-dimensional (2D) gel western analysis, suggests the presence of other members of GS_1 genes that have not yet been characterized (Temple et al., 1996). The study also involves a phylogenetic analysis of the different GS_1 genes that have been described in the literature in relation to the GS_1 genes described. Based on genomic Southern analysis and analysis of hybrid-select translation (HST) products, it appears that there are three distinct classes of GS_1 in soybean and each class has two members. Although the two members of each gene class show homology in the coding region and the 3′-untranslated region (UTR), there are differences in the expression pattern suggesting differential evolution of the promoters of orthologous GS_1 gene members. In an attempt to understand the regulatory mechanism underlying the expression of the different GS_1 genes, the promoter regions of the two *gln-γ* genes have been isolated and subjected to comparative sequence analysis with each other and with the promoter region of the *gln-γ* gene of French bean (Shen et al., 1992).

Gln synthetase (GS) is the key enzyme in N metabolism and it catalyzes the synthesis of Gln from glutamic acid, ATP, and NH_4^+. There are two major isoforms of GS in plants, a cytosolic form (GS_1) and a chloroplastic form (GS_2). In leaves, GS_2 functions to assimilate ammonia produced by nitrate reduction and photorespiration, and GS_1 is the major isoform assimilating NH_3 produced by all other metabolic processes, including symbiotic N_2 fixation in the nodules. GS_1 is encoded by a small multigene family in soybean (*Glycine max*), and cDNA clones for the different members have been isolated. Based on sequence divergence in the 3′-untranslated region, three distinct classes of GS_1 genes have been identified (α, β, and γ). Genomic Southern analysis and analysis of hybrid-select translation products suggest that each class has two distinct members. The α forms are the major isoforms in the cotyledons and young roots. The β forms, although constitutive in their expression pattern, are ammonia inducible and show high expression in N_2-fixing nodules. The γ1 gene appears to be more nodule specific, whereas the γ2 gene member, although nodule enhanced, is also expressed in the cotyledons and flowers. The two members of the α and β class of GS_1 genes show subtle differences in the expression pattern. Analysis of the promoter regions of the γ1 and γ2 genes show sequence conservation around the TATA box but complete divergence in the rest of the promoter region. It was postulated that each member of the three GS_1 gene classes may be derived from the two ancestral genomes from which the allotetraploid soybean was derived (Fig. 9.8) (Morey et al., 2002).

9.3 Biological Nitrogen Fixation

Recently, the recognition, host specificity and initial nodulation processes between leguminous plants and compatible rhizobia has been rapidly uncovered. There are many good reviews about this aspect so the outline can be briefly introduced.

1) Rhizobia live heterotrophically depending on the organic matter in soil, and they don't fix atmospheric N_2. The host legume roots excrete species specific isoflavonoid compounds. Daizein and genistein are two major isoflavonoids released from soybean roots.

2) Compatible rhizobia, usually species *Bradyrhizobium japonicum* for soybean, recognize the isoflavonoid released from host legume, and NOD genes are expressed by specific isoflavonoid signals to make NOD factor. The NOD factor is a lipochitine oligosaccharide with some modification. The structures of NOD factors are different among rhizobia species and only compatible NOD factor can induce nodule formation in host plants with very low concentrations. The cell division in the cortex restores to prepare nodule formation.

3) Rhizobia move to the host roots and proliferate near the root surface.

4) Rhizobia attach to the extending root hair.

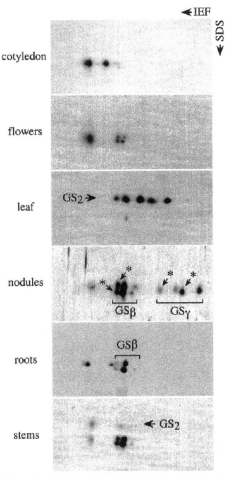

Fig. 9.8 Glutamine synthetase isoforms in different organs of soybean.

5) Then, root hair entraps the rhizobia by root hair curling. Host plant makes the infection thread, which has a tunnel like structure, and rhizobia can enter into the roots through it. Finally rhizobia are released into the proliferating nodule meristem cells. One or several rhizobia are enclosed in PBM (peribacteroid membrane) or symbiosome membrane in synonym.

6) Plant cell division and rhizobium proliferation occur and nodule structure develops (Fig. 9.9).

7) Nodule vascular bundles connect to the root vascular bundles and nodules and roots exchange materials through phloem and xylem. Nodules are formed and bacteroid, a symbiotic state of rhizobia, start to fix N_2. In the case of soybean nodules, nodule organogenesis completed in the initial stage when nodule diameter is about 1 mm and further nodule growth is mostly depending on the cell expansion rather than cell proliferation.

Recently, gene expression analysis can be applied for model legume *Lotus japonicus* by cDNA array analysis (Kouchi et al., 2004). Using the cDNA array of 18,144 non-redundant expressed sequence tags (ESTs) isolated from *L. japonicus*, and the expression of 1,076 genes was significantly accelerated during the successive stages after infection of compatible rhizobia, *Mesorhizobium loti*. These include 32 nodulin and nodulin-homolog genes, as well as a number of genes involved in the catabolism of photosynthates and assimilation of fixed nitrogen. The gene expression profile in early stages of rhizobium-legume interaction was considerably different from that in subsequent nodule development. A number of genes involved in the defense responses to pathogens and other stresses were induced

Fig. 9.9 Nodules in soybean.

abundantly in the infection process, but their expressions were suppressed during subsequent nodule formation. The genome sequencing information and genome resources in model legumes, *L. japonicas* and *Medicago truncathula* have been available for many aspects of legume studies (Sato et al., 2007).

In soybean, N derived from the atmosphere (NDFA) via BNF can range from 0 to 98% of the total N uptake, representing 0 to 337 kg N ha^{-1} (Salvagiotti et al., 2008), depending on rhizobia activity. However, N removal from the system (i.e., by seed N) is determined by different factors that affect seed yield and N harvest index (NHI; seed N uptake to total N uptake). In a recent study, Tamagno et al. (2017) showed that NHI in soybean grown in three different regions ranged from 44 to 91% (at R7–R8 growth stages), with yields ranging from ~ 1 to 8 Mg ha^{-1}. The previous information related to NDFA contribution and NHI calculation for soybeans portrays the complexity of estimating N contribution of soybeans to the rotation. A review study summarizing 108 scientific papers published from 1966 to 2006 documented an average NDFA contribution of 50 to 60% (Salvagiotti et al., 2008). Comparable NDFA estimations were documented in Argentina: 60% (ranging 46–71%) (Collino et al., 2015) and up to 80% in less fertile soils in Brazil (Alves et al., 2003). The review by Salvagiotti et al. (2008) suggested that the NDFA contribution was not sufficient for high-yielding soybeans (> 7 Mg ha^{-1}). Salvagiotti et al. (2009) showed a slight increase in seed yield in crops that yielded more than 5 Mg ha^{-1} when N was supplied without affecting the N$_2$ fixation process.

The "N-gap" between crop N uptake and N supplied by N$_2$ fixation has not yet been estimated at varying yield and NDFA levels. A better understanding of the so-called N-gap could allow development of potential N management strategies to further boost soybean yields and profitability. These strategies should take into account the trade-off between inorganic N fertilizer and fixed N (Salvagiotti et al., 2008). High-yielding soybean environments should be accompanied by high N$_2$ fixation activity, as suggested by van Kessel and Hartley (2000). Collino et al. (2015) showed greater N$_2$ fixation as yield potential increased. However, if N$_2$ fixation contribution remains constant (or increasing less than proportional to yield) at increasing productivity levels, then mineral soil N contribution should meet the crop demand. Consequently, N could be mined from the soil producing negative partial N balance and affecting soil health (Table 9.2 and Fig. 9.10).

Table 9.2 Descriptive statistics of the meta-database relative to soybean yield (adjusted to 13% moisture), Plant N content at the end of the season (dry basis), N contribution from biological N fixation (N$_2$ fixation) at the end of the season in absolute values and expressed in relative terms, N derived from the atmosphere, NDFA %, in aboveground biomass.

Parameter	n	Mean	SD
Seed yield (Mg ha^{-1})	733	3.1	1.4
Plant N content (kgha^{-1})	733	245	108
N$_2$ fixation (all N) (kgha^{-1})	733	137	82
N$_2$ fixation (no N) (kgha^{-1})	473	142	78
NDFA% (all N)	733	56	21
NDFA% (no N)	473	58	19

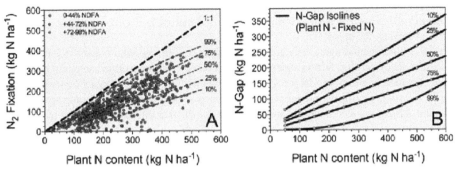

Fig. 9.10 N$_2$ fixation and N-Gap relations with plant N content.

In Argentina, soybean inoculated with strain E109, without N-fertilizer, achieved 6,000 kg/ha (Hungria et al., 2006a), and in Londrina, Brazil, soybean yielded 5,890 kg/ha again without N-fertilizer (Zotarelli, 2000). Still in Brazil, a survey of studies using dilution, δ^{15} ^{15}N isotopic dilution, δ^{15}N, N balance and the N-ureide technique indicated that the N derived from BNF ranged from 69 to 94%, with estimated rates reaching 300 kg N/ha (Hungria et al., 2005). The contributions of BNF may be considerably higher, as root N was not considered in these studies; it may represent 30 to 35% of total plant N (Peoples and Herridge, 2000; Khan et al., 2002).

Figure 9.11 shows one example of a combination of twenty field trials, with no response to 30 kg N/ha as starter N. Experiments have also been performed to confirm that five extra benefits would be obtained by the application of N-fertilizer throughout the plant-growth cycle, because doubts were raised about the capacity of BNF to provide N after flowering, due to nodule senescence (Hungria et al., 2006b; Mendes et al., 2008). Again, no benefits were observed with the supplementary dose of 50 kg of N/ha at the R2 or R4 stages. In the experiments shown in Fig. 9.11, BNF contribution evaluated by the N-ureide technique dropped from 84% on the reinoculated treatment to 77 and 78% when N was applied at R2 and R4, respectively, and to 44% with the application of 200 kg of N/ha (Hungria et al., 2006b).

First in Argentina and now in Brazil in the last few years, there has been a preference for cultivars of indeterminate habit, in which vegetative growth occurs mainly after flowering. Concerns have been raised about a decrease of BNF after flowering, as two thirds of the plant's growth occurs after this period. Then again, in a series of experiments performed for the last two years, it has consistently been observed that these cultivars also do not need any N-fertilizer supply. It is noteworthy also that the inhibition of nodulation and BNF, resulting in less yield by application of chemical N was stronger with cultivars

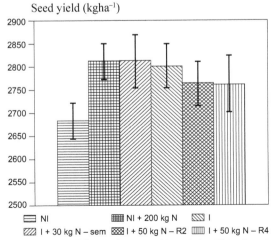

Fig. 9.11 Seed yield of soybean under different nitrogen and inoculation (I) treatments (NI – no inoculation).

of the indeterminate type. Finally, in a comparison of twenty experiments with cultivars of short- and long-growth cycles, Hungria et al. (2006b) also reported that there was no benefit in yield by adding N-fertilizer at sowing, R2 or R4.

Several environmental conditions are limiting factors to the growth and activity of the N_2-fixing plants. A principle of limiting factors states that "the level of crop production can be no higher than that allowed by the maximum limiting factor" (Brockwell et al., 1991). In the *Rhizobium*-legume symbiosis, which is a N_2-fixing system, the process of N_2 fixation is strongly related to the physiological state of the host plant. Therefore, a competitive and persistent rhizobial strain is not expected to express its full capacity for nitrogen fixation if limiting factors (e.g., salinity, unfavorable soil pH, nutrient deficiency, mineral toxicity, temperature extremes, insufficient or excessive soil moisture, inadequate photosynthesis, plant diseases, and grazing) impose limitations on the vigor of the host legume (Brockwell et al., 1991; Peoples et al., 1995; Thies et al., 1995).

Typical environmental stresses faced by the legume nodules and their symbiotic partner (*Rhizobium*) may include photosynthate deprivation, water stress, salinity, soil nitrate, temperature, heavy metals, and biocides (Walsh, 1995). A given stress may also have more than one effect, e.g., salinity may act as a water stress, which affects the photosynthetic rate, or may affect nodule metabolism directly. The most problematic environments for rhizobia are marginal lands with low rainfall, extremes of temperature, acidic soils of low nutrient status, and poor water-holding capacity (Bottomley, 1991). Populations of *Rhizobium* and *Bradyrhizobium* species vary in their tolerance to major environmental factors; consequently, screening for tolerant strains has been pursued (Keyser et al., 1993). Biological processes (e.g., N_2 fixation) capable of improving agricultural productivity while minimizing soil loss and ameliorating adverse edaphic conditions are essential.

A pot experiment was carried out in order to determine the effects of soil water contents on BNF. Four different soil water contents (25%, 50%, 75% and 100% of water holding capacity) were adjusted either every 3 days or just after plants indicate wilting point. Non-inoculated pots were added to experiment as a control. The results revealed that BNF is affected by different levels of soil water content. The mechanism of this effect would not be the direct effect of water, but the side effect of water on soil oxygen content; therefore, an aeration capability.

Table 9.3 Total nitrogen amount of whole plant (mg N).

| | | \% of water holding capacity | | | | |
		25	50	75	100	Average
Near wilting point	−bacteria	9.6	10.2	12.9	14.7	11.8
	+bacteria	22.6	24.2	27.1	20.4	23.6
	Average	16.1	17.2	20.0	17.6	17.7
Constant period	−bacteria	8.2	12.3	13.9	15.6	12.5
	+bacteria	15.5	22.0	27.9	25.1	22.6
	Average	11.9	17.2	20.9	20.3	17.6

NO_3^- transport pathway into soybean nodule was investigated by tungstate and $^{15}NO_3^-$ tracers (Mizukoshi et al., 1995). There are several possible pathways of NO_3^- transport into soybean nodules. First, the NO_3^- absorbed from the lower part of roots is transported through the xylem and supplied to the nodules. Second, NO_3^- is supplied by phloem from shoots, which was transferred from xylem to phloem in stems or leaves. Third, NO_3^- is directly absorbed from nodule surface. Forth, NO_3^- absorbed from the adjacent root part is transported cell to cell from root cortex to nodule cortex via the symplastic pathway. The importance of symplastic transport of C unloaded from nodule phloem and N from infected cells was suggested in soybean nodules (Brown et al., 1995).

Tungstatre (WO_4^{2+}) was used as an anion tracer, and the distribution of tungsten (W) in the roots and nodules was examined by electron probe X-ray microanalysis (EPMA). At 3 days after 1 mM WO_4^{2+} treatment in culture solution, accumulation of W in the roots cortex was observed while the W movement into root nodules was not detected. It was suggested that external anions cannot be readily transported into nodules via apoplastic pathway. In contrast, when 1.7 mM of $^{15}NO_3^-$ was supplied to the solution for 1 day, an appreciable amount of NO_3^- and ^{15}N was detected in nodule cortex, although a little was distributed in the infected region. In another experiment, $^{15}NO_3^-$ solution was supplied to one large nodule through a cheese cloth wrapping the nodule. These results suggest that NO_3^- can be absorbed from nodule surface (epidermal cells or loosely packed lenticel cells), then it is transported from cell to cell into cortex via simplistic pathway. Arrese-Igor et al. (1998) reported that nitrate accumulation was 5 times higher in cortex than infected region of the soybean nodules supplied with 10 mM NO_3^- for 8 days. The nitrate treatment did not cause free nitrite accumulation in nodules in 8 days.

The bacteroid has NR activity when soybean plants are cultivated without nitrate. The characteristics of nitrate respiration of isolated soybean bacteroids were reported (Ohyama and Kumazawa, 1987). Under anaerobic conditions, the respiratory CO_2 evolution was very low. When 1 mM NO_3^- was added, the CO_2 evolution increased by 18 times. The CO_2 evolution associated with NO_3^- reduction was inhibited by the addition of 1 mM DNP (2,4-dinitrophenol) or 1 mM $HgCl_2$, but not by 1 mM KCN. On the other hand, aerobic respiration atpO^2 = 0.2 was severely depressed by 1 mM KCN and 1 mM $HgCl_2$.

Schematic representation of partnership between a diazotrophic bacterial cell and a nodulating plant cell during symbiotic nitrogen fixation suggests that Rhizobia induce the formation of nodules on legumes using either Nod factor-dependent or Nod factor-independent processes. In the Nod factor-dependent strategy, plants release signals, such as flavonoids, that are perceived by compatible bacteria in the rhizosphere. This activates the nodulation (nod) genes of rhizobia, which, in turn, synthesize and release bacterial signals, mainly lipochitooligosaccharides (LCOs) (Nod factors), which trigger early events in the nodulation process. Synthesis of the Nod factors backbone is controlled by the canonicalnodABC genes, which are present in all rhizobia, but a combination of other nodulation genes (nod, nol, ornoe) encode the addition of various decorations to the core structure. In the Nod factor-independent process, bacteria enter in the plant via cracks in the epidermis. Accumulation of cytokinin synthesized by the bacteria in these infection zones might trigger nodule organogenesis. In the mature nodule, bacteria progressively experience lower oxygen concentrations and differentiate into bacteroids, fixing diffused nitrogen gas using their nitrogenase enzyme complex. NH_3 produced by nitrogenase from the bacteria (nif, fix, and cytochromebd) can be incorporated into amino acids via the glutamine synthetase-glutamate synthase (GS-GOGAT) pathway. NH_3 can also diffuse through the bacterial membrane and be transported to the plant cytoplasm via ammonia transporters (e.g., AmtB), where it is assimilated into nitrogen compounds (amino acids, proteins, and alkaloids) in exchange for food molecules, e.g., glucose, amino acids, and other saccharides. The plant provides amino acids to the bacterial cell and in return the bacterial cell cycles amino acids back to the plant for asparagine synthesis. Other nutrients have to be made available for the microbe, including phosphorus, sulfur, molybdenum, and cobalt (Fig. 9.12).

Inside newly formed legume nodules, rhizobia differentiate and depend on carbon sources derived from the plant in order to sustain metabolism, including nitrogen fixation. Plant metabolism is altered to support this energy demand. Genes involved in metabolic pathways, like glycolysis, photosynthesis, amino acid biosynthesis, purine and redox metabolism, and metabolite transport, are all upregulated during symbiosis (Colebatch et al., 2004; Benedito et al., 2010). The primary metabolite is sucrose, which is produced in the aerial parts of plants and travels through phloem to the root nodule, where it is catabolized (Kouchi and Yoneyama, 1984). In nodule cells, sucrose is cleaved reversibly to UDP-glucose and fructose by sucrose synthase and irreversibly to glucose and fructose by invertase (Craig et al., 1999; Horst et al., 2007). Hexoses subsequently enter glycolysis, which is upregulated transcriptionally in the nodules (Colebatch et al., 2004), to produce phosphoenolpyruvate, which, in turn, is converted to dicarboxylic acids. Several studies have shown that carbonic anhydrase, phosphoenolpyruvate carboxylase, and malate dehydrogenase are upregulated during nodule development, which directs carbon flow toward malate (Udvardi and Poole, 2013). The exchange of metabolites between the plant and bacteroids does not happen freely but is facilitated by specialized transporters. Analysis of the genomic inventory of *Medicago truncatula* transporters revealed that a wide range of transporters are induced during nodule development (Benedito et al., 2010). Among these are genes encoding putative sugar transporters, amino acid transporters, and sulfate transporters (Limpens et al., 2013). In Rhizobium-legume symbiosis, carbon is specifically supplied to the bacteroids in the form of dicarboxylic acids, such

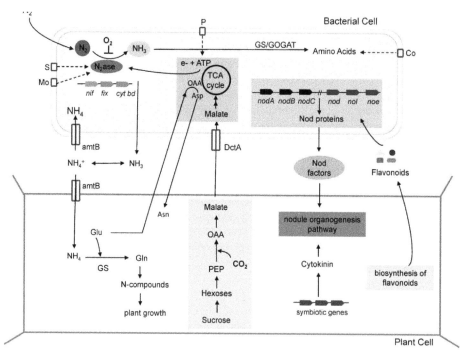

Fig. 9.12 Schematic representation of partnership between a diazotrophic bacterial cell and a nodulating plant cell during symbiotic nitrogen fixation. Asn, asparagine; Asp, aspartate; KG, alpha ketoglutarate; AmtB, ammonia transporter; Co, cobalt; cyt bd, cytochromebd; DctA, dicarboxylate transporter; Glu, glutamate; Gln, glutamine; GOGAT, glutamate synthase; GS, glutamine synthetase; HCO_3, bicarbonate; Mo, molybdenum; NH_3, ammonia; N2ase, nitrogenase; Nod factors, nodulation factors; NFR, Nod factor receptor; OAA, oxaloacetate; P, phosphorus; S, sulfur.

as malate (Udvardi et al., 1988). After crossing the symbiosome membrane that separates the bacteroids from the plant cell cytoplasm, dicarboxylates are taken up by DctA, a transporter of the major facilitator super family (Yurgel and Kahn, 2004). Dicarboxylic acids are assimilated by gluconeogenesis or catabolized via enzymes of the tricarboxylic acid (TCA) cycle to provide the reductant and ATP required for nitrogen fixation (Finan et al., 1988; 1991).

In legume-Rhizobium symbiosis, ammonia produced by nitrogenase is delivered to the plant cell as NH_4^+ and/or NH_3. Ammonia in its neutral lipophilic form probably crosses the bacteroid membranes via diffusion. The bacterial NH_4^+ transporter AmtB, which transports NH_4^+ in the opposite direction (i.e., into the bacteroid), is repressed in bacteroids, ensuring that NH_3 lost from the cell is not recovered by the bacterium but rather is taken into the plant cytoplasm. After entering the symbiosome space between the bacteroid and the symbiosome membrane, ammonia is protonated to NH_4^+ because of the acidic environment there (Day et al., 2001). In the next step, ammonium crosses the symbiosome membrane and enters the cytoplasm of the infected plant cell, where it is rapidly assimilated into organic form. Two possible pathways exist for ammonium transport across the symbiosome membrane: One through an NH_3 channel (Niemietz and Tyerman, 2000), and the other through a cation channel that transports K^+, Na^+, and NH_4^+ (Tyerman et al., 1995). Once inside the plant cell, ammonia is assimilated into amino acids mainly by the action of GS, glutamate synthase (GOGAT), and aspartate aminotransferase. The expression of genes encoding these enzymes is induced during nodule development (Colebatch et al., 2004). Interestingly, nodulin 26, which can transport NH_3 (Hwang et al., 2010), interacts physically with cytosolic GS that is responsible for the assimilation of ammonia to glutamine (Masalkar et al., 2010). Several other genes encoding aquaporin-like proteins that potentially transport ammonia are induced in infected cells of *Medicago truncatula* nodules (Limpens et al., 2013). The symbiosome membrane NH_4^+/K^+ channels have not yet been identified genetically.

Zhang et al. (2014) have extensively discussed recent research that was carried out in order to elucidate the molecular basis for the sensitivity of soybean to cold stress. The genome-wide expression analysis of miRNAs in response to chilling was particularly helpful as it revealed the miRNAs that are involved in response of mature nodules to cold stress. These findings have concluded that miRNAs are involved in the protection against chilling injury in mature soybean nodules.

Sulfur (S) is an essential element for SNF, although it has not been studied much in this context. Nitrogenase, the bacterial enzyme responsible for the reduction of N, is a complex [Fe-S] enzyme (Rees and Howard, 2000) that is produced in relatively high amounts by N-fixing bacteroids (Gaude et al., 2004). NifS is a cysteine desulfurase, which uses l-cysteine for the specific mobilization of S for maturation of nitrogenase, and accumulates only under N-fixing conditions (Zheng et al., 1993; Johnson et al., 2005). The *Lotus japonicus Sst1* gene, which is expressed in a nodule-specific manner, encodes a sulfate transporter that is essential for SNF (Krusell et al., 2005). The SST1 protein appears to reside on the symbiosome membrane (Wienkoop and Saalbach, 2003) and is thought to transport sulfate from the plant cell cytoplasm to the bacteroids (Krusell et al., 2005). In legume nodules, the content of (homo) glutathione, a S-containing antioxidant, correlates strongly with nitrogenase activity and is able to modulate the efficiency (El Msehli et al., 2011).

Changes in whole-plant S-metabolism associated with nodule formation and SNF were studied by comparing sulfate and thiol content and APR activity of nodules, roots, stems, and leaves of *L. japonicus* plants inoculated with wild-type or $\Delta nifA$ and $\Delta nifH$ mutants of *Mesorhizobium loti*, and of uninoculated plants. Both mutant strains of *M. loti* form ineffective (Fix⁻) nodules with no nitrogenase activity, in contrast to Fix⁺ nodules containing wild-type rhizobia. In nodules harboring the $\Delta nifA$ strain, infected cells contain undifferentiated rhizobia, whereas $\Delta nifH$ nodules contain well-differentiated, albeit ineffective bacteroids (Fotelli et al., 2011).

Sulfate levels varied significantly within and between organs, depending on the symbiotic and N-fixing status of plants (Fig. 9.13). Sulfate levels were lowest in nodules of N-fixing plants containing wild-type bacteria and significantly higher in nodules containing either of the *nif* mutants. Sulfate levels were several times higher in roots, stems, and leaves than in nodules of plants inoculated with wild-type bacteria. Sulfate levels were significantly higher in roots of inoculated than of uninoculated plants, whereas sulfate levels in stems of inoculated plants were lower than of uninoculated plants. Interestingly, sulfate levels in each organ were highest for nodulated plants that were unable to fix N due to the presence of Fix-bacteria (Kalloniati et al., 2015) (Fig. 9.13).

The possible role of H_2S in the symbiosis between soybean (*Glycine max*) and rhizobium (*Sinorhizobium fredii*) was investigated. Results demonstrated that exogenous H_2S donor (sodium hydrosulfide, NaHS) treatment promoted soybean growth, nodulation and nitrogenase (Nase) activity. Western blotting analysis revealed that the abundance of nitrogenase component nifH was increased by

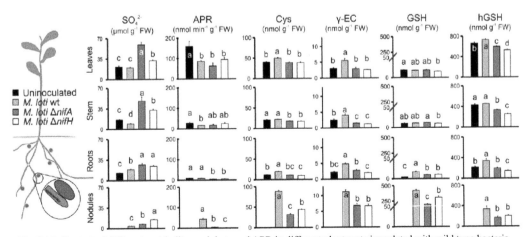

Fig. 9.13 Content of different metabolites, sulphate and APR in different plant parts inoculated with wild type bacteria.

NaHS treatment in nodules. Quantitative real-time PCR data showed that NaHS treatment up-regulated the expressions of symbiosis-related genes *nodC* and *nodD* of *S. fredii*. Besides, expression of soybean nodulation marker genes including early nodulin 40 (*GmENOD40*), ERF required for nodulation (*GmERN*), nodulation signaling pathway2b (*GmNSP2b*) and nodulation inception genes (*GmNIN1a*, *GmNIN2a* and *GmNIN2b*) were up-regulated. Moreover, the expressions of glutamate synthase (*GmGS*), nitrite reductase (*GmNiR*), ammonia transporter (*GmSAT1*), and *nifH* involved in nitrogen metabolism were up-regulated in NaHS-treated soybean roots and nodules. Together, results suggested that H$_2$S may act as a positive signaling molecule in soybean-rhizobia symbiosis.

Ammonia is known to be the initial product of nitrogenase. Bergersen (1965) observed that after the exposure of ^{15}N$_2$ to soybean nodules for 1 min, more than 90% of the fixed N in soluble fraction was detected as ammonia. The *Km* value (Michaelis constant) of N$_2$ for nitrogen fixation by purified nitrogenase was 8–16%, and that in the detached nodule in the air is 5% in the air and that in solution was 0.029 mole m^{-3} (Bergersen, 1999). Until the late 1970s, no direct evidence as to how the fixed N in bacteroids is transported to the host plant cytosol was available, nor how it is metabolized to translocation forms of N to the shoots. Until 1970, the fixed ammonia had been considered to be initially combined with 2-OG producing Glu catalyzed by glutamate dehydrogenenase (GDH) in plants and microbes. Not only ammonia produced by nitrogen fixation, the ammonia absorbed in the roots and reduced from nitrate was also believed to be assimilated by GDH enzyme.

GDH catalyzes the reaction as follows (Layzell, 1990):

NH$_4^+$ + 2-oxoglutarate + 2 e– → glutamate

In roots or leaves, the enzyme is located in mitochondria or chloroplast, and the electron donor may be either NADH or NADPH.

After discovering a new enzyme glutamate synthase (GOGAT) in *Aerobacter aerogenes* (Tempest et al., 1970), it is confirmed that ammonia can be assimilated via alternative of GDH, via glutamine synthetase (GS) and GOGAT pathway.

Wolk et al. (1976) demonstrated that the fixed ammonia is assimilated through GS/GOGAT pathway in nitrogen fixing blue green algae, *Anabaena cylindrica* by using ^{13}N as a tracer.

GS catalyzes the following reaction:

NH$_4^+$ + glutamate + ATP → glutamine + ADP + Pi

This reaction requires divalent cation, such as Mg^{2+}, Mn^{2+} or Co^{2+} as a cofactor. There are a number of isoforms of GS: GS1 (cytosolic) and GS2 (chloroplast) forms of leaves, GSr root cytosolic form, and GSn nodule cytosolic form.

GOGAT catalyzes the following reaction:

glutamine + 2-oxoglutarate + 2 e– → 2 glutamate

The electron donor is ferredoxin (Fd) or NADH in higher plants. In leaves, the most activity is Fd-dependent enzyme, and NADH-GOGAT locates in the plastid of non-photosynthetic tissues.

The net reaction of GS/GOGAT pathway is as follows:

NH$_4^+$ + 2-oxoglutarate + 2 e– + ATP → glutamate + ADP + Pi

Compared with GDH reaction, GS/GOGAT pathway requires one extra ATP as a substrate, and it means GS/GOGAT needs more energy than GDH. Although the higher cost of the ammonia assimilation by GS/GOGAT than GDH, the GS with lower *Km* for ammonia has an advantage to assimilate ammonia at low concentrations in cells before reaching toxic level.

To elucidate the initial ammonia assimilation pathway in soybean nodules, the intact nodules attached to the upper part of the roots were exposed to ^{15}N$_2$ gas for 21 min (Ohyama and Kumazawa, 1978). ^{15}N$_2$ gas was prepared from ^{15}N labeled ammonium sulfate then mixed with He and O$_2$ (^{15}N$_2$:He:O$_2$ = 1:7:2) (Ohyama and Kumazawa, 1981c). The 21 min of ^{15}N$_2$ exposure was followed by non-labeled conditions for 29 min (chase period), and the nodules were harvested at 2, 4, 6, 10, 15, 20, 25, 30, 35, 40, 50 min after starting $^{15}N_2$ exposure. The fresh nodules were rapidly homogenized with 80% ethanol, and ammonia, amino acids and ureides (allantoin and allatoic acid) were extracted.

The incorporation of ^{15}N into various nitrogen compounds was determined by the optical emission spectrometry (Ohyama and Kumazawa, 1979b; Ohyama, 1982; Ohyama et al., 2004; FNCA, 2006). This suggests that there are two or more compartments of ammonia in nodules, and one of which may be the ammonia pool directly derived from nitrogen fixation. The size is relatively small, less than 1% of total ammonia in nodules, but the turnover rate is very rapid in a few min. Among amino acids, glutamine gave the highest ^{15}N abundance until 10 min of ^{15}N supply. When the ^{15}N abundance of amido-N and amino-N of glutamine was separately measured, the ^{15}N was rapidly incorporated into amido-N, then amino-N after a few min lag-time. Following glutamine, glutamic acid and alanine increased ^{15}N abundance. After changing gas phase from $^{15}N_2$ to non-labeled conditions, ammonia and glutamine showed the immediate decrease of ^{15}N abundance, this indicates the characteristics of primary assimilatory products. On the other hand, the ^{15}N abundance of glutamate alanine continue to increase for a few min after changing to non-labeled conditions. ^{15}N was relatively rapidly incorporated into ureides (the sum of allantoin and allatoic acid) in 10 min of $^{15}N_2$ exposure, although the ^{15}N abundance did not decrease during chase period. This is the first experimental evidence that ureides are synthesized actively in soybean nodules from fixed nitrogen. The incorporation of ^{15}N was faster in allantoin than allatoic acid, suggesting that allatoic acid is formed from allnatoin in nodules, and not *vice versa* (Ohyama and Kumazawa, 1978). The ^{15}N was slowly incorporated into asparagine, although the time lag was longer than ureides.

Soybean plants absorb inorganic N from the roots, and they can fix atmospheric N_2 in the nodules associated with soil bacteria rhizobia. It shows a model of nutrients and water flow via xylem and phloem in soybean plants. Soybean roots absorb water and nutrients in soil solution, and they are transported to the shoots via xylem vessels by the transpiration and root pressure. The fixed N in nodule is also transported to the shoots via xylem. On the other hand, photoassimilates (mainly sucrose), amino acids (Asn, etc.), and minerals (potassium, etc.) are transported from leaves to the apical buds, roots, nodules, and pods via the phloem by osmotic pressure or protoplasmic streaming.

The distribution of radioactivity shows xylem flow (Sato et al., 1999) or phloem flow (Fujikake et al., 2003) and shows the positron imaging of the distribution of radioactivity in nodulated soybean (T 202) after 1 hour of $^{13}NO_3^-$ supply to the root medium (Sato et al., 1999). All parts of the roots exhibited the highest radioactivity (red), and stems and first trifoliolate leaf were relatively high (yellow). The radioactivity was not observed in the nodules, although they are attached in the roots. The positron imaging of distribution of radioactivity in nodulated soybean (cv. Williams) and after ^{11}C-labeled CO_2 was exposed to the first trifoliolate leaf, and the radioactivity was monitored after 2 hours (Fujikake et al., 2003), the highest radioactivity was shown in the $^{11}CO_2$-fed leaf (red) and stems (red) with apical bud (red) and root (yellow) and nodules (red). No radioactivity was observed in the primary leaf and other matured leaves. Nodules showed a higher radioactivity than that in the roots.

9.4 Macromolecules

Protein is a polymer or a complex of polymers of 20 amino acids in higher plants and plays an essential role on metabolism as enzymes, storage proteins, and structure components of the cells. The 20 amino acids consist of alanine, arginine, asparagine, aspartic acid, cysteine, glutamic acid, glutamine, glycine, histidine, isoleucine, leucine, lysine, methionine, phenylalanine, proline, serine, threonine, tyrosine, tryptophan, and valine. Enzyme is a kind of protein that catalyzes a specific chemical reaction in plant cells, and regulation of enzyme synthesis and the activity are essential for maintaining life and growth.

Nucleic acids, deoxyribonucleic acid (DNA), and ribonucleic acid (RNA) are a polymer of purine bases (adenine and guanine) and pyrimidine bases (thiamine, cytosine for DNA, and uracil, cytosine for RNA) with pentose (2-deoxyribose for DNA and ribose for RNA) and phosphate. DNA serves as a template of mRNA, and the mRNA is translated into amino acid sequences of protein. Purine base contains 4 N atoms in a molecule, and they are derived from two glutamines, one aspartic acid, and one glycine. Pyrimidine base contains 2 N atoms in a molecule, and they are derived from one glutamine and one aspartic acid. Amino acids are the precursors of most of N compounds in plants.

In soybean plants, ureides, allantoin, and allantoic acid are mainly used for transport of N in addition to amino acids. Ureides have 4N and 4C atoms in a molecule, and it is considered to be more efficient to transport N than asparagine (2N and 4C) in terms of carbon economy.

Table 9.4 Concentration of amino acid, ureides, nitrate, ammonium, and other 80% ethanol soluble fraction of soybean organs (μgN/gDW).

Organ	Total amino acid-N	Ureides-N	Nitrate-N	Ammonium-N	Others-N	Total soluble-N
Nodules	519	483	147	132	4179	5460
Roots	147	19	32	39	477	714
Stems (UP)	270	143	56	36	418	923
Stems (LP)	134	62	32	23	376	627
Leaves (UP)	96	35	33	46	622	832
Leaves (LP)	96	36	43	46	630	851
Pods (UP)	776	1529	95	167	655	3222
Pods (LP)	415	916	32	124	1281	2768
Seeds (UP)	1243	82	91	119	8635	10170
Seeds (LP)	982	82	69	171	6326	7630

It shows the total amino acid-N, ureides-N, nitrate-N, ammonium-N, and others in 80% ethanol soluble fraction of each organ. Both hydrophilic and hydrophobic low-molecular weight compounds, such as sugars, amino acids, ureides, organic acids, and chlorophyll.

N_2 gas is diffused into central symbiotic region of the nodule and reduced into ammonia by the enzyme "nitrogenase" in the bacteroid, a symbiotic form of rhizobia. Most of the ammonia produced by N_2 fixation is rapidly excreted into the cytosol through the peribacteroid membrane (PBM) of the infected cell.

Based on the time course experiment with $^{15}N_2$ feedings in the nodulated intact soybean plants, the ammonia produced by nitrogen fixation is initially assimilated into amide group of Gln with Glu by the enzyme glutamine synthetase (GS) (Ohyama and Kumazawa, 1978; 1980). Then, Gln and 2-OG produce two moles of Glu by the enzyme glutamate synthase (GOGAT). Some part of Gln is used for purine base synthesis, and uric acid is transported from the infected cells to the adjacent uninfected cells in the central symbiotic region of nodule. Uric acid is catabolized into allantoin and allantoic acid in the uninfected cells and then transported to the shoot through xylem vessels in the roots and stems. A small portion of fixed N was assimilated into alanine and Glu in the bacteroids, but it was not by GS/GOGAT pathway (Ohyama and Kumazawa, 1980).

A small portion of fixed N is transported as amino acid-like asparagine (Asn) in addition to ureides, but the percentage is about 10–20% of total fixed N. A small amount of ureides is synthesized from NO_3^- in soybean nodules (Ohyama and Kumazawa, 1979). Nitrate in culture solution can be absorbed from nodule surface (Mizukoshi et al., 1995) and assimilated in the cortex of the nodules. Nitrate absorbed from lower part of roots was not readily transported to the nodules attached to the upper part of the root system (Sato et al., 1999) (Fig. 9.14).

The mature soybean seed consists of a seed coat surrounding a large embryo. Seed coat has a hilum (seed scar), and there is a tiny hole (micropyle) at the end of the hilum. The tip of the hypocotyl radical axis is located just below the micropyle. There is a main vain at the dorsal part of a pod, and nutrients, such as sugar and ureides and amino acids, are transported through the vain. Seed coat has a funiculus connecting main vain and hilum. Nutrients are transported through vascular bundles in seed coat; however, the vascular systems are not directly connected to the cotyledons. The cotyledons are cultured in a seed coat cavity. Therefore, nutrients are excreted to the cavity, and cotyledons absorb them by themselves.

As shown in Table 9.4, young pods contained a high concentration of ureides both in the upper and the lower pods. The high accumulation of ureides in the pods may be due to the fact that ureides are tentatively stored in the pods before transporting to the seeds. Seeds contain a high concentration of amino acids, especially Asn and GABA.

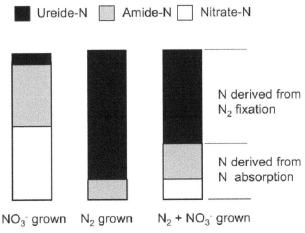

Fig. 9.14 Ureide, amide and nitrate grown under different N sources.

Figure 9.15 shows the changes in ureide-N and amino acid-N in seeds and pods of nodulated and non-nodulated soybean (Ohtake et al., 1996). The concentrations of ureides in the pods of nodulated soybean were high at 1st September and decreased after 15th September. The ureide-N concentration stayed low in the pods of non-nodulated soybean plants. Amino acid-N concentrations were similar between nodulated and non-nodulated soybeans and decreased linearly from 1st September to 10th October at maturing stage. On the other hand, the ureide-N concentrations in the seeds of nodulated and non-nodulated plants were constantly low. The amino acid-N concentrations were similar between nodulated and non-nodulated soybeans, decreased from 1st September to 22nd September, and then constant until maturity at 10th October (Ohyama et al., 2017) (Fig. 9.16).

Seed protein content is important for both feed and food utilisation of soybean. In soybeans grown in Central Europe, considerable variation in protein content was due to seasonal influences, as demonstrated in different experiments from a breeding programme. In soybean genotypes of early maturity groups, average to high protein content (range 399–476 g kgÿ$^{-1}$) was found in years with high air temperature and moderate rates of rainfall during the seed-filling period, whereas seed protein content was drastically reduced (range 265–347 g kgÿ$^{-1}$) in seasons of insufficient nitrogen fixation or higher amounts of precipitation during seed filling. In a set of 60 genotypes, protein content was increased both by late nitrogen fertilisation before the onset of seed filling and by inoculation of seed with nitrogen-fixing rhizobia. Despite the high degree of environmental modification, genetic variation of seed protein content

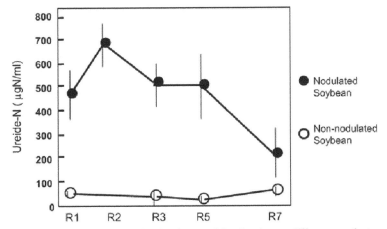

Fig. 9.15 Ureide nitrogen in nodulated and non-nodulated soybean at different growth stages.

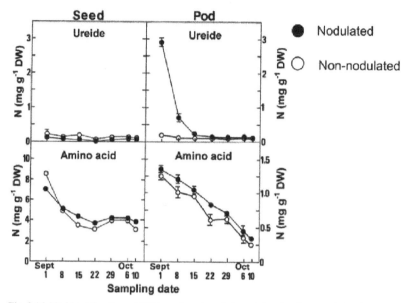

Fig. 9.16 Ureide and amino acid N in pods and seeds in nodulated and non-nodulated soybean.

was considerable, and genotype environment interaction was of low magnitude. Therefore, selection of early maturing soybean genotypes with improved seed protein content appears to be feasible and is only limited by the moderately negative correlation between protein content and seed yield.

In 1993 and 1994, low rates of precipitation and high temperatures clearly favoured protein instead of oil synthesis, whereas the extremely low protein content in 1996 was probably due to insufficient nitrogen availability, as nearly no nodule formation could be observed in that particular season.

Moreover, in 1996 the lowest temperature during the seed-filling period was observed, which also might have affected seed protein content. In 1995, 1996 and 1998, the formation of an enhanced oil content was promoted by high amounts of rainfall during the seed-filling period (R3 to R8 stages). A moderately negative correlation between seed protein content and grain yield was observed in nearly all seasons (Table 9.5), whereas the correlation between protein content and time to maturity was positive in high-protein seasons and negative in low-protein seasons. The relationship between protein and oil content was consistently negative for all growing seasons (Table 9.5). In the exceptionally low-protein season (1996), the early maturing cv Ultra was higher in protein content than high-protein cvs Proto and Apache (Fig. 9.16). In this season of low nitrogen availability, high amounts of rainfall occurred late in August and in September, when cv Ultra had already reached full maturity (data not shown). In genotypes of later maturity, oil synthesis was enhanced by late water availability, which reduced the protein fraction

Table 9.5 Phenotypic coefficients of correlation (r) between seed protein content and agronomic characters in soybean performance trials from different growing seasons.

Year	Grain yield	Time to maturity	Seed weight	Oil content
1993	+0.150	+0.400**	+0.126	−0.902**
1994	−0.327*	+0.584**	−0.006	−0.827**
1995	−0.578**	−0.265*	−0.274*	−0.850**
1996	−0.493**	−0.412*	−0.339*	−0.597**
1997	−0.425**	−0.217	−0.027	−0.899**
1998	−0.573**	−0.296*	+0.158	−0.690**

*, ** Significant at the 0.05 and 0.01 levels, respectively.

of the seed as a consequence. This view is also supported by the finding of a clearly negative correlation between protein content and time to maturity in 1996 (Vollmann et al., 2000) (Table 9.6).

The ranking of genotypes by protein content appeared to be rather similar for the different nitrogen regimes applied. Moreover, genotypes were ranked similarly across years (Spearman rank correlation coefficient rs = 0.78), which was also confirmed by an analysis of variance, in which genotype treatment as well as genotype year interactions were of a much lower magnitude than was the effect of genotype on protein content (Table 9.6).

Table 9.6 Degrees of freedom (Df) and mean squares from the analysis of variance for seed quality parameters in 60 soybean genotypes and three nitrogen treatments of the protein ranking experiment.

Source of variation	Df	Mean square		
		Seed weight	Protein content	Oil content
Replication within year	2	747.9	1912.0	724.6
Year	1	265397.8***	347723.6***	304033.1***
Treatment	2	9298.5***	42657.5***	8803.1**
Year × treatment	2	5216.4**	27828.8***	4581.9**
Error(a)	4	113.8	233.4	154.9
Genotype	59	3804.0***	3333.3***	1321.8***
Genotype × year	59	473.7***	431.3***	153.1***
Genotype × treatment	118	127.1*	244.9**	107.6***
Genotype × year × treatment	118	154.1**	262.2**	121.7***
Error(b)	354	98.7	172.4	63.8

*, **, *** Significant (F-test) at the 0.05, 0.01 and 0.001 levels, respectively.

Most parts of N stored in matured seeds are storage proteins in the cotyledons (Fig. 9.17A) (Ohtake et al., 1997). Soybean seed storage protein consists of glycinin and β-conglycinin (Fig. 9.17B). The glycinin is the hexamer, which has acidic and basic subunits. The β-conglycinin is the trimer, which has α', α, and β subunits.

Plant seeds accumulate proteins, oils, and carbohydrates because these nitrogen and carbon reserves are necessary for early seed germination and seedling growth (for review, see Weber et al., 2005). These reserve components are synthesized during an extended phase of seed development, loosely termed seed filling. Seed filling is the period when rapid metabolic and morphological (size, weight, and color) changes occur, encompassing cellular processes that include cell expansion and the early stage of desiccation (Rubel et al., 1972; Mienke et al., 1981; Agrawal and Thelen, 2006). Seed filling is also the period that

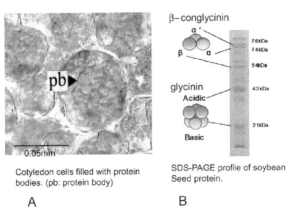

Cotyledon cells filled with protein bodies. (pb: protein body)

SDS-PAGE profile of soybean Seed protein.

A   B

Fig. 9.17 Photograph of protein bodies (A) and seed protein profile (B).

largely determines the relative levels of storage reserves in seeds. The relative proportion of storage components in seeds varies dramatically among different plant species. For example, soybean (*Glycine max*) seed contains approximately 40% protein and 20% oil (Hill and Breidenbach, 1974; Ohlrogge and Kuo, 1984). In contrast, seed of oilseed rape (*Brassica napus*; also called rapeseed or canola) contains approximately 15% protein and 40% oil (Norton and Harris, 1975). To gain insight into the complex process of seed development, identification of genes and proteins and their dynamic expression profiles during seed filling are beginning to provide a framework for more in-depth comparative studies.

A high-throughput proteomic approach was employed in order to determine the expression profile and identity of hundreds of proteins during seed filling in soybean (*Glycine max*) cv Maverick. Soybean seed proteins were analyzed at 2, 3, 4, 5, and 6 weeks after flowering using two-dimensional gel electrophoresis and matrix-assisted laser desorption ionization time-of-flight mass spectrometry. This led to the establishment of high-resolution proteome reference maps, expression profiles of 679 spots, and corresponding matrix-assisted laser desorption ionization time-of-flight mass spectrometry spectra for each spot. Database searching with these spectra resulted in the identification of 422 proteins representing 216 non-redundant proteins. These proteins were classified into 14 major functional categories. Proteins involved in metabolism, protein destination and storage, metabolite transport, and disease/defense were the most abundant. For each functional category, a composite expression profile is presented in order to gain insight into legume seed physiology and the general regulation of proteins associated with each functional class. Using this approach, an overall decrease in metabolism-related proteins versus an increase in proteins associated with destination and storage was observed during seed filling. The accumulation of unknown proteins, sucrose transport and cleavage enzymes, cysteine and methionine biosynthesis enzymes, 14-3-3-like proteins, lipoxygenases, storage proteins, and allergenic proteins during seed filling is also discussed. A user-intuitive database (http://oilseedproteomics.missouri.edu) was developed in order to give access to these data for soybean and other oilseeds currently being investigated.

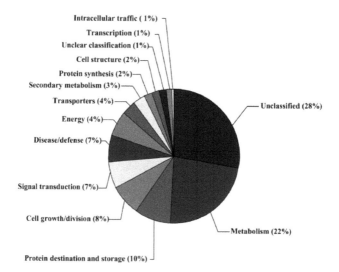

Legumes are able to access atmospheric di-nitrogen (N_2) through a symbiotic relationship with rhizobia that reside within root nodules. In soybean, following N_2 fixation by the bacteroids, ammonia is finally reduced in uninfected cells to allantoin and allantoic acid (Smith and Atkins, 2012). These ureides present the primary long-distance transport forms of nitrogen (N), and are exported from nodules via the xylem for shoot N supply. Transport of allantoin and allantoic acid out of nodules requires the function of ureide permeases (UPS1) located in cells adjacent to the vasculature (Collier and Tegeder, 2012; Pelissiel et al., 2004). It was expressed a common bean *UPS1* transporter in cortex and endodermis cells of soybean nodules and found that delivery of N from nodules to shoot, as well as seed set, was significantly increased. In addition, the number of transgenic nodules was increased and symbiotic N_2 fixation per nodule was elevated, indicating that transporter function in nodule N export is a limiting step in bacterial N acquisition. Furthermore, the transgenic nodules showed considerable increases in nodule N assimilation, ureide synthesis, and metabolite levels. This suggests complex adjustments of nodule N metabolism and partitioning processes in support of symbiotic N_2 fixation. It was proposed that the transgenic *UPS1* plants display metabolic and allocation plasticity to overcome N_2 fixation and seed yield limitations. Overall, it is demonstrated that transporter function in N export from nodules is a key step for enhancing atmospheric N_2 fixation and nodule function and for improving shoot N nutrition and seed development in legumes (Carter and Tegeder, 2016) (Figs. 9.17 and 9.18 and 9.19 and 9.20).

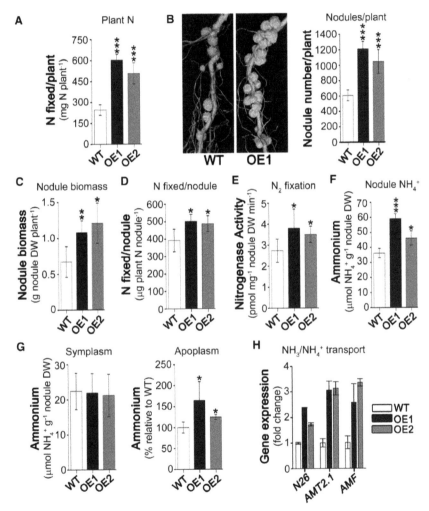

Fig. 9.18 (A) N fixed per plant based on the total amount of elemental N in shoot, roots, and nodules of *UPS1* overexpressors (OE1 and OE2) and WT plants (n = 7 plants). (B) Left: Nodulated roots of *UPS1*-OE1 and WT plants. Right: Nodule number (n = 6 plants). (C) Nodule biomass (n = 6 plants). DW, dry weight. (D) N fixed per nodule calculated from the total plant N content and nodule number (n = 7 plants). (E) Analysis of nodule nitrogenase activity using an acetylene reduction assay (n = 7 plants; for each plant, three pools of ten nodules were measured). (F and G) Analysis of NH_3/NH_4^+ levels in (F) whole nodules and (G) nodule symplasm and apoplasm (n = 4 pools of nodules; pools are from two plants each). (H) Expression analysis of genes involved in NH_3/NH_4^+ transport using qPCR. Results are shown as fold change compared to WT, and are presented as the mean of three technical repetitions. Results presented here are from one growth set but are representative of a minimum of two independently grown sets of plants. The data are mean ± SD. *** $p < 0.001$, ** $p < 0.01$, * $p < 0.05$.

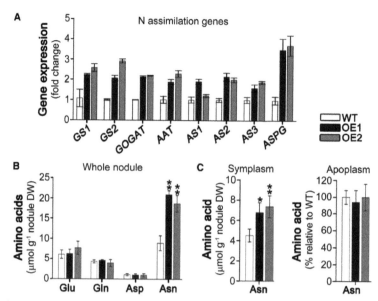

Fig. 9.19 (A) Expression of genes involved in N assimilation in *UPS1* overexpressors (OE1 and OE2) using qPCR. Results are shown as fold change compared to WT, and are presented as the mean of three technical repetitions. For gene and primer information. (B and C) Amino acid levels in (B) whole nodules and (C) nodule symplasm and apoplasm (n = 4 pools of nodules; pools are from two plants each). Results presented here are from one growth set but are representative of a minimum of two independently grown sets of plants. The data are mean ± SD. ** $p < 0.01$, * $p < 0.05$.

Ruffel et al. (2008) reported the transcriptome analysis of split nodulated roots and shoot comparing various N sources in *Medicago trancatula*. The effects of nitrate on gene expression in the nodules of *Medicago truncatula* were recently published in a study that used RNA-sequencing transcriptome analysis (Cabeza et al., 2014). After 4- and 28-h nitrate treatments, the expression of 127 genes for the nodule-specific cysteine-rich peptides and leghemoglobin, as well as the genes related to iron allocation and mitochondrial ATP synthesis, were all downregulated, while the expression of genes related to nitrate transport and nitrate reduction were upregulated. In that research work, however, gene expression in the roots was not analyzed, and the mechanism of nitrogen inhibition may not be the same as for *Glycine max*, as it could differ among legume species and determinate (soybean) and indeterminate (Medicago trancatula) types of nodules. The present study is the first report of transcriptome and metabolome analyses that compare the roots and nodules in relation to the nitrate inhibition of nitrogen-fixing soybean nodules.

The number of legume nodules is systemically regulated by the previous rhizobial infections in order to avoid excessive nodulation, and this phenomenon is called the autoregulation of nodulation (AON) (Caetano-Anolles and Gresshoff, 2011; Reid et al., 2011; Soyano and Kawaguchi, 2014). Several lines of hypernodulation or supernodulation mutants that lacked AON have been isolated and exhibited profuse nodulation compared with parent lines (Delves et al., 1986; Gremaud and Harper, 1989; Akao and Kouchi, 1992; Ohyama et al., 1993). A reciprocal shoot and root grafting experiment, by using a hypernodulation mutant line and the wild-type, revealed that AON was controlled by the shoot genotype, and not the roots (Akao and Kouchi, 1992). It is interesting that all hypernodulation mutant lines so far showed partial tolerance to nitrate for their nodule formation, thus, a common mechanism may be shared between the nitrate inhibition of nodulation and AON (Delves et al., 1986; Gremaud and Harper, 1989; Akao and Kouchi, 1992; Ohyama et al., 1993). Genes responsible for AON, such as LjHAR1, have been identified in the model legume *Lotus japonicus* and GmNARK in the soybean have been identified (reviewed by Soyano et al., 2014). Small peptides, named CLAVATA3/EMBRYO SURROUNDING REGION-RELATED (CLE) peptides, may be candidates of the infection signal from the roots to the shoot, based on similarities between the CLAVATA1 and CLE peptide signaling in *Arabidopsis thaliana*,

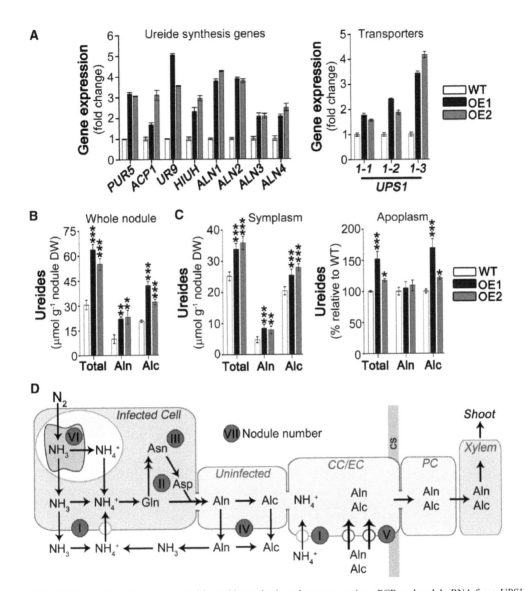

Fig. 9.20 (A) Expression of genes involved in ureide synthesis and transport using qPCR and nodule RNA from *UPS1* overexpressors (OE1 and OE2) and WT plants. Results are shown as fold change compared to WT, and are presented as the mean of three technical repetitions. For gene and primer information. (B and C) Total ureides, allantoin, and allantoic acid levels of (B) whole nodules and (C) nodule symplasm and apoplasm (n = 4 pools of nodules; pools are from two plants each). Results presented here are from one growth set but are representative of a minimum of two independently grown sets of plants. The data are mean ± SD. *** p < 0.001, ** p < 0.01, * p < 0.05. (D) Model of key regulatory steps in *UPS1*-OE soybean nodules suggesting metabolic and allocation plasticity in support of increased N_2 fixation and nodule-to-shoot N supply. Seven (I–VII) regulatory events downstream of bacteroid function are proposed and discussed. (I) Ammonium levels within the host cells are kept at steady state to prevent potential NH_4^+ toxicity and N_2 fixation inhibition. (II) Glutamine and aspartate levels are kept constant to avoid downregulation of N_2 fixation and N assimilation. (III) Excess organic N is transiently stored as asparagine. (IV) Symplasmic and apoplasmic levels of allantoin and allantoic acid are adjusted to facilitate increased ureide export from nodules and to potentially prevent downregulation of N_2 fixation. (V) Movement of ureides from the apoplasm of uninfected cells toward the vascular bundles is controlled by UPS1 plasma membrane transporters located in cortex and endodermis cells. (VI and VII) The increased N demand of *UPS1*-OE shoots is met by both (VI) elevated N_2 fixation per nodule and (VII) an increase in nodule number. Arrows with circles indicate transporters. CC, cortex cell; EC, endodermis cell; PC, parenchyma cell; Asn, asparagine; Asp, aspartate.

which controls shoot meristem development (Reid et al., 2011; Soyano and Kawaguchi, 2014) in the form of arabinosylated glycopeptides (Okamoto et al. 2013). Collectively, these findings suggest that an inhibitor is synthesized in the shoot when NARK proteins receive the CLE peptides; the shoot-derived unknown inhibitor is then transported from the shoot to the roots, where it prevents the differentiation and development of the nodule primordia. However, there is no direct evidence to link the nitrate inhibition and AON.

A microarray-based transcriptome analysis and the metabolome analysis were carried out for the roots and nodules of hydroponically grown soybean plants treated with 5 mM of nitrate for 24 h and compared with control without nitrate. Gene expression ratios of nitrate vs. the control were highly enhanced for those probesets related to nitrate transport and assimilation and carbon metabolism in the roots, but much less so in the nodules, except for the nitrate transport and asparagine synthetase. From the metabolome analysis, the concentration ratios of metabolites for the nitrate treatment vs. the control indicated that most of the amino acids, phosphorous-compounds and organic acids in roots were increased about twofold in the roots, whereas, in the nodules, most of the concentrations of the amino acids, P-compounds and organic acids were reduced, while asparagine increased exceptionally. These results may support the hypothesis that nitrate primarily promotes nitrogen and carbon metabolism in the roots, but mainly represses this metabolism in the nodules (Ishikawa et al., 2018) (Fig. 9.21).

(A) Upregulated over x4

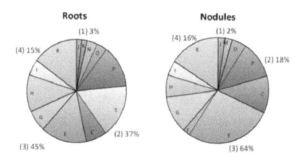

(B) Downregulated over x4

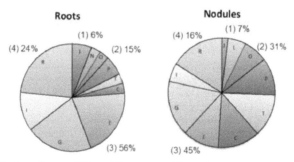

Fig. 9.21 Functional distribution of genes differentially expressed in soybean roots and nodules by the 5-mM nitrate treatment. (A) Upregulated genes > fourfold in the roots and nodules of soybean plants after 24 h of the 5 mM of the nitrate treatment compared with control. (B) Downregulated genes > fourfold in the roots and nodules of soybean plants after 24 h of the 5 mM of the nitrate treatment compared with control. Functional categories: (1) Information storage and processing, J: translation, ribosomal structure and biogenesis, K: transcription, L: DNA replication, recombination, and repair. (2) Cellular processes, D: cell division and chromosome partitioning, M: cell envelope biogenesis, outer membrane, N: cell motility and secretion, O: post-translational modification, protein turnover, chaperones, P: inorganic ion transport and metabolism, T: signal transduction mechanisms. (3) Metabolism, C: energy production and conversion, E: amino acid transport and metabolism, F: nucleotide transport and metabolism, G: carbohydrate transport and metabolism, H: coenzyme transport and metabolism, I: lipid metabolism. (4) Poorly characterized proteins, R: general function prediction only.

References

Agrawal, G.K. and J.J. Thelen. 2006. Large scale identification and quantification profiling of phosphoproteins expressed during seed filling in oilseed rape. Mol. Cell Proteomics 5: 2044–2059.

Akao, S. and H. Kouchi. 1992. A supernodulating mutant isolated from soybean cultivar Enrei. Soil Sci. Plant Nutr. 38: 183–187.

Arrese-Igor, C., E.M. Gonzalez, A.J. Gordon, F.R. Minchin, L. Galvez, M. Royuela, P.M. Cabrerizo and P.M. Aparicio-Tejo. 1999. Sucrose synthase and nodule nitrogen fixation under drought and other environmental stresses. Symbiosis 27: 189–212.

Atwell, B.J. 1992. Nitrate and ammonium as nitrogen sources for lupins prior to nodulation. Plant Soil 139: 247–251.

Baroniya, S.S., S. Kataria, G.P. Pandey and K.N. Guruprasad. 2014. Growth, photosynthesis and nitrogen metabolism in soybean varieties after exclusion of the UV-B and UV-A/B components of solar radiation. The Crop Journal 2: 388–397.

Becana, M. and J.I. Sprent. 1987. Nitrogen fixation and nitrate reduction in the root nodules of legumes. Physiol. Plant. 70: 757–765.

Benedito, V.A., H. Li, X. Dai, M. Wandrey, J. He, R. Kaundal, I. Torres-Jerez, S.K. Gomez, M.J. Harrison, Y. Tang, P.X. Zhao and M.K. Udvardi. 2010. Genomic inventory and transcriptional analysis of *Medicago truncatula* transporters. Plant Physiol. 152: 1716–1730.

Bergersen, F.J. 1965. Ammonia—an early stable product of nitrogen fixation by soybean nodules. Aust. J. Biol. Sci. 18: 1–9.

Bergersen, F.J. 1999. Delivery of N_2 to bacteroids in simulated soybean nodule cells: Consideration of gradients of concentration of dissolved N_2 in cell walls, cytoplasm, and symbiosomes. Protoplasma 206: 137–142.

Bottomley, P. 1991. Ecology of *Rhizobium* and *Bradyrhizobium*. pp. 292–347. *In*: Stacey, G., R.H. Burris and H.J. Evans (eds.). Biological Nitrogen Fixation. New York, NY: Chapman & Hall.

Brockwell, J., A. Pilka and R.A. Holliday. 1991. Soil pH is a major determinant of the numbers of naturally-occurring *Rhizobium meliloti* in non-cultivated soils of New South Wales. Aust. J. Exp. Agric. 31: 211–219.

Brown, S.M., K.J. Oparka, J.I. Sprent and K.B. Walsh. 1995. Symplastic transport in soybean root nodules. Soil Biol. Biochem. 27: 387–399.

Caetano-Anolles, G. and P.M. Gresshoff. 1991. Plant genetic control of nodulation. Ann. Rev. Microbiol. 45: 345–382.

Cafaro La Menza, N., J.P. Monzon, J.E. Specht and P. Grassini. 2017. Is soybean yield limited by nitrogen supply? F. Crop. Res. 213: 204–212.

Campbell, W.H. 1999. Nitrate reductase structure, function and regulation: Bridging the gap between biochemistry and physiology. Annu. Rev. Plant Physiol. Plant Mol. Biol. 50: 277–303.

Carter, A.M. and M. Tegeder. 2016. Increasing nitrogen fixation and seed development in soybean requires complex adjustments of nodule nitrogen metabolism and partitioning processes. Current Biology 26: 2044–2051.

Colebatch, G., G. Desbrosses, T. Ott, L. Krusell, O. Montanari, S. Kloska, J. Kopka and M.K. Udvardi. 2004. Global changes in transcription orchestrate metabolic differentiation during symbiotic nitrogen fixation in *Lotus japonicus*. Plant J. 39: 487–512.

Collier, R. and M. Tegeder. 2012. Soybean ureide transporters play a critical role in nodule development, function and nitrogen export. Plant J. 72: 355–367.

Collino, D.J., F. Salvagiotti, A. Perticari, C. Piccinetti, G. Ovando, S. Urquiaga and R.W. Racca. 2015. Biological nitrogen fixation in soy-bean in Argentina: Relationships with crop, soil, and meteorological factors. Plant Soil 392: 239–252.

Craig, J., P. Barratt, H. Tatge, A. Déjardin, L. Handley, C.D. Gardner, L. Barber, T. Wang, C. Hedley, C. Martin and A.M. Smith. 1999. Mutations at the rug4locus alter the carbon and nitrogen metabolism of pea plants through an effect on sucrose synthase. Plant J. 17: 353–362.

Day, D.A., B.N. Kaiser, R. Thomson, M.K. Udvardi, S. Moreau and A. Puppo. 2001. Nutrient transport across symbiotic membranes from legume nodules. Aust. J. Plant Physiol. 28: 669–676.

Delhon, P., A. Gojon, P. Tillard and L. Passama. 1995a. Diurnal regulation of NO_3^- uptake in soybean plants. I. Changes in NO_3^- influx, efflux, and N utilization in the plant during the day/night cycle. J. Exp. Bot. 46: 1585–1594.

Delhon, P., A. Gojon, P. Tillard and L. Passama. 1995b. Diurnal changes of NO_3^- uptake in soybean plants II. Relationship with accumulation of NO_3^- and asparagine in the roots. J. Exp. Bot. 46: 1595–1602.

Delves, A.C., A. Mathews, D.A. Day, A.S. Carter, B.J. Carroll and P.M. Gresshoff. 1986. Regulation of the soybean-Rhizobium nodule symbiosis by shoot and root factors. Plant Physiol. 82: 588–590.

El Msehli, S., A. Lambert, F. Baldacci-Cresp, J. Hopkins, E. Boncompagni, S.A. Smiti, D. Hérouart and P. Frendo. 2011. Crucial role of (homo)glutathione in nitrogen fixation in *Medicago truncatula* nodules. New Phytol. 192: 496–506.

Finan, T.M., I. Oresnik and A. Bottacin. 1988. Mutants of Rhizobium meliloti defective in succinate metabolism. J. Bacteriol. 170: 3396–3403.

Finan, T.M., E. McWhinne, B. Driscoll and R.J. Watson. 1991. Complex symbiotic phenotypes result from gluconeogenic mutations in Rhizobium meliloti. Mol. Plant Microbe Interact 4: 386–392.

FNCA Biofertilizer manual. 2006. ISBN 4-88911-301-0 c0550, http://www.fnca.mext.go.jp/english/index.html.

Fotelli, M.N., D. Tsikou, A. Kolliopoulou, G. Aivalakis, P. Katinakis, M.K. Udvardi, H. Rennenberg and E. Flemetakis. 2011. Nodulation enhances dark CO_2 fixation and recycling in the model legume *Lotus japonicus*. J. Exp. Bot. 62: 2959–2971.

Fujikake, H., A. Yamazaki, N. Ohtake, K. Sueyoshi, S. Matsuhashi, T. Ito, C. Mizuniwa, T. Kume, S. Hashimoto, N.S. Ishioka, S. Watanabe, A. Osa, T. Sekine, H. Uchida, A. Tsuji and T. Ohyama. 2003. Quick and reversible inhibition of soybean root nodule growth by nitrate involves a decrease in sucrose supply to nodules. Journal of Experimental Botany 54: 1379–1388.

Gaude, N., H. Tippmann, E. Flemetakis, P. Katinakis, M. Udvardi and P. Dörmann. 2004. The galactolipid digalactosyldiacylglycerol accumulates in the peribacteroid membrane of nitrogen-fixing nodules of soybean and Lotus. J. Biol. Chem. 279: 34624–34630.

Gremaud, M.F. and J.E. Harper. 1989. Selection and initial characterization of partially nitrate tolerant nodulation mutants of soybean. Plant Physiol. 89: 169–173.

Guruprasad, K.N., S. Bhattacharjee, S. Kataria, S. Yadav, A. Tiwari, S.S. Baroniya, A. Rajiv and P. Mohanty. 2007. Growth enhancement of soybean (*Glycine max*) upon exclusion of UV-B and UV-A components of solar radiation characterization of photosynthetic parameters in leaves. Photosynth. Res. 94: 299–306.

Hill, J.E. and R.W. Breidenbach. 1974. Proteins of soybean seeds. II. Accumulation of the major protein components during seed development and maturation. Plant Physiol. 53: 747–751.

Horst, I., T. Welham, S. Kelly, T. Kaneko, S. Sato, S. Tabata, M. Parniske and T.L. Wang. 2007. TILLING mutants of Lotus japonicas reveal that nitrogen assimilation and fixation can occur in the absence of nodule-enhanced sucrose synthase. Plant Physiol. 144: 806–820.

Hoshi, S. 1982. Nitrogen fixation, growth and yield of soybean. pp. 5–33. *In*: Nitrogen Fixation in Root Nodules, Japanese Society of Soil Science and Plant Nutrition (ed.). Hakuyusha Publishers, Japan.

Hungria, M., J.C. Franchini, R.J. Campo and P.H. Graham. 2005. The importance of nitrogen fixation to soybean cropping in South America. pp. 25–42. *In*: Werner, D. and W.E. Newton (eds.). Nitrogen Fixation in Agriculture, Forestry, Ecology, and the Environment. Dordrecht: Springer.

Hungria, M., R.J. Campo, I.C. Mendes and P.H. Graham. 2006a. Contribution of biological nitrogen fixation to the N nutrition of grain crops in the tropics: The success of soybean [*Glycine max* L. Merr.] in South America. pp. 43–93. *In*: Singh, R.P., N. Shankar and P.K. Jaiwal (eds.). Nitrogen Nutrition and Sustainable Plant Productivity. Houston, 1 Texas: Studium Press, LLC.

Hungria, M., J.C. Franchini, R.J. Campo, C.C. Crispino, J.Z. Moraes and R.N.R. Sibaldelli et al. 2006b. Nitrogen nutrition of soybean in Brazil: Contributions of biological N_2 fixation and of N-fertilizer to grain yield. Can. J. Plant Sci. 86: 927–939.

Hwang, J.H., S.R. Ellingson and D.M. Roberts. 2010. Ammonia permeability of the soybean nodulin 26 channel. FEBS Lett. 584: 4339–4343.

Ishikawa, S., Y. Ono, N. Ohtake, K. Sueyoshi, S. Tanabata and T. Ohyama. 2018. Transcriptome and metabolome analyses reveal that nitrate strongly promotes nitrogen and carbon metabolism in soybean roots, but tends to repress it in nodules. Plants (Basel) 7(2): 32.

Johnson, D.C., D.R. Dean, A.D. Smith and M.K. Johnson. 2005. Structure, function, and formation of biological iron-sulfur clusters. Annu. Rev. Biochem. 74: 247–281.

Kalloniati, C., P. Krompas, G. Karalias et al. 2015. Nitrogen-fixing nodules are an important source of reduced sulfur, which triggers global changes in sulfur metabolism in Lotus japonicus. Plant Cell. 27(9): 2384–2400.

Kaschuk, G., T.W. Kuyper, P.A. Leffelaar, M. Hungria and K.E. Giller. 2009. Are the rates of photosynthesis stimulated by the carbon sink strength of rhizobial and arbuscular mycorrhizal symbioses? Soil Biol. Biochem. 41: 1233–1244.

Kessel, C.V. and C. Hartley. 2000. Agricultural management of grain legumes; has it led to an increase in nitrogen fixation? Field Crop. Res. 65: 165–181.

Keyser, H.H., P. Somasegaran and B.B. Bohlool. 1993. Rhizobial ecology and technology. pp. 205–226. *In*: Blaine Metting, F. (ed.). Soil Microbial Ecology: Applications in Agricultural and Environmental Management. New York, N.Y: Marcel Dekker, Inc.

Khan, D.F., M.B. Peoples, P.M. Chalk and D.F. Herridge. 2002. Quantifying below-ground nitrogen of legumes 2. A comparison of ^{15}N and non-isotopic methods. Plant Soil 239: 277–289.

Kouchi, H. and T. Yoneyama. 1984. Dynamics of carbon photosynthetically assimilated in nodulated soy bean plants under steady-state conditions 2. The incorporation of 13C into carbohydrates, organic acids, amino acids and some storage compounds. Ann. Bot. 53: 883–896.

Kouchi, H., K. Shimomura, S. Hata, A. Hirota, G.-J. Wu, H. Kumagai, S. Tajima, N. Suganuma, A. Suzuki, T. Aoki, M. Hayashi, T. Yokoyama, T. Ohyama, E. Asamizu, C. Kuwata, D. Shibata and S. Tabata. 2004. Large-scale analysis of gene expression profiles during early stages of root nodule formation in a model legume, Lotus japonicus. DNA Research 11: 263–274.

Krusell, L. et al. 2005. The sulfate transporter SST1 is crucial for symbiotic nitrogen fixation in *Lotus japonicus* root nodules. Plant Cell 17: 1625–1636.

Lambers, H. 1983. 'The functional equilibrium', nibbling on the edges of a paradigm. Netherl. J. Agric. Sci. 31: 305–311.

Layzell, D.B. 1990. N$_2$ fixation, NO$_3^-$ reduction and NH$_4^+$ assimilation. *In*: Dennis, D.T. and D.H. Turpin (eds.). Plant Physiology, Biochemistry and Molecular Biology. Longman Scientific Technology (Essex, UK).

Limpens, E., S. Moling, G. Hooiveld, P.A. Pereira, T. Bisseling, J.D. Becker and H. Küster. 2013. Cell- and tissue-specific transcriptome analyses of *Medicago truncatula* root nodules. PLoS One 8: e64377.

Marsolier, M.C., G. Debrosses and B. Hirel. 1995. Identification of several soybean cytosolic glutamine synthetase transcripts highly or specifically expressed in nodules: Expression studies using one of the corresponding genes in transgenic *Lotus corniculatus*. Plant Mol. Biol. 27: 1–15.

Medic, J., C. Atkinson and C.R. Hurburgh. 2014. Current knowledge in soybean composition. JAOCS, J. Am. Oil Chem. Soc. 91: 363–384.

Mendes, I.C., M. Hungria and M.A.T. Vargas. 2003. Soybean response to starter nitrogen and Bradyrhizobium inoculation on a cerrado oxisol under no-tillage and conventional tillage systems. Rev. Bras. Cienc. Solo 27: 81–87.

Menza, N.C.La., J.P. Monzon, J.E. Specht and P. Grassini. 2017. Is soybean yield limited by nitrogen supply? Field Crops Research 213: 204–212.

Mienke, D.W., J. Chen and R.N. Beachy. 1981. Expression of storage-protein genes during soybean seed development. Planta 153: 130–139.

Mizukoshi, K., T. Nishiwaki, N. Ohtake, R. Minagawa, T. Ikarashi and T. Ohyama. 1995. Nitrate transport pathway into soybean nodules traced by tungstate and ^{15}NO$_3^-$. Soil Sci. Plant Nutr. 41: 75–88.

Morey, K.J., J.L. Ortega and C. Sengupta-Gopalan. 2002. Cytosolic glutamine synthetase in soybean is encoded by a multigene family, and the members are regulated in an organ-specific and developmental manner. Plant Physiol. 128(1): 182–193.

Mourtzinis, S. et al. 2018. Soybean response to nitrogen application across the United States: A synthesis-analysis. F. Crop. Res. 215: 74–82.

Niemietz, C.M. and S.D. Tyerman. 2000. Channel-mediated permeation of ammonia gas through the peribacteroid membrane of soybean nodules. FEBS Lett. 465: 110–114.

Norton, G. and J.F. Harris. 1975. Compositional changes in developing rape seed (*Brassica napus* L.). Planta. 123: 163–174.

Ohlrogge, J.B. and T.M. Kuo. 1984. Control of lipid synthesis during soybean seed development: Enzymic and immunochemical assay of acyl carrier protein. Plant Physiol. 74: 622–625.

Ohtake, N., M. Suzuki, Y. Takahashi, T. Fujiwara, M. Chino, T. Ikarashi and T. Ohyama. 1996. Differential expression of β-conglycinin genes in nodulated and non-nodulated isolines of soybean. Physiology Plant. 96: 101–110.

Ohtake, N., Y. Ikarashi, T. Ikarashi and T. Ohyama. 1997. Distribution of mineral elements and cell morphology in nodulated and non-nodulated soybean seeds. Bulletin of the Faculty of Agriculture, Niigata University. 49: 93–101.

Ohyama, T. and K. Kumazawa. 1978. Incorporation of ^{15}N into various nitrogenous compounds in intact soybean nodules after exposure to ^{15}N$_2$ gas. Soil Sci. Plant Nutr. 24: 525–533.

Ohyama, T. and K. Kumazawa. 1979. Assimilation and transport of nitrogenous compounds originated from ^{15}N$_2$ fixation and NO$_3^-$ absorption. Soil Science and Plant Nutrition 25: 9–19.

Ohyama, T. and K. Kumazawa. 1979b. Assimilation and transport of nitrogenous compounds originating from ^{15}N$_2$ fixation and ^{15}NO$_3^-$ absorption. Stable Isotopes, Proceedings of the Third International Conference 327–335 (Academic Press).

Ohyama, T. and K. Kumazawa. 1980. Nitrogen assimilation in soybean nodules I. The role of GS/GOGAT system in the assimilation of ammonia produced by N$_2$ fixation. Soil Science and Plant Nutrition 26: 109–115.

Ohyama, T. and K. Kumazawa. 1980. Nitrogen assimilation in soybean nodules II. ^{15}N$_2$ assimilation in bacteroid and cytosol fractions of soybean nodules. Soil Science and Plant Nutrition 26: 205–213.

Ohyama, T. and K. Kumazawa. 1981c. A simple method for the preparation, purification and storage of ^{15}N$_2$ gas for biological nitrogen fixation studies. Soil Sci. Plant Nutr. 27: 263–265.

Ohyama, T. 1982. Emission spectrometric ISN analysis of amino acids. Radioisotopes 31: 212–221 (in Japanese).

Ohyama, T. 1983. Comparative studies on the distribution of nitrogen in soybean plants supplied with N$_2$ and NO$_3^-$ at the pod filling stage. Soil Sci. Plant Nutr. 29: 133–145.

Ohyama, T. and K. Kumazawa. 1987. Characteristics of nitrate respiration of isolated soybean bacteroids. Soil Sci. Plant Nutr. 33: 69–78.

Ohyama, T., K. Saito and N. Kato. 1989a. Assimilation and transport of nitrate, nitrite, and ammonia absorbed by nodulated soybean plants. Soil Sci. Plant Nutr. 35: 9–20.

Ohyama, T., K. Saito and N. Kato. 1989b. Diurnal rhythm in nitrate absorption by roots of soybeans (*Glycine max*). Soil Sci. Plant Nutr. 35: 33–42.

Ohyama, T., N. Kato and K. Saito. 1989c. Nitrogen transport in xylem of soybean plant supplied with ^{15}NO$_3^-$. Soil Sci. Plant Nutr. 35: 131–137.

Ohyama, T., J.C. Nicholas and J.E. Harper. 1993. Assimilation of ^{15}N$_2$ and ^{15}NO$_3^-$ by partially nitrate-tolerant nodulation mutants of soybean. J. Exp. Bot. 44: 1739–1747.

Ohyama, T., K. Tewari, S.A. Latif, S. Ruamrungsri, S. Komiyama, S. Ito, A. Yamazaki, K. Sueyoshi and N. Ohtake. 2004. Direct analysis of ^{15}N abundance of Kjeldahl digested solution by emission spectrometry. Bull. Facul. Agric. Niigata Univ. 57: 43–50.

Ohyama, T., N. Ohtake, K. Sueyoshi, Y. Ono, K. Tsutsum, M. Ueno, S. Tanabata, T. Sato and Y. Takahashi. 2017. Amino acid metabolism and transport in soybean plants. pp. 171–196. *In*: Amino Acid—New Insights and Roles in Plant and Animal. In Tech Inc.

Okamoto, S., H. Shinohara, T. Mori, Y. Matsubayashi and M. Kawaguchi. 2013. Root-derived CLE glycopeptides control nodulation by direct binding to HAR1 receptor kinase. Nat. Commun.

Pélissier, H.C., A. Frerich, M. Desimone, K. Schumacher and M. Tegeder. 2004. PvUPS1, an allantoin transporter in nodulated roots of French bean. Plant Physiol. 134: 664–675.

Peoples, M.B., J.K. Ladha and D.F. Herridge. 1995. Enhancing legume N$_2$ fixation through plant and soil management. Plant Soil. 174: 83–101.

Peoples, M.B. and D.F. Herridge. 2000. Quantification of biological nitrogen fixation in agricultural systems. pp. 519–524. *In*: Pedrosa, F.O., M. Hungria, M.G. Yates and W.E. Newton (eds.). Nitrogen Fixation: From Molecules to Crop Productivity. Dordrecht: Kluwer Academic Press.

Rees, D.C. and J.B. Howard. 2000. Nitrogenase: Standing at the crossroads. Curr. Opin. Chem. Biol. 4: 559–566.

Reid, D.E., B.J. Ferguson, S. Hayashi, Y.H. Lin and P.M. Gresshoff. 2011. Molecular mechanisms controlling legume autoregulation of nodulation. Ann. Bot. 108: 789–795.

Roche, D., S.J. Temple and C. Sengupta-Gopalan. 1993. Two classes of differentially regulated glutamine synthetase genes are expressed in the soybean nodule: A nodule specific class and a constitutively expressed class. Plant Mol. Biol. 22: 971–983.

Rotundo, J.L., S.L. Naeve and J.E. Miller-Garvin. 2016. Regional and temporal variation in soybean seed protein and oil across the United States. Crop Sci. 56: 797.

Rubel, A., R.W. Rinne and D.T. Canvin. 1972. Protein, oil, and fatty-acid in developing soybean seeds. Crop Sci. 12: 739–741.

Ruffel, S., S. Freixes, S. Balzergue, P. Tillard, C. Jeudy, M.L. Martin-Magniette, M.J. van der Merwe, K. Kakar, J. Gouzy, A.R. Fernie et al. 2008. Systemic signaling of the plant nitrogen status triggers specific transcriptome responses depending on the nitrogen source in *Medicago truncatula*. Plant Physiol. 146: 2020–2035.

Salon, C. et al. 2001. Grain legume seed filling in relation to nitrogen acquisition: A review and prospects with particular reference to pea. Agronomie 21: 539–552.

Salvagiotti, F., K.G. Cassman, J.E. Specht, D.T. Walters, A. Weiss and A. Dobermann. 2008. Nitrogen uptake, fixation and response to fertilizer N in soybeans: A review. F. Crop. Res. 108: 1–13.

Salvagiotti, F., J.E. Specht, K.G. Cassman, D.T. Walters, A. Weiss and A. Dobermann. 2009. Growth and nitrogen fixation in high-yielding soybean: Impact of nitrogen fertilization. Agron. J. 101: 958–970.

Sato, T., N. Ohtake, T. Ohyama, N.S. Ishioka, S. Watanabe, A. Osa, T. Sekine, H. Uchida, A. Tsuji, S. Matsushashi, T. Ito and T. Kume. 1999. Analysis of nitrate absorption and transport in non-nodulated and nodulated soybean plants with $^{13}NO_3^-$ and $^{15}NO_3^-$. Radioisotopes 48: 450–458.

Sato, S., Y. Nakamura, E. Asamizu, S. Isobe and S. Tabata. 2007. Genome sequencing and genome resources in model legumes. Plant Physiol. 144: 588–593.

Sharifi, M. and B.J. Zebarth. 2006. Nitrate influx kinetic parameters of five potato cultivars during vegetative growth. Plant Soil 288: 91–99.

Shen, W.-J., M.S. Williamson and B.G. Forde. 1992. Functional analysis of the promoter region of a nodule-enhanced glutamine synthetase gene from *Phaseolus vulgaris* L. Plant Mol Biol. 19: 837–846.

Smith, P.M.C. and C.A. Atkins. 2002. Purine biosynthesis. Big in cell division, even bigger in nitrogen assimilation. Plant Physiol. 128: 793–802.

Soyano, T. and M. Kawaguchi. 2014. Systemic regulation of root nodule formation. pp. 73–88. *In*: Ohyama, T. (ed.). Advances in Biology and Ecology of Nitrogen Fixation. InTech: Rijeka, Croatia.

Soyano, T., H. Hirakawa, S. Sato, M. Hayashi and M. Kawaguchi. 2014. Nodule Inception creates a long-distance negative feedback loop involved in homeostatic regulation of nodule organ production. Proc. Natl. Acad. Sci., USA 111: 14607–14612.

Stefan, H. and N. Christian. 2002. Biomass production and N$_2$ fixation of five Mucuna pruriens varieties and their effect on maize yield in the forest zone of Cameroon. J. Plant Nutr. Soil Sci. 165: 101–109.

Tamagno, S. and I.A. Ciampitti. 2017. Seed yield and biological N fixation for historical soybean genotypes. Kansas Field Research Report.

Tamagno, S., G.R. Balboa, Y. Assefa, P. Kovács, S.N. Casteel, F. Salva-giotti, F.O. García, W.M. Stewart and I.A. Ciampitti. 2017. Nutrient partitioning and stoichiometry in soybean: A synthesis-analysis. F. Crop. Res. 200: 18–27.

Tempest, D.W., J.L. Meers and C.M. Brown. 1970. Synthesis of glutamate in Aerobacter aerogenes by a hitherto unknown route. Biochem. J. 117: 405–407.

Temple, S.J., J. Heard, G. Ganter, K. Dunn and C. Sengupta-Gopalan. 1995. Characterization of a nodule-enhanced glutamine synthetase from alfalfa: Nucleotide sequence, *in situ* localization and transcript analysis. Mol. Plant-Microbe Interact. 8: 218–227.

Temple, S.J., S. Kunjibettu, D. Roche and C. Sengupta-Gopalan. 1996. Total glutamine synthetase activity during soybean nodule development is controlled at the level of transcription and holoprotein turnover. Plant Physiol. 112: 1723–1733.

Thies, J.E., P.L. Woomer and P.W. Singleton. 1995. Enrichment of Bradyrhizobium spp. populations in soil due to cropping of the homologous host plant. Soil Biol. Biochem. 27: 633–636.

Tyerman, S.D., L.F. Whitehead and D.A. Day. 1995. A channel-like transporter for NH_4 on the symbiotic interface of N_2-fixing plants. Nature 378: 629–632.

Udvardi, M. and P.S. Poole. 2013. Transport and metabolism in legume-rhizobia symbioses. Annu. Rev. Plant Biol. 64: 781–805.

Udvardi, M.K., G.D. Price, P.M. Gresshoff and D.A. Day. 1988. A dicarboxylate transporter on the peribacteroid membrane of soybean nodules. FEBS Lett. 231: 36–40.

Vollmann, J., C. Fritz, H. Wagentristi and P. Ruckenbauer. 2000. Environmental and genetic variation of soybean seed protein content under Central European growing conditions. Journal of the Science of Food and Agriculture 80: 1300–1306.

Weber, H., L. Borisjuk and U. Wobus. 2005. Molecular physiology of legume seed development. Annu. Rev. Plant Biol. 56: 253–279.

Wolk, C.P., J. Thomas and P.W. Shaffer. 1976. Pathways of nitrogen metabolism after fixation of ^{13}N-labelled nitrogen gas by cyanobacterium, *Anabaena cylindrica*. J. Biol. Chem. 256: 5027–5034.

Yurgel, S.N. and M.L. Kahn. 2004. Dicarboxylate transport by rhizobia. FEMS Microbiol. Rev. 28: 489–501.

Zhang, S., Y. Wang, K. Li, Y. Zou, L. Chen and X. Li. 2014. Identification of cold-responsive miRNAs and their target genes in nitrogen-fixing nodules of soybean. Int. J. Mol. Sci. 15: 13596–13614.

Zheng, L., R.H. White, V.L. Cash, R.F. Jack and D.R. Dean. 1993. Cysteine desulfurase activity indicates a role for NIFS in metallocluster biosynthesis. Proc. Natl. Acad. Sci., USA 90: 2754–2758.

Zotarelli, L. 2000. Balanço de nitrogênio na rotação de culturas em sistemas de plantio direto e convencional na região de Londrina, PR. Universidade Federal Rural do Rio de Janeiro, Seropédica, Brazil. 164 pp. [Ph.D. Thesis].

CHAPTER 10
Lipid Metabolism

Generally, oilseeds are composed of two parts: The kernel, which is main part, and the seed covering that encloses the kernel and is called the husk or tegument. The kernel is comprised of two parts: The embryo and the endosperm. Lipase activity is investigated during seed germination where it has the maximum value (Hellyer et al., 1999; Paques et al., 2006). Triacylglycerols is stored in oleosomes and comprise in range from 20 to 50% of dry weight. As germination proceeds, triacylglycerols are hydrolyzed to produce the energy required for the synthesis of sugars, amino acids (mainly asparagine, aspartate, glutamine and glutamate) and carbon chains required for embryonic growth (Quettier and Eastmond, 2009).

Lipid level and lipase activity were studied in various germinating seeds. It showed that β-oxidation takes place 4 days after germination of Castor bean seeds (Hutton and Stumpf, 1969). The major hydrolytic enzymes concerned with the lipid metabolism during germination are the lipases which catalyze the hydrolysis of ester carboxylate bonds and release fatty acids and organic alcohols (Pereira et al., 2003; Leal et al., 2002) and the reverse reaction (esterification) or even various transesterification reactions (Freire and Castilho, 2008). The ability of lipases to catalyze these reactions with great efficiency, stability and versatility makes these enzymes highly attractive from a commercial point of view.

The induction of lipase activity during germination might be dependent on factors from embryo (Tavener and Laidman, 1972). An early study of Shoshi and Reevers (1974) showed the presence of two lipases in the endosperm of Castor bean seed, acid lipase in dry seed and alkaline lipase during germination. On the other hand, storage tissues of all the oilseeds except Castor bean contained only lipase activity, which increased during germination (Anthony et al., 1978).

Because sucrose is the substrate for lipid biosynthesis in developing seed and the end product of lipid degradation, it might be primarily considered as a regulatory factor in studying the mechanisms of lipid metabolism (Quettier and Eastmond, 2009; Baud et al., 2008).

Storage lipid mobilization in germinating seeds begins with hydrolysis of triacylglycerols in oleosomes by lipases into free fatty acids and glycerol. Then, fatty acids undergo β-oxidation in peroxisomes. Next, glyoxylate cycle will proceed partially in the peroxisome and partially in the cytoplasm. Three of the five enzymes of the glyoxylate cycle (citrate synthase, isocitrate lyase and malate synthase) are located in peroxisomes, the two other enzymes (aconitase and malate dehydrogenase) operate in the cytoplasm (Pracharoenwattana and Smith, 2008). Succinate is transported from peroxisome to mitochondria and converted to malate via the Krebs cycle. Malate, in turn, after transport to the cytoplasm, is converted to oxaloacetate. Finally, gluconeogenesis and the synthesis of sugars proceed; these are the processes which are a form of carbon transport, especially in germinating seeds (Quettier and Eastmond, 2009; Borek and Ratajczak, 2010).

Glyoxylate cycle has been known to play a crucial role in lipid degradation in oilseeds, whereas stored lipid is converted into glucose, the main respiratory substrate during germination and, hence, seedling establishment (Eastmond et al., 2000). Seed imbibition triggers highly increase in oxygen consumption, reflecting the enhancement of oxidation of produced carbohydrates from the glyoxylate cycle (Muscolo et al., 2007). Alongside to glyoxylate cycle, the OPPP operates where a number of enzymes and intermediates participate the two pathways (ap Rees, 1980). It functions to provide the cell with NADPH for biosynthetic reactions and appears to be important in the regulation of germination (Perino and Come, 1991).

The action of the two glyoxylate cycle enzymes, isocitrate lyase (ICL) and malate synthase (MS), that bypass the decarboxylation steps of the TCA cycle are essential in oilseed germination. Two moles of acetyl-CoA are introduced with each turn of the cycle, resulting in the synthesis of one mole of the four-carbon compound succinate transported from the glyoxysome into the mitochondrion and converted into malate via TCA cycle. This malate is then exported to cytosol in exchange for succinate and is converted to oxalacetate. PEP-CK catalyzes the conversion of oxaloacetate to phosphoenolpyruvate, and this fuels the synthesis of soluble carbohydrates necessary for germination (Muscolo et al., 2007).

After germinating, oil-containing seeds metabolize stored triacylglycerols by converting lipids to sucrose. Plants are not able to transport fats from the endosperm to the root and shoot tissues of the germinating seedling, so they must convert stored lipids to a more mobile form of carbon, generally sucrose. This process involves several steps that are take place in different cellular compartments: Oleosomes, gly-oxysomes, mitochondria, and cytosol.

The conversion of lipids to sucrose in oilseeds is triggered by germination and begins with the hydrolysis of triacylglycerols stored in the oil bodies to free fatty acids, followed by oxidation of the fatty acids to produce acetyl-CoA (Fig. 10.1).

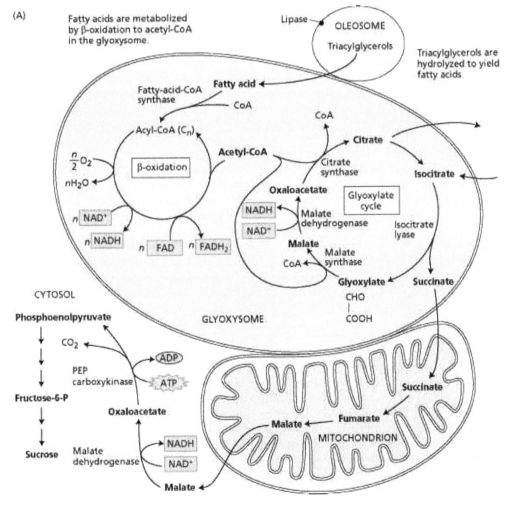

Fig. 10.1 Sucrose formation from lipid.

10.1 Seed Germination

In soybean seeds, lipids are stored in the form of triacyl-glycerol (TAG) in which fatty acids are esterified to glycero-lipids. Fatty acids vary in type and content according to genetic and environmental factors (Heppard et al., 1996) that affect their nutritional values and processing properties. Soybean lipids are composed of five fatty acid species: Palmitic acid (C16:0), stearic acid (C18:0) of saturated fatty acids, and oleic acid (C18:1), lin-oleic acid (C18:2) and linolenic acid (C18:3) of unsaturated fatty acids (Yadav, 1996). The lipids in soybean seeds typically contain 10% linolenic acid, which is a polyunsaturated fatty acid with one more double bond than linoleic acid. Linolenic acid is considered unstable and responsible for the development of off-flavors because it is more easily oxidized than other unsaturated fatty acids (Frankel et al., 1980). In the oil industry, the oxidative stability of edible oil is improved by hydrogenating soybean oil to reduce linolenic acid levels (Pantalone et al., 1997a); however, trans-isomer fatty acids are produced during hydrogenation, which are believed to increase the risk of heart disease. Consequently, a major breeding aim for soybean oil has been to reduce the amount of linolenic acid in soybean oil. On the other hand, some studies have shown that linolenic acid plays important roles in maintaining brain, nerve and retina functions (Okuyama, 1990; Chalon, 2006); hence, the physiological importance of linolenic acid has recently being noted.

Soybean contains about 20% oil in seeds and is, therefore, currently an important oil seed crop. Considering the fatty acid composition which constitutes the soybean lipid, the concentration of linolenic acid is less than 10%. On the other hand, wild soybean contains about twice the linolenic acid concentration and half as much lipid content as that of soybean. Based on these differences in the lipid content and linolenic acid concentration between soybean and wild soybean, a genetic study on the fatty acid concentration and lipid content using *G. max* × *G. soja* populations was carried out. These traits showed normal distribution and were highly heritable in F2 and RIL populations; moreover; a negative correlation was shown between these traits. In addition, a QTL for the lipid content and linolenic acid concentration was detected in the same position near SSR marker Satt384 on LG E, suggesting that the factors controlling the lipid content might be partly shared by those of the linolenic acid concentration. Further elucidation of the regulatory aspects of these traits will provide information that could lead to the improvement of fatty acid composition and lipid content of soybean.

One of the major constraints in soybean cultivation is the non-availability of high vigour seeds at the time of sowing. Soybean seeds undergo rapid loss of vigour and viability during storage, which is more pronounced in sub-tropical conditions. Seed longevity is greatly influenced by storage conditions, such as relative humidity and temperature; the lowering of these parameters significantly increases the storage life of seeds (Sauer et al., 1992). Deterioration of seed during storage is manifested as a reduction in percent germination while those seeds that do germinate, produce weak seedlings, which ultimately affect the growth and yield of crop plant (Tekrony et al., 1993). Contents of soluble carbohydrates generally decline with aging of seed (Petruzelli and Taranto, 1989; Sharma et al., 2005) and this decline might result in limited availability of respiratory substrates for germination. Depletion of disaccharides may lessen the protective effects of sugars on structural integrity of membrane (Crowe et al., 1984). The lipid related changes of seeds during storage revealed decline in phospholipids and polyunsaturated fatty acids leading to marked decline in seed vigour (Priestley and Leopold, 1983). An increase in necrosis in cotyledons and substantial reduction in total germination was reported in soybean seeds stored at high temperature by Falivene et al. (1980). Soybean seeds stored at controlled temperature (15–20°C) had a higher percentage germination than those stored at ambient temperature (Somchai, 1999; Sharma et al., 2006).

Loss in seedling vigour is reported to precede the loss of seed viability in a number of crops, including soybean (Dharamlingam and Basu, 1990). Similar results of decreased germination percentage and seedling vigour index with aging were reported by Singh and Dadlani (2003) and Gupta and Aneja (2004) in soybean. The decrease in % germination and vigour index can be attributed to DNA degradation with aging, which leads to impaired transcription causing incomplete, or faulty enzyme synthesis essential for earlier stages of germination.

The dry matter content of cotyledons from germinating soybean seeds stored under different conditions increased with the increase in storage period up to 180 days (Fig. 10.2) and increase in dry

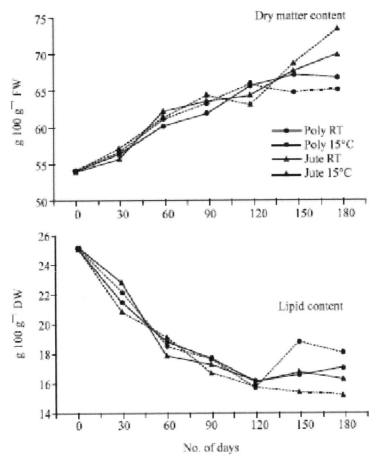

Fig. 10.2 Changes in dry matter and lipid content in cotyledons of germinating soybean seeds during storage. Each point represents the mean of three replications. CD (p > 0.05) for DOSx PMxT is 0.19 for lipid content. (DOS: days of storage, PM: packing material and T: temperature).

matter content in cotyledons of germinating soybean seeds stored in jute bags at 15°C or RT was more as compared to those stored in polythene bags beyond 120 DOS. The total lipid content decreased initially up to 120 DOS in all the treatments and slightly increased thereafter in seeds stored in polythene bags at RT or 15°C (Fig. 10.2).

The phospholipid content increased up to 90 DOS followed by a decrease with further increase in storage period up to 180 DOS, irrespective of the packing and temperature conditions (Table 10.1). The glycolipid content did not vary much in germinating seeds stored in polythene or jute bags with the storage period, however, the decrease in glycolipid content was more at 15°C in both the packings as compared to RT. On germination, the free fatty acid content in cotyledons of germinating soybean seeds increased with the increase in storage period from 30 to 180 days in all the treatments. The sterol content increased up to 90 DOS and then decreased with the increase in storage period irrespective of the packings and temperature conditions. The increase in sterol content during the initial 90 days of storage was notably more in the seeds stored at RT than those stored at 15°C in polythene bags.

Data on lipid composition has revealed that lipid degradation marginally decreased triglyceride content during storage and increased free fatty acids, sterol and phospholipid content suggesting that the lipase present in the seeds remains active and alters membrane integrity (Bernal Lugo and Leopold, 1992). The damage to membrane integrity has also been indicated by the decrease in sucrose content during initial period of storage.

Table 10.1 Changes in lipid composition in cotyledons of germinating soybean seeds during storage.

Days of storage (DOS)	Polythene bags		Jute bags	
	Room temperature	15°C	Room temperature	15°C
Phospholipids (g 100 g⁻¹ oil)				
30	0.9	0.8	0.8	0.9
60	1.1	1.1	1.2	1.1
90	1.3	1.3	1.3	1.3
120	1.1	0.9	0.9	1.0
150	0.8	0.8	0.6	0.6
180	0.6	0.6	0.6	0.5
Glycolipid content (g 100 g⁻¹ oil)				
30	1.5	1.4	1.2	1.3
60	1.3	1.2	1.4	1.1
90	1.2	1.0	1.1	0.9
120	1.5	1.5	1.7	1.2
150	1.4	1.3	1.3	1.1
180	1.2	0.9	1.1	0.8
Free fatty acids (g 100 g⁻¹ oil)				
30	1.1	1.4	1.2	1.4
60	1.0	1.1	1.0	1.1
90	1.4	1.4	1.3	1.3
120	1.8	1.8	1.7	1.7
150	2.2	1.9	1.9	2.0
180	2.1	1.9	1.9	2.6
Sterol (g 100 g⁻¹ oil)				
30	8.2	7.5	7.6	7.8
60	8.8	8.2	8.8	7.9
90	9.7	9.3	9.6	9.5
120	9.1	8.9	9.1	9.0
150	7.5	6.8	6.5	6.7
180	6.4	5.9	5.7	5.3
CD (p < 0.05)	**Phospholipids**	**Glycolipids**	**Free fatty acids**	**Sterols**
DOS X PM	0.05	0.08	0.09	0.19
DOS X T	NS	0.08	0.09	0.19
PM X T	0.03	0.05	0.05	0.11
DOS X PM X T	0.07	0.12	0.13	0.28

The values are procured on the 7th day of germination. PM-packing material; T-temperature; NS-non significant, Values at zero DOS: PL-0.71, GL-1.41, FFA-1.5 and Sterol-6.6 g 100 g⁻¹ oil.

The possible role of lipid peroxidation in seed deterioration was investigated during natural aging and accelerated aging of seeds of edible soybean (*Glycine max* [L.], Merr. cv. Kaohsiung Selection No.1). Natural aging was achieved by sealing the seeds in aluminum foil bags coated with polyethylene and storing the seeds at room temperature for 3 to 12 months. Accelerated aging was obtained by incubating the seeds at 45°C and close to 100% relative humidity for 3 to 12 days, after which the seeds were air dried to their original moisture level (8%). The results indicate that both natural and accelerated aging enhanced lipid peroxidation, as germination was depressed. Aging also inhibited the activity of superoxide dismutase, catalase, ascorbate peroxidase and peroxidase. The changes in germination and physiological activities, expressed as a function of aging duration, were somewhat similar in the two aging treatments.

Vegetative oil derivatives tend to deteriorate due to hydrolytic and oxidative reactions. These reactions especially hydrolysis of lipids result in the accumulation of free fatty acids (FFA) as well as mono- and diglycerides. Oxidation, auto-oxidation and lipoxygenase induced the presence of high FFA content which will cause later off-flavor in oil seeds. The source of primary catalysis which initiates lipid peroxidation are free radicals which contain one or more unpaired electrons (Halliwell, 2006), as active forms of oxygen, i.e., the superoxide radical (^2O), hydrogen peroxide (H_2O_2) and the hydroxyl radical (OH°) which can react with biological molecules. In plant cell, many experimental results indicated the production of reactive oxygen species where has been shown in peroxisome. The major by-product of the peroxisomal metabolism is H_2O_2. The respiratory of mitochondria generally generated through electron leakage to oxygen where under normal physiological conditions ca. 1–2% of the oxygen used by the mitochondria is transformed into H_2O_2 (Puntarulo et al., 1988). Moreover, ROS can be enhanced in response to a range of abnormal conditions, including exposure to biotic and abiotic stresses (Rhoads, 2006). The rate of ROS generation in plant cells can bring about extensive oxidative damage (Halliwell and Gutteridge, 2006). These active forms of oxygen may initiate many reactions on polyunsaturated fatty acids leading to the oxidative degradation of lipids or lipid peroxidation which causes various types of cellular damage (Bailey, 2004). Several comprehensive reviews have identified free radical-mediated lipid peroxidation, enzyme inactivation or protein degradation, disruption of cellular membranes and damage to genetic (nucleic acids) integrity as major causes of seed ageing (McDonald, 1999).

In soybean, the axes of such aged seeds contain high levels of malondialdehyde (MDA); a product of the peroxidation of unsaturated fatty acids. The levels of linoleic and linolenic acids in a polar lipid (phospholipid) fraction decrease during aging and more dramatically during post-aging deterioration (Robert et al., 1980). Moreover, a significant increase of MDA in the embryo axes and cotyledon of aged peanut seed was identified by Sung and Jeng (1994). Nevertheless, Walters (1998) found that the relationship between the increase of TAGs peroxidation and decline of viability during aging is rarely clear because decreased un-saturation and increased free fatty acids are usually detected after seeds dies.

The ability of seeds to withstand stress might be related to their ability to scavenge active oxygen species in order to avoid among others lipid peroxidation (Leprince et al., 1993) resulting in formation of free oxygen radicals (Wilson and McDonald, 1986). Foyer et al. (1991) indicated that plants have evolved antioxidant systems to protect cellular membranes and organelles from damaging effects of active oxygen species (AOS). Cellular damage caused by lipid peroxidation might be prevented or reduced by these systems. It involves protective mechanisms with free radical and peroxide-scavenging enzymes, such as superoxide dismutase (SOD, IC 1.15.1.1), catalase (CAT, EC 1.11.1.6), and enzymes of the ascorbate glutathione cycle (Foyer et al., 1994). This oxide reduction cycle involves the enzymes ascorbate peroxidase (APX, EC 1.11.1.11) and glutathione reductase (GR, EC 1.6.4.2), which can react with H_2O_2 and neutralize the activity of AOS (Asada, 1992; Foyer et al., 1991). Superoxide dismutases are generally considered as key enzymes in the regulation of intracellular concentrations of superoxide radicals and peroxides, these products can react in the Haber-Weiss reaction to form hydroxyl radicals, leading to lipid peroxidation (Gutteridge and Halliwell, 1990; Bailey et al., 1996). In addition, APX reduce H_2O_2 to water with ascorbate as electron donor (Asada, 1992). Peroxidase (POD) and CAT are involved in the removal of hydrogen peroxide (Fridovich, 1986). It is well known that CAT and APX play an important role in preventing oxidative stress by catalyzing the reduction of H_2O_2, however, it has been shown that seed germinability might be related to the efficiency of free radical scavenging because this preventing may affect seed storability and vigour. In peanut, Jeng and Sung (1994) evaluated the effect of accelerated aging on germinability and several physiological characteristics related to peroxidation in the seed of two cultivars. The results indicated that accelerated aging inhibited seed germination, seedling growth and the activity of SOD, POX, APX and LOX. Baily et al. (1998) has shown that lipid peroxidation resulted in losses of free radical scavenging, which is thought to be involved in deterioration of sunflower seeds during accelerated aging. It was also characterized by a decrease in the activities of CAT and GR. The results suggest that seed germinability might be related to the antioxidant defense systems which play a key role in seed storability and vigour. Investigation about enzymatic changes during storage also indicated that the activity of GR, CAT and APX decreased and were also correlated with the seed vigour. However, the activity of SOD and POX remained unchanged (Murthy et al., 2002).

Tocopherols or vitamin E are an important class of lipid-soluble compounds with antioxidant activities that are synthesized only by plants. Assembly of phospholipids and TAG occurs primarily in the endoplasmic reticulum (ER) and shares a common biosynthetic precursor phosphatidic acid (PA). Membrane phospholipid synthesis is active in young and green tissues (Li et al., 2015), however, in developing oilseeds, phospholipid metabolism is overwhelmingly directed to TAG accumulation (Bates et al., 2007; 2009), through two major pathways: The diacylglycerol acyltransferase (DGAT)-mediated Kennedy pathway that uses acyl-CoA and the diacylglycerol (DAG) to generate TAG, and phospholipid:diacylglycerol acyltransferase (PDAT)-mediated pathway that uses phosphatidylcholine (PC) and DAG to produce TAG (Bates et al., 2009). Overexpression of an A-type phospholipase, *pPLAIIIδ*, enhanced TAG production in Arabidopsis and Camelina seeds (Li et al., 2013; 2015). A-type phospholipase, PLA catalyzing the hydrolysis of PC to generate a free fatty acid and lysoPC (LPC), can also affect TAG production (Li et al., 2013; 2015). Acyl-CoA:LPC acyltransferase (LPCAT) can modify PC saturation by introducing an acyl-CoA into a new PC (Bates et al., 2012; Wang et al., 2012).

De novo PC biosynthesis from choline by the actions of choline/ethanolamine kinase (CEK), choline-phosphate cytidylyltransferase (CCT), and DAG cholinephosphotransferase (DAG:CPT) is also important for phospholipid and TAG metabolism (Lin et al., 2015). The interconversion between PC and DAG by PC:DAG cholinephosphotransferase (PDCT, also ROD1) can significantly affect TAG biosynthesis and unsaturation through above mentioned pathways (Bates et al., 2012; Lu et al., 2009; Wang et al., 2014). PDCT transfers phosphocholine from PC to DAG actively during oil seed development, and edits TAG composition using PC that is also extensively modified by fatty acid desaturases (FADs) on their acyl chains (Lu et al., 2009; Hu et al., 2012; Wickramarathna et al., 2015). In the ER, the partitioning of PA, PC, and DAG precursors for TAG or phospholipid biosynthesis may be controlled by a unexplored complex network. For instance, it has been estimated that more than 70% PC-derived DAG is used to synthesize TAG in flax seeds (Pan et al., 2013). Thus, PC plays multiple roles in TAG and phospholipid biosynthesis by recycling or incorporation of the newly synthesized fatty acids in TAG acyl editing (Bates et al., 2007).

PLD may contribute to diurnal cycling of PA, and PC acyl editing and significantly affect PC pools and TAG production (Yang et al., 2017). A study demonstrated that PC derived DAG is the major source for TAG synthesis in *Camelina* seeds overexpressing *PLDζ1* and *PLDζ2* (Yang et al., 2017).

Recently, a genome-wide association study suggested a *PLDα* gene as a key locus affecting oil biosynthesis in soybean (Fang et al., 2017; Manan et al., 2017). Studies indicated that PLDα1 is involved in the mature seed aging and deterioration while stored under high temperature and humidity (Devaiah et al., 2007; Lee et al., 2011). Two *PLDα-RNAi* knockdown (*PLDα1KD*) soybean lines had altered unsaturation fatty acids in both phospholipids and TAG when grown in Kansas in the United States (Lee et al., 2011), however, it is still not understood why *PLDα1KD* seeds have such changes in fatty acyl chains saturation. Although the anti-deterioration effects of *PLDα1* mutation on naturally or artificially aging Arabidopsis and soybean seeds have been reported (Devaiah et al., 2007; Lee et al., 2011; 2012), the underlying molecular mechanism by which *PLDα1* mutation affects lipids metabolism and seed quality has not been explored (Lee et al., 2011; 2012). In particular, the way in which the PLDα1 mutation affects soybean developing seeds under heat and humidity stresses is unknown.

High temperature and humidity stresses occur frequently in southern China amid the rising global temperature in recent years and cause soybean pre-harvest deterioration and significant losses. High temperature and humidity are major problems and limiting factors in soybean production area, and they not only caused soybean yield loss, but also adversely affected seed storage and reduced nutrition (Shu et al., 2015). In the Mid-Yangzi River region of China, soybean seed development and ripening occur during the season of high precipitations (~ more than 150 mm), high temperature (~ 34–38°C), and high humidity (~ 75–80%) usually from the late June to the early September. The soybean seeds grown in these regions usually have pre-harvest deterioration, rapidly losing seed vigor and nutrition quality, and are more vulnerable to pathogens during storage (Shu et al., 2015; Wang et al., 2012). A better understanding of the mechanism by which high temperatures and humidity impact developing soybean seeds will help design effective genetic strategies to improve soybean tolerance to these environmental stress conditions. Here, *PLDα1KD* soybean lines were generated the *PLDα1KD* soybean performance was investigated in such stressful environments.

Transgenic soybean lines with knockdown of phospholipase Dα1 (*PLDα1KD*) were generated to study *PLDα1*'s effects on lipid metabolism and seed vigor under high temperature and humidity conditions. Under such stress, as compared with normal growth conditions, *PLDα1KD* lines showed an attenuated stress-induced deterioration during soybean seed development, which was associated with elevated expression of reactive oxygen species scavenging genes when compared with wild-type control. The developing seeds of *PLDα1KD* had higher levels of unsaturation in triacylglycerol (TAG) and major membrane phospholipids, but lower levels of phosphatidic acid and lysophospholipids compared with control cultivar. Lipid metabolite and gene expression profiling indicates that the increased unsaturation on phosphatidylcholine (PC) and enhanced conversion between PC and diacylglycerol (DAG) by PC:DAG acyltransferase underlie a basis for increased TAG unsaturation in *PLDα1KD* seeds. Meanwhile, the turnover of PC and phosphatidylethanolamine (PE) into lysoPC and lysoPE was suppressed in *PLDα1KD* seeds under high temperature and humidity conditions. *PLDα1KD* developing seeds suffered lighter

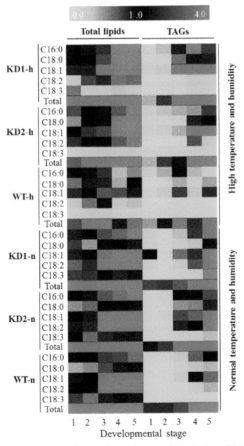

Fig. 10.3 Heatmap analyses on fatty acid compositions of total lipids and TAGs from *PLDα1KD* and wild-type developing seeds. The values represented are the mean ± SD from at least three independent repeats. KD1-h, KD2-h and WT-h: *PLDα1* knockdown line 1, 2 and wild-type jack, respectively, under high temperature and humidity which used thick lines; KD1-n, KD2-n and WT-n:*PLDα1* knockdown line 1, 2 and wild-type Jack, respectively, under normal temperature and humidity. 1, 2, 3, 4, and 5 indicate different developing stages of seeds, corresponding to fresh weights as described previously. The values are from the mean ± SD (*n* = 3).

oxidative stresses than did wild-type developing seeds in the stressful environments. *PLDα1KD* seeds contain higher oil contents and maintained higher germination rates than the wild-type seeds.

Totally, 764 non-redundant proteins were identified, slightly more than those identified in rice seeds (He et al., 2011). However, there were 257 proteins that were annotated as putative or unknown proteins, which may be ascribed to the fact that the soybean is a newly sequenced species. These unknown proteins were blasted in NCBI database in order to obtain their homologs in other plant species. The best matches with probability > 90% were selected. Based on this, 206 proteins from the putative/unknown proteins were assigned with different biological functions. To better understand the physiological status of the imbibied seeds, the 764 identified proteins were classified into 14 functional categories (Fig. 10.4) according to MapMan ontology. These categories are metabolism related proteins (215), protein biosynthesis and destination (183), RNA related proteins (28), DNA related proteins (5), redox regulation (36), stress response (55), signaling (38), transporting proteins (28), cell structure related proteins (48), development related proteins (22), storage proteins (33), hormone related (4), metal handling proteins (1), miscellaneous enzymes (15) and the function unassigned proteins (53).

The three categories with the greatest number of proteins are metabolism-related proteins, proteins related to protein biosynthesis and destination and cell cycle and organization-related proteins. Those for germinating rice seeds were metabolism, protein biosynthesis and destination and signaling proteins groups, respectively (He et al., 2011), however, the most abundant group is the storage proteins in soybean, rather than the metabolism related proteins in rice. These proteins accounted for 42.8% of the total proteins in terms of abundance. Since proteins are the most abundant reserves in soybean seeds, it is reasonable that there are more storage proteins in soybean than in rice. The major storage proteins in soybean seed are 11S globulin (beta-conglycinin) and 7S globulin, whereas those in rice are glutelin and globulin family proteins. It has been reported that the degradation of these storage proteins might help to nourish the germinating soybean seeds and young seedlings (Kim et al., 2011). Another obvious

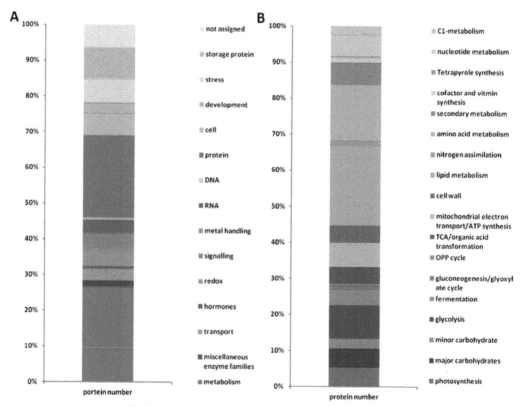

Fig. 10.4 Protein functional categories (A) and metabolism (B).

difference between the soybean and rice lists is the redox regulation proteins group. There were 36 redox regulation proteins identified in germinating soybean seeds, accounting for less than 0.5% of the total proteins in terms of abundance, whereas 25 such proteins accounted for more than 8% in the germinating rice seeds. How the detoxification of ROS happened in soybean seed germination might be an interesting question.

Lipids and proteins are the two major reserves in soybean seeds. Mobilization of these two reserves should be very important for the soybean germination and the ensuing seedling growth. This is supported by the fact that the lipid and amino acid metabolism related proteins were the two major sub-groups in the metabolic protein group (Fig. 10.4B). Fifty nine lipids metabolic proteins were detected, much more than those detected in rice seeds. Most of the enzymes were lipoxygenase (LOX), which indicated that the oils in soybean seeds might be degraded through a LOX-dependent pathway. LOXs are non-heme iron-containing dioxygenases which catalyze the oxidation of polyunsaturated fatty acids by adding of molecular oxygen at C9 and C13 of the acyl chain in linolenic or linoleic acid (Brash, 1999). It was reported that there are four major branches of LOX pathway: (a) The peroxygenase (POX) or hydroperoxide isomerase pathway, (b) The hydroperoxide dehydratase (AOS) pathway, (c) The hydroperoxide isomerase (HPL) pathway, (d) The divinyl ether (DES) pathway (Feussner and Wastenack, 2002). The detection of some P450 monooxygenases implied that the AOS branch might be the major pathway for lipids degradation during soybean germination. Some reports also showed that LOXs were storage proteins in soybean vegetative tissues and seeds, and might be associated with defense response (Porta and Rocha-Sosa, 2002). Besides LOXs and P450 monooxygenases, a full set of fatty acid beta-oxidation enzymes were also detected. The beta-oxidation along with glyoxylate cycle and TCA cycle will help to degrade the lipids completely.

The distribution of fatty acids in dry and in germinating seeds of three soybean (*Glycine max* (L.) Merr.) varieties was studied. Palmitic, oleic, linoleic, and linolenic acids were homogeneously distributed throughout the seed, while stearic acid was slightly higher near the root-shoot axis. With a possible small adjustment for stearic acid, therefore, the analysis of a piece of soybean seed taken at a maximum distance from the root-shoot axis would have given an accurate estimate of fatty acid proportions of the whole seed in the material studied. Practically no significant change in fatty acid proportions during germination up to 6 days was observed, which suggests that the utilization of fatty acids is random. Thus, a sample could have been take any time during this germination period of 6 days to determine the fatty acid proportions of the mature (dry) seed.

Although soybean seeds contained much more fatty acids than rice, the changes in fatty acids content were similar in these two crops (Fig. 10.5). The contents of fatty acid experienced a sharp decrease and then decreased slowly in the following stages. The results also showed that C18:1 and C18:2 were the major fatty acids in both rice and soybean seeds. The content of C18:2 fatty acids are much higher than C18:1 in soybean seeds, while more C18:1 fatty acids were detected in rice seeds.

Oleosomes are subcellular lipid reservoirs (oil bodies) present in all plant seeds and found primarily in their cotyledons (embryonic seed leaves) and embryonic axis (seedling's root) (Yoshida et al., 2003). Owing to the initial absence of photosynthesis in germination, nearly all energy for initial development in plants comes from lipids via lipolysis of TAGs by surface-bound lipases, β-oxidation in glyoxysomes and catabolism in mitochondria (Graham, 2008). As a result, the density of oleosomes, and correspondingly of oleosins, is initially quite high; for instance, oleosins constitute nearly 10% of the total protein mass in *Arabidopsis thaliana* seeds (Hsieh and Huang, 2004). The high levels of oleosin can be understood from its important role as an emulsifier, helping to maintain small oil bodies with a high surface-to-volume ratio for augmented lipolysis by surface-localized lipases (McClements and Li, 2010; Eastmond, 2006). Although previous studies have shown that oleosin disappears from oil bodies during germination (Chen et al., 2014; Deruyffelaere et al., 2015) and that oil bodies fuse when oleosin is genetically suppressed (Siloto et al., 2006; Miquel et al., 2014), it is unknown if oil bodies grow or shrink during unperturbed, native germination and how this correlates to oleosin levels. Recent work has shown that oleosins are degraded prior to lipid mobilization from oil bodies via a ubiquitination–proteasome.

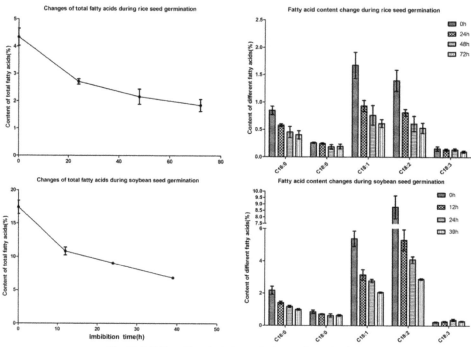

Fig. 10.5 Fatty acid content change during seed germination.

10.2 Plant Organs

In plants, the pathways for lipid biosynthesis and oil accumulation have been studied and the genes related to fatty acid biosynthesis have been characterized. There are several key genes in the process of fatty acid biosynthesis. One is *ACCase*, encoding acetyl CoA carboxylase in the first key step of fatty acid biosynthesis, and malonyl-CoA is produced (Ohlrogge and Jaworski, 1997). The second one is *KASIII*, which encodes 3-ketoacyl-ACP synthase III to catalyze the formation of a 4-carbon product (Clough et al., 1992; Jackowski and Rock, 1987). The carbon number of fatty acid is increased by two in acyl chain, and elongation of the acyl chain from six to 16 carbon molecules is catalyzed by an enzyme named KAS1 (Shimakata and Stumpf, 1982). Without KAS1, FA contents would be sharply reduced, and plant growth and development would be strongly affected (Wu and Xue, 2010). The genes related to FA biosynthesis, such as *Pl-PKβ1* (*pyruvate kinase*), *PDHE1α* (*pyruvate dehydrogenase E1 alpha subunit*), *BCCP2* (*acetyl-CoA carboxylase*), *ACP1* (*acyl carrier protein*), and *KAS1,* have a similar expression pattern to *WRI1* (*WRINKLED1*), and the FA biosynthesis-related genes were up-regulated in the *WRI1*-overexpressing plants (Ruuska et al., 2002). WRI1 is an AP2-type transcription factor (TF) with two AP2 DNA-binding domains, and it appears to be a master regulator of *FAS* (fatty acid synthesis) genes in expression level. There is a specific sequence motif AW-box in the promoter regions of the *FAS* genes, and WRI1 binds to this motif in *Arabidopsis*. Overexpression of *WRI1* enhanced the oil content in transgenic *Arabidopsis* (Liu et al., 2010), and maize (Shen et al., 2010; Pouvreau et al., 2011). In Castor bean, there are WRI1 binding consensus sites in the promoter region of *RcBCCP2* and *RcKAS1*, and RcWRI1 possibly binds to these sites to play a pivotal role in fatty acid biosynthesis (Tajima et al., 2013). Overexpression of a single transcription factor gene *WRI* effect on the oil content (Shintani et al., 1997; Dehesh et al., 2001).

Transcription factors can regulate expression of genes involved in a wide range of plant processes and have a cascade amplification effect (Riechmann and Meyerowitz, 1988). Therefore, transcription factors are the promising targets to improve oil contents in plants. Several candidate transcription factors involved in fatty acid biosynthesis and accumulation have been characterized, including *WRI18* (Masaki et al., 2005; Baud et al., 2007), and *LEC2* (*leafy cotyledon2*) (Santos-Mendoza et al., 2008) in *Arabidopsis*. *WRI1* is a target of LEC2 (Baud et al., 2007). The transcription factors regulating fatty acid contents have been identified from soybean in the lab. Two Dof-type (DNA-binding one zinc finger) genes, *GmDof4* and *GmDof11*, were found to increase the content of total fatty acids in their transgenic *Arabidopsis* seeds by activating the ACCase and ACSL long-chain-acyl CoA synthetase genes, respectively (Wang et al., 2007). Through microarray analysis, a MYB-type gene, *GmMYB73*, was identified and this gene can suppress the expression of *GL2* (*GLABRA 2*), a negative regulator of oil accumulations (Liu et al., 2014). Overexpression of *GmMYB73* enhanced lipid contents in seeds of transgenic *Arabidopsis* through release of GL2-inhibited *PLDα1* (phospholipase D) expression (Liu et al., 2014; Shen et al., 2006; Shi et al., 2012). Overexpression of *GmbZIP1* (Shen et al., 2006) also enhanced lipid content and oil accumulation by regulating two sucrose transporter genes, *SUC1* and *SUC5*, and three cell-wall invertase genes, *cwINV1*, *cwINV3* and *cwINV6* (Song et al., 2013). Recently, through RNA-seq analysis, gene co-expression networks have been identified for soybean seed trait regulation and *GmNFYA* (*nuclear transcription factor Y alpha*) is found to enhance seed oil contents in transgenic *Arabidopsis* plants (Lu et al., 2016).

A *DREB*-type transcription factor gene, *GmDREBL*, has been characterized for its functions in oil accumulation in seeds. The gene is specifically expressed in soybean seeds. The GmDREBL is localized in the nucleus and has transcriptional activation ability. Overexpression of *GmDREBL* increased the fatty acid content in the seeds of transgenic *Arabidopsis* plants. GmDREBL can bind to the promoter region of *WRI1* to activate its expression. Several other genes in the fatty acid biosynthesis pathway were also enhanced in the *GmDREBL*-transgenic plants. The *GmDREBL* can be up-regulated by *GmABI3* and *GmABI5*. Additionally, overexpression of *GmDREBL* significantly promoted seed size in transgenic plants compared to that of WT plants. Expression of the *DREBL* is at higher level on the average in cultivated soybeans than that in wild soybeans. The promoter of the *DREBL* may have been subjected to selection during soybean domestication. Results demonstrate that *GmDREBL* participates in the regulation of fatty acid accumulation by controlling the expression of *WRI1* and its downstream genes, and manipulation of the gene may increase the oil contents in soybean plants. Study provides novel insights into the function of *DREB*-type transcription factors in oil accumulation in addition to their roles in stress response.

Longevity is conferred by the ability to stabilize the biological entity for long periods of time by the formation of an amorphous highly viscous, solid-like matrix (i.e., a glassy state) in the cells that suspends integrated metabolic activities and severely slows down deteriorative reactions (Walters et al., 2005; 2010; Buitink et al., 2000). Seed longevity is also attributed to a range of protective compounds (Sano et al., 2016; Leprince et al., 2017), including non-reducing soluble sugars (sucrose (Suc) and raffinose (Raf) family oligosaccharides), RFO (Salvi et al., 2016; Zinsmeister et al., 2016) and a set of late embryogenesis abundant (LEA) proteins and heat shock proteins (HSP). Together with sugars, both types of proteins act as chaperones and molecular shields in order to prevent protein denaturation and membrane destabilization during drying and in the dry state. Longevity is also conferred by antioxidant mechanisms that limit oxidation of lipids, proteins and nucleic acids during storage, such as glutathione (Nagel et al., 2015 and references therein), tocopherols (Sattler et al., 2004), flavonoids that are present in the seed coat (Debeaujon et al., 2000) and lipocalins (Boca et al., 2014). Several repair mechanisms also contribute to longevity when they are activated during seed imbibition to fix damage that occurred to proteins and DNA during storage (Oge et al., 2008; Waterworth et al., 2010). Next to protection and repair, an impaired degradation of chlorophyll appears to negatively affect longevity (Zinsmeister et al., 2016; Nakajima et al., 2012). The presence of chlorophyll is considered as an indicator of immaturity, but how it affects longevity remains unsolved.

To be commercially successful, crop seeds should be harvested when longevity reaches its maximum (Finch-Savage and Bassel, 2016; Leprince et al., 2017). In legumes, longevity is progressively acquired during seed maturation from seed filling onwards (Leprince et al., 2017). In soybean, there exist conflicting data as to whether seed longevity reaches a maximum at seed filling (Marcos-Filho, 2016; Gillen et al., 2012) or later, during maturation (Zanakis et al., 1994). Delaying harvest to obtain maximum longevity increases the risk of exposing mature seeds to rapid deterioration in the field due to high humidity and temperature (Gillen et al., 2012; Zanakis et al., 1994).

Plants contain phosphoglycerides of the same types as other eukaryotes. In the extra-plastidic membranes, phosphatidylcholine and phosphatidylethanolamine are major components with smaller amounts of phosphatidylinositol and phosphatidylserine (Harwood, 1980) (Table 10.2). Phosphorylated derivatives of phosphatidylinositol which, as in animals, have signalling functions, are found in small amounts, mainly in the plasma membrane (Drobak, 2005; Im et al., 2007). Diphosphatidylglycerol (cardiolipin) is confined to the inner mitochondrial membrane (Bligny and Douce, 1980) where it is a major component (Table 10.2). Phosphatidylglycerol, in contrast to animal tissues, is a major constituent because it is the only significant phospholipid in chloroplast (plastid) thylakoids. Due to the dominance of chloroplasts in green tissues, phosphatidylglycerol is often present in comparable amounts to phosphatidylcholine (Table 10.2).

Table 10.2 Acyl lipid composition of selected plant tissues.

	\% total lipids						
	PC	**PE**	**PI**	**PG**	**MGDG**	**DGDG**	**SQDG**
Leaf lipids	10	5	3	8	40	28	6
Mito. lipids — outer	68	24	5	2	n.d.	n.d.	n.d.
Mito. lipids — innera	29	50	2	1	n.d.	n.d.	n.d.
PM lipids	32	46	19	tr.	n.d.	n.d.	n.d.
Thylakoid lipids	2	tr.	n.d.	10	48	31	8
Root microsomes	35	28	14	n.d.	n.d.	n.d.	n.d.
Cyanobacteria	n.d.	n.d.	n.d.	19	56	14	11

Abbreviations: DGDG, digalactosyldiacylglycerol; MGDG, monogalactosyldiacylglycerol; PC, phosphatidylcholine; PE, phosphatidylethanolamine; PG, phosphatidylglycerol; PI, phosphatidylinositol; PM, plasma membrane; SQDG, sulphoquinovosyldiacylglycerol; n.d., none detected; tr., trace (< 0.5); Mito., mitochondrial.

One aspect of membrane composition that sets plants and algae apart from other eukaryotes are their chloroplast thylakoids. The major lipids of these are three glycosylglycerides—monogalactosyldiacylglycerol (MGDG), digalactosyldiacylglycerol (DGDG) and sulphoquinovosyldiacylglycerol (SQDG, the plant sulpholipid). Interestingly, all oxygen-evolving photosynthetic organisms, including cyanobacteria, contain rather similar thylakoid membrane compositions, with about 45% MGDG, 29% DGDG, 7% SQDG and 9% phosphatidylglycerol in different leaves (see Table 10.2). The reason for this is unclear, although some hypotheses have been put forward (Jones, 2007; Domonkos et al., 2013). Because of its prevalence in thylakoids, MGDG is the most abundant membrane lipid on earth, although many textbooks fail to even acknowledge its existence.

Sphingolipids are found in plants but, again, with their own distinctiveness compared to animals and yeast. Of particular interest, is their enrichment in membrane rafts (Cacas et al., 2012). Plant sphingolipids are summarised well in (Lynch and Dunn, 2004) and their metabolism is updated in (Markham et al., 2013). There are also a number of novel lipids in plants which may be present in appreciable quantities in certain tissues. These would include molecules like *N*-acyl-phosphatidylethanolamine and sterol glycosides (Stumpf and Conn, 1980). Plant sterols themselves are dominated by β-sitosterol and stigmasterol (rather than cholesterol in animals). However, a whole host of minor compounds, including cholesterol, are found widely (Stumpf and Conn, 1980; Hemmerlin et al., 2012).

The long-chain (16 or 18C) saturated fatty acids are converted to unsaturated acids by the action of desaturases. While the desaturation of stearate to oleate is catalysed by a soluble stearoyl-ACP Δ9-desaturase (in the stroma) (Hildebrand et al., 2005), other plant desaturases use complex lipid substrates. For such reactions, both the desaturase proteins and the substrates are found in membranes.

Given the prevalence of stearate as a product of plant fatty acid synthase, it is not surprising that the main unsaturated fatty acids in plants are the 18C molecules—oleate, linoleate, and α-linolenate.

Soybean (*Glycine max*) is one of the most important oilseed crops, contributing to 59% of all the world oilseed production in 2014 (Soystats International, 2015), and is one of the world's most widely used and healthy edible oils. In addition to this, the industrial products and uses for soybean oil are becoming increasingly popular and diverse.

Soybean oil, like most edible oils, is composed of palmitic (C16:0), stearic (C18:0), oleic (C18:1), linoleic (C18:2) and linolenic (C18:3) acids. Oleic, linoleic and linolenic acids are 18 carbon unsaturated fatty acids, containing one, two and three cis double bonds interrupted by a methylene group, respectively. The double bond positions in the acyl chain from the carboxyl terminal are 9 in C18:1, 9 and 12 (or 6, counting from the methyl terminal) in C18:2, and 9, 12 and 15 (or 3) in C18:3. Oleic acid is also referred to as monounsaturated fatty acid, while the linoleic and linolenic acids as polyunsaturated fatty acids (Yadav, 1996).

Soybean oil contains about 11% palmitic, 4% stearic, 24% oleic, 54% linoleic and 7% linolenic acids (Kinney, 1996). The quality of the oil fraction varies considerably among these sources and it depends on the fatty acid composition and, particularly, on the proportion of unsaturated fatty acids, mainly oleic, linoleic and linolenic acids (Somerville and Browse, 1991). Due to high levels of polyunsaturated fatty acids, the quality of soybean oil is not ideal for industrial purposes, mainly due to its low oxidative stability. Currently, chemical hydrogenation is the industrial process used to increase the oxidative stability of the soybean oil (Hildebrand and Collins, 1998). However, this process also generates significant amounts of trans fatty acids, which have been linked to heart problems in animals and humans (Yadav, 1996). For this reason, there is a considerable interest on the genetic modification of soybean oil composition, by traditional breeding or by the use of molecular biology techniques. These modifications could avoid the production of the undesirable trans fatty acids and also produce oils with better nutritional and functional attributes (Kinney, 1996).

After harvesting at the full maturity stage, R8, the dry weight of a soybean seed consists of the following elements: Oil (20%), protein (40%), carbohydrates (30%), crude fiber (5%) and ash (5%) (Fehr and Caviness, 1977). Typically, soybean oil consists of approximately: 13% palmitic (C16:0), 4% stearic (C18:0), 20% oleic (C18:1), 55% linoleic (C18:2) and 8% linolenic (C18:3) acid at 13% moisture (Pham et al., 2010; 2011). In addition to these five major FAs, numerous minor FAs that may also have commercial value could be found in soybean oil, such as myristic acid (C14:0), arachidic acid (C20:0), behenic acid (C22:0) or erucic acid (C22:1) (Jokic et al., 2013). Despite the fact that seed development is a highly regulated process, dry matter accumulation at seed filling is affected by both genetic and environmental factors, leading to changes of oil and protein concentrations of crops. In soybean, the FA accumulation during seed maturation takes place in a short period of about 4 to 6 weeks, as opposed to those of other oil plants, such as olive, oil palm or avocado. Soybean is, therefore, sensitive to stressful conditions during their short seed filling period, and this makes them susceptible to incurring permanent changes in oil content and FA profile as well as crop quality and productivity (Wang and Frei, 2011).

Oil and protein concentrations of crops are sensitive to both genetic and environmental factors. The major stress factors that have been investigated are: Drought, salinity, ozone and heat. The observed effects are variable and depend on the stress type, crop species, and experimental conditions, but some typical patterns can be characterized. A decrease in the lipid concentration has been reported in almost every study involving crops grown under unfavourable conditions. By contrast, these stresses usually stimulate higher protein concentration in the harvested fraction of crops, with only a few studies showing no effect or lower protein concentration (Wang and Frei, 2011).

The proportions of oil, protein, and carbohydrate in soybean (*Glycine max*) seeds influence their value, and the control of their accumulation has been studied extensively. Maternally supplied substrates (Pipolo et al., 2004) and seed genotype (Hernandez-Sebastia et al., 2005) determine the oil and protein levels in the seed. Although the fatty acid composition of soybeans has been successfully engineered (Damude and Kinney, 2007), molecular attempts to modify the proportions of oil and protein have resulted in only a few successes for related legumes (Rolletschek et al., 2007). In part, this reflects the complexity of metabolic networks (Egli, 1998) and the uncertain relationship between seed composition and seed metabolism.

10.3 Lipid Metabolism in Developing Seeds

The biosynthesis of seed storage oils containing the five major FAs occurs primarily in two subcellular compartments. FA biosynthesis occurs in the plastid of cells and involves the cyclic condensation of two-carbon units, in which acetyl coenzyme A (acetyl-CoA) is the precursor. When conjugated to the acyl carrier protein (ACP), the FA chain is referred to as acyl-ACP. The first committed step in the pathway is the synthesis of malonyl-CoA from acetyl-CoA and CO_2 by the enzyme acetyl-CoA carboxylase (Li-Beisson et al., 2013). In the following step, some 16:0-ACP is released from the FA synthase machinery, but most molecules that are elongated to 18:0-ACP are efficiently converted to 18:1-ACP by a desaturase enzyme (Fig. 10.6) depicts the biosynthesis of the five common FAs present in the oil of annual oil crops and the main enzyme steps involved. The first three FAs (C16:0, C18:0 and C18:1) are produced by *de novo* synthesis and desaturation in the plastids (Li-Beisson et al., 2013). Elongation and desaturation are carried out while the FAs are attached to an acyl carrier protein (ACP). After removal of the ACP group by acyl-ACP thioesterases (FatA or FatB), the FAs are exported from the plastid and incorporated into the cytosolic acyl-CoA by the action of an acyl-CoA synthetase (ACS). 18:1 is then acylated onto the membrane lipid phosphatidylcholine (PC), mainly by the action of the lysophosphatidylcholine acyltransferase (LPCAT) (Li-Beisson et al., 2013). Further desaturations of the 18:1 to 18:2 and 18:3 are catalysed by FA desaturase 2 (FAD2) and FAD3 while the acyl substrates are acylated to PC. Storage

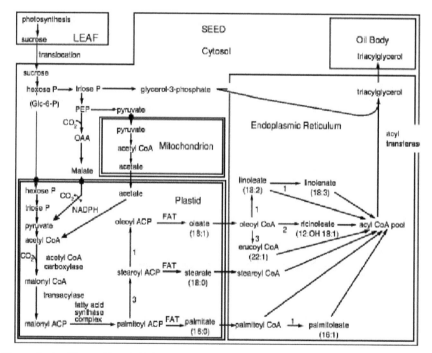

Fig. 10.6 The synthesis of fatty acids and oils in developing seeds involves the participation of enzymes in several cellular compartments, the cytosol, mitochondrion, plastid, and ER, the latter becoming modified to form the oil body; numbered stages require the following enzymes: (1) FADs, (2) Hydroxylase, (3) Elongase Complex; Other enzymes occur in the ER of some species to produce rarer fatty acids, e.g., by Epoxidation, Acetylenation, or Methylation. Biochemical reactions in the ER are intimately associated with its membrane, as are the final stages of TAG production, when fatty acids from the Acyl-CoA Pool are added sequentially by acyltransferases to glycerol-3-P. ER, Endoplasmic Reticulum; FAD, Fatty Acid Desaturases; Glc-6-P, Glucose-6-Phosphate; PEP, Phosphoenol Pyruvate; OAA, Oxaloacetate; ACP, Acyl Carrier Protein; CoA, Coenzyme A; FAT, Fatty Acyl Thioesterase.

TAGs are synthesized by the Kennedy pathway (Fig. 10.7) in developing seeds. The enzymes involved are probably located in the endoplasmic reticulum (ER) and act by the sequential acylation of the sn-1, -2 and -3 positions of glycerol-3-phosphate, with the removal of the phosphate group occurring before the final acylation step. The distribution of acyl groups on the glycerol backbone is often non-random because of the substrate selectivity of the acyltransferases for different FAs (Li-Beisson et al., 2013). In detail, TAGs can be formed through three sequential acyl-CoA-dependent acylations of the glycerol backbone beginning with sn-glycerol-3-phosphate. The acylation of sn-glycerol-3-phosphate is catalyzed by acyl-CoA:sn-glycerol-3-phosphate acyltransferase (GPAT). The second acylation is catalyzed by acyl-CoA:lyso-phosphatidic acid acyltransferase (LPAAT). After removal of the phosphate group to generate sn-1,2-diacylglycerol (sn-1,2-DAG), the final acyl-CoA-dependent acylation is catalyzed by acyl-CoA:diacylglycerol acyltransferase (DGAT) to form TAG (Li-Beisson et al., 2013).

Most of the polyunsaturated fatty acids production in developing seeds, occurs via desaturation of oleic and linoleic acids catalyzed by desaturases in the smooth endoplasmatic reticulum (Somerville and Browse, 1991). The substrate for the desaturases is PC. The fatty acids linked to PC, which may become unsaturated, can subsequently be incorporated into storage triacylglycerol (TAG) molecules. The linoleic and linolenic acid levels in the oil depend on their biosynthesis rate and availabilities (Yadav, 1996). The polyunsaturated fatty acids can become available through two distinct mechanisms: (a) reversible reaction, by which cholinephosphotransferase (CPT, CDP-choline:1,2-diacylglycerolcholine phosphotransferase, EC 2.7.8.2) converts phosphatidylcholine (PC) containing polyunsaturated fatty acids into diacylglycerols (DAGs), which can be used for oil synthesis via diacylglycerol acyltransferase (DAGAT) and (b) reversible reaction, by which acyl-CoA:lysophosphatidylcholine acyltransferase (LPCAT, EC 2.3.1.23) catalyses the exchange of acyl groups, generally between oleoyl-CoA and a polyunsaturated acyl group linked to position 2 of PC.

LPCAT catalyzes the reversible reaction between acyl groups of acyl-CoA cytoplasmatic pool and unsaturated acyl group linked to position 2 of PC. The reaction equilibrium is towards PC synthesis, i.e., lysophosphatidylcholine acylation, since it results in cleavage of an energy-rich thioester bond (Stymne and Stobart, 1984). The permutation between acyl groups is dictated by: (a) velocity and specificity of exchange of acyl groups between acyl-CoA and PC; (b) unsaturation rate of fatty acids in PC and (c) activity and specificity of LPCAT.

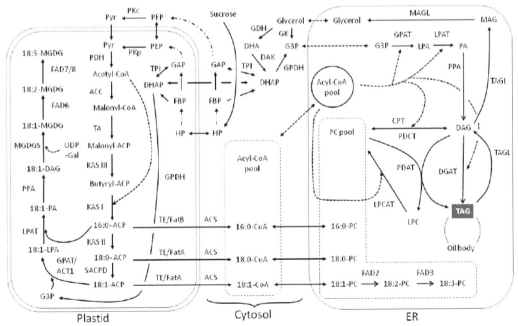

Fig. 10.7 Lipid metabolic pathway.

CPT catalyzes the reversible exchange between the PC pool and DAGs. Due to reaction reversibility, PC can act as a precursor of highly unsaturated molecular species of DAGs in seeds that accumulate polyunsaturated fatty acids in the oil fraction. It is a key enzyme in the metabolism of oilseeds, thus, its activity and regulation mechanisms are essential for understanding the fatty acid distribution for lipid synthesis (Vogel and Browse, 1996).

Oleoyl-CoA was a preferential substrate in the acyl-CoA pool for LPCAT activity, while stearyl-CoA was completely excluded. It showed that, during exchange of acyl groups, oleoyl-CoA enters position 2 of PC, liberating linoleate, which is preferentially used in the acylation of position 2 of glycerol 3-phosphate. Thus, this enzyme regulates the type of acyl groups constituting the TAGs, which accumulate in developing seeds. The exchange of acyl groups between acyl-CoA and PC is a major step for regulating quality of polyunsaturated fatty acids in the acyl-CoA pool for oil synthesis in developing safflower seeds.

Up until few years ago, soybean produced more oil than any other crop plant, despite the fact that it is grown primarily for protein. Even today, soybean accounts for about 22% of the world production of oils and fats. Therefore, it was important to study oil accumulation in this crop. The soybean embryos used in a study have been shown to be an excellent system for studying transgenic and physiological influences on resource partitioning and have proven to be a very predictive model for seeds (Ttruong et al., 2013).

A detailed description of the lipid synthesis, compartment wise, can now be outlined below (Fig. 10.7).

In the flux control experiments, [1-^{14}C] acetate was used to label fatty acids and [U-^{14}C] glycerol for incorporation into the backbone of complex lipids during assembly (Block B reactions). These two precursors are virtually specific for each type of incorporation (> 96%), as demonstrated for other plant oil tissues (Ramli et al., 2013; Tang et al., 2012). The distribution of radioactivity into lipid classes during the linear period of incorporation (4 h) shows that, of the non-polar lipids, only TAG and diacylglycerol (DAG) were significantly labeled, while phosphatidylcholine (PC) contained the bulk of radiolabel amongst the polar lipids. The latter is indicative of a cycling of carbon flux between DAG and PC, as expected from the high activity of the "acyl editing" reactions in soybean (Bates et al., 2009). Since, the soybean cultures are non-photosynthetic and mimic the developing seed metabolism of chloroplast lipids, such as MGDG, and, consequently, their labeling was minor. The relatively small accumulation of radioactivity in the Kennedy pathway intermediates, phosphatidate and, especially, lysophosphatidate, compared to DAG attests to the important control exerted by the final enzyme of the Kennedy pathway, DGAT, in soybean.

The Kennedy Pathway produces triacylglycerol (TAG), an oil storage molecule in plants. The Pathway consists of four enzymes: Glycerol-3-phosphate acyltransferase (G3PAT; EC 2.3.1.15), lysophosphatidic acid acyltransferase (LPAAT; EC 2.3. 1.51), phosphatidic acid phosphatase (PAP; EC 3.1.3.4), and diacylglycerol acyltransferase (DGAT; EC 2.3.1.20).

New evidence suggests there are alternate pathways that also produce triacylglycerol (Chapman and Ohlrogge, 2012).

G3PAT catalyzes the first reaction in the Pathway. The fatty acid preference of various isoforms depends on plant species and subcellular location (Frentzen, 1993). The fatty acid preference can also be influenced by growing temperature, as was shown in a study on safflower G3PAT (Ichihara, 1984). Overexpression of a G3PAT from either safflower or *Escherichia coli* increases the total seed oil concentration of arabidopsis (Jain et al., 2000). LPAAT catalyzes the second reaction in the Pathway. LPAAT usually prefers unsaturated fatty acids, but the level of that preference is species specific. Overexpression of yeast LPAAT increases the total oil concentration in soybean seed and the concentration of both oil and very-long chain fatty acids in arabidopsis and Brassica napus seed (Rao and Hildebrand, 2009).

DGAT1 catalyzes the final reaction in the Kennedy Pathway. The substrate preferences of DGAT1s vary with temperature in activity assays: for example, a canola DGAT1s prefers oleic acid at 24°C and erucic acid at 32 and 40°C (Cao and Huang, 1987). In general, DGAT1 utilizes linoleic, palmitic, and oleic acids (Cao and Huang, 1986). Expression of DGAT1 is evident in all aerial tissues of Brassica napus, but is more tissue specific in arabidopsis (Lu et al., 2003). In arabidopsis, DGAT1 also shows a developmentally dependent expression pattern (Lu et al., 2003). DGAT1 expression in soybean seed peaks 45 d after flowering (DAF) (Li et al., 2010). DGAT2 and PDAT are also capable of catalyzing this final reaction, but DGAT1 is the predominant enzyme in soybean (Li et al., 2010; Chapman and Ohlrogge, 2012).

The air temperature during seed fill can influence the final fatty acid profile of soybean seed. For example, when temperature increases, the level of polyunsaturated fatty acids decreases, with the level of linolenic acid being the most affected (Hou et al., 2006). It has also been shown that the levels of linoleic and linolenic acids decrease with increasing temperature, but oleic acid increases (Wolf et al., 1982).

Since temperature influences the fatty acid composition, this study used controlled environment conditions to grow the genotypes of interest. The main oil storage molecule, triacylglycerol (TAG), can be created by the Kennedy Pathway. The objective of this study was to determine the impact of growing temperature on the expression of the Kennedy Pathway genes, glycerol-3-phosphate acyltransferase (G3PAT), lysophosphatidic acid acyltransferase (LPAAT), and diacylglycerol acyltransferase 1 (DGAT1), in developing seed of four soybean genotypes with altered fatty acid composition using quantitative polymerase chain reaction. The three growing temperatures were: high, 30°C day/25°C night; normal, 25°C day/20°C night; and low, 20°C day/15°C night. The expression of G3PAT steadily declined following 15 d after flowering (DAF), suggesting that it is likely to be more highly expressed earlier in development than was measured in the study. As a result, the expression of G3PAT did not correspond to fatty acid accumulation. LPAAT expression coincided with accumulation of oleic acid (18:1D9) and linolenic acid (18:3D9,12,15) in a temperature-dependent manner. The expression of DGAT1 corresponded to accumulation of linoleic acid (18:2D9,12), which varied among the soybean genotypes, indicating a genotypic effect on the expression of DGAT1. This study suggests that the expression of the acyltransferase enzymes of the Kennedy Pathway influences the fatty acid composition in seed of four altered fatty acid soybean genotypes.

Single manipulation used the addition of oleate. We felt that this was entirely appropriate for soybean which accumulates this fatty acid as a major component of its oil (~ 25%) and uses oleate to produce the main fatty acid, linoleate (~ 50%). Calculation of changes induced by the addition of oleate gave group flux controls for Block A and Block B of 0.63 and 0.37, respectively. In fact, the data showed that oleate

Table10.3 Pair-wise comparison LS means for temperature of significant differences in fatty acid composition of developing seeds of four soybean genotypes grown at three temperatures. LS means could not be estimated for RG7 at low temperature because samples at 15 d after flowering were not available for evaluation. Comparisons are within a column for each genotype. Least significant difference (LSD) was based on arithmetic means.

Genotype	Temperature	Palmitic (g kg^{-1})	Stearic (g kg^{-1})	Oleic (g kg^{-1})	Linoleic (g kg^{-1})	Linolenic (g kg^{-1})
RG2	Low	62a	41a	215a	527a	156a
	Normal	62a	37a	203a	536a	161a
	High	56a	42a	292b	505a	106b
	LSD	41.4	20.7	31.0	45.0	23.3
RG7	Low	-	-	-	-	-
	Normal	128a	53a	185a	491a	143a
	High	129a	104b	160a	489a	118a
	LSD	12.6	44.8	56.2	37.3	25.9
RG10	Low	136a	42ab	157a	570ab	95a
	Normal	135a	41a	153a	599a	73ab
	High	139a	50b	236b	520b	55b
	LSD	4.3	7.3	33.5	40.0	23.0
SV64-53	Low	134a	47a	165a	472a	181a
	Normal	132a	49a	217b	480a	122b
	High	130a	54a	210ab	492a	114b
	LSD	9.5	7.6	41.0	34.9	25.7

a, b LS means with the same letter designation are not significantly different.

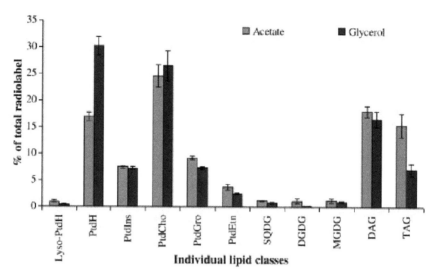

Fig. 10.8 Labeling of individual polar lipids in soybean cultures from [1-14C] acetate and [U-14C] glycerol. Abbreviation: DAG: diacylglycerol; DGDG: digalactosyldiacylglycerol; MGDG: monogalactosyldiacylglycerol; PDAT: phospholipid diacylglycerol acyltranferase; PtdCho: Phosphatidyl choline; PtdEtn: phosphatidylethanol amine; PtdGro: phosphatidylglycerol; SQDG: Sulphoquinovosyldiacylglycerol; TAG: Triacylglycerol; PtdH: phosphatidic acid; Lyso-PtdH: Lyso phosphatidic acid, etc.

reduced labelling of fatty acids from [1-14C] acetate and enhanced that of lipids from [U-14C] glycerol. This is most simply interpreted as product inhibition (by oleate) of the fatty acid biosynthesis block while constraints caused by limitation in fatty acid supply are alleviated by oleate addition. Product inhibition may be similar to the reduction of acetyl-CoA carboxylase activity by oleoyl-ACP observed in oilseed rape seeds (Andre et al., 2012).

10.4 Seed Development

Much work has been done in legumes and in particular on Arabidopsis, which strongly implicates metabolite and hormone responsive pathways as key contributors (Gibson, 2004). Soybean Prestorage or morphogenesis begins with the fertilization of the first flower, follows on to the completion of embryogenesis, and ends once pod development has been achieved. Prestorage includes GS R1–R4. Moreover, the zygote undergoes extensive cell divisions, and resembles the globular heart stage. This cell differentiation subsequently results in the tissue types required to form the root-shoot axis (Berger, 2003) and large cotyledon where oil, protein and starch reserves are localized during seed maturation. In the early stage of embryogenesis, the embryo is supported by a temporary organ called a suspensor, which provides a connection for the embryo to the surrounding nutrient-providing tissues. Measurements of endogenous hormone concentrations during morphogenesis have shown that cytokinins (CKs), abscisic acid (ABA), gibberellin (GA) and indole-3-acetic acid (IAA) are all transiently high and significantly active (Bewley et al., 2012). Tissue culture studies involving Phaseolus (common bean) have shown that the addition of exogenous GA can substitute.

A generalized graph showing the relative levels of water, dry weight (DW), and hormones during the stages of seed development for a detached suspensor is in promoting embryonic growth, suggesting that the suspensor may normally provide GAs as well as nutrients to the developing embryo. Similarly, other studies involving a focus on either exogenous hormone addition, genetic responses or exudates from tissue culture all suggest that the roles of GAs and CKs are primarily nutritive. IAA has shown to play a major role in establishing the embryonic body-plan via effects on apical-basal polarity/pattern formation and vascular development (Vogler and Kuhlemeier, 2003). ABA can act to prevent seed abortion and promote embryo growth during the early embryogenesis (Frey et al., 2004). Despite the low levels of

ABA generally detected during early embryogenesis, the ABA biosynthetic pathway is apparently active at this stage. In agreement, high ABA levels have been found in the pedicel/placento-chalazal complex of maize kernels (Jones and Brenner, 1987). CKs have been implicated in a number of processes, including support of suspensor function, significant promotion of embryonic growth to reduce seed abortion, and enhancement of grain filling and seed yield via the promotion of cell division, especially within the cotyledons (Zalewski et al., 2010). In dicots such as soybean, prestorage cell division is critical as it dictates the total number of cells that will exist within, and in doing so lays down the ground work for cell enlargement during maturation. Moreover, once the number of embryonic cells has been defined by the key contributors, the seed cotyledon will enlarge and accumulate the important constituents based upon the available number of cells, assimilate supply, and regulatory signals. Accumulation of oils/FAs and proteins occurs throughout cell enlargement and is central to cotyledon development. Inside the cells of cotyledons, oil is stored in small discrete oil bodies in the form of triacylglycerols (TAGs) (Ohlrogge and Kuo, 1984). It is believed that the more intracellular volume is available, the more space oil bodies can occupy. However, this limited available intracellular space must be shared between both protein bodies and TAGs. Thus, it is well-known that the production of TAGs and protein bodies is inversely correlated (Chung et al., 2003).

Following the first phase, the reserve accumulation is the next critical period in soybean seed production. Soybean seed value is determined in this phase as lipid bodies and proteins are synthesized and stored throughout development stage of R5 until the end of R6. This is one of the last two phases of embryonic development and is sometimes collectively referred to as "maturation". At that time, seeds acquire the ability to survive desiccation and become ready to initiate growth of the next generation, independent of the maternal plant. Seed maturation begins when developing embryos cease growth by cell division; this coincides with an increase in seed ABA, a hormone which induces expression of a cyclin-dependent kinase inhibitor (ICK1) that could lead to cell cycle arrest at the G1/S transition. As demonstrated in the Arabidopsis seed model, ABA, classically associated with seed maturation, is produced first in maternal tissues and later in the embryo (Karssen et al., 1983). Maternal ABA, synthesized in the seed coat and translocated to the embryo, promotes its growth and prevents abortion (Frey et al., 2004). A major increase in ABA levels occurs during the maturation phase corresponds to the positive regulation of a number of genes for seed reserves. The middle stage of seed development is a period of massive reserve accumulation and cell enlargement as cells fill with protein and lipid bodies (Harada, 1997). Multiple seed mass and composition studies on "Williams 79" soybean seeds by Dornbos and McDonald (1986) demonstrated that stages R5 and R7 corresponded to seed filling initiation and physiological maturity, respectively. Between those phases, water content (% fresh weight—FW) declines steadily, although the total amount of water per embryo is still increasing. The most abundant hormone at this stage is ABA, which reaches peak levels during the period of maximal seed weight gain. In the late-developmental stage, ABA induces dormancy and inhibits germination in the matured seeds by upregulating its own levels and down-regulating GA synthesis (Wilkinson et al., 2010). During the final phase of seed development, embryos become desiccation tolerant, lose water, and become relatively metabolically inactive. A decrease in the ABA level during the desiccation phase is also expected to result from decreased ABA synthesis (Audran et al., 1998).

Lipids accumulate as triacylglycerides that are found in oil storage bodies surrounded by the protein oleasin or occasionally as oil droplets in the cytosol. Predominant fatty acids in triacylglycerides are palmitate (16:0), stearate (18:1), linoleate (18:2) and linolenate (18:3). In soybean, cell division in the seed is completed at an early stage of development (20–25 DAF) while the embryo is still quite small. The major increase in seed size which occurs between 25 to 60 days after flowering (DAF) is brought about through enlargement of pre-existing cells. The majority of oil, protein and carbohydrate synthesis and storage occur during this period by simultaneous partitioning of the photosynthates among those three major reserves (Ohlrogge and Kuo, 1984). It was reported that, by 26 DAF, starch, lipid and protein bodies were present in the cytosol of soybean cotyledons. As the seed developed, the cells of the cotyledons became packed with the lipids, protein and starch bodies. However, the starch bodies disappeared just prior to maturation. Developing soybean seeds contained 5% oil at 25 DAF. The oil

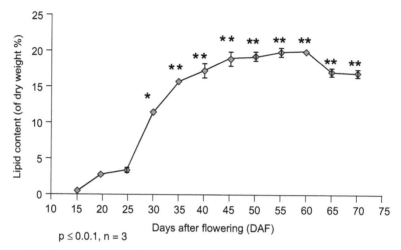

Fig. 10.9 Changes in lipid content during seed development. Lipid content was determined every 5 days. One star stands for the discrepancy is significant and double stars stand for the discrepancy is extremely significant.

percentage increased slightly to around 20% by 40 DAF and remained essentially constant during the remaining period of seed development (Fig. 10.9).

10.5 Abiotic Stresses

Consequences of exposure to abiotic stresses include various physiological changes in crop plants, such as: Alterations in the photosynthetic gas exchange and assimilate translocation (Morgan et al., 2004), altered water uptake and evapotranspiration, effects on nutrient uptake and translocation, antioxidant reactions (Apel and Hirt, 2004), programmed cell death (Kangasjari et al., 2005), and altered gene expression and enzyme activity. These exposures are likely to have numerous effects on the chemical composition of crops and, consequently, the quality of agricultural products.

Oil and protein concentrations of crops are sensitive to both genetic and environmental factors. The major stress factors that have been investigated are: Drought, salinity, ozone and heat. The observed effects are variable and depend on the stress type, crop species, and experimental conditions, but some typical patterns can be characterized. A decrease in the lipid concentration has been reported in almost every study involving crops grown under unfavorable conditions. The FA profile of soybean oil is a fundamental quality attribute. Genotype is the main determinant of FA composition, but environmental factors such as climate conditions have been linked to variations in oil quality and yield. The majority of the studies reported decreases in the lipid concentration when crops were grown under stressful conditions. Liu et al. (2013) indicated UV-B radiation reduced total biomass and seed yield per plant. These losses were mainly attributed to the change of pod number per plant and seed size. In a report on seed development gene expression, Fatihi et al. (2013) indicated that a reduced seed size is primarily associated with reduced TAG content in the embryos of Arabidopsis. In case of the drought stressed crops, almost all studies reported a decrease in the lipid concentration of the harvested products compared to that of the sufficiently watered plants. It is important to note that, in seeds of annual crops, such as soybean and sunflower, oil accumulates at a high rate during a short period of time (between 30 and 45 days). On the other hand, in olive fruit—similar to those of oil palm and avocado—oil accumulates principally in the mesocarp at a low rate, i.e., over a long period (100 to 140 days). Thus, it is possible that greater opportunities for recovery to normal values after a high-temperature event might exist in olives. A similar trend towards declining oil concentration was seen under salinity and heat stress, for which only a few studies reported increases or no effects on lipid concentrations. In contrast, ozone stress seemed to be an exception, as the available studies reported either no effect, or even an increase in lipid concentration (Wang and Frei, 2011). Temperature effects on seed growth (Wardlaw et al., 2002) are well documented in annual crops, including oil-seed species. Seed oil concentration decreased in response to high temperatures during the period of oil synthesis (Roudanini et al., 2003). Processes indirectly linked to oil synthesis, such as photosynthesis or respiration, could also be simultaneously modulating the oil concentration. Photosynthesis of both leaves and fruit are likely negatively affected by exposure to high temperatures. Increases in leaf temperature above 32°C in growth chambers resulted in a decline in photosynthetic rate (Nambara and Marion-Poll, 2005). The high temperature stress reduces the duration of seed filling period via accelerated leaf senescence and, consequently, oil accumulation is stopped before fulfilling seed oil capacity, when the seed is ready for desiccation. The environmental stresses not only change the oil contents of oil crops but also affect oil composition. A general trend that indicated an increase in the saturation level of the oil fraction due to various abiotic stresses has been reported (Wang and Frei, 2011). The proportion of polyunsaturated FAs (PUFA) in soybean oil dropped considerably under heat stress (Dornbos and Mullen, 1992). These decreases in PUFA (especially linoleic acid, C18:2) were consistently accompanied by increases in the proportion of oleic acid (C18:1) (Pham et al., 2010; 2011). FA composition varies depending on the timing of the high temperature event. In soybean, as well as in sunflower, lower latitudes leading to the increase of temperature have been associated with high C18:1 oils (Taarit et al., 2010). In addition, it has been demonstrated that the differences in night temperatures are a better indicator of the changes in FA composition than daily average temperatures in annual oil-seed crops (sunflower and soybean). The observed changes in FA composition are believed to be a result of the activity of enzymes involved in lipid synthesis and conversion.

FA synthesis in oil seeds starts its early steps in the plastids and then C18:1 as the main product of plastidal lipid synthesis is exported to the cytosol. The enzyme activity, in which oleate desaturase (OD) moderates the cytosolic desaturation of C18:1 to form PUFA (i.e., C18:2), is believed to be an explanation for shifts in the C18:1/C18:2 ratio in several crops under various types of stress: Salinity, drought, and heat (Hernandez et al., 2009). Numerous studies have demonstrated the temperature dependence of this enzyme (Esteban et al., 2004). In sunflower, the highest OD activity was observed at 20°C and its activity dropped considerably at higher temperatures. On the other hand, in safflower, OD was more heat stable and maintained its full activity up to 30°C. Two factors, including (i) the heat stability of the enzyme, and (ii) the effects of temperature on the internal oxygen concentration of seeds, which are key regulators of OD activity (Rolletschek et al., 2007), have been proposed in order to explain this temperature-dependent decline in enzyme activity. Besides enzymatic desaturation of FAs, transport of plastidal FAs to the cytosol is potentially affected by environmental stresses. It is generally considered that the common stress factors including drought, heat and tropospheric ozone result in an increased protein concentration in wheat grains and soybean seeds.

10.6 Molecular Aspects

Soybean is an important economic crop and provides oil and proteins for human and animals. Increasing the FA (fatty acid) contents and improving the oil quality are closely related to our daily life. So far, numerous efforts have been made to meet the needs of human food and industry production by changing the fatty acid content in seeds. However, the extracted fatty acids from the existing oil plants are far from enough, therefore, traditional breeding methods and transgenic approaches manipulating fatty acid biosynthesis pathways are used to increase oil content in soybean. In plants, the pathways for lipid biosynthesis and oil accumulation have been studied and the genes related to fatty acid biosynthesis have been characterized. There are several key genes in the process of fatty acid biosynthesis. One is *ACCase*, encoding acetyl CoA carboxylase in the first key step of fatty acid biosynthesis, and malonyl-CoA is produced (Ohlrogge and Jaworski, 1997). The second one is *KASIII*, which encodes 3-ketoacyl-ACP synthase III to catalyze the formation of a 4-carbon product (Clough et al., 1992). The carbon number of fatty acid is increased by two in acyl chain, and elongation of the acyl chain from six to 16 carbon molecules is catalyzed by an enzyme named KAS1 (Shimakata and Stumpf, 1982). Without KAS1, FA contents would be sharply reduced, and plant growth and development would be strongly affected (Wu and Xue, 2010). The genes related to FA biosynthesis, such as *Pl-PKβ1* (*pyruvate kinase*), *PDHE1α* (*pyruvate dehydrogenase E1 alpha subunit*), *BCCP2* (*acetyl-CoA carboxylase*), *ACP1* (*acyl carrier protein*), and *KAS1* have a similar expression pattern to *WRI1* (*WRINKLED1*), and the FA biosynthesis-related genes were up-regulated in the *WRI1*-overexpressing plants (Ruuska et al., 2002). WRI1 is an AP2-type transcription factor (TF) with two AP2 DNA-binding domains (Cernac and Benning, 2004), and it appears to be a master regulator of *FAS* (fatty acid synthesis) genes in expression level. There is a specific sequence motif AW-box in the promoter regions of the *FAS* genes, and WRI1 binds to this motif in *Arabidopsis* (Maeo et al., 2009). Over expression of *WRI1* enhanced the oil content in transgenic *Arabidopsis* (Liu et al., 2010) and maize (Pouvreau et al., 2011). In castor bean, there are WRI1 binding consensus sites in the promoter region of *RcBCCP2* and *RcKAS1*, and RcWRI1 possibly binds to these sites to play a pivotal role in fatty acid biosynthesis (Tajima et al., 2013). Overexpression of a single transcription factor gene *WRI* can increase the seed oil contents, while manipulating a single fatty acid biosynthesis gene had only very limited effect on the oil content (Dehesh et al., 2001).

Transcription factors can regulate expression of genes involved in a wide range of plant processes and have a cascade amplification effect (Riechmann and Meyerowitz, 1998). Therefore, transcription factors are the promising targets to improve oil contents in plants. Several candidate transcription factors involved in fatty acid biosynthesis and accumulation have been characterized, including *WRI1* (Baud et al., 2007) and *LEC2* (*leafy cotyledon2*) (Santos-Mendoza et al., 2008) in *Arabidopsis. WRI1* is a target of LEC2. The transcription factors regulating fatty acid contents have been identified from soybean in our lab. Two Dof-type (DNA-binding one zinc finger) genes *GmDof4* and *GmDof11* were found to increase the content of total fatty acids in their transgenic *Arabidopsis* seeds by activating the ACCase and ACSL (long-chain-acyl CoA synthetase) genes, respectively (Wang et al., 2007). Through microarray analysis, a MYB-type gene, *GmMYB73*, was identified and this gene can suppress the expression of *GL2* (*GLABRA 2*), a negative regulator of oil accumulations (Liu et al., 2014). Overexpression of *GmMYB73* enhanced lipid contents in seeds of transgenic *Arabidopsis* through the release of GL2-inhibited *PLDα1* (phospholipase D) expression (Shi et al., 2012). Overexpression of *GmbZIP123* also enhanced lipid content and oil accumulation by regulating two sucrose transporter genes, *SUC1* and *SUC5*, and three cell-wall invertase genes, *cwINV1*, *cwINV3* and *cwINV6* (Song et al., 2013).

Recently, through RNA-seq analysis, gene co-expression networks have been identified for soybean seed trait regulation and *GmNFYA* (*nuclear transcription factor Y alpha*) is found to enhance seed oil contents in transgenic *Arabidopsis* plants (Lu et al., 2016). A DREB-type (dehydration-responsive element-binding) transcription factor gene *GmDREBL*, was cloned and found to increase the seed lipid content in the transgenic plants. GmDREBL directly activates the expression of *WRI1* to promote fatty acid accumulation.

Transcriptomic analyses of RHA1 grown under conditions of N-limitation and N-excess revealed 1,826 dysregulated genes. Genes whose transcripts were more abundant under N-limitation included those involved in ammonium assimilation, benzoate catabolism, fatty acid biosynthesis and the methylmalonyl-CoA pathway. Of the 16 *atf* genes potentially encoding diacylglycerol *O* acyltransferases, *atf8* transcripts were the most abundant during N-limitation (~ 50-fold more abundant than during N-excess). Consistent with Atf8 being a physiological determinant of TAG accumulation, a Δ*atf8* mutant accumulated 70% less TAG than wild-type RHA1 while *atf8* over expression increased TAG accumulation 20%.

Triacylglycerol (TAG) is the main storage lipid in plant seeds and the major form of plant oil used for food and increasingly, for industrial and biofuel applications. Several transcription factors, including FUSCA3 (At3, g26790, FUS3), are associated with embryo maturation and oil biosynthesis in seeds.

In plants, the oil formation process is composed of four steps: Fatty acid *de novo* synthesis, acyl elongation and editing, triacylglycerol (TAG) assembly and oil drop formation (Bates et al., 2013), and each step involves many genes. It has been suggested that the number of genes involved in lipid signalling and membrane lipid synthesis were two to three-fold higher, and 63% more genes involved in the plastid *de novo* fatty acid synthesis in soybean than Arabidopsis. Many single-member enzymes in Arabidopsis have multiple homologs in soybean (Schmutz et al., 2010). For instance, there is only one gene (At2g30200) encoding malonyl-CoA:ACP malonyltransferase (MCMT) but two genes (Glyma.11G164500 and Glyma.18G057700) encoding MCMT in soybean. In total, there are 1127 putative acyl lipid related genes in the soybean genome (Schmutz et al., 2010). Even though the in-depth study of fatty acid metabolism in model plant Arabidopsis lays a foundation for the study of lipid synthesis in other plants (Baud and Lepiniec, 2009), the fatty acid metabolic pathways in soybean remain to be elucidated.

The whole genome expression profile plays an important role in exploring candidate genes and investigating complex metabolisms (Wang et al., 2009). Some lipid-related candidate genes have been identified by the transcriptome analysis of Brassica napus pods (Xu et al., 2015). However, traditional differential gene expression analysis only compares the transcriptional changes between two samples each time, while the relationships between genes are not investigated. Genes with similar expression patterns may be co-regulated, functionally related or in the same pathway (Zhang and Horvath, 2005; Horvath and Dong, 2008). Gene co-expression network (GCN) analysis can simultaneously analyze the gene expression data of all samples in order to effectively identify functionally co-expressed genes. It is especially suitable for the study of complex large-scale gene expression data, such as different developmental stages of the same tissue (El-Sharkawy et al., 2015), different organs or tissues (Singh et al., 2017), responses at different time points after abiotic stress (Hopper et al., 2016) and pathogen infection (Li et al., 2011). GCN analysis includes three steps (Van Dam et al., 2018): First, the relationships between genes are determined by various measurements, such as Pearson's correlation. Second, the associations of genes are used to construct a network, where the genes are connected with each other and each node represents a gene and each edge indicates the strength of the relationship. Third, the co-expressed genes are identified using the available clustering methods, such as k-means or hierarchal clustering.

With the development of GCN analysis, it is suggested that a threshold for Pearson's correlation coefficient should be set in order to determine the existence of a network (Butte and Kohane, 2000). To pick the appropriate threshold, Zhang and Horvath proposed a new framework for 'soft' thresholding that weighs each connection, which is called the weighted correlation network (Zhang and Horvath, 2005), and constructs the co-expression network more consistently with the scale-free network distribution and biological significance.

Several groups have studied the transcriptome profiles in developing soybean seeds to analyze the patterns of differently expressed genes (DEGs) during soybean seed development (Chen et al., 2012; Jones and Vodkin, 2013), identifying the transcript sequence polymorphisms (including SNPs, small Indels and large deletions) among soybean varieties differing in oil content and composition (Goettel et al., 2014; 2016), screening candidate genes related to oil synthesis based on gene expression patterns among different tissues or in seeds of developmental stages (Liu et al., 2014), and understanding the genetic basis underlying soybean domestication (Gao et al., 2018). These studies have contributed greatly to dissecting the oil synthesis process in soybean seeds. However, the dynamic changes in transcriptome related to oil accumulation during soybean seed development are still not well understood, and the correlation between transcriptomic changes with seed oil content needs to be explored (Table 10.4 and Fig. 10.10).

Table 10.4 Differentially expressed genes between seeds of 35 DAF and 15 DAF, 55 DAF and 15 DAF, and 65 DAF and 15 DAF.

Contrast groups	Total genes	Co-expressed genes	Distinct genes	Up-regulated genes		Down-regulated genes	
				Known	Unknown	Known	Unknown
Seeds of 35 DAF vs 15 DAF	11592	9905	1687	419	261	7314	3598
Seeds of 55 DAF vs 15 DAF	16594	9905	6689	646	358	10203	5378
Seeds of 65 DAF vs 15 DAF	16255	9905	6350	529	321	10246	5159

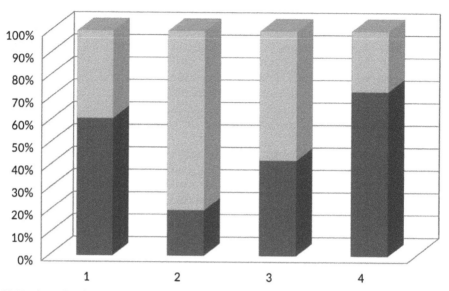

Fig. 10.10 Numbers of differently expressed genes (DEGs) in comparison with the soybean developing seeds between adjacent developmental stages were determined using the thresholds of FDR_0.05 and log2|Fold change| _ 1. The log2|Fold change| value of the pairwise comparison "a DAF vs. b DAF" was calculated by the formula: log2|Fold change| = log2(FPKM_b DAF)-log2(FPKM_a DAF). Up-regulated (black); Down-regulated (white).

References

Andre, C., R.P. Haslam and J. Shanklin. 2012. Feedback regulation of plastidic acetyl-CoA carboxylase by 18:1-acyl carrier protein in *Brassica napus*. Proceedings of National Academy of Sciences, U.S.A. 109: 10107–10112.

Anthony, H.C., A.H.C. Huang and R.A. Moreau. 1978. Lipases in the storage tissues of peanut and other oil seeds during germination. Planta. 141: 111–116.

Apel, K. and H. Hirt. 2004. Reactive oxygen species: metabolism, oxidative stress, and signal transduction. Annual Review of Plant Biology 55: 373–399.

Asada, K. 1992. Ascorbate peroxidase—a hydrogen scavenging enzyme in plants. Physiol. Plant 85: 235–241.

Audran, C., C. Borel, A. Frey, B. Sotta, C. Meyer, T. Simonneau and A. Marion-Poll. 1998. Expression studies of the zeaxanthin epoxidase gene in Nicotiana plumbaginifolia. Plant Physiology 118: 1021–1028.

Bailly, C., A. Benamar, F. Corbineau and D. Come. 1998. Free radical scavenging as affected by accelerated ageing and subsequent priming in sunflower seeds. Physiol. Plant. 104: 646–652.

Bailly, C. 2004. Active oxygen species and antioxidants in seed biology. Seed Sci. Res. 14: 93–107.

Bates, P.D., J.B. Ohlrogge and M. Pollard. 2007. Incorporation of newly synthesized fatty acids into cytosolic glycerolipids in pea leaves occurs via acyl editing. J. Biol. Chem. 282: 31206–16.

Bates, P.D., T.P. Durrett, J.B. Ohlrogge and M. Pollard. 2009. Analysis of acyl fluxes through multiple pathways of triacylglycerol synthesis in developing soybean embryos. Plant Physiology 150: 55–72.

Bates, P.D., A. Fatihi, A.R. Snapp, A.S. Carlsson, J. Browse and C. Lu. 2012. Acyl editing and headgroup exchange are the major mechanisms that direct polyunsaturated fatty acid flux into triacylglycerols. Plant Physiol. 160: 1530–9.

Bates, P.D., S. Stymne and J. Ohlrogge. 2013. Biochemical pathways in seed oil synthesis. Curr. Opin. Plant Biol. 16: 358–364.

Baud, S., B. Dubreucq, M. Miquel, C. Rochat and L. Lepiniec. 2008. Storage reserve accumulation in *Arabidopsis*: Metabolic and developmental control of seed filling. *Arabidopsis* Book American Society of Plant Biology 6: e0113.

Baud, S. and L. Lepiniec. 2009. Regulation of *de novo* fatty acid synthesis in maturing oilseeds of Arabidopsis. Plant Physiol. Biochem. 47: 448–455.

Berger, F. 2003. Endosperm: The crossroad of seed development. Current Opinion of Plant Biology 6: 42–50.

Bernal, L.I. and A.C. Leopold. 1992. Changes in soluble carbohydrates during seed storage. Plant Physiol. 98: 1207–1210.

Bewley, J.D., K. Bradford and H. Hilhorst. 2012. Seeds: Physiology of Development, Germination and Dormancy. Springer-Verlag, New York, USA.

Bligny, R. and R. Douce. 1980. A precise localisation of cardiolipin in plant cells. Biochim. Biophys. Acta 617: 254–263.

Boca, S., F. Koestler, B. Ksas, A. Chevalier, J. Leymarie, A. Fekete et al. 2014. Arabidopsis lipocalins AtCHL and AtTIL have distinct but overlapping functions essential for lipid protection and seed longevity. Plant Cell Environ. 37: 368–81.

Bongi, G. and S.P. Long. 1987. Light-dependent damage to photosynthesis in olive leaves during chilling and high temperature stress. Plant, Cell & Environment 10: 241–249.

Borek, S. and L. Ratajczak. 2010. Storage lipids as a source of carbon skeletons for asparagine synthesis in germinating seeds of yellow lupine (*Lupinus luteus* L.). Journal of Plant Physiology 167: 717–724.

Brash, A.R. 1999. Lipoxygenases: Occurrence, functions, catalysis, and acquisition of substrate. J. Biol. Chem. 274: 23679–23682.

Butte, A.J. and I.S. Kohane. 2000. Mutual information relevance networks: Functional genomic clustering using pairwise entropy measurements. Biocomputing 1999: 418–429.

Buitink, J., O. Leprince, M.A. Hemminga and F.A. Hoekstra. 2000. Molecular mobility in the cytoplasm: An approach to describe and predict lifespan of dry germplasm. Proc. Natl. Acad. Sci., U.S.A. 97: 2385–90.

Cacas, J.-L., F. Furt, M. Le Guedard, J.-H. Schmitter, C. Bure et al. 2012. Lipids of plant membrane rafts. Prog. Lipid Res. 51: 272–299.

Cao, Y.-Z. and A.H.C. Huang. 1987. Acyl coenzyme A preference of diacylglycerol acyltransferase from the maturing seeds of *Cuphea*, maize, rapeseed, and Canola. Plant Physiol. 84: 762–765.

Cernac, A. and C. Benning. 2004. WRINKLED1 encodes an AP2/EREB domain protein involved in the control of storage compound biosynthesis in *Arabidopsis*. Plant Journal 40: 575–585.

Chalon, S. 2006. Omega-3 fatty acids and monoamine neurotransmission. Prostaglandins Leukot. Essent. Fatty Acids 75: 259–269.

Chapman, K.D. and J.B. Ohlrogge. 2012. Compartmentation of triacylglycerol accumulation in plants. Journal of Biological Chemistry 287: 2288–2294.

Chen, H., F.W. Wang, Y.Y. Dong, N. Wang, Y.P. Sun, X.Y. Li, L. Liu, X.D. Fan, H.L. Yin, Y.Y. Jing et al. 2012. Sequence mining and transcript profiling to explore differentially expressed genes associated with lipid biosynthesis during soybean seed development. BMC Plant Biol. 12: 122.

Chen, Y., L. Zhao, Y. Cao, X. Kong and Y. Hua. 2014. 24 kDa and 18 kDa oleosins are not only hydrolyzed in extracted soybean oil bodies but also in soybean germination. J. Agric. Food Chem. 62: 956–965.

Chung, J., H.L. Babka, G.L. Graef, P.E. Staswick, D.J. Lee, P.B. Cregan, R.C. Shoemaker and J.E. Specht. 2003. The seed protein, oil, and yield QTL on soybean linkage group I. Crop Sci. 43: 1053–1067.

Clough, R.C., A.L. Matthis, S.R. Barnum and J.G. Jaworski. 1992. Purification and characterization of 3-Ketoacyl-Acyl carrier protein synthase-iii from Spinach—a condensing enzyme utilizing acetyl-coenzyme-a to initiate fatty-acid synthesis. Journal of Biological Chemistry 267: 20992–20998.

Crowe, L.M., R. Mourdian, J.H. Crowe, S.A. Jackson and C. Womersly. 1984. Effects of carbohydrates on membrane stability at lower water activities. Biochem. Biophys. Acta 769: 141–150.

Damude, H.G. and A.J. Kinney. 2007. Engineering oilseed plants for a sustainable, land-based source of long chain polyunsaturated fatty acids. Lipids 42: 179–185.

Debeaujon, I., K.M. Leon-Kloosterziel and M. Koornneef. 2000. Influence of the testa on seed dormancy, germination, and longevity in Arabidopsis. Plant Physiol. 122: 403–13.

Dehesh, K., H. Tai, P. Edwards, J. Byrne and J.G. Jaworski. 2001. Over expression of 3-ketoacyl-acyl-carrier protein synthase IIIs in plants reduces the rate of lipid synthesis. Plant Physiology 125: 1103–1114.

Deruyffelaere, C., I. Bouchez, H. Morin, A. Guillot, M. Miquel, M. Froissard, T. Chardot and S. D'Andrea. 2015. Ubiquitin-mediated proteasomal degradation of oleosins is involved in oil body mobilization during post-germinative seedling growth in *Arabidopsis*. Plant Cell Physiol. 56: 1374–1387.

Devaiah, S.P., X. Pan, Y. Hong, M. Roth, R. Welti and X. Wang. 2007. Enhancing seed quality and viability by suppressing phospholipase D in *Arabidopsis*. Plant J. 50: 950–7.

Dharamlingam, C. and R.N. Basu. 1990. Maintenance of viability and vigour in sunflower (*Helianthus annus* L.). Seed Res. 18: 15–25.

Dornbos, D.L. and M.B. McDonald. 1986. Mass and composition of developing soybean seeds at five reproductive growth stages. Crop Science 26: 624–630.

Dornbos, D.L. and R.E. Mullen. 1992. Soybean seed protein and oil contents and fatty acid composition adjustments by drought and temperature. Journal of American Oil Chemical Society 69: 228–231.

Drobak, B.K. 2005. Inositol-containing lipids: Roles in cellular signalling. pp. 303–328. *In*: Murphy, D.J. (ed.). Plant Lipids: Biology, Utilisation and Manipulation. Blackwell Publishing, Oxford.

Eastmond, P.J., V. Germain, P.R. Lange, J.H. Bryce, S.M. Smith and I.A. Graham. 2000. Post germinative growth and lipid catabolism in oilseeds lacking the gloxylate cycle. Proceedings of the National Academy of Sciences of the United States of America 97: 5669–5674.

Eastmond, P.J. 2006. Sugar-Dependent1 encodes apatatin domain triacylglycerol lipase that initiates storage oil breakdown in germinating Arabidopsis seeds. Plant cell Online 18: 665–675.

Egli, D.B. 1998. Seed Biology and the Yield of Grain Crops. USA, New York: CAB International.

El-Sharkawy, I., D. Liang and K. Xu. 2015. Transcriptome analysis of an apple (*Malus domestica*) yellow fruit somatic mutation identifies a gene network module highly associated with anthocyanin and epigenetic regulation. J. Exp. Bot. 66: 7359–7376.

Esteban, A.B., M.D. Sicardo, M. Mancha and J.M. Martínez-Rivas. 2004. Growth temperature control of the linoleic acid content in safflower (*Carthamus tinctorius*) seed oil. Journal of Agriculture and Food Chemistry 52: 332–336.

Falivene, S.M.P., M.A.C. de Miranda and L.D.A. de Almeida. 1980. Temperature and the occurrence of cotyledons necrosis in soybean. Revista-Brasileira-Desementes 2: 43–51.

Fang, C., Y. Ma, S. Wu, Z. Liu, Z. Wang, R. Yang et al. 2017. Genomewide association studies dissect the genetic networks underlying agronomical traits in soybean. Genome Biol. 18: 161.

Farese Jr, R.V. and T.C. Walther. 2009. Lipid droplets finally get a little R-E-S-P-E-C-T. Cell 139: 855–860.

Fatihi, A., A.M. Zbierzak and P. Dörmann. 2013. Alterations in seed development gene expression affect size and oil content of Arabidopsis seeds. Plant Physiology 163: 973–985.

Fehr, W.R. and C.E. Caviness. 1977. Stages of soybean development. Cooperative Extension Service, Agriculture and Home Economics Experiment Station, Iowa State University, Ames, Iowa.

Feussner, I. and C. Wasternack. 2002. The lipoxygenase pathway. Annu. Rev. Plant Biol. 53: 275–297.

Finch-Savage, W.E. and G.W. Bassel. 2016. Seed vigour and crop establishment: Extending performance beyond adaptation. J. Exp. Bot. 67: 567–91.

Foyer, C.H., M. Leiandais, C. Galap and K.J. Kunert. 1991. Effects of elevated cytosolic glutathione reductase activity on the cellular glutathione pool and photosynthesis in leaves under normal and stress conditions. Plant Physiol. 97: 863–872.

Foyer, C.H., M. Leiandais and K.J. Kunert. 1994. Photooxidative stress in plants. Physiol. Plant. 92: 696–717.

Frankel, E.N. 1980. Soybean oil flavor stability. Handbook of Soy Oil Processing and Utilization. St. Louis, American Soybean Association and American Oil Chemists' Society. pp. 229–244.

Freire, G.D.M. and F.L. Castilho. 2008. Lipases em Biocatálise. *In*: Bon et al. (org) (ed.). Enzimas em biotecnologia: Produção, Aplicação e Mercado. Rio de Janeiro: Interciência.

Frey, A., B. Godin, M. Bonnet, B. Sotta and A. Marion-Poll. 2004. Maternal synthesis of abscisic acid controls seed development and yield in *Nicotiana plumbaginifolia*. Planta 218: 958–964.

Fridovich, I. 1986. Biological effects of the superoxide radical. Arch. Biochem. Biophys. 147: 1–11.

Gao, H., Y. Wang, W. Li, Y. Gu, Y. Lai, Y. Bi and C. He. 2018. Transcriptomic comparison reveals genetic variation potentially underlying seed developmental evolution of soybeans. J. Exp. Bot. 69: 5089–5104.

Gibson, L.R. and R.E. Mullen. 1996. Soybean seed quality reductions by high day and night temperature. Crop Science 36: 1615.

Gibson, S.I. 2004. Sugar and phytohormone response pathways: navigating a signalling network. Journal of Experimental Botany 55: 253–264.

Gillen, A.M., J.R. Smith, A. Mengistu and N. Bellaloui. 2012. Effects of maturity and *Phomopsis longicolla* on germination and vigor of soybean seed of near-isogenic lines. Crop Sci. 52: 2757–66.

Goettel, W., E. Xia, R. Upchurch, M.L. Wang, P.Y. Chen and Y.Q. An. 2014. Identification and characterization of transcript polymorphisms in soybean lines varying in oil composition and content. BMC Genom. 15: 299.

Goettel, W., M. Ramirez, R.G. Upchurchand and Y.Q. An. 2016. Identification and characterization of large DNA deletions affecting oil quality traits in soybean seeds through transcriptome sequencing analysis. Appl. Genet. 129: 1577–1593.

Graham, I.A. 2008. Seed storage oil mobilization. Annu. Rev. Plant Biol. 59: 115–142.

Gupta, A. and K.R. Aneja. 2004. Seed deterioration in soybean varieties during storage-physiological attributes. Seed Res. 32: 26–32.

Halliwell, B. and J.M.C. Gutteridge. 1984. Oxygen toxicity, oxygen radicals, transition metals and disease. Biochemical Journal 219: 1–14.

Halliwell, B. and J.M.C. Gutteridge. 2006. Free Radicals in Biology and Medicine, Ed 4. Clarendon Press.

Halliwell, B. 2006. Reactive species and antioxidants. Redox biology is a fundamental theme of aerobic life. Plant Physiol. 141: 312–322.

Harada, J.J. 1997. Seed maturation and control of germination. pp. 545–592. *In*: Cellular and Molecular Biology of Plant Seed Development. Kluwer Academic, Dordrecht, Netherland.

Harwood, J.L. 1980. Plant acyl lipids: Structure, distribution and analysis. pp. 1–50. *In*: Stumpf, P.K. and E.E. Conn (eds.). The Biochemistry of Plants. vol. 4, Academic Press, New York.

He, D., C. Han, J. Yao, S. Shen and P. Yang. 2011. Constructing the metabolic and regulatory pathways in germinating rice seeds through proteomic approach. Proteomics 11: 2693–2713.

Hellyer, S.A., I.C. Chandler and J.A. Bosley. 1999. Can the fatty acid selectivity of plant lipases be predicted from the composition of the seed triglyceride? Biochemica Biophysica Acta 1440(2-3): 215.

Hemmerlin, A., J.L. Harwood and T. Bach. 2012. A raison d'être for two distinct pathways in the early steps of plant isoprenoid biosynthesis? Prog. Lipid Res. 51: 95–148.

Heppard, E.P., A.J. Kinney, K.L. Stecca and G.H. Miao. 1996. Developmental and growth temperature regulation of two different microsomal omega-6 desaturase genes in soybeans. Plant Physiol. 110: 311–319.

Hernandez-Sebastia, C., F. Marsolais, C. Saravitz, D. Israel, R.E. Dewey and S.C. Huber. 2005. Free amino acid profiles suggest a possible role for asparagine in the control of storage product accumulation in developing seeds of low- and high protein soybean lines. Journal of Experimental Botany 56: 1951–1963.

Hernández, M.L., M.N. Padilla, M. Mancha and J.M. Martínez-Rivas. 2009. Expression analysis identifies FAD2-2 as the olive oleate desaturase gene mainly responsible for the linoleic acid content in virgin olive oil. J. Agric. Food Chem. 57: 6199–6206.

Hildebrand, D. and G. Collins. 1998. Manipulation of linolenate and other fatty acids in soybean oil. Soybean Research Documents Online. Available: http://www.ag.uiuc.edu/~stratsoy/research.

Hildebrand, D.F., K. Yu, C. McCracken and S.S. Rao. 2005. Fatty acid manipulation. pp. 67–102. *In*: Murphy, D.J. (ed.). Plant Lipids: Biology, Utilisation and Manipulation. Blackwell Publishing, Oxford.

Hopper, D.W., R. Ghan, K.A. Schlauch and G.R. Cramer. 2016. Transcriptomic network analyses of leaf dehydration responses identify highly connected ABA and ethylene signaling hubs in three grapevine species differing in drought tolerance. BMC Plant Biol. 16: 118.

Horvath, S. and J. Dong. 2008. Geometric interpretation of gene coexpression network analysis. PLoS Comput. Biol. 4: e1000117.

Hou, G., G.R. Ablett, K.P. Pauls and I. Rajcan. 2006. Environmental effects on fatty acid levels in soybean seed oil. J. Am. Oil Chem. Soc. 83: 759–763.

Hsieh, K. and A.H.C. Huang. 2004. Endoplasmic reticulum, oleosins, and oils in seeds and tapetum cells. Plant Physiol. 136: 3427–3434.

Hu, Z., Z. Ren and C. Lu. 2012. The phosphatidylcholine diacylglycerol cholinephosphotransferase is required for efficient hydroxy fatty acid accumulation in transgenic Arabidopsis. Plant Physiol. 158: 1944–54.

Hutton, D. and P.K. Stumpf. 1969. Fat metabolism in higher plant. Characterisation of β-oxidation system from maturing and germinating caster bean seeds. Plant Physiology 44: 508–516.

Im, Y.J., I.Y. Perera, I. Brglez, A.J. Davis et al. 2007. Increasing plasma membrane phosphatidylinositol(4,5)bisphosphate biosynthesis increases phosphoinositide metabolism. Plant Cell 19: 1603–1616.

Ichihara, K. 1984. sn-Glycerol-3-phosphate acyltransferase in a particulate fraction from maturing safflower seeds. Arch. Biochem. Biophys. 232: 685–698.

Jackowski, S. and C.O. Rock. 1987. Acetoacetyl-Acyl carrier protein synthase, a potential regulator of fatty-acid biosynthesis in bacteria. J. Biol. Chem. 262: 7927–7931.

Jain, R.K., K. Coffey, A. Kumar and S.L. MacKensie. 2000. Enhancement of seed oil content by expression of glycerol-3-phosphate acyltransferase genes. Biochem. Soc. Trans. 28: 958–961.

Jeng, T.L. and J.M. Sung. 1994. Hydration effect on lipid peroxidation and peroxide-scavenging enzymes activity of artificially aged peanut seed. Seed Sci. Technol. 22: 531–539.

Jokic, S., R. Sudar, S. Svilovic, S. Vidovic, M. Bilic, D. Velic and V. Jurkovic. 2013. Fatty acid composition of oil obtained from soybeans by extraction with supercritical carbon dioxide. Czech Journal of Food Science 31: 116–125.

Jones, R.J. and M.L. Brenner. 1987. Distribution of abscisic acid in maize kernel during grain filling. Plant Physiology 83: 905–909.

Jones, S.I. and L.O. Vodkin. 2013. Using RNA-Seq to profile soybean seed development from fertilization to maturity. PLoS ONE 8: e59270.

Kangasjärvi, J., P. Jaspers and H. Kollist. 2005. Signalling and cell death in ozone-exposed plants. Plant, Cell and Environment 28: 1021–1036.

Karssen, C.M., D.L.C. Brinkhorst-Van der Swan, A.E. Breekland and M. Koornneef. 1983. Induction of dormancy during seed development by endogenous abscisic acid: studies on abscisic acid deficient genotypes of *Arabidopsis thaliana* (L.) *Heynh*. Planta 157: 158–165.

Kim, H.T., U.K. Choi, H.S. Ryu, S.J. Lee and O.S. Kwon. 2011. Mobilization of storage proteins in soybean seed (*Glycine max* L.) during germination and seedling growth. Biochim. Biophys. Acta 1814: 1178–1187.

Kinney, A.J. 1996. Development of genetically engineered soybean oils for food applications. Journal of Food Lipids 3: 273–292.

Leal, M.C.M., M.C. Cammarota, D.M.G. Freire and J.G.L. Sant'Anna. 2002. Hydrolytic enzymes as coadjuvants in the anaerobic treatment of dairy waste waters. Brazilian Journal of Chemical Engineering 19(2): 175.

Lee, J., R. Welti, W.T. Schapaugh and H.N. Trick. 2011. Phospholipid and triacylglycerol profiles modified by PLD suppression in soybean seed. Plant Biotechnol. J. 9: 359–372.

Lee, J., R. Welti, M. Roth, W.T. Schapaugh, J. Li and H.N. Trick. 2012. Enhanced seed viability and lipid compositional changes during natural ageing by suppressing phospholipase Dalpha in soybean seed. Plant Biotechnol. J. 10: 164–173.

Leprince, O., A. Pellizzaro, S. Berriri and J. Buitink. 2017. Late seed maturation: Drying without dying. J. Exp. Bot. 68: 827–41.

Li-Beisson, Y., B. Shorrosh, F. Beisson, M.X. Andersson, V. Arondel, P.D. Bates, S. Baud, D. Bird, A. Debono, T.P. Durrett, R.B. Franke, I.A.Graham, K. Katayama, A.A. Kelly, T. Larson, J.E. Markham, M. Miquel, I. Molina, I. Nishida, O. Rowland, L. Samuels, K.M. Schmid, H. Wada, R. Welti, C. Xu, R. Zallot and J. Ohlrogge. 2013. Acyl-lipid metabolism. *Arabidopsis* Book 11: e0161.

Li, M., S.C. Bahn, C. Fan, J. Li, T. Phan, M. Ortiz et al. 2013. Patatinrelated phospholipase pPLAIIIdelta increases seed oil content with long chain fatty acids in Arabidopsis. Plant Physiol. 162: 39–51.

Li, M., F. Wei, A. Tawfall, M. Tang, A. Saettele and X. Wang. 2015. Overexpression of patatinrelated phospholipase AIIIdelta altered plant growth and increased seed oil content in camelina. Plant Biotechnol. J. 13: 766–78.

Li, C., Z. Guan, D. Liu and C.R. Raetz. 2011. Pathway for lipid A biosynthesis in *Arabidopsis thaliana* resembling that of *Escherichia coli*. Proc. Natl. Acad. Sci., U.S.A. 108: 11387–11392.

Li, R., K. Yu and D.F. Hildebrand. 2010. DGAT1, DGAT2 and PDAT expression in seeds and other tissues of epoxy and hydroxyl fatty acid accumulating plants. Lipids 45: 145–157.

Lin, Y.C., Y.C. Liu and Y. Nakamura. 2015. The choline/ethanolamine kinase family in Arabidopsis: Essential role of CEK4 in phospholipid biosynthesis and embryo development. Plant Cell. 27: 1497–511.

Liu, B., X. Liu, Y.S. Li and S.J. Herbert. 2013. Effects of enhanced UV-B radiation on seed growth characteristics and yield components in soybean. Field Crop Research 154: 158–163.

Liu, J., W. Hua, G. Zhan, F. Wei, X. Wang and G. Liu. 2010. Increasing seed mass and oil content in transgenic Arabidopsis by the overexpression of WRI1-like gene from *Brassica napus*. Plant Physiology and Biochemistry 48: 9–15.

Liu, Y.F., Q.T. Li, X. Lu, Q.X. Song, S.M. Lam, W.K. Zhang, B. Ma, Q. Lin, W.Q. Man, W.G. Du, G.H. Shui, S.Y. Chenand and J.S. Zhang. 2014. Soybean GmMYB73 promotes lipid accumulation in transgenic plants. BMC Plant Biology 14: 73.

Lu, C.L., S.B. de Noyer, D.H. Hobbs, J. Kang, Y. Wen, D. Krachtus and M.J. Hills. 2003. Expression pattern of diacylglycerol acyltransferase-1, an enzyme involved in tria-cylglycerol biosynthesis in *Arabidopsis thaliana*. Plant Mol. Biol. 52: 31–41.

Lu, C., Z. Xin, Z. Ren, M. Miquel and J. Browse. 2009. An enzyme regulating triacylglycerol composition is encoded by the ROD1 gene of Arabidopsis. Proc. Natl. Acad. Sci., U.S.A. 106: 18837–4.

Lu, X., Q.T. Li, Q. Xiong, W. Li, Y.D. Bi, Y.C. Lai, X.L. Liu, W.Q. Man, W.K. Zhang, B. Ma, S.Y. Chen and J.S. Zhang. 2016. The transcriptomic signature of developing soybean seeds reveals genetic basis of seed trait adaptation during domestication. Plant Journal 86: 530–544.

Lynch, D.V. and T.M. Dunn. 2004. An introduction to plant sphingolipids. New Phytol. 161: 677–702.

Maeo, K., T. Tokuda, A. Ayame, N. Mitsui, T. Kawai, H. Tsukagoshi, S. Ishiquros and K. Nakamura. 2009. An AP2-type transcription factor, WRINKLED1, of *Arabidopsis thaliana* binds to the AW-box sequence conserved among proximal upstream regions of genes involved in fatty acid synthesis. Plant Journal 60: 476–487.

Masaki, T. et al. 2005. ACTIVATOR of Spo(min):: LUC1/WRINKLED1 of a *Arabidopsis thaliana* transactivates sugar-inducible promoters. Plant Cell Physiol. 46: 547–556.

Manan, S., M.Z. Ahmad, G. Zhang, B. Chen, B.U. Haq, J. Yang and J. Zhao. 2017. Soybean LEC2 regulates subsets of genes involved in controlling the biosynthesis and catabolism of seed storage substances and seed development. Front Plant Sci. 8: 1604.

Marcos-Filho, J. 2016. Seed physiology of cultivated plants. 2nd Edition. Londrina: AssociacÉão Brasileira de Tecnologia de Sementes-ABRATES.

McClements, D.J. and Y. Li. 2010. Review of *in vitro* digestion models for rapid screening of emulsion-based systems. Food Funct. 1: 32–59.

McDonald, M.B. 1999. Seed deterioration: Physiology, repair and assessment. Seed Science and Technology 27: 177–237.

Miquel, M. et al. 2014. Specialization of oleosins in OB dynamics during seed development in Arabidopsis thaliana seeds. Plant Physiol. 164: 1866–1878.

Murthy, U.M.N., Y.H. Liang, P.P. Kumar and W.Q. Sun. 2002. Non-enzymatic protein modification by the Maillard reaction reduces the activities of scavenging enzymes in *Vigna radiata*. Physiologia Plantarum 115: 213–220.

Muscolo, A., M. Sidari, C. Mallamaci and E. Attinà. 2007. Changes in germination and glyoxylate and respiratory enzymes of *Pinus pinea* seeds under various abiotic stresses. Journal of Plant Interactions 2(4): 273–279.

Nagel, M., I. Kranner, K. Neumann, H. Rolletschek, C.E. Seal, L. Colville et al. 2015. Genome-wide association mapping and biochemical markers reveal that seed ageing and longevity are intricately affected by genetic background and developmental and environmental conditions in barley. Plant, Cell Environ. 38: 1011–22.

Nakajima, S., H. Ito, R. Tanaka and A. Tanaka. 2012. Chlorophyll b reductase plays an essential role in maturation and storability of Arabidopsis seeds. Plant Physiol. 160: 261–73.

Nambara, E. and A. Marion-Poll. 2005. Abscisic acid biosynthesis and catabolism. Annual Review of Plant Biology 56: 165–185.

Oge, L., G. Bourdais, J. Bove, B. Collet, B. Godin, F. Granier et al. 2008. Protein repair L-Isoaspartyl methyltransferase1 is involved in both seed longevity and germination vigor in *Arabidopsis*. Plant Cell 20: 3022–37.

Ohlrogge, J.B. and T.M. Kuo. 1984. Control of lipid synthesis during soybean seed development: enzymic and immunochemical assay of acyl carrier protein. Plant Physiology 74: 622–625.

Ohlrogge, J.B. and J.G. Jaworski. 1997. Regulation of fatty acid synthesis. Annual Review of Plant Physiology 48: 109–136.

Okuyama, H. 1990. Does food affect brain functions? Protein, Nucleic Acid and Enzym. 35: 75–79.

Pan, X., R.M. Siloto, A.D. Wickramarathna, E. Mietkiewska and R.J. Weselake. 2013. Identification of a pair of phospholipid:diacylglycerol acyltransferases from developing flax (*Linum usitatissimum* L.) seed catalyzing the selective production of trilinolenin. J. Biol. Chem. 288: 24173–88.

Pantalone, V.R., G.J. Rebetzke, J.W. Burton and R.F. Wilson. 1997a. Genetic regulation of linolenic acid concentration in wild soybean Glycine soja accessions. J. Am. Oil Chem. Soc. 74: 159–163.

Paques, F.W. and G.A. Macedo. 2006. Lipases de Látex Vegetais: Propriedades e Aplicações Industriais: A review. Química Nova. 29(1): 93.

Pereira, E.P., G.M. Zanin and H.F. Castro. 2003. Immobilization and catalytic properties of lipase on chitosan for hydrolysis and etherification reactions. Brazilian Journal of Chemical Engineering 20(4): 343.

Perino, C. and D. Come. 1991. Physiological and metabolic study of the germination phases in apple embryo. Seed Science and Technology 19: 1–14.

Petruzelli, L. and G. Taranto. 1989. Wheat ageing: The contribution of embryonic and non-embryonic lesions to loss of seed viability. Physiol. Plant. 76: 289–294.

Pham, A.T., J.D. Lee, J.G. Shannon and K.D. Bilyeu. 2010. Mutant alleles of FAD2-1A and FAD2-1B combine to produce soybeans with the high oleic acid seed oil trait. BMC Plant Biology 10: 195.

Pham, A.T., J.D. Lee, J.G. Shannon and K.D. Bilyeu. 2011. A novel FAD2-1 A allele in a soybean plant introduction offers an alternate means to produce soybean seed oil with 85% oleic acid content. Theoretical and Applied Genetics 123: 793–802.

Pipolo, A.E., T.R. Sinclair and G.M.S. Camara. 2004. Protein and oil concentration of soybean seed cultured *in vitro* using nutrient solutions of differing glutamine concentration. Annuals of Applied Biology 144: 223–227.

Porta, H. and M. Rocha-Sosa. 2002. Plant lipoxygenases. Physiological and molecular features. Plant Physiol. 130: 15–21.

Pouvreau, B., S. Baud, V. Vernoud, V. Morin, C. Py, G. Gendrot, J.P. Pichon, J. Rouster, W. Paul and P.M. Rogowsky. 2011. Duplicate maize wrinkled1 transcription factors activate target genes involved in seed oil biosynthesis. Plant Physiology 156: 674–686.

Pracharoenwattana, I. and S.M. Smith. 2008. When is a peroxisome not a peroxisome? Trends in Plant Science 13: 522–525.

Puntaluro, S., R.A. Sanchez and A. Boveris. 1988. Hydrogen peroxide metabolism in soybean embryonic axes at the onset of germination. Plant Physiol. 86: 626–630.

Quettier, A.L. and P.J. Eastmond. 2009. Storage oil hydrolysis during early seedling growth. Plant Physiology and Biochemistry 47: 485.

Ramli, U.S., D.S. Baker, P.A. Quant and J.L. Harwood. 2002. Control mechanisms operating for lipid biosynthesis differ in oil palm and olive callus cultures. Biochemistry Journal 364: 385–391.

Rao, S.S. and D. Hildebrand. 2009. Changes in oil content of transgenic soybeans expressing the yeast SLC1 gene. Lipids 44: 945–951.

ap Rees, T. 1980. Integration of pathways of synthesis and degradation of hexose phosphates. pp. 1–42. *In*: Preiss, J. (ed.). The Biochemistry of Plants. London: Academic Press.

Rhoads, D.M., A.L. Umbach, C.C. Subbaiah and J.N. Siedow. 2006. Mitochondrial reactive oxygen species. Contribution to oxidative stress and interorganellar signaling. Plant Physiol. 141: 357–366.

Riechmann, J.L. and E.M. Meyerowitz. 1998. The AP2/EREBP family of plant transcription factors. Biological Chemistry 379: 633–646.

Robert, R., C. Stewart and J.D. Bewley. 1980. Lipid peroxidation associated with accelerated aging of soybean axes. Plant Physiol. 65: 245–248.

Rolletschek, H., L. Borisjuk, A. Sánchez-García, C. Gotor, L.C. Romero, J.M. Martínez-Rivas and M. Mancha. 2007. Temperature dependent endogenous oxygen concentration regulates microsomal oleate desaturase in developing sunflower seeds. Journal of Experimental Botany 58: 3171–3181.

Rolletschek, H., T.H. Nguyen, R.E. Hausler et al. 2007. Antisense inhibition of the plastidial glucose-6-phosphate/phosphate translocator in Vicia seeds shifts cellular differentiation and promotes protein storage. Plant Journal 51: 468–484.

Rondanini, D., R. Savin and A.J. Hall. 2003. Dynamics of fruit growth and oil quality of sunflower (*Helianthus annuus* L.) exposed to brief intervals of high temperature during grain filling. Field Crop Research 83: 79–90.

Ruuska, S.A., T. Girke, C. Benning and J.B. Ohlrogge. 2002. Contrapuntal networks of gene expression during *Arabidopsis* seed filling. Plant Cell 14: 1191–1206.

Salvi, P., S.C. Saxena, B.P. Petla, N.U. Kamble, H. Kaur, P. Verma et al. 2016. Differentially expressed galactinol synthase(s) in chickpea are implicated in seed vigor and longevity by limiting the age induced ROS accumulation. Sci. Rep. 6: 35088.

Sano, N., L. Raijou, N.M. North, I. Debeaujon, A. Marion-Poll and M. Seo. 2016. Staying alive: Molecular aspects of seed longevity. Plant Cell Physiol. 57: 660–74.

Santos-Mendoza, M., B. Dubreucq, S. Baud, F. Parcy, M. Caboche and L. Lepiniec. 2008. Deciphering gene regulatory networks that control seed development and maturation in *Arabidopsis*. Plant Journal 54: 608–620.

Sattler, S.E., L.U. Gilliland, M. Magallanes-Lundback, M. Pollard and D. DellaPenna. 2004. Vitamin E is essential for seed longevity, and for preventing lipid peroxidation during germination. Plant Cell 16: 1419–32.

Sauer, D.B., C.M. Meronuck and C.M. Christensen. 1992. Microflora. *In*: Sauer, D.B. (ed.). Storage of Cereal Grains and their Products. American Association of Cereal Chemists, USA.

Sharma, S., P. Virdi, S. Gambhir and S.K. Munshi. 2005. Changes in soluble sugar content and antioxidant enzymes in soybean seeds stored under different storage conditions. Ind. J. Agric. Biochem. 18: 9–12.

Sharma, S., S. Gambhir and S.K. Munshi. 2006. Effect of temperature on vigour and biochemical composition of soybean seed during storage. J. Res. Punjab Agric. Univ. 43: 29–33.

Shen, B., K.W. Sinkevicius, D.A. Selinger and M.C. Tarczynski. 2006. The homeobox gene GLABRA2 affects seed oil content in *Arabidopsis*. Plant Mol. Biol. 60: 377–387.

Shen, B. et al. 2010. Expression of ZmLEC1 and ZmWRI1 increases seed oil production in maize. Plant Physiol. 153: 980–987.

Shi, L., V. Katavic, Y.Y. Yu, L. Kunst and G. Haughn. 2012. Arabidopsis glabra2 mutant seeds deficient in mucilage biosynthesis produce more oil. Plant Journal 69: 37–46.

Shimakata, T. and P.K. Stumpf. 1982. Isolation and function of Spinach leaf beta-Ketoacyl-[Acyl-Carrier-Protein] synthases. Proceedings of National Academy of Science, U.S.A. 79: 5808–5812.

Shintani, D., K. Roesler, B. Shorrosh, L. Savage and J. Ohlrogge. 1997. Antisense expression and overexpression of biotin carboxylase in tobacco leaves. Plant Physiol. 114: 881–886.

Shoshii, M. and H. Reevers. 1974. Lipase activity in castor been endosperm during germination. Plant Physiology 54: 23–28.

Shu, Y., Y. Tao, S. Wang, L. Huang, X. Yu, Z. Wang et al. 2015. GmSBH1, a homeobox transcription factor gene, relates to growth and development and involves in response to high temperature and humidity stress in soybean. Plant Cell Rep. 34: 1927–37.

Siloto, R.M.P., K. Findlay, A. Lopez-Villalobos, E.C. Yeung, C.L. Nykiforuk and M.M. Moloney. 2006. The accumulation of oleosins determines the size of seed oil bodies in *Arabidopsis*. Plant Cell Online 18: 1961–1974.

Singh, V.K., M.S. Rajkumar, R. Garg and M. Jain. 2017. Genome-wide identification and co-expression network analysis provide insights into the roles of auxin response factor gene family in chickpea. Sci. Rep. 7: 10895.

Singh, K.K. and M. Dadlani. 2003. Effect of packaging on vigour and viability of soybean seed during ambient storage. Seed Res. 31: 27–32.

Somerville, C. and J. Browse. 1991. Plant lipids: Metabolism, mutants and membranes. Science 252: 80–87.

Song, Q.X., Q.T. Li, Y.F. Liu, F.X. Zhang, B. Ma, W.K. Zhang, W.Q. Man, W.G. Du, G.D. Wang, S.Y. Chen and J.S. Zhang. 2013. Soybean GmbZIP123 gene enhances lipid content in the seeds of transgenic *Arabidopsis* plants. Journal of Experimental Botany 64: 4329–4341.

Somchai, P.O. 1999. Quality and storability of soybean seed after conditioning. www.grad.emu.ac.th/abstract/1999/agi/abstract/agi990007.html.

SoyStats International. 2015. World Oilseed Production 2014. Available: http://soystats.com/international-world-oilseed-production/.

Stumpf, P.K. and E.E. Conn (eds.). 1980. Plant Biochemistry. vol. 4, Academic Press, New York.

Stymne, S. and A.K. Stobart. 1984. Evidence for the reversibility of the acyl-CoA lysophosphatidyl choline acyltransferase in microsomal preprations from developing safflower (*Carthamus tinctorious* L.) cotyledons and rat liver. Biochem. Journal 223: 305–314.

Sung, J.M. and T.L. Jeng. 1994. Lipid peroxidation and peroxide-scavenging enzymes associated with accelerated aging of peanut seed. Physiol. Plant. 91: 51–55.

Taarit, M.B., K. Msaada, K. Hosni and B. Marzouk. 2010. Essential oil composition of sage (*Salvia officinalis* L.) leaves under NaCl stress. Food Chemistry 119: 951–956.

Tajima, D., A. Kaneko, M. Sakamoto, Y. Ito, N.T. Hue, M. Miyaaki, Y. Ishibashi, T. Yussa and M. Iwaya-Inoue. 2013. Wrinkled 1 (WRI1) homologs, AP2-type transcription factors involving master regulation of seed storage oil synthesis in castor bean (*Ricinus communis* L.). American Journal of Plant Sciences 4: 333–339.

Tang, M., I.A. Guschina, P. O'Hara, A.R. Slabas, P.A. Quant, T. Fawcett and J.L. Harwood. 2012. Metabolic control analysis of developing oilseed rape (*Brassica napus* cv Westar) embryos shows that lipid assembly exerts significant control over oil accumulation. New Phytol. 196: 414–426.

Tavener, R.J.A. and D.L. Laidman. 1972. The induction of triglyceride metabolism in the germinating wheat grain. Phytochemistry 11: 981–987.

Tekrony, D.M., C. Nelson, D.B. Egli and G.M. White. 1993. Predicting soybean seed germination during warehouse storage. Seed Sci. Technol. 21: 127–137.

Truong, Q., K. Koch, J.M. Yoon, J.D. Everard and J.V. Shanks. 2013. Influence of carbon to nitrogen ratios on soybean somatic embryo (cv. Jack) growth and composition. Journal of Experimental Botany 64: 2985–2995.

Van Dam, S., U. Vosa, A. van der Graaf, L. Franke and J.P. de Magalhaes. 2018. Gene co-expression analysis for functional classification and gene-disease predictions. Brief. Bioinform. 19: 575–592.

Vogel, G. and J. Browse. 1996. Cholinephosphotransferase and diacylglycerolacyltransferase: Substrate specificities a key branch point in seed lipid metabolism. Plant Physiology 110: 923–931.

Vogler, H. and C. Kuhlemeier. 2003. Simple hormones but complex signalling. Current Opinion in Plant Biology 6: 51–56.

Walters, C. 1998. Understanding the mechanisms and kinetics of seed ageing. Seed Science Research 8: 223–244.

Walters, C., L.M. Wheeler and J.M. Grotenhuis. 2005. Longevity of seeds stored in a genebank; species characteristics. Seed Sci. Res. 15: 1–20.

Walters, C., D. Ballesteros and V.A. Vertucci. 2010. Structural mechanics of seed deterioration: Standing the test of time. Plant Sci. 179: 565–73.

Wang, H.Y., J.H. Guo, K.N. Lambert and Y. Lin. 2007. Developmental control of *Arabidopsis* seed oil biosynthesis. Planta 226: 773–783.

Wang, X.M. and D.F. Hildebrand. 1988. Biosynthesis and regulation of linolenic acid higher plants. Plant Physiology and Biochemistry 26: 777–792.

Wang, Y. and M. Frei. 2011. Stressed food—The impact of abiotic environmental stresses on crop quality. Agriculture, Ecosystem and Environment 141: 271–286.

Wang, Z., M. Gerstein and M. Snyder. 2009. RNA-Seq: A revolutionary tool for transcriptomics. Nat. Rev. Genet. 10: 57–63.

Wang, L., H. Ma, L. Song, Y. Shu and W. Gu. 2012. Comparative proteomics analysis reveals the mechanism of preharvest seed deterioration of soybean under high temperature and humidity stress. J. Proteomics. 75: 2109–27.

Wang, L., W. Shen, M. Kazachkov, G. Chen, Q. Chen, A.S. Carlsson et al. 2012. Metabolic interactions between the Lands cycle and the Kennedy pathway of glycerolipid synthesis in Arabidopsis developing seeds. Plant Cell 24: 4652–69.

Wang, L., M. Kazachkov, W. Shen, M. Bai, H. Wu and J. Zou. 2014. Deciphering the roles of Arabidopsis LPCAT and PAH in phosphatidylcholine homeostasis and pathway coordination for chloroplast lipid synthesis. Plant J. 80: 965–76.

Wardlaw, I.F., C. Blumenthal, O. Larroque and C.W. Wrigley. 2002. Contrasting effects of chronic heat stress and heat shock on kernel weight and flour quality in wheat. Functional Plant Biology 29: 25.

Waterworth, W., G. Masnavi, R.M. Bhardwaj, Q. Jiang, C.M. Bray and C.E. West. 2010. A plant DNA ligase is an important determinant of seed longevity. Plant J. 63: 848–60.

Wickramarathna, A.D., R.M.P. Siloto, E. Mietkiewska, S.D. Singer, X. Pan and R.J. Weselake. 2015. Heterologous expression of flax phospholipid:diacylglycerol cholinephosphotransferase (PDCT) increases polyunsaturated fatty acid content in yeast and Arabidopsis seeds. BMC Biotechnol. 15: 63.

Wilkinson, S. and W.J. Davies. 2010. Drought, ozone, ABA and ethylene: New insights from cell to plant to community. Plant Cell Environment 33: 510–525.

Wilson, D.O. and M.B. McDonald. 1986. The lipid peroxidation model of seed ageing. Seed Science and Technology 14: 269–300.

Wu, G.Z. and H.W. Xue. 2010. *Arabidopsis* beta-Ketoacyl-[Acyl Carrier Protein] synthase I is crucial for fatty acid synthesis and plays a role in chloroplast division and embryo development. Plant Cell 22: 3726–3744.

Xu, H.M., X.D. Kong, F. Chen, J.X. Huang, X.Y. Lou and J.Y. Zhao. 2015. Transcriptome analysis of Brassica napus pod using RNA-Seq and identification of lipid-related candidate genes. BMC Genom. 16: 858.

Yadav, N.S. 1996. Genetic modification of soybean oil quality. pp. 127–188. *In*: Verma, D.P.S. and R.C. Shoemaker (eds.). Soybean Genetics, Molecular Biology and Biotechnology. CAB International, New York, U.S.A.

Yang, W., G. Wang, J. Li and P.D. Bates. 2017. Phospholipase Dzeta enhances diacylglycerol flux into triacylglycerol. Plant Physiol. 174: 110–23.

Yoshida, H., Y. Hirakawa, C. Murakami, Y. Mizushina and T. Yamade. 2003. Variation in the content of tocopherols and distribution of fatty acids within soya bean seeds (*Glycine max* L.). J. Food Compos. Anal. 16: 429–440.

Zalewski, W., P. Galuszka, S. Gasparis, W. Orczyk and A. Nadolska Orczyk. 2010. Silencing of the HvCKX1 gene decreases the cytokinin oxidase/dehydrogenase level in barley and leads to higher plant productivity. Journal of Experimental Botany 61: 1839–1851.

Zanakis, G.N., R.H. Ellis and R.J. Summerfield. 1994. Seed quality in relation to seed development and maturation in 3 genotypes of soyabean (*Glycine max*). Expl. Agric. 30: 139–156.

Zhang, B. and S. Horvath. 2005. A general framework for weighted gene co-expression network analysis. Stat. Appl. Genet. Mol. Biol. 4: 17.

Zinsmeister, J., D. Lalanne, E. Terrasson, E. Chatelain, C. Vandecasteele, B.L. Vu et al. 2016. ABI5 is a regulator of seed maturation and longevity in legumes. Plant Cell 28: 2735–54.

CHAPTER 11
Plant Growth Regulators

Plant regulators are organic compounds which, in small amounts, somehow modify a given physiological plant process and rarely act alone, as the action of two or more of these compounds is necessary to produce a physiological effect. Gibberellins (GAs) play an essential role in many aspects of plant growth and development, such as seed germination (Maske et al., 1997), stem elongation and flower development (Yamaguchi and Kamiya, 2000). They are extensively used to manipulate flower formation and fruit set in horticultural plants. When applied at the pre-blooming stage, GAs decrease the number of flowers and fruit set, probably by increasing vegetative mass which, in turn, shares the photoassimilates with the fruit (Birnberg and Brenner, 1987).

Cytokinins (CK) are known to stimulate or inhibit a great number of physiological processes. For soybean, however, great variability can be observed in reported results. These dissimilar results primarily occur because of differences between concentrations utilized and differences in the physiological stages at which these products are applied. CK can be utilized in a variety of applications, from the treatment of seeds (Riedell et al., 1985) to applications during flowering (Dyer et al., 1986), and the same is true of GA.

11.1 Seed Germination

Plants are characterized by producing various types of growth regulators that differ in their chemical structure and physiological action. They include auxins, cytokinins (CK), gibberellins (GA), abscisic acid (ABA), ethylene (ET), salicylic acid (SA), jasmonates (JA), brassinosteroids (BR) and strigolactones. Each of ABA, SA, JA and ET is found to play an essential role in mediating plant defense response against stresses. During the early phase of seed germination, a decrease in JA and SA contents and an increased level of auxins were recorded in *Arabidopsis* seeds. Both JA and SA were shown to act as negative regulators of seed germination. Auxins are considered to be regulators of the seed germination process in a crosstalk with GAs, ABA, and ET. The brassinosteroids' signal could stimulate germination by decreasing the sensitivity to ABA.

A variety of cellular processes in plants are under the control of phytohormones which play key roles and coordinate various signal transduction pathways during abiotic-stress response. Seed imbibitions resulted in an activation of GA biosynthesis and response pathways with the production of the bioactive GAs. Then, GAs stimulated the genes encoding for enzymes such as endo-β-1,3 glucanase and β-1,4 mannan endohydrolase which hydrolyze the endosperm and alleviate the inhibitory effects of ABA on embryo growth potential. These results indicate the antagonistic relation between each of ABA and GA which interpret the presence of high GA and low ABA levels in seeds under favourable environmental conditions and a reverse ratio under unfavorable conditions. Thus, the crosstalk relation between seed dormancy and germination is balanced by GA-ABA ratio, a key mechanism for cope early abiotic-stress conditions.

ABA inhibits water uptake by preventing cell wall loosening of the embryo, thereby reducing embryo growth potential. GAs are involved in the direct enhancement of embryo growth during the late phase. GAs repress the ABA effect during the early and the late phases of germination through stimulation of genes expression encoding cell wall loosening, resulting in remodeling of enzymes, such as α-expansins, in the early phase of germination.

Light and cold act together to break dormancy of imbibed seeds and promote seed germination by increasing GAs levels. A rapid decrease of ABA endogenous content during Phase II is one of many factors that influence the successful completion of germination. Significant leakage of cellular solutes due to initial imbibition indicates cellular membranes damage caused by rehydration. In addition, drying and rapid seed dehydration processes influence DNA integrity. Seeds have developed a number of repair mechanisms during seed germination, including the repair of membranes, as well as proteins and DNA.

Under stress conditions, phytohormones play a crucial role via responsive protein mediated stress. C1-(cysteine rich protein family) domain containing proteins that play a part in plant hormone-mediated stress responses were found. In addition, 72 responsive proteins mediating stress are identified in *Arabidopsis* that contained all three unique signature domains. Many hydrolytic enzymes' biosynthesis and activity are influenced by GA_3 in wheat and barley. Catalase and ascorbate peroxidase activity showed a significant improvement in wheat SA- and GA-primed wheat seeds compared to unprimed.

Seed germination and post-germinative growth are critical stages during the plant life-cycle. The phytohormone abscisic acid (ABA) delays seed germination, whereas gibberellin (GA) promotes this process, and both hormones are key regulators involved in seed germination processes (Shu et al., 2016b). Consistently, mutants with dysfunctional ABA and GA biosynthesis, and catabolic or signaling transduction pathways always show an altered germination phenotype. In *Arabidopsis*, the ABA biosynthesis mutants nced6, nced5, nced3, and aba2 (Seo et al., 2006; Frey et al., 2012), and ABA signaling mutants abi3 (Koornneef et al., 1989), abi4 (Shu et al., 2013), and abi5 (Piskurewicz et al., 2008) had a faster germination phenotype compared to wild type, whereas overexpression of *Nicotiana* ABA2 or *Arabidopsis* ABI4 resulted in a delayed-germination phenotype (Frey et al., 1999; Shu et al., 2013). However, the seeds of GA-deficiency mutants, ga1 and ga2, failed to germinate unless there was an exogenous supply of GA (Lee et al., 2002; Shu et al., 2013), while the mutants that are defective in GA2-oxidases (GA2ox), which deactivate bioactive GA, have a faster germination phenotype (Yamauchi et al., 2007). Further, the ratio between the levels of active GA/ABA is also a key determinant of seed germination (Meng et al., 2016). Taken together, both ABA and GA levels, along with signaling, play important roles in the regulation of seed germination.

Seed germination is a key stage during a plant's life-cycle, and the germination process is determined by diverse environmental cues, such as the availability of suitable levels of light, water and oxygen, as well as the presence of endogenous phytohormones (Oracz and Karpinski, 2016). Abscisic acid (ABA) promotes seed dormancy and, thus, inhibits seed germination, while gibberellins (GAs) release seed dormancy and promote seed germination. These are the key hormonal regulators of seed dormancy and germination, and have been very well studied in past decades (Shu et al., 2016).

Another important phytohormone, auxin, is involved in almost all stages of plant development, including root growth, apical dominance, fruit growth and response to environmental signals (Wang and Estelle, 2014; Strader and Zhang, 2016). Auxin also interacts with ABA to regulate plant drought and osmotic stress responses in Chinese kale (*Brassica oleracea* var. *alboglabra*) (Zhu et al., 2016). In addition, ABA and auxin signaling are also mediated by the type B Gγ subunit SlGGB1 in tomato, with silencing of *SlGGB1* affecting seed development, lateral root growth and fruit shape formation (Subramaniam et al., 2016). Recently, auxin was demonstrated to be a regulator of several phytohormone signaling pathways, including ABA and GA, and, as a result, regulated tomato fruit formation (Ren and Wang, 2016).

Regarding seed germination processes, earlier studies showed that exogenous application of the auxin indole-3-acetic acid (IAA) delayed wheat seed germination and inhibited pre-harvest sprouting on mother plants (Ramaih et al., 2003), while exogenous auxin treatment also inhibited seed germination under salt stress conditions in *Arabidopsis* (Park et al., 2011), although the mechanisms underlying the inhibitory effects of auxin on seed germination are yet to be elucidated (Fig. 11.1).

A recent study revealed that auxin induced seed dormancy by enhancing ABA signal transduction, identifying auxin as a promoter of seed dormancy (Liu et al., 2013). Detailed investigation showed that *ABI3* is required for auxin-activated seed dormancy. The auxin-responsive transcription factors, *ARF10*

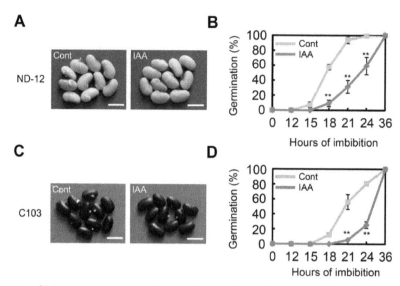

Fig. 11.1 Exogenous IAA treatment represses soybean seed germination under dark conditions. Healthy and elite soybean seeds (cultivars ND-12 and C-103) were incubated on two layers of filter paper in Petri dishes. The concentration of IAA used was 1 μM, and the equivalent amount of ultrapure water was added as control (Cont). The germination rates under dark conditions were recorded using a safe green light. Quantitative analysis of germination rates is shown in the right panels. The representative images (21 hours after sowing) are shown (left panels). (A, B) for cultivar ND-12; (C, D) for cultivar C-103. Bar in panel A and C = 10 mm. The average percentages of four repeats ± standard error are shown. ** Difference is significant at the 0.01 level.

and *ARF16*, indirectly promote *ABI3* transcription, and consequently maintain seed dormancy levels and repress germination (Liu et al., 2013). However, there are some important scientific questions which remain unanswered. For instance, does auxin regulate ABA biosynthesis during seed germination, and, if so, how does that happen? Furthermore, the relationship between auxin and GA biosynthesis is still not fully clear.

Currently, seed priming techniques include osmopriming (soaking seeds in osmotic solutions as PEG or in salt solutions), hydropriming (soaking seeds in predetermined amounts of distilled water or limiting imbibition periods), and hormone priming (seed are treated with plant growth regulators) which are more commonly studied in laboratory conditions, and thermopriming (physical treatment achieved by pre-sowing of seeds at different temperature that improve germination vigor under adverse environmental conditions) and matric priming (mixing seeds with organic or inorganic solid materials and water in definite proportions and, in some cases, adding chemical or biological agents) (Paparella et al., 2015; Jisha et al., 2013). Hydropriming and osmopriming with large-sized priming molecules cannot permeate cell wall/membrane so water influx would be the only external factor affecting priming. The determination of suitable priming technique is dependent mainly on plant species, seed morphology and physiology.

On the other hand, salts and hormone priming affect not only the seed hydration but also other germination-related processes due to absorption of exogenous ions/hormones, consequently confusing the effects of imbibition *versus* that of ions/hormones.

Improvement germination performance of primed seeds may be considered a result of advanced metabolism processes (Soeda et al., 2015) including enhancement each of the efficiency of respiration (Sun et al., 2011) and antioxidant activity (Chen and Arora, 2011), initiation of repairing processes (Balestrazzi et al., 2011) and alteration of phytohormonal balance (El-Araby et al., 2006). Also, improvement of germination performance may be linked to higher expressions of genes and proteins involved in water transport, cell wall modification, cytoskeletal organization, and cell division and increases in protein synthesis potential, post-translational processing capacity, and targeted proteolysis have been linked to the advanced germination of primed seeds (Kubala et al., 2015; Hilhorst, 2007).

ABA pre-treated seeds showed a reduction in germination that may be attributed to metabolic deviation, limiting the available energy and changes in metabolomics or to the modulating of the endogenous ABA level (Fincher, 1989). On the contrary, GA$_3$ seed treatment has not affect seed germination substantially. It is documented that GA$_3$ have a stimulatory effect on germination and the associated enzymes (Liu et al., 2013). Also, auxin (IAA) is documented to regulate seed dormancy and plant shade avoidance syndrome that adversely affects seedling development and crop yield (Procko et al., 2014). Cytokinin pre-treatment may act as auxins in promoting seed germination by antagonizing the inhibitory effect of ABA on the germination process. However, it was found that cytokinin antagonizes the inhibitory effect of ABA on post-germinating growth of *Arabidopsis* through the stimulation of ABI5 protein degradation (Guan et al., 2014).

In seed physiology, exogenous H$_2$O$_2$ increased ABA catabolism by enhancing the expression of CYCP707A genes, played a major role in ABA catabolism and enhanced gibberellic acid (GA) biosynthesis genes in *Arabidopsis* dormant seeds (Liu et al., 2010). ROS regulated the expression of ethylene response factor ERF1, a component of the ethylene signalling pathway in sunflower seed germination (Oracz et al., 2009). In barley dormant seed, H$_2$O$_2$ enhanced GA synthesis genes, such as GA20ox1, rather than repressing ABA signalling in embryo (Bahin et al., 2011). Recently, researchers have also shown that ROS regulate the induction of a-amylase through gibberellin–ABA signalling in barley aleurone cells (Ishibashi et al., 2012) (Fig. 11.2).

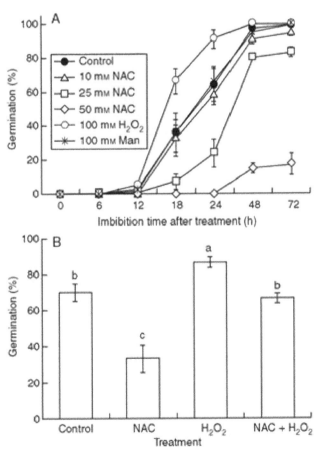

Fig. 11.2 The effects of hydrogen peroxide (H$_2$O$_2$) and N-acetylcysteine (NAC) on soybean seed germination. (A) Change of germination with time of soybean seeds treated with distilled water (control), and with 10, 25, or 50 mM NAC, 100 mM H$_2$O$_2$ or 100 mM mannitol (Man). (B) Germination of soybean seeds treated with distilled water (control), 25 mM NAC, 100 mM H$_2$O$_2$ or 25 mM NAC + 100 mM H$_2$O$_2$ for 24 h. Bars with different letters differ significantly (P, 0.05, Tukey's test, n ¼ 5).

In soybean seeds, ROS are produced in the embryonic axis during germination (Puntarulo et al., 1988; 1991), and the production and scavenging of ROS during ageing were related to vigour and cell death, respectively, during accelerated ageing of the embryonic axis (Tian et al., 2008). In addition, low temperatures led to oxidative stress and lipid peroxidation caused by ROS in the embryonic axis (Posmyk et al., 2001).

During a study seed priming with T4 (hydration with GA3 100 ppm 12 hr) and T5 (hydration with 0.5% KNO$_3$ 12 hr) higher germination was exhibited, i.e., 87.75 and 86.5%, respectively, over control as compared to other treatments. Remaining treatments also contributed improved germination and field emergence in comparison to control. Seed priming with only water (T1) could not significantly contribute for optimum plant stand and other quality studies. Gu, Gong Ping (2001) reported that the dry dressing of freshly harvested seed of soybean showed significant beneficial effects on germinability over the untreated control. Significant variation in root length, shoot length and seedling vigour index were observed due to different seed priming treatments. Highest Root and shoot length (42.75 cm), vigour index I (3740.13) and vigour index II (126.59) were recorded in T4; lowest (36.95 cm, 3075.60 and 93.22, respectively) were recorded in T1. The above explanation regarding the superiority of GA3 (100 ppm) (T4), CaCl$_2$ (2.0%) (T3) primed seeds is plausible for the root length, shoot length and vigour index as a higher positive correlation exists among them (Table 11.1).

Table 11.1 Effect of seed priming on germination (%) and root and shoot length (cm).

Treatment	Germination (%)	Root+shoot length (cm)
T1: Untreated (control)	83.00	36.95
T2: Hydration with distilled water (12 hrs)	84.00	39.92
T3: Hydration with CaCl$_2$ (2%) (12 hrs)	85.00	40.71
T4: Hydration with 100 ppm GA$_3$ (12 hrs)	87.33	42.75
T5: Hydration with 0.5% KNO$_3$ (12 hrs)	87.00	40.01
T6: Hydration with KCl (0.5%) (12 hrs)	85.33	40.93
T7: Hydration with water + Bavistin @ 3.0 g/kg of seed	86.67	41.85
T8: Hydration with IAA@ 80 ppm (12 hrs)	85.67	41.99
CD at 5%	1.98	1.81

Both seed germination and seedling establishment are essential for subsequent plant development. It is worth noting that both processes are powered by the energy that is stored in the seed itself (Chen and Thelen, 2010).

The mitochondrial FAD-dependent glycerol-3-P dehydrogenase ubiquinone oxidoreductase (FAD-GPDH) pathway has been proposed to be involved in the breakdown of fatty acids and glycerol in plant seeds (Huang, 1975). Several elegant studies demonstrated that the FAD-GPDH cascade is particularly important in oilseed plants, by which the hydrolysis of triacylglycerol releases free fatty acids and glycerol, with the fatty acids and glycerol then being converted to sugars, which support the seed germination and seedling establishment processes (Theodoulou and Eastmond, 2012). The *Arabidopsis SDP6* (*Sugar-Dependent 6*) gene encodes FAD-dependent glycerol-3-P dehydrogenase (FAD-G3P), and, although *sdp6* mutant seeds are able to germinate, the seedlings exhibit a marked arrested growth phenotype in the absence of exogenous sucrose during the seedling establishment stage (Quettier et al., 2008). This means that the transition from glycerol and fatty acids to sucrose is significantly impaired in *sdp6* seeds, with the exogenous supply of sucrose fully rescuing the arrested growth phenotype of the *sdp6* mutant (Quettier et al., 2008).

Another key gene, *SDP1* (*Sugar-Dependent 1*), encoding a patatin-like domain-containing triacylglycerol lipase, also plays a key role during post-germinative growth (Eastmond, 2006). In *sdp1*, the hydrolysis of triacylglycerol is blocked, resulting in *sdp1* seeds also exhibiting an arrested growth phenotype in the absence of sucrose, mimicking the *sdp6* phenotype (Eastmond, 2006). Altogether, both the hydrolysis of triacylglycerol and the conversion of fatty acids and glycerol to sugars are important for the successful powering of seed germination and seedling establishment, with the sugar supply allowing the young seedlings to achieve the photosynthetic autotrophic state (Graham, 2008; Kelly et al., 2011).

Numerous studies have demonstrated that, compared with the cereal crop seeds, including rice, wheat, and maize, soybean seeds contain much higher oil and fatty acid content (Teng et al., 2017). During storage, seed respiration, a catabolic reaction, utilizes glucose and other biomolecules (principally oils and fatty acids), and, as a consequence, significantly shortens seed longevity and decreases the rates of seed germination and seedling establishment, even causing soybean seeds which had been stored for long periods to be incapable of germination (Barros et al., 2017; Munz et al., 2017).

The ability of aged soybean seed to germinate decreases markedly as storage time increases; interestingly, the loss of germination potential also correlates with a decline in RNA integrity in soybean seeds following prolonged storage (Fleming et al., 2017). Another study demonstrated that, during natural aging processes, phospholipase Dα (PLDα) affected the soybean seed phospholipid and triacylglycerol profiles, suggesting that suppression of PLDα activity in soybean seed has the potential to improve seed quality during long-term storage (Lee et al., 2012). A decline in the germination ability of aged soybean seed significantly constrains soybean production, as it results in poor germination and seedling emergence from farm-saved seeds in the field.

It is reported that diethyl aminoethyl hexanoate (DA-6), a plant growth regulator, increases germination and seedling establishment from aged soybean seeds by increasing fatty acid metabolism and glycometabolism. Phenotypic analysis showed that DA-6 treatment markedly promoted germination and seedling establishment from naturally and artificially aged soybean seeds. Further analysis revealed that DA-6 increased the concentrations of soluble sugars during imbibition of aged soybean seeds. Consistently, the concentrations of several different fatty acids in DA-6-treated aged seeds were higher than those in untreated aged seeds. Subsequently, quantitative PCR analysis indicated that DA-6 induced the transcription of several key genes involved in the hydrolysis of triacylglycerol to sugars in aged soybean seeds. Furthermore, the activity of invertase in aged seeds, which catalyzes the hydrolysis

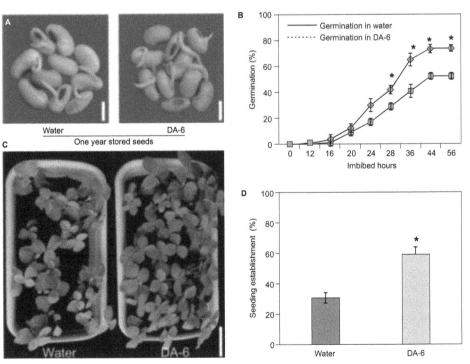

Fig. 11.3 Improving germination and seedling establishment of soybean with DA-6.

of sucrose to form fructose and glucose, increased following DA-6 treatment. Taken together, DA-6 promotes germination and seedling establishment from aged soybean seeds by enhancing the hydrolysis of triacylglycerol and the conversion of fatty acids to sugars (Fig. 11.3).

It is reported that NaCl delays soybean seed germination by negatively regulating gibberellin (GA) while positively mediating abscisic acid (ABA) biogenesis, which leads to a decrease in the GA/ABA ratio. This study suggests that fluridone (FLUN), an ABA biogenesis inhibitor, might be a potential plant growth regulator that can promote soybean seed germination under saline stress. Different soybean cultivars, which possessed distinct genetic backgrounds, showed a similar repressed phenotype during seed germination under exogenous NaCl application. Biochemical analysis revealed that NaCl treatment led to high MDA (malondialdehyde) levels during germination and the post-germinative growth stages. Furthermore, catalase, superoxide dismutase, and peroxidase activities also changed after NaCl treatment. Subsequent quantitative Real-Time Polymerase Chain Reaction analysis showed that the transcription levels of ABA and GA biogenesis and signaling genes were altered after NaCl treatment. In line with this, phytohormone measurement also revealed that NaCl considerably down-regulated active GA1, GA3, and GA4 levels, whereas the ABA content was up-regulated, thereby reducing ratios such as GA1/ABA, GA3/ABA, and GA4/ABA. Consistent with the hormonal quantification, FLUN partially rescued the delayed-germination phenotype caused by NaCl-treatment. Altogether, these results demonstrate that NaCl stress inhibits soybean seed germination by decreasing the GA/ABA ratio, and that FLUN might be a potential plant growth regulator that could promote soybean seed germination under salinity stress.

Changes in antioxidant compounds and antioxidant activity were studied in soybean seeds during germination with applied gibberellic acid (GA_3) in soaking water. GA_3 concentrations in soaking water were set up as 0; 0.001; 0.01; 0.1; 1 and 10 mg/L. The soaking process was carried out at ambient temperature for 12 hours prior germination. Soaked soybeans were germinated in dark condition, at 25°C for 0, 12, 23, 36, 48, 60 and 72 hours. The total phenolic content (TPC), total flavonoid content (TFC), vitamin C and α-tocopherol contents as well as antioxidant activity assayed by DPPH radical-scavenging activity in terms of IC50 of germinated soybean were determined. It was found that the increase of germination time caused an increase in the TPC, TFC, vitamin C, α-tocopherol and antioxidant activity, with most of them tending to reach their maximum values after a period of 60 hours of germination. In addition, GA_3 has a profound effect upon the antioxidant potentials and it caused a significant enhancement in the production of antioxidant compounds when compared to control. GA_3 1 mg/L in soaking solution was proposed as the optimum concentration to apply for the soaking process.

11.2 Plant Growth

Plant growth regulators, when applied in very small quantities, influence plant growth. Several reports indicated that the application of a growth regulator improved plant growth and yield. The 2, 3, 5-triiodobenzoic acid (TIBA) is a well-known plant growth regulator. The application of TIBA in soybean resulted in higher grain yield (Pankaj kumar et al., 2001). GA_3 enlarged the length of stem and flower number plant-1. GA_3 accelerated stem elongation and bud development. Kinetin increased the fresh weight by increasing stem diameter in morning glory but reduced shoot length (Chaudhry and Khan, 2000). Salicylic acid is an endogenous signaling molecule that has several functions, particularly the suppression of germination and growth, intervention with root absorption, reduced leaf abscission and transpiration (Ashraf et al., 2010; Hayat et al., 2010). Application of salicylic acid also significantly increased root dry weight. Salicylic acid application to soybean and corn promoted dry weight and leaf area of plants (Khan et al., 2003) (Tables 11.2 and 11.3).

A study was conceptualized and executed with the prime objective of studying the effect of chlormequat chloride, NAA, Mepiquat chloride and Brassinosteroids on morphological, physiological and biochemical parameters of soybean. The field trial was conducted following randomized block design with nine treatments replicated thrice. The basic material for the present investigation consisted of soybean cv. JS-335 and two growth promoting (NAA and Brassinosteroid) and growth retarding substances (chlormequat chloride and mepiquat chloride). These growth regulators were sprayed at flower initiation stage. The Morphophysiological parameters, namely, plant height, number of branches, number

Table 11.2 Effect of different plant growth regulators at different stages on plant height (cm) and branches per plant of soybean.

Treatment	Plant height (A1S)	Plant height (A2S)	Plant height (H)	Branches (A1S)	Branches (A2S)	Branches (H)
Control	38.24	48.63	61.08	2.55	4.01	4.56
SA @ 50 ppm	36.16	45.92	55.79	2.80	4.59	5.37
IAA @ 100 ppm	42.38	55.53	67.46	2.21	3.16	3.63
CCC @ 250 ppm	32.16	41.16	49.68	3.04	4.98	5.87
MC @ 500 ppm	35.91	44.87	54.49	2.60	4.10	4.67
ABA @ 10 ppm	37.67	47.57	57.77	2.24	3.20	3.74
TIBA @ 50 ppm	41.12	50.82	65.94	2.68	4.28	5.20
GA$_3$ @ 100 ppm	43.37	56.72	69.22	2.15	3.05	3.38
MH @ 1000 ppm	32.60	41.75	51.15	2.91	4.77	5.58
CD at 5%	2.70	4.70	3.38	0.26	0.41	0.66

A1S = After 1st spray; A2S = After 2nd spray; H = at harvest.

Table 11.3 Effect of different plant growth regulators at different stages on total dry weight and leaf area of soybean.

Treatment	TDM (g/plant) A1S	TDM (g/plant) A2S	TDM (g/plant) harvest	Leaf area (dm²/plant) A1S	Leaf area (dm²/plant) A2S	Leaf area (dm²/plant) harvest
Control	5.77	9.63	11.84	3.97	6.70	8.44
SA @ 50 ppm	5.82	9.71	12.23	4.21	7.11	9.24
IAA @ 100 ppm	5.93	10.02	12.72	6.42	11.42	15.53
CCC @ 250 ppm	7.92	14.09	19.16	4.68	8.34	11.07
MC @ 500 ppm	7.85	13.97	18.99	5.51	9.92	13.29
ABA @ 10 ppm	6.91	11.88	15.91	4.1	6.93	8.93
TIBA @ 50 ppm	8.11	14.43	19.62	6.19	11.01	14.75
GA$_3$ @ 100 ppm	6.76	11.62	15.10	6.50	11.57	15.73
MH @ 1000 ppm	7.16	12.31	16.49	4.76	8.52	11.45
CD at 5%	0.44	0.82	1.44	0.95	2.65	2.26

of trifoliates per plant, dry matter accumulation in leaf, stem and reproductive parts, LAI, CGR and RGR was observed to increase significantly with the application of NAA (20 ppm) and brassinosteroid (25 ppm). However, it decreased with the application of chlormequat chloride and mepiquat chloride. Biochemical parameters, namely, chlorophyll content, was observed to increase significantly with application of NAA (20 ppm), brassinosteroid (25 ppm), mepiquat chloride 5%, AS (5%) and chlormequat chloride 50% SL at different concentrations compared to control and water spray, whereas fluorescence emission and photosynthetic rate were noticed to be non-significant. A significant increase in the seed protein content was also noticed with the application of NAA (20 ppm), brassinosteroid (25 ppm), mepiquat chloride 5% AS (5%) and chlormequat chloride at different concentrations, compared to control and water spray. In conclusion, the study revealed that superiority of NAA (20 ppm) treatment for majority of the morphological, physiological and biochemical parameters at different growth stages, compared to other growth regulators and control treatments studied in the investigation for rabi soybean (Tables 11.4 and 11.5).

2,3,5-triiodobenzoic acid (TIBA) and succinic acid-2,2-dimethylhydrazide (daminozide) are well-known plant growth regulators. TIBA is an inhibitor of basipetal auxin transport (Geldner et al., 2001). Daminozide inhibits the action of tryptamine oxidase in the endogenous indolylacetic acid biosynthesis

Table 11.4 Effect of different growth regulators on leaf area index, crop growth rate (CGR) (g. m^{-2}.d^{-1}) and relative growth rate (RGR) (g.g^{-1}d^{-1}) in rabi soybean.

Treatments		LAI						CGR					RGR				
		Days after emergence						Days after emergence					Days after emergence				
		15	30	45	60	75	At harvest	15–30	30–45	45–60	60–75	75–Harvest	15–30	30–45	45–60	60–75	75–Harvest
T1	Chlormequat chloride 50% SL (137.5 g a.i/ha)	0.14	0.75	2.4	4.52	3.43	2.34	2.03	12.3	21.57	4.36	0.62	0.1167	0.1194	0.06	0.0075	0.001
T2	Chlormequat chloride 50% SL (162.5 g a.i/ha)	0.13	0.73	2.23	4.32	3.07	2.23	2.23	11.59	23.58	4.62	1.01	0.1163	0.111	0.0649	0.0077	0.0016
T3	Chlormequat chloride 50% SL (187.5 g a.i/ha)	0.16	0.72	2.13	4.26	3.05	2.15	2.42	9.79	19.63	4.03	1.94	0.1115	0.097	0.0621	0.0078	0.0035
T4	Chlormequat chloride 50% SL (375 g a.i/ha)	0.14	0.71	2.06	4.18	2.91	2.04	2.4	7.81	18.54	3.35	0.26	0.1171	0.0871	0.067	0.0072	0.0005
T5	Alpha naphthylacetic acid (NAA) 20 ppm	0.14	0.81	2.96	4.77	3.81	2.87	2.24	12.97	25.61	9.13	1.98	0.119	0.1173	0.0646	0.0133	0.0026
T6	Mepiuat chloride 5% AS (5%)	0.14	0.74	2.36	4.43	3.22	2.28	2.19	11.31	23.92	4.07	1.84	0.1146	0.1104	0.0665	0.0068	0.0029
T7	Brassinosteroid (25 ppm)	0.15	0.76	2.63	4.71	3.73	2.79	2.02	13.04	24.76	4.45	1.97	0.1162	0.123	0.0637	0.007	0.0029
T8	Water	0.14	0.71	2.06	4.12	2.84	2.01	2.11	8.39	15.73	2.65	0.23	0.1236	0.098	0.0596	0.0063	0.0005
T9	Control	0.13	0.59	1.42	3.32	2.48	1.89	1.87	7.83	15.47	2.58	0.13	0.1188	0.1	0.062	0.0064	0.0003
	Mean	0.14	0.72	2.25	4.29	3.17	2.29	2.17	10.56	20.98	4.36	1.11	0.1171	0.107	0.0634	0.0078	0.0017
	Sed	0.08	0.03	0.17	0.1	0.16	0.29	0.49	0.32	0.59	0.59	0.02	0.0016	0.0064	0.0098	0.0007	0.0005
	CD (0.05)	NS	0.06	0.37	0.21	0.35	NS	NS	0.68	1.26	1.27	0.04	0.0036	0.0136	NS	0.0015	0.001

Table 11.5 Effect of different growth regulators on chlorophyll (SPAD-502), fluorescence, photosynthetic rate values, seed protein (%) and nitrogen harvest index in rabi soybean.

Treatments		SPAD-502 VALUE	Fo	Fm	Fv	Fv/Fm	Photosynthetic rate (μ mol CO_2 m^{-2})	Seed protein (%)	Nitrogen harvest index
T1	Chlormequat chloride 50% SL (137.5 g a.i/ha)	32.74	59.67	173	115.91	0.67	17.5	41.8	17.06
T2	Chlormequat chloride 50% SL (162.5 g a.i/ha)	32.6	61	179.3	116.54	0.65	16.44	42.17	17.31
T3	Chlormequat chloride 50% SL (187.5 g a.i/ha)	32.53	62	184.3	123.48	0.67	16.98	42.63	17.78
T4	Chlormequat chloride 50% SL (375 g a.i/ha)	31.3	62.67	182.3	122.14	0.67	16.36	41.2	16.49
T5	Alpha naphthyl acetic acid (NAA) 20 ppm	33.74	65.33	188	125.96	0.67	15.52	42.47	17.66
T6	Mepiuat chloride 5% AS (5%)	32.6	62.67	184.3	121.63	0.66	16	41.7	17.05
T7	Brassinosteroid (25 ppm)	33.03	65.67	188	124.08	0.66	16.36	42	17.29
T8	Water	30.26	63.67	179.67	120.37	0.67	15.82	40.43	16.44
T9	Control	30.16	61.67	177.67	117.26	0.66	15.88	40.2	16.26
	Mean	32.11	62.71	181.84	120.82	0.66	16.32	19.62	16.73
	Sed	0.3	1.78	6.91	4.52	0.29	1.64	0.14	1.26
	CD (0.05)	0.63	NS	NS	NS	NS	NS	0.31	NS

(Castro and Appezzato-da-Glória, 1993). According to Rademacher (2000), daminozide is an inhibitor of 2-oxoglutarate dioxygenase and blocks the formation of gibberellic acid (GA). The TIBA can promote changes in vegetative and reproductive characteristics of soybean. The height reduction is associated with the shortening of the internode of the plant. The application of TIBA in soybean resulted in higher grain yield (Pankaj et al., 2001). In contrast, different doses of TIBA applied in soybean resulted in no significant differences in the number of grains and pods and weight of 100 grains (Cato and Castro, 2006).

Soybean genotypes grown in sub-tropical climates may exhibit lodging. The plant lodging is influenced by soil type and fertility level, sowing date, latitude and altitude of the location, plant population and conditions of crop development.

Plant regulators and herbicides can prevent or reduce plant lodging. This study aimed to verify the effects of the growth regulators TIBA and daminozide on vegetative growth and yield of soybean cultivar CD 214 RR. The experiment was carried out at a field in randomized block design with four replications in a factorial scheme. The A factor was represented by the combination of regulators TIBA and daminozide and its concentrations, and the Factor B was seven times the evaluation of injury and plant height or eight times the evaluation of lodging. In the range of doses used, the application of daminozide resulted in greater injury to soybean plants than TIBA. The smaller plant height was achieved by the application of 6 g ha^{-1} of TIBA and 1200 g ha^{-1} of daminozide. Treatments with daminozide (100 g ha^{-1}) and TIBA (10 g ha^{-1}) stood out due to the reduced lodging of soybean plants. Grain weight increased linearly when the levels of TIBA increased. There was a negative correlation between lodging and grain yield and a positive correlation between plant height and lodging. There was also a negative correlation between injury caused by the application of plant regulators and lodging (Fig. 11.4).

A pot experiment was carried out in order to study the effects of GA$_3$ and cytokinin on the vegetative growth of the soybean. GA$_3$ (50 mg L^{-1}) was applied as seed treatment, leaving plants with water application as control. GA$_3$ (100 mg L^{-1}) and cytokinin (30 mg L^{-1}) were sprayed on leaves at the physiological stage V3/V4, and 15 days after, cytokinin (30 mg L^{-1}), again as foliar spray. Seed treatment decreased plant emergence and initial soybean root growth, but, as the season progressed, differences in root growth disappeared; plants were shorter, and presented a decrease in the number of nodes, in stem diameter, in leaf area and in dry matter yield. Conversely, foliar application of GA$_3$ led to an increase in plant height, first node height and stem diameter. Leaf area and dry matter production also increased as a result of GA$_3$ foliar application. There was no effect of exogenous gibberellin and cytokinin on the

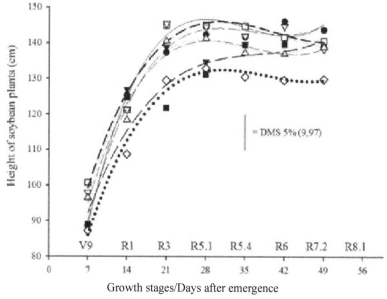

Growth stages/Days after emergence

Fig. 11.4 Changes in plant height of soybean at different growth stages.

number of soybean leaves, number of stem branches and root dry matter. Joint application of gibberellin and cytokinin tended to inhibit gibberellin effects. Cytokinin applied to leaves during soybean vegetative growth was not effective in modifying any of the evaluated plant growth variables (Table 11.6).

The levels of different cytokinins, indole-3-acetic acid (IAA) and abscisic acid (ABA) in roots of *Glycine max* [L.] Merr. cv. Bragg and its supernodulating mutant nts382 were compared. Forty-eight hours after inoculation with *Bradyrhizobium*, quantitative and qualitative differences were found in the root's endogenous hormone status between cultivar Bragg and the mutant nts382. The six quantified cytokinins, ranking similarly in each genotype, were present at higher concentrations (30–196% on average for isopentenyl adenosine and dihydrozeatin riboside, respectively) in mutant roots. By contrast, the ABA content was 2-fold higher in Bragg, while the basal levels of IAA [0.53 µmol (g DW))$^{-1}$, on average] were similar in both genotypes. In 1 mM NO$_3^-$ fed Bragg roots 48 h post-inoculation, IAA, ABA and the cytokinins isopentenyl adenine, and isopentenyl adenosine quantitatively increased with respect to uninoculated controls. However, only the two cytokinins increased in the mutant (Table 11.7).

High NO$_3^-$ (8 mM) markedly reduced root auxin concentration, and neither genotypic differences nor the inoculation-induced increase in auxin concentration in Bragg was observed under these conditions. Cytokinins and ABA, on the other hand, were minimally affected by 8 mM NO$_3^-$. Root IAA/cytokinin and ABA/cytokinin ratios were always higher in Bragg, relative to the mutant, and responded to inoculation (mainly in Bragg) and nitrate (both genotypes). The overall results are consistent with the auxin-burst-control hypothesis for the explanation of autoregulation and supernodulation in soybean. However, they are still inconclusive with respect to the inhibitory effect of NO$_3^-$.

Cytokinins are a class of plant-specific hormones that play a central role during the cell cycle and influence numerous developmental programs. Because of the lack of biosynthetic and signaling

Table 11.6 Results of plant emerged, plant height, first node height, stem diameter, leaf area, total dry matter and length root of soybean plants cv. IAC 17 under seed treatment with GA$_3$ and leaf application of GA$_3$ and CK.

Treatment	Plant emerged	Plant height (cm)	F.N.H.[1] (cm)	Stem diameter (mm)	Leaf area (cm^2)	T.D.M.[2] (g/m)
First evaluation						
No GA$_3$	8.4a*	17.1a	−2.2a	237a	1.35a	184
With GA$_3$	5.8b	17.2a	−2.1a	237a	1.24a	94b
VC%	8.91	12.12	−14.08	15.07	19.45	33.62
Second evaluation						
No GA$_3$	66.5bc	5.9b	9.3a	1951a	14.06a	199a
No GA$_3$ With GA$_3$	98.5a	6.1b	9.1a	1984a	14.52a	255a
GA$_3$ + CK	101.4a	5.5b	9.5a	2183a	16.41a	266a
No GA$_3$	55.0c	8.6b	7.2b	1328a	8.09a	190a
With GA$_3$ With GA$_3$	104.0a	11.5a	8.6ab	2166a	16.22a	294a
GA$_3$ + CK	86.6ab	9.0ab	7.6b	1476a	10.18a	199a
VC%	12.73	23.41	7.57	20.86	28.79	31.52
Third evaluation						
No GA$_3$	108.8cde	5.1a	13.3a	4957a	40.73a	278a
With GA$_3$	139.8abcd	6.5a	12.8a	5070a	39.79a	289a
No GA$_3$ GA$_3$ + CK	148.3ab	5.6a	13.6a	5010a	48.99a	382a
No GA$_3$ CK	113.3cde	5.6a	12.8a	5141a	41.26a	307a
With GA$_3$ CK	155.0a	5.3a	14.1a	6094a	48.90a	393a
GA$_3$ + CK CK	152.3a	7.4a	13.4a	5759a	46.65a	357a
No GA$_3$	90.8a	9.1a	11.8a	4404a	37.83a	315a
With GA$_3$	156.0a	10.0a	13.4a	5326a	45.54a	295a
With GA$_3$ GA$_3$ + CK	142.5abc	8.1a	13.1a	5122a	44.06a	302a
No GA$_3$ CK	104.0de	7.4a	12.9a	5189a	39.31a	315a
With GA$_3$ CK	146.0abc	8.8a	13.5a	5858a	47.86a	416a
GA$_3$ + CK CK	143.3abc	8.4a	13.0a	5240a	41.08a	313a
VC%	11.52	28.27	11.30	20.47	17.40	35.25

* Averages followed by the same letter do not differ; Tukey test, $P = 0.05$. [1] First node height. [2] Total dry matter.

Table 11.7 Effect of inoculation and NO₃ (mM) on cytokinin content (pmol.gDW⁻¹) in roots of soybean cv. Bragg and its supernodulating mutant nts382. Roots were harvested 48 h after inoculation (12 DAP). Cytokinins were extracted, immunopurified, separated by HPLC and quantified by ELISA. Data, means of eight replicates (coming from two experiments, four replicates per experiment) were subjected to analysis of variance and compared using the LSD test. Z, zeatin; DHZ, dihydrozeatin; ZR, zeatin riboside; DHZR, dihydrozeatin riboside; iP, isopentenyl adenine; iPA, isopentenyl adenosine.

Genotype	NO₃	Inocu.	Z	DHZ	ZR	DHZR	iP	iPA
Bragg	1	−	64	37	28	15	47	36
	1	+	56	46	28	15	139	105
	8	−	59	28	27	17	59	42
	8	+	61	45	30	19	77	59
Nts382	1	−	119	55	51	45	112	43
	1	+	84	48	44	62	179	124
	8	−	119	50	42	41	79	52
	8	+	126	66	50	47	145	93

mutants, the regulatory roles of cytokinins are not well understood. Cytokinin oxidase expression was genetically engineered in transgenic tobacco plants in order to reduce their endogenous cytokinin content. Cytokinin-deficient plants developed stunted shoots with smaller apical meristems. The plastochrone was prolonged, and leaf cell production was only 3–4% that of wild type, indicating an absolute requirement of cytokinins for leaf growth. In contrast, root meristems of transgenic plants were enlarged and gave rise to faster growing and more branched roots. These results suggest that cytokinins are an important regulatory factor of plant meristem activity and morphogenesis, with opposing roles in shoots and roots.

Soybean plants differentiate abundant floral buds, but a large proportion abort during development (Abernethy et al., 1977); the pod-set percentage ranged from 20 to 40% (Jiang and Egli, 1993). Prevention of the abortion might result in an increase in the number of pods and seeds, and thereby lead to an increase in grain yield (Kokubun, 2011). It was observed that abortion occurred most frequently at initial stages of pro embryo development after fertilization.

The abortion was amplified by unfavourable environments, including water deficiency (Saitoh et al., 1999), suboptimum solar radiation and temperature. Previous studies suggested two putative factors controlling the abortion: Availability of photosynthate or nutrients for pod development (Heitholt et al., 1986a; 1986b) and availability of certain hormones (Kokubun, 2011).

Within individual racemes, the pod-set percentage in basal flowers was considerably higher than that in distal ones (Kokubun and Honda, 2000), which appeared to be associated with the intra-raceme variation of endogenous cytokinin content (Kokubun and Honda, 2000). In addition, the application of synthetic cytokinin to racemes enhanced the pod formation (Nonokawa et al., 2007). These lines of evidence suggest that cytokinin plays a promotive role in pod setting in soybean.

The application of synthetic cytokinin (6-benzylaminopurine, BA) to racemes of soybean genotype IX93-100 at 7 days after anthesis (DAA) enhanced pod-set percentage of the florets at the 5th position and above (numbered from the base on rachis). The endogenous cytokinin (transzeatin riboside) content of individual florets was measured at the 1, 3, 5, 7th position every 3 days after anthesis. Cytokinin was detected only from the florets at 9 DAA, and the content was higher in the more proximal florets while it became negligible in the 7th floret. These results suggest that an increase in the amount of cytokinin in individual florets might enhance the pod setting of the florets positioned at the middle or distal part within the raceme (Fig. 11.5).

The critical role of phytohormones in the formation and abortion of reproductive organs in soybean was clearly recognized when Huff and Dybing (1980) observed that extracts from flowers and young pods applied to growing flowers accelerated flower abortion. They then applied a lanolin paste containing either indoleacetic acid (IAA), giberellin (GA) or 6-benzylaminopurine (BA) to the growing raceme, and found that IAA enhanced the abortion rate, as did the extract, whereas GA and BA did not. These results indicated that IAA plays a crucial role in increasing the abortion rate, although there was a conflicting report indicating that IAA delays the abortion (Oberholster et al., 1991).

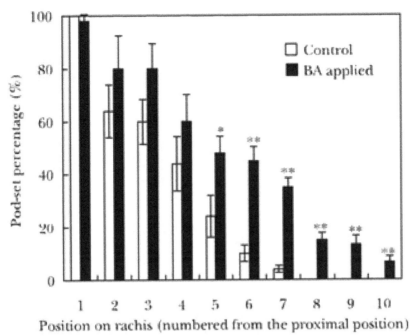

Fig. 11.5 Pod setting on rachis as influenced by BA applied.

Among phytohormones, cytokinins are considered to play a vital role in floral development in soybean. There have been numerous reports showing that the application of BA to racemes reduced seed abortion and increased pod number (Nagel et al., 2001). A limited number of reports support the notion that cytokinins are produced by the root system and transported to the shoot, where they are involved in the regulation of shoot development (Wareing et al., 1968). To examine the hypothesis that cytokinins produced in the root system are transported to the shoot, Heindle et al. (1982) collected root pressure exudates from detopped roots, and analyzed the forms and quantities of cytokinins in these exudates. Using high-performance liquid chromatography, they isolated and quantitated several forms of cytokinins: Zeatin, zeatin riboside, and their dihydro derivatives, dihydrozeatin and dihydrozeatin riboside. Their results indicated that cytokinin fluxes were independent of exudate flux, and that the ribosides accounted for the majority of the observed transport. In a later study, they also found a peak in cytokinin concentration during the period from the beginning of anthesis until 9 days after initial anthesis (Heindl et al., 1982; Carlson et al., 1987). This period corresponds to a stage in which most flowers are destined to either initiate or abort floral structure. Based on these findings, they concluded that cytokinins exported from the root may function in the regulation of reproductive growth in soybean.

The magnitude of the rate of flower abortion was observed to vary with the position on the plant. For example, it was higher in the branches, in the lower part of the main stem and in the top nodes of the main stem (Gai et al., 1984). Within individual racemes, flowers on the distal portions on the rachis exhibited a higher probability of abortion than those on the basal portions (Wiebold and Paciera, 1990).

As described above, the role of cytokinins in promoting development of floral structures in soybean had been clarified by several studies (Carlson et al., 1987). However, the nature of ontogenetic changes in the cytokinin content of racemes, and the relationship between the location of cytokinin on the plant and pod abortion remained unknown until 1990s. For precise analyses of intra-raceme variation in pod set and cytokinin contents, soybean genotype IX93-100, which has long racemes (approximately 10 cm, depending on the environmental conditions), was used as plant material.

Studies on the ontogenetic changes in cytokinins detected in different portions in racemes to determine whether a relationship exists between cytokinin concentration and pod-set probability at a specific floral position on a raceme (Kokubun and Honda, 2000). In an experiment using genotype IX93-100 grown in an environmentally controlled chamber (30/20°C day/night temperature, 15 h day length, 600 μmol mol^{-2} s^{-1} photosynthetic photon flux density (PPFD)), it was found that the total amount of cytokinin in racemes peaked one to two weeks after the first flowering event on a raceme, when pod development initiated. Within individual racemes, the total amount of cytokinin was greater at more proximal floral positions, as was the probability of pod set (Fig. 11.6). Removal of proximal flowers increased both cytokinin content and pod-set probability at middle positions on the raceme. Thus, cytokinin content in racemes was closely associated with pod-set probability within individual racemes. Each flower on a raceme initiated a pod 4 to 7 days after anthesis. During this period, a fertilized ovule develops embryo and endosperm, and cells are dividing actively (Carlson and Lersten, 1987). Microscopic observation showed that most of the abscised flowers are in the proembryo stage, which occurs several days after anthesis (Abernethy et al., 1977). Abernethy et al. (1977) speculated that the frequent occurrence of abscission at this stage was due to the reduced level of a cell division mediating factor.

The biosynthetic pathway for cytokinins has not been fully elucidated in soybean. In a study, the dominant forms of cytokinins were identified to be cis-zeatin riboside (c-ZR) and isopentenyladenosine (iPA), which differed from the forms detected in exudates previously reported by Heindl et al. (1982) and Carlson et al. (1987). In chickpeas, cis-isomers were found to be predominant in seeds (Emery et al., 1998), while trans-isomers are the more commonly reported forms in higher plants (McGaw and Burch, 1995; Prinsen et al., 1997). Further studies are currently in progress to identify the forms of cytokinins detected both in exudates and racemes in laboratory.

Plant hormones often do not act alone, but in conjunction with or in opposition to each other in such a manner that the final state of plant development reflects the net effect of the interplay of two or more hormones. Regarding the effects of hormones other than cytokinins on soybean floral development, there have not been many studies (Huff and Dybing, 1980; Oberholster et al., 1991). Views on the role of IAA were conflicting, which could be a reflection of genotypic or cultural differences in those experiments (Fig. 11.7).

Using the same plant material (IX93-100) grown in pots and in the field, it was noted that the auxin (IAA) concentration in racemes was high for a long period from pre-anthesis to ca. 10 days following the anthesis (DAA) of the first flower on a raceme, but the cytokinin concentration remained elevated for a shorter period, with a peak at 9 DAA (Nonokawa et al., 2007) (Fig. 11.7). The two phytohormones are

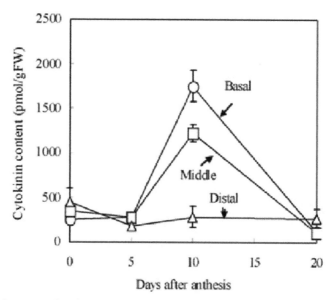

Fig. 11.6 Changes in the amount of cytokininis in various portions (basal, middle and distal) of racemes during reproductive development in soybean. Portions: 1–3 (basal), 4–6 (Middle), and 7 and above (Distal), numbered from the most basal portion on the rachis. Adapted from Kokubun and Honda (2000).

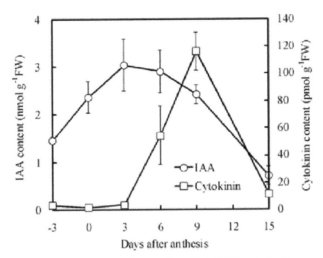

Fig. 11.7 Changes in the endogenous concentration of IAA and cytokinin (*t*-ZR equivalent) in racemes during reproductive development of soybean plant. Racemes were sampled for analysis at intervals before and after anthesis. Values represent the mean ± SE (*n* = 6). Adapted from Nonokawa et al. (2007).

located primarily at different positions within a raceme; the IAA concentration was higher in the distal portion of racemes, whereas the cytokinin concentration was higher in the basal portions of racemes. IAA application to racemes reduced the number of flowers and pods throughout the reproductive stage. In contrast, the effect of cytokinin (BA) application varied depending on the growth stage: Application of BA at around 7 DAA significantly increased the pod-set percentage, while, at other stages, BA application reduced pod set. Thus, the concentrations of the two endogenous hormones changed in a different manner, with cytokinins exerting a positive effect, and auxin exerting a negative effect on pod set, depending on the growth stage.

The long raceme soybean genotype IX93-100 was grown in pots and in the field. The auxin IAA concentration in racemes was high for a long period from preanthesis to 9 DAA of the first flower of raceme, but cytokinin concentration was high for a short period, with a peak at 9 DAA. In pot-grown plant, IAA applied to racemes tended to reduce the number of flowers and pods. BA applied to racemes before anthesis tended to reduce the number of flowers and pods and that applied around 7 DAA significantly increased the pod set percentage. However, these effects of IAA and BA application were slight in field grown plants (Table 11.8).

The results showed that severe drought stress significantly decreased pod set up to 40% and the critical stage for pod abortion was 3–5 days after anthesis (DAA), when cell division was active in the ovaries. Drought at later stages, when pod filling had begun, reduced seed size but had no significant effect on pod set. Pod water potential decreased with drought, however, pod turgor was maintained at similar level to the well-watered controls. ABA concentration increased significantly in the xylem sap, leaves, and pods of drought stressed plants. Xylem-borne ABA and leaf ABA were seemingly the source of ABA accumulated in the drought-stressed pods (Fig. 11.8).

Ovule formation is a complex developmental process in plants, with a strong impact on the production of seeds. Ovule primordia initiation is controlled by a gene network, including components of the signalling pathways of auxin, brassinosteroids and cytokinins. By contrast, gibberellins (GAs) and DELLA proteins, the negative regulators of GA signaling, have never been shown to be involved in ovule initiation. Molecular and genetic evidence that points to DELLA proteins as novel players in the determination of ovule number in *Arabidopsis* and in species of agronomic interest, such as tomato and rapeseed, adding a new layer of complexity to this important developmental process. DELLA activity correlates positively with ovule number, acting as a positive factor for ovule initiation. In addition, ectopic expression of a dominant DELLA in the placenta is sufficient to increase ovule number. The role

Table 11.8 Effects of IAA and BA application on the number of flowers, pod set percentage, number of pods and grain yield at the applied nodes. IAA and BA were applied to racemes at intervals before and after anthesis.

		Time of application (DAA)				
		Control	−7	0	7	14
No. of flowers per nodes	IAA	63.0a	64.3a	51.3a	51.3a	
	BA	63.0a	58.8a	54.7a	64.0a	64.3a
Pod set (%)	IAA	22.8a	18.7ab	17.4b	15.1b	
	BA	22.8a	20.5c	29.4ab	31.9a	23.5ab
No. of pods per nodes	IAA	14.2a	12.0ab	8.7bc	7.7c	
	BA	14.2ab	12.5b	15.7ab	19.7a	15.3ab
Grain yield per nodes (g)	IAA	5.4a	4.1ab	3.1b	2.9b	
	BA	5.4ab	3.7b	5.3ab	6.7a	5.9ab

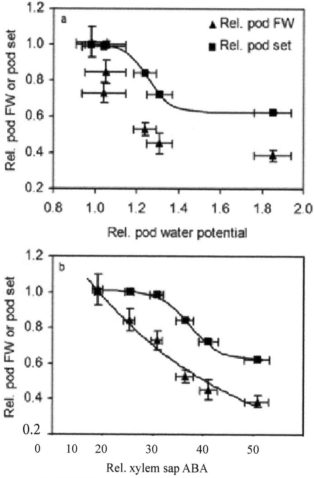

Fig. 11.8 Pod set and xylem sap ABA under water stress.

of DELLA proteins in ovule number does not appear to be related to auxin transport or signaling in the ovule primordia. Possible crosstalk between DELLA proteins and the molecular and hormonal network controlling ovule initiation is also known.

11.3 Yield

Soybean is a grain legume crop. Soybean plays an important role throughout the different countries of the world as food and feed. It provides oil as well as protein to the living beings. This very useful crop is grown in many countries, but land coverage is highest in United States of America. It is noted that the productivity of soybean is related to the number of seeds per square meter and the weight of the seeds. Hence, high number of seeds and high seed weight are important for higher productivity. Medium duration and medium height plants are most favored. Average production is about 2.5 t/ha. The minimum age at which plants can be induced to flower varies with species and with environmental conditions. Flowering of soybeans occurs most rapidly under short day conditions. It had also been shown that soybean yields tend to reach a maximum at populations of 50,000 to 1,00,000 per acre. Drought stress is the most important limiting factor at the initial phase of plant growth and establishment. Soybean is particularly sensitive to the lack of moisture during the blooming process (growth stages R1 and R2) and during the legume and seed growing processes (growth stages R3–R6). Plant growth regulators play important roles in plant growth and development, but little is known about the roles of plant growth regulators in improving the yield components and seed qualities of soybean. Endogenous plant growth regulators determine many growths and development processes, ultimately manifesting yield components and yield. Plant growth regulators are known to enhance the source-sink relationship and stimulate the translocation of photo-assimilates, thereby helping in effective flower formation, fruit and seed development and ultimately enhancing productivity of the crop. Growth regulators can improve the physiological efficiency, including photosynthetic ability, and can enhance the effective partitioning of accumulates from source to sink in the field crops (Solamani et al., 2001). Soybean plants differentiate abundant floral buds, but most of them fail to grow pods and abort during development. It is noted that cytokinins play a role in regulating flower and pod development in soybean (Reese et al., 1995). Carlson et al. (1987) and Nooden et al. (1990) have demonstrated a correlation between endogenous levels of cytokinins and the level of flower abortion and set.

Application of exogenous benzyladenine (BA) to individual racemes has been shown to prevent abortion of flowers and/or pods (Reese et al., 1995). Abortion of a significant percentage of flowers and pods commonly occurs in soybean (Abernethy et al., 1997; Dybing et al., 1986). The aborted flowers have generally been pollinated and fertilized, but tend to grow more slowly than the flowers that set (Huff and Dybing, 1980). Most proximal flowers set while most distal flowers abort, removal of proximal flowers can induce set of distal flowers (Wiebold, 1990). Compared with setting flowers, the sink intensity is lower in aborting flowers, as early as 3 day after anthesis (Brun and Betts, 1984). Conflicting data suggest that nutritional availability (carbohydrates and nitrogen) may play a role(s) in the number of flowers formed and the total number of mature soybean fruits and seeds that develop (Hayati et al., 1995). Research findings indicate that the concentration of endogenous auxin and cytokinin in racemes changes in a different manner, and that cytokinins have a positive, and auxin a negative effect on pod setting when respective hormones are applied to racemes after the anthesis stage. Thus, in soybean, the pod formation, development and filling, i.e., both number of pods and the growth of pods is determined by the assimilate supply and the balance of endogenous plant growth regulators. Furthermore, the exogenous application and time of application of plant growth regulators have some influence on the manifestation of yield components and yield of soybean. Hence, it is interesting to understand the responses of plant growth regulators applied exogenously.

11.3.1 Auxins

IAA has been found to increase the plant height, number of leaves per plant, and fruit size, with a consequent enhancement of seed yield in groundnut (Lee, 1990), cotton (Kapgate et al., 1989), cowpea (Khalil and Mandurah, 1989) and rice (Kaur and Singh, 1991). It also increases the flowering, fruit set, the total dry matter of crops (Gurdev and Saxena, 1991). It is also known that auxin suppresses axillary bud outgrowth (Shimuzu-Sato et al., 2009). In soybean, it was noted that both tryptophan and IAA-induced root nodules are formed. Additionally, these extracellular hormones also increased shoot and

Table 11.9 Influence of IAA on development and seed yield of soybean.

Treatment	Flowers Per plant	Pods per plant	Pod set (%)	Seeds per plant	Seed yield (g per plant)	100-seed weight (g)	Seed yield (t ha⁻¹)
Control	16.78	12.67	64.06	26.56	1.69	6.41	0.67
100 ppm	24.67	19.11	71.97	39.67	3.56	8.82	1.42
200 ppm	22.00	16.78	69.50	36.44	3.30	8.94	1.32

root dry weight as well as soybean yields. Highest root nodules number and soybean yield were observed from the treatment of 1.0 ppm tryptophan applied in early planting. It seemed that higher concentrations of tryptophan or IAA are required when applied at early planting, presumably because its concentration decreases prior to the root hairs being formed (Sudadi, 2012). Sarkar et al. (2002), in an experiment with soybean cv. BS-3 when sprayed at three different times with two concentrations (100 and 200 ppm) of indole acetic acid (IAA) noted that 100 ppm IAA produced increase in plant height, number of flowers, number of pods, percent of fruit set, number of seeds per plant, seed yield per plant and seed yield (t/ha), as compared to control (Table 11.9).

The growth regulator, napthaleneacetic acid (NAA), delayed the flowering of soybean, hence, delayed maturity. Chaudhuri et al. (1980) observed that NAA delayed the senescence of rice so that translocation of assimilates from source to sink is more and hence the yield. Fruiting was also delayed in NAA-treated plants. NAA-treated plants utilized more material in the vegetative parts of the plant. Fewer early pods were set on these plants. Dhakne et al. (2015) noted that application of 40 ppm of NAA increased the yield of soybean. In an experiment, the effects of the application of 1-napthaleneacetic acid (NAA) solutions of 0, 10, 25, 50 and 100 ppm to soybean SJ2 and SJ4 was studied using three applications of each dose at vegetative stage, flowering stage and pod setting stage. The result indicated that the spraying of 10 ppm NAA resulted in a 15% increase in the number of pods per plant and gave a 26% increase in seed yield. Application of higher dose of NAA to soybean plant would obtain bigger seed size and higher protein content but lower oil content. Soybean SJ4 showed higher response to the application of NAA than SJ2, in that the yield, seed size, seed protein and fat content of SJ4 are greater than that of SJ2 (Phanophot et al., 1986). Merlo et al. (1987) also reported that NAA application on soybean at flowering increased branches per plant and average pod weight. 100 seeds weight were increased with 20 ppm of NAA (Ravikumar and Kulkarni, 1998). Deotale et al. (1998) studied the effect of NAA on growth parameter of soybean and obtained the highest values for plant height, number of leaves per plant, number of branches per plant, leaf area, dry matter, days to maturity and seed yield with 100 ppm. It is also noted that NAA at 40 ppm foliar spray improves total chlorophyll content, nitrate reductase activity, soluble protein and thereby pod yield in soybean (Senthil, 2003). In 3 soybean varieties, application of NAA at 50 ppm increased the number of pods per plant, number of seeds per pod, 100 seed weight, harvest index, and grain yield significantly in relation to the control (Kalyankar et al., 2007). Khaswa et al. (2014) observed that application of 100 ppm of NAA at 30 and 65 DAS increased grain yield over control.

The work revealed that seed yield and yield components were significantly increased by NAA at 20 ppm over other treatments and control. The application of brassinosteroid at 25 ppm, mepiquat chloride 5% AS and chlormequat chloride applied at 187.5 g a.i/ha, 162.5 g a.i/ha and 137.5 g a.i/ha also resulted in higher seed yield, compared to control and water spray. The study revealed the superiority of NAA (20 ppm) treatment for majority of the morphological, physiological, growth, biochemical, quality and yield parameters at different growth stages, in addition to seed yield, compared to other growth regulator and control treatments. Suggesting the use of NAA will have many positive benefits on yield and yield attributing characters in *rabi*-soybean (Table 11.10).

TIBA is known for its antiauxin physiological activities. TIBA appeared to have an effect similar to shortening the photoperiod or removing the auxin, producing young leaves. It is hypothesized that TIBA slows auxin production or action in young leaves and meristems of the plants, therefore lowering the competition of the vegetative parts of the plant for growth materials. An increase in seed yield was shown for TIBA treated plots at stage 2 in the range of 15 to 120 ppm. Higher levels were injurious to

Table 11.10 Effect of plant growth regulators on yield and yield attributes in *Rabi*-soybean.

Treatments		No. of seeds/pod	No. of pods/ plant	100 seed weight (g)	Seed yield (kg.ha⁻¹)
T1	Chlormequat chloride 50% SL (137.5 g a.i/ha)	2.73	29.66	13.75	1472.33
T2	Chlormequat chloride 50% SL (162.5 g a.i/ha)	2.78	30.33	13.76	1491.00
T3	Chlormequat chloride 50% SL (187.5 g a.i/ha)	2.89	31.86	14.08	1502.66
T4	Chlormequat chloride 50% SL (375 g a.i/ha)	2.25	25.42	13.51	1449.66
T5	Alpha naphthyl acetic acid (NAA) 20 ppm	3.05	34.66	14.27	1563.33
T6	Mepiuat chloride 5% AS (5%)	2.80	31.00	13.86	1504.66
T7	Brassinosteroid (25 ppm)	2.94	32.33	14.26	1548.66
T8	Water	2.19	26.29	13.45	1437.66
T9	Control	2.06	25.66	13.85	1411.33
	Mean	2.63	29.69	13.87	1486.81
	Sed	0.07	0.63	2.47	6.09
	CD (0.05)	0.15	1.34	NS	12.98

the plants. It facilitates better conditions for pod set and fruit growth. Soybean genotypes grown in sub-tropical climates may exhibit lodging. Treatment with TIBA (10 gha⁻¹) under field conditions reduced lodging to soybean plants. Grain weight increased linearly when the levels of TIBA increased. There was a negative correlation between lodging and grain yield and a positive correlation between plant height and lodging. There was also a negative correlation between injury caused by the application of plant regulators and lodging (Buzzello et al., 2013). Chung and Kim (1989) investigated the effect of TIBA on soybean cultivar Hwang Keumkong. TIBA reduced stem length and lodging but increased stem diameter, podding rate, number of pods and seeds per plant and seed yield. TIBA was effective to healthy growth and to increase seed yield. Optimum treatment method for healthy plant growth and higher seed yield was 2–3 times spray with 5 days interval from 6 leaf stage (V6) of soybean plants. Soybean seed yield in the plot of TIBA treatment with 3 times from 6 leaf stage was 20% higher both in early and ordinary seeding field than those of non-treatment plots. Kumar et al. (2002) noted that seed yield was significantly higher with TIBA (50 ppm) application, which was associated with higher number of pods and seeds per plant in soybean genotypes (Table 11.11).

A field experiment was conducted at VNMKV, Parbhani, MH during June 2017 to Oct 2017 in a Randomized Block Design (RBD) based on three replications with a view to find out the influence of different plant growth regulators applied at different stages on the growth and yield of soybean cv. Application of plant growth regulators at flower initiation stage (35 DAS) and pod initiation stage (50 DAS) on plant showed significant effect on plant height, number of branches plant⁻¹, chlorophyll content (SPAD value), Dry weight plant⁻¹ and seed yield of soybean. Results showed that spraying of GA₃ @ 100 ppm produced the higher plant height (43.37 cm, 56.72 cm, 69.22 cm) and higher leaf area

Table 11.11 Influence of TIBA on yield attributes and seed yield.

Treatment	Plant height (cm)		Biomass (g.plant⁻¹)		No. of pods per plant		No. of seeds per plant		Seed yield (g per plant)	
	V1	V2	V1	V2	V1	V2	V1	V2	V1	V2
Control	30.86	64.44	16.67	17.20	33.80	49.00	68.03	61.64	10.21	9.12
TIBA 50 ppm	25.16	58.89	20.08	19.22	43.69	64.56	93.49	83.93	13.37	12.27
TIBA 100 ppm	23.78	56.72	19.16	18.95	38.80	58.00	86.92	79.00	12.25	11.31

V1 = MACS124; V2 = JS335.

Table 11.12 Effects of different growth regulators on chlorophyll (SPAD value), protein content and seed yield in soybean.

Treatment	SPAD value A1S	SPAD value A2S	Seed protein (%)	Seed yield (tha⁻¹)
Control	39.44	41.78	38.83	1.685
SA @ 50 ppm	43.08	44.09	41.44	1.814
IAA @ 100 ppm	41.13	42.97	39.18	1.750
CCC @ 250 ppm	48.89	51.72	42.11	1.953
MC @ 500 ppm	46.98	48.56	40.82	2.010
ABA @ 10 ppm	44.76	46.69	41.24	1.861
TIBA @ 50 ppm	43.83	44.91	41.22	2.124
GA₃ @ 100 ppm	41.77	43.48	39.18	1.712
MH @ 1000 ppm	48.17	50.83	41.54	1.916
CD at 5% P	0.81	0.92	NS	0.207

observed (6.5 dm², 11.57 dm², 15.53 dm²), Spraying of CCC @ 250 ppm produced the higher number of branches (3.04, 4.98, 5.87) and same in chlorophyll content and protein content and also gave the highest dry weight (19.16 g plant⁻¹) of soybean at harvest. Results also revealed that Seed yield was highest in spraying of TIBA @ 50 ppm (21.24 q/ha) at par with CCC @ 250 ppm and MC 500 ppm, as compared to other growth regulators. So, TIBA acts an important role for increasing soybean yield, when it is applied at 35 DAS and 50 DAS (Table 11.12).

11.3.2 Gibberellins

Gibberellins (GA₃) constitute a group of tetracyclic diterpenoids involved in plant growth and development. Gibberellic acid (GA₃) is a well-known phytohormone and has numerous physiological effects on plants, including seed germination, growth, stem elongation, leaf expansion, photosynthesis, flowering and cell expansion (Taiz and Zeiger, 2010; Yuan and Xu, 2001). Exogenous application of GA₃ to plants causes the increase in activities of many key enzymes, like carbonic anhydrase (CA), nitratereductase (NR) (Aftab et al., 2010) and ribulose–1, 5 biophosphate carboxylase/oxygenase (RuBPCO) (Yuan and Xu, 2001). Results indicate that seeds primed with GA₃@100 ppm for 12 hr had significantly higher germination percentage over untreated control of soybean. The seed priming significantly influenced the seed yield and yield contributing characters of soybean cv. JS-9305 showing to the corresponding favorable improvement in number of pods per plant, number of seeds per pod, test weight, seed yield and biological yield (Agawane and Parhe, 2015) (Table 11.13).

Upadhyay and Ranjan (2015) observed that application of GA₃ (20 ppm) at bud initiation and 50% flowering of soybean cv. Harit Soya increased the biological yield and seed yield along with test weight and harvest index (Table 11.14).

Field experiments with different levels (0, 125, 250 and 375 ppm) of GA₃ on soybean genotypes M11 and L17 when applied showed that interactions between different levels of GA₃ and the soybean genotypes had a significant effect on pod number per plant, seeds per pod, 1000 seed weight and economic yield of soybean (Azizi et al., 2012).

In a pot experiment, Leite et al. (2003) studied the effects of GA₃ on soybean at the rate of 100 mgl⁻¹ as foliar spray on leaves at the physiological stage V3/V4 and 15 days after. Foliar application of GA₃ led to an increase in plant height, first node height and stem diameter. Leaf area and dry matter production also increased as a result of GA₃ foliar application. The effect of gibberellic acid on the growth and yield components of soybean cultivar (Clark 63) was studied for its influence when applied to the different stages of growth. Gibberellin proved to be potential in increasing the yield components, like the number of nodes pods and seeds, which resulted in an increase in yield per hectare. Soaking the seed with

Table 11.13 Effect of seed priming on germination, yield attributes and seed yield.

Treatment	Germination (%)	Pods per plant	Test weight (g)	Seed yield (t per ha)
Control	83.00	94	12.95	2.29
Hydration with 100 ppm GA$_3$ (12 h)	87.33	160	14.18	2.89

Table 11.14 Effect of GA$_3$ on seed yield of soybean.

Treatment	Test weight (g)	Seed yield (g.plant^{-1})	Biological yield (g.plant^{-1})	Harvest index (%)
Control	128	12.0	35.6	33.73
GA$_3$ (20 ppm) spray	161	21.1	45.1	46.66

gibberellic acid with a concentration of 10 ppm before seeding then spraying with the same substance during the vegetative and flowering stages was proven to give the highest yield, producing 2.62 tons per hectare, which has a significant difference over the control, which was 1.72 tons per hectare. The growth substance, however, did not significantly accelerate and increase germination, plant height, fertility of nodes and weight of 1000 seeds (Domingo, 1981). When plants of soybean cv. BS-3 were sprayed at three different times with two concentrations (100 and 200 ppm) of gibberellic acid, GA$_3$ at 100 ppm had regulatory effect to enhance the plant height, number of branches, number of leaves, leaf area per plant, number of flowers, number of pods, percentage of fruit set, number of seeds per plant, seed yield per plant, 1000 seed weight and seed yield (t ha^{-1}) (Sarkar et al., 2002) (Table 11.15).

The study to know the effect of GA$_3$ applications on physiological and productive parameters an experiment was performed with soybean cultured in the field for 3 crop seasons and in the greenhouse for 1 crop season sprayed at the R2 physiological stages and 7 days later at the rate 300 gl^{-1}. GA$_3$-treated plants had longer shoots. Although there were no differences in number of pods, GA$_3$ treated plants had a lower number of seeds per pod. GA$_3$ spray increased oil content but reduced seed proteins (Travaglia et al., 2009).

The results exhibited that seeds primed with GA$_3$ @ 100 ppm (T4), 0.5% KNO$_3$ (T5) recorded significantly higher germination percentage, i.e., 87.33% and 87.00%, respectively, over the untreated control T1 (83.00%). The treatments (T4) GA$_3$ 100 ppm 12 hr (77.19), (T8) Hydration with IAA @ 80 ppm 12 hr (76.55) and (T3) Hydration with CaCl$_2$ (2.0%) 12 hr (76.22) maintained the optimum plant stand at harvest over untreated control (T1). This likely contributed to boosting economic yield in soybean cultivar, JS-9305. The seed priming significantly influenced the seed yield and yield contributing characters of soybean. The seed priming treatments (T4) GA$_3$ 100 ppm 12 hr, (T7) hydration with water + Bavistin 3.0 g/kg were found to be effective for improvement in dry matter content of seedling (g) in soybean variety JS-9305. The treatments (T4) 100 ppm GA$_3$ 12 hr (2078.00 g/plot), (T8) hydration with IAA 80 ppm 12 hr (2008.67 g/plot), (T5) Hydration with 0.5 per cent KNO$_3$ 12 hr (1991.00 g/plant) seed yield per plot, respectively, over the untreated control T1 (1647.67 g/plot), showing the corresponding favourable improvement in number of pods per plant, number of seeds per pod, test weight (g), seed yield per plot (gm), seed yield per Ha (q), biological yield (g) and numerical harvest index (%) (Table 11.16).

Table 11.15 Effect of GA$_3$ on yield components and seed yield of soybean.

Treatment	No. of flowers per plant	No. of pods per plant	Percent fruit set	No. of seeds per plant	1000 seed weight (g)	Seed yield per plant (g)
Control	16.78	12.67	64.00	26.56	64.10	1.69
GA$_3$ 100 ppm	35.44	26.00	77.64	54.22	107.60	5.87
GA$_3$ 200 ppm	29.78	22.00	71.33	46.78	96.30	4.52

Table 11.16 Effect of seed priming on seed yield (q/ha) and harvest index.

Treatment	Seed yield (q/ha)	Harvest index (%)
T1: Untreated (control)	22.88	36.61
T2: Hydration with distilled water (12 hrs)	25.17	38.01
T3: Hydration with CaCl$_2$ (2.0%) (12 hrs)	26.95	39.98
T4: Hydration with 100 ppm GA$_3$ (12 hrs)	28.86	40.88
T5: Hydration with 0.5 per cent KNO$_3$ (12 hrs)	27.65	40.60
T6: Hydration with KCl (0.5%) (12 hrs)	25.32	38.03
T7: Hydreation with water (12 hrs) + Bavistin @ 3.0 g/kg of seed	27.56	39.24
T8: Hydration with IAA @ 80 ppm (12 hrs)	27.90	40.63
Mean	26.54	39.25
CD at 5%	1.30	NS

11.3.3 Cytokinins

Studies indicate that soybean seed yield is more decisively determined by the number of pods than by other components of production (Yashima et al., 2005). The amount of flowers that give rise to the pods until reaching maturity is a key factor for giving high yields. Although soybean flowers are produced abundantly, a large number of flowers and young pods abort naturally (Nonokawa et al., 2012). Some researches show that, in normal conditions, the abscission of the reproductive structures of soybean can vary between 20 and 82% of the total number of the number of flower produced (Peterson et al., 2005). The mechanisms responsible for flower and pod fixing are completely established. According to Dario et al. (2005), the application of growth regulators could raise the productivity above levels established until now. Researches have pointed out the use of plant growth regulators to reduce the pod abortion (Nonokawa et al., 2012; Passos et al., 2011). Abortion prevention may result in an increased number of pods and seeds, thus leading to gain in seed yield (Nonokawa et al., 2012). Exogenous application of cytokinin to raceme tissues of soybean has been shown to stimulate flower production and prevent flower abortion (Nagel et al., 2001). In the greenhouse, application of 3.4×10^{-7} moles of 6-benzylaminopurine resulted in a 79% increase in seed yield compared with controls. Results of field trials showed much greater variability within treatments, with consistent, but non-significant increases in seed number and total yields of about 3%. Data suggest that cytokinin levels play a significant role in determining total yield in soybeans, and that increasing cytokinin concentrations in certain environments may result in increased total seed production (Nagel et al., 2001).

In a pot experiment on soybean by Leite et al. (2003), cytokinin (30 mgl^{-1}) was sprayed on leaves at the physiological stage V3/V4 and 15 days after. Cytokinin applied to leaves during vegetative growth was not effective in modifying any of the plant growth variables. In a field experiment with soybean genotypes with determinate and semideterminate nature, Kumar et al. (2002) studied the effect of plant growth regulators. Cytokinin improved all the parameters and the effect was seen more in determinate genotype. The seed yield was 10 to 15% higher with kinetin, which was associated with higher number of pods and seeds per plant (Table 11.17).

Dhakne et al. (2015) also noted that application of Kinetin 40 ppm recorded significantly higher yield and net return in soybean. Cytokinin has long been known to have an important involvement in controlling shoot branching (Leyser, 2003). Cytokinin promotes axillary bud out growth (Shimizu-Sato et al., 2009). In soybean raceme, cytokinin concentration was high for a short period with a peak at 9 days after anthesis. Again, cytokinin concentration was higher in the basal portions of racemes. In contrast, 6-benzylaminopurine (BA) applied to racemes before anthesis tended to reduce the number of flowers and pods, and that applied around 7 days after anthesis significantly increased the pod set percentage (Nonokawa et al., 2007). Genetic improvement of synthesis and transport of endogenous cytokinins from the root system, via conventional breeding or molecular approaches, may strengthen pod set capacity

Table 11.17 Effect of kinetin on yield attributes and yield of soybean.

Treatment	Biomass (g per plant)		No. of pods per plant		No. of seeds per plant		Seed yield (g per plant)	
	V1	V2	V1	V2	V1	V2	V1	V2
Control	16.67	17.20	33.80	49.00	68.03	61.46	10.21	9.12
Kinetin 25 ppm	17.83	18.37	34.10	52.40	73.96	68.32	11.43	10.44
Kinetin 50 ppm	18.48	19.30	34.53	54.60	74.21	68.47	11.57	10.60

V1 = MACS124; V2 = JS335.

of agriculturally significant genotypes. Clarification of the physical and chemical properties of the rhizosphere optimizing synthesis of endogenous cytokinins in roots should improve pod set. In soybean cultivation, there seems to be a link between exogenous benzyadenine and reduction of flower and pod abortion (Nagel et al., 2001). It was observed that benzyladenine application reduced pod abortion in the lower, middle and upper third of the canopy of soybean. Soybean plants treated with benzyladenine showed higher yields than control plants. The highest productivity was obtained in soybean plants treated with a concentration of 300 mgl⁻¹. Benzyladenine application at the end of flowering is a promising management practice for soybean cultivation (Borges et al., 2014).

11.3.4 Abscisic Acid

The results obtained in the experiments with soybean grown in field conditions support the idea that ABA enhances yield by a combination of factors. Therefore, foliar application (300 mgl⁻¹) of ABA may be an alternative tool for enhancing yield of short-cycle soybean, since it gives relief to temporary situations of water stress, such as the stress that happens in the hours of maximum irradiance, where an imbalance between water transpiration and absorption it is frequently produced. ABA seems to improve a combination of factors that contribute to increasing the number of lateral roots and the density of the radial system, to protect the photosynthetic apparatus, to keep the stomata conductance more stable over the time, and to enhance carbon allocation and partitioning to the seeds (Reinoso et al., 2011). Chung and Kim (1989) reported that when ABA was sprayed on soybean plants stem length and lodging decreased, however, stem diameter, podding rate, number of pods per plant and seed yield increased with 2–3 times spray at 6 days interval at 6 leaf stage (V6). The study to investigate the effect of ABA application on physiological and productive parameters with soybean cultured in field for 3 crop seasons and in the greenhouse for 1 season at the rate of 300 mgl⁻¹ sprayed at V7 and R2 physiological stages revealed that ABA treated plants had greater dry weight of aerial parts and root density and also leaf area and chlorophyll content. ABA application increased soybean yield by enhancing carbon allocation and partitioning to the seeds (Travaglia et al., 2009).

When the effect of S-abscisic acid application on soybean seed yield in response to water stress was evaluated, the results indicated that the application of 2-ppm S-ABA by foliar spray at reproductive stages increased soybean seed yield, seed number and pod number under optimum and mild water stress conditions, while under severe water stress conditions, the application of S-ABA was not effective to increase soybean seed yield. The application of S-ABA also increased the water use efficiency of soybean plants (Kamal et al., 1998). A study on the effect of ABA application on pod formation and seed yield of soybean plants was carried out. The results indicated that the application of 1 and 10 ppm ABA foliar spray before the flowering (V7) and flowering stages promoted pod set, filled pod number, pod respiration, glucose and fructose contents in leaf, petiole and pod shell and yield of soybean plants. However, the application of 100 ppm ABA tended to inhibit pod respiration, glucose and fructose contents and grain yield of soybean plants (Takahashi et al., 1996). The effect of exogenous ABA on rate of sucrose uptake by soybean embryos was evaluated in an *in vitro* system. In addition, the concentrations of endogenous ABA in seeds of three soybean Plant Introduction (PI) lines, differing in seed size, were compared to their seed growth rates. ABA (10⁻⁷ molar) stimulated *in vitro* sucrose uptake in soybean cv. Clay embryos removed from the plants and grown in a controlled environment chamber, but not in embryos removed from field grown plants of the three PI lines. However, the concentration of ABA in seeds of three field grown PI lines correlated well with their *in situ* seed growth rates and *in vitro* (14C) sucrose uptake rates. Across genotypes, the concentration of ABA in seeds peaked at 8.5 micrograms per gram fresh weight, corresponding to the time of most rapid seed growth rate and declined to 1.2 micrograms per gram at physiological maturity. Seeds of the large-seeded genotype maintained an ABA concentration at least 50% greater than that of the small seeded genotype throughout the latter half of the seed filling. A higher concentration of ABA was found in seed-coats and cotyledons than in embryonic axes. Seed coats of the large seeded genotype always had a higher concentration of ABA than seed coats of small seeded line. It is suggested that this higher, concentration of ABA in seed coats of the large seeded genotypes stimulate sucrose unloading into the seed-coat apoplast and that ABA in cotyledons may enhance sucrose uptake by the cotyledons (Schussler et al., 1984). Variation in light intercepted during and after seed initiation has been found to be a major environmental determinant of soybean seed size. Investigation on the influence of light enrichment and shading on seed growth rate, effective filling, cotyledon cell number, cell volume and endogenous ABA concentrations of cotyledons/testas during seed filling of soybean has been undertaken. Even an indeterminate Group 0 soybean was subjected to light reduction and enrichment treatments from the beginning of pod formation until final harvest for two years in Massachusetts. Higher rates of seed growth, greater seed dry weight, and higher cotyledon cell number, along with a significant lowering of endogenous ABA levels in testa and cotyledon with shade. The level of ABA in cotyledon during seed development was significantly correlated with seed growth rates only under shade treatments (Liu et al., 2006). Under shade stress, seedling height and first inter node length increased, stem diameter decreased, abscisic acid (ABA) and zeatin (ZT) concentration decreased, while indole acetic acid (IAA) and gibberellins 3 (GA3) concentration increased (Table 11.18). Furthermore, branch numbers, pod number of branches and seed number of branches increased. Branch yield did not reduce significantly under shade stress, which was related to the decrease of ABA and IAA. Based on the results, the decrease in soybean yield under shade and drought stresses was mainly due to yield reduction of the main stem (Zhang et al., 2011).

Drought and shade stress have effects on auxin, cytokinins and abscisic acid concentrations (Davies, 2010). Water stress at the seeding and at the flowering stage reduces soybean yield by 20 and 46% (Shou et al., 1991), respectively, due to decreased photosynthetic rate, stomatal conductance and transpiration rate of soybean (Ohashi et al., 2006; Vuetal., 2001). Moreover, shade treatments have been shown to reduce yield and seeds per plant. The effects of PGRs including benzyladenine (BA), uniconazole (S180), brassinolide (Br) and abscisic acid (ABA) on leaf water potential (Ψ), Chlorophyll (Chl), photosynthetic rate (Pn), PSII photochemical efficiency (Fv/FM) and seed yield of soybean cv. Keng 5 were studied under water deficit. PGRs were applied at R1 of 50, 100, 0.1 and 50 mgl^{-1} for BA, S180, Br and ABA, respectively. Two levels of soil moisture stress were applied at R1. The result indicated that water deficit decreased biomass of stems and leaves, and induced significant yield loss.

PGRs treatments increased soybean yields both under well-watered and water-deficit conditions, with the exception of BA under water-deficit. It was noted that PGRs treatments minimized the yield loss caused by water deficit (Zhang et al., 2004) (Table 11.19).

Table 11.18 Effects of drought and shade on plant hormone of soybean.

Treatment		ABA (ng g⁻¹FW)	IAA (ng g⁻¹FW)	GA₃ (ng g⁻¹FW)	ZT (ng g⁻¹FW)
2009					
HI	WW	26.62c	47.36c	227.29bc	52.66a
	MD	358.55a	22.15d	205.64c	38.08b
LI	WW	14.55c	117.07a	411.81a	49.01a
	MD	285.79b	78.08b	278.74b	33.46b
2010					
HI	HW	24.50c	79.60b	128.29c	37.45a
	LW	207.41a	61.86b	104.01c	30.86ab
LI	HW	15.57c	136.06a	300.26a	32.87a
	LW	125.12b	76.55b	274.58b	24.61b

Table 11.19 Effect of water deficit and plant growth regulators on biomass and seed yield of soybean.

Treatment	Stems + Leaves biomass (g per plant)		Seed yield (g per plant)	
	Well-watered	Water deficit	Well-watered	Water deficit
CK	8.94	6.17	3.14	2.27
BA	9.34	6.54	3.35	2.48
S180	9.32	7.01	3.71	2.79
Br	10.20	7.18	3.51	2.83
ABA	11.80	6.61	3.96	2.78
Mean	9.91	6.70	3.53	2.63

CK = Check; Well-watered (–0.02 MPa); Water deficit (–0.06 MPa).

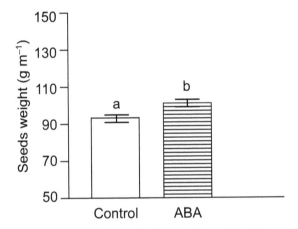

Fig. 11.9 Seed weight of soybean as influenced by ABA.

Given that photosynthetic activity by crop canopy declines gradually during the effective grain filling period and current photosynthesis (rather than remobilization of stored carbohydrate) is considered to be main source for seed growth in soybean (Liu et al., 2006), supporting a regular photosynthesis could be the cause of the higher carbohydrate amounts found at the flowering period in shoots of the ABA-treated plants as compared to controls; the difference in shoots disappears at harvest because of an increased carbohydrate remobilization (21%) to the seed in the ABA-treated plants (Travaglia et al., 2009). These results confirm the participation of ABA in promoting source to sink transport of assimilates during the stage of seed filling.

The number of ripped pods in the ABA-treated soybean plants was similar to the controls; however, it was observed that, during the first periods of development, the pods in ABA treated plants were bigger, a difference that disappeared at maturity (data not shown). The weight of seeds per m^2 was significantly higher in the ABA-treated plants (Fig. 11.9); the seeds maintain the same protein content and higher oil concentration as compared with the controls (Fig. 11.9). Thus, the treatment with ABA did not affect the quality of seeds, an important characteristic since the seed quality is one of the key aspects for agriculture success.

11.3.5 Cycocel

The plant growth retardant Cycocel (2-Chloroethyl, trimethyl ammonium chloride) has been used to check the abscission of flower and modify the crop canopy for improving the yield in gram (Bangal et al., 1982), pigeopea (Vikhi et al., 1983) and soybean (Singh et al., 1987). Grewal et al. (1993) reported that cycocel improves the translocation of photosynthates. More protein content stored in the seeds might be due to improvement of translocation of photosynthates to the seeds. In a three-year study conducted during 2006–2008, when cycocel at the rate of 500 ppm was applied as foliar spray at flower initiation, pod initiation and flower initiation + pod initiation of growth, there was an increase in chlorophyll content and carotenoids in leaves as well as increase in seed yield (Devi et al., 2011) (Table 11.20).

Kumar et al. (2002) also noted a higher seed yield of soybean with the application of 250 ppm of CCC, which was associated with higher number of pods and seeds per plant (Table 11.21).

A field experiment was conducted during *kharif*, 2009 at College of Agriculture, University of Agricultural Sciences, Dharwad to study the growth, development, physiology, yield and quality of soybean (*Glycine max* (L.) Merrill) as influenced by plant growth regulators (PGRs). The experiment was laid out in factorial randomized block design, replicated thrice with different plant growth regulators, namely, Progibb (20, 40 and 60 ppm), CCC (500 and 1000 ppm) and TIBA (100 and 200 ppm), as foliar spray, with two varieties (KHSb-2 and JS-335).

The results on various yields and yield attributes indicated that all the yield contributing characters, i.e., seed yield per plant, number of seeds per plant, number of pods per plant, pod weight per plant and

Table 11.20 Effect of bioregulators on yield of soybean (average for three years).

Characters	Bioregulators	FI	PI	FI+PI	Mean
Seed yield (t/ha)	SA @ 50 ppm	1.45	1.41	1.68	1.51
	Ethrel @ 200 ppm	1.69	1.68	1.88	1.75
	Cycocel @ 500 ppm	1.22	1.19	1.27	1.23
	Control	1.08	1.05	1.09	1.07
	Mean	1.36	1.33	1.48	
	CD(0.05)A 0.11				
	CD(0.05)B 0.10				
	CD(0.05)AxB NS				
Harvest index (%)	SA @ 50 ppm	38.01	37.80	39.60	38.47
	Ethrel @ 200 ppm	42.94	43.16	42.07	42.72
	Cycocel @ 500 ppm	44.51	41.69	42.52	42.91
	Control	35.90	33.16	36.89	35.31
	Mean	40.34	38.95	40.27	
	CD(0.05) A 0.03				
	CD(0.05) B NS				
	CD(0.05) AXB NS				
Straw yield (tha⁻¹)	SA @ 50 ppm	2.38	2.32	2.57	2.42
	Ethrel @ 200 ppm	2.23	2.21	2.59	2.34
	Cycocel @ 500 ppm	1.53	1.60	1.61	1.58
	Control	1.92	2.12	2.18	2.07
	Mean	2.02	2.06	2.24	
	CD(0.05)A 181				
	CD(0.05)B 157				
	CD(0.05)AxB NS				

A = Bioregulators; B = Stages of application (FI = Flower initiation; PI = Pod initiation ; FI + PI = Both flower initiation and pod initiation).

Table 11.21 Effect of cycocel on yield of soybean (average of three years).

Treatment	Biomass (g per plant)		No. pods per plant		No. of seeds per plant		Seed yield (g per plant)	
	V1	V2	V1	V2	V1	V2	V1	V2
Control	16.67	17.20	33.80	49.00	68.03	61.46	10.21	9.12
CCC 250 ppm	19.99	20.93	37.13	55.60	76.43	72.45	12.06	11.22

V1 = MACS124; V2 = JS335.

Table 11.22 Influence of plant growth regulators on yield in soybean Seed yield per plant (g) Seed yield (q ha⁻¹).

Treatment	Seed yield per plant			Seed yield		
	V1	V2	Mean	V1	V2	Mean
Progibb 20 ppm	8.85	10.05	9.45	29.5	33.6	31.6
Progibb 40 ppm	9.40	10.62	10.01	31.5	35.3	33.4
Progibb 60 ppm	9.65	11.11	10.38	32.5	37.1	34.8
CCC 500 ppm	10.05	11.21	10.63	33.5	37.5	35.5
CCC 1000 ppm	9.71	10.72	10.22	32.3	35.7	34.0
TIBA 100 ppm	9.80	10.73	10.27	32.9	36.2	34.5
TIBA 200 ppm	9.61	10.62	10.12	31.7	35.3	33.5
Control	8.48	9.34	8.91	28.2	31.2	29.7
Mean	9.44	10.55	10.00	31.5	35.2	33.3
CD (0.05) treatment	0.22	0.64		0.17	0.48	

V1 = KHSb2; V2 = JS335.

100-seed weight, increased significantly due to PGRs. Among the treatments, CCC (500 ppm) was found to be very effective in increasing the seed yield, followed by Progibb (60 ppm) and TIBA in both the genotypes. JS-335 recorded significantly higher yield than KHSb-2 (Table 11.22).

11.3.6 Ethrel

Ethylene released from ethrel (2-chloroethyl phosphonic acid) could possibly be utilized for promoting growth, as Abbas (1991) has shown that early pod development is related to higher ethylene levels, thus, less flower and pod shedding, thereby reducing abscission and improving pod set. Ethrel is reported to have induced an increase in cell division in tomato fruits, resulting in greater fruit production and yield (Atta-Aly et al., 1999). A study revealed that application of Ethrel 200 ppm at flower initiation + pod initiation gave higher vegetative growth and yield in soybean (Devi et al., 2011) (Table 11.23).

11.3.7 Salicylic Acid

Salicylic acid ($C_7H_6O_3$) is an endogenous growth regulator of phenolic nature, which participates in the regulation of physiological processes in plant, such as stomal closures, ion uptake, inhibition of ethylene biosynthesis, transpiration and stress tolerance (Khan et al., 2003; Shakirova et al., 2003). Foliar application of salicylic acid exerted a significant effect on plant growth metabolism when applied at physiological concentration and, thus, acted as one of the plant growth regulating substances (Kalarani et al., 2002). Salicylic acid increased the number of flowers, pods per plant and seed yield of soybean (Gutierrez-Coronado et al., 1998). In a soybean cv. JS 335, when salicylic acid was applied at 50 ppm as foliar spray at flower initiation, pod initiation and flower initiation + pod initiation, seed yield increased (Devi et al., 2011) (Table 11.24).

In an experiment on BARI soybean by Khatun et al. (2016), when salicylic acid was sprayed on leaves and plants at the vegetative stage, flower initiation stage and pod initiation stage, flower + pod initiation stages in pot experiment under field condition gave the highest number of seeds pod⁻¹, harvest index, small size seed, protein and moisture content in seed (1.60, 39.06%, 19.47%, 44.45% and 12.91%, respectively). Salicylic acid application at flower and pod initiation stages showed the highest yield attributes and maximum protein content.

Kinetin spray produced the maximum 100-seed weight (11.58 g). Application of growth regulators at vegetative stage produced the highest stover yield (6.46 g plant⁻¹), flower initiation stage gave the larger size seed (59.09%), pod initiation stage showed the maximum pod length (2.43 cm), highest

Table 11.23 Effect of ethrel on yield of soybean.

Treatment		Seed yield (t per ha)	
Spray time	Flower initiation	Pod initiation	Flower initiation + Pod initiation
Control	1.08	1.05	1.09
Ethrel 200 ppm	1.69	1.68	1.88

Table 11.24 Effect of salicylic acid on yield of soybean.

Treatment		Seed yield (t per ha)	
Spray time	Flower initiation	Pod initiation	Flower initiation + pod initiation
Control	1.08	1.05	1.09
SA 50 ppm	1.45	1.41	1.68

Table 11.25 Interaction effect of different hormone and their time of application on yield parameters of soybean.

Treatment combination	Pods per plant	Pod length (cm)	Seeds per pod	100-seed weight (g)	Stover yield (g per plant)	Biological Yield (g per plant)	Harvest index (%)
H0S1	19.60	2.00	1.12	8.33	6.92	8.79	21.85
H0S2	20.40	2.16	1.24	9.73	6.30	8.17	22.11
H0S3	28.60	2.50	1.48	9.26	6.64	10.44	28.41
H0S4	25.60	2.32	1.36	11.29	5.50	8.43	33.67
H1S1	32.80	2.10	1.28	9.27	6.83	10.14	30.77
H1S2	25.60	2.60	1.60	7.99	5.96	9.49	37.07
H1S3	28.00	2.52	1.76	9.43	5.00	8.22	35.78
H1S4	24.00	2.32	1.76	16.79	4.71	11.10	52.76
H2S1	27.80	2.22	1.54	11.02	5.42	9.94	42.62
H2S2	29.80	2.36	1.32	8.81	4.57	7.06	31.46
H2S3	23.00	2.36	1.20	11.31	6.80	8.98	21.91
H2S4	41.00	2.56	1.68	9.05	4.48	9.43	52.51
H3S1	37.60	2.42	1.76	9.98	6.69	11.20	39.99
H3S2	32.60	2.48	1.52	11.77	6.45	11.16	42.10
H3S3	20.40	2.32	1.52	13.73	6.53	9.75	32.34
H3S4	20.00	2.34	1.36	10.85	5.89	9.83	34.66
LSD(0.05)	17.44	0.438	0.394	3.682	2.412	4.015	16.710

H0 = Control (water), H1 = Salicylic acid (50 ppm), H2 = GA3 (100 ppm), H3 = Kinetin (500 ppm). S1 = Vegetative stage at 25 DAS, S2 = Flower initiation at 40 DAS, S3 = Pod initiation at 50 DAS and S4 = Flower + Pod initiation stage at 40 and 50 DAS.

moisture content in seed (13.50%) and spray at flower + pod initiation stage produced the maximum 100-seed weight (12.00 g), harvest index (43.42%), medium size seed (32.53%), and protein content in seed (44.31%). Among the treatment combinations, the application of salicylic acid at flower and pod initiation stage showed the highest yield attributes and maximum protein content in relation to the other growth regulators (Table 11.25).

References

Abbas, S. 1991. Biosynthesis pathways as control points in ethylene regulated flower and fruit drop and seed absorption in chickpea. Proceedings Grain Legumes Indian Society of Genetics and plant Breeding, IARI, New Delhi.

Abernethy, R.H., R.G. Palmer, R. Shibles and I.C. Anderson. 1977. Histological observations on abscising and retained soybean flowers. Canadian Journal of Plant Science 57: 713–716.

Aftab, T., M.M.A. Khan, M. Idrees, M. Naeem and A.S. Moinuddin. 2010. Salicylic acid acts as potent enhancer of growth, photosynthesis and artemisinin production in *Artemisia annua* L. Journal of Crop Science and Biotechnology 13: 183–188.

Agawane, R.B. and S.D. Parhe. 2015. Effect of seed priming on crop growth and seed yield of soybean (*Glycine max* (L.) Merill). The Bioscan 10: 265–270.

Ashraf, M., N.A. Akram, R.N. Arteca and M.R. Foolad. 2010. The physiological, biochemical and molecular roles of brassinosteroids and salicylic acid in plant processes and salt tolerance. Critical Reviews Plant Sci. 29(3): 162–190.

Atta-aly, M.A., G.S. Riad, Z. Ets Lacheene and A.S. Beltagy. 1999. Early application of ethrel extends tomato fruit cell division and increase fruit size and yield with ripening delay. Journal of Plant Growth Regulator 18: 15–25.

Azizi, K., J. Moradii, S. Heidari, A. Khalili and M. Feizian. 2012. Effect of different concentrations of gibberellic acid on seed yield and yield components of soybean genotypes in summer intercropping. International Journal of Agricultural Sciences 2: 291–301.

Bahin, E., C. Bailly, B. Sotta, I. Kranner, F. Corbineau and J. Leymarie. 2011. Crosstalk between reactive oxygen species and hormonal signaling pathways regulates grain dormancy in barley. Plant Cell and Environment 34: 980–993.

Balestrazzi, A., C. Macovei and D. Carbonera. 2011. Seed imbibition in *Medicago truncatula* Gaertn. Expression profiles of DNA repair genes in relation to PEG-mediated stress. Journal of Plant Physiology 168: 706–713.

Bangal, D.B., S.N. Deshmukh and V.A. Patil. 1982. Note on the effects of growth regulators and urea on yield attributes of gram (*Cicer arietinum*). Legume Research 5: 54–56.

Barros, J.A.S., J.H.F. Cavalcanti, D.B. Medeiros, A. Nunes-Nesi, T. Avin-Wittenberg, A.R. Fernie and W.L. Araújo. 2017. Autophagy deficiency compromises alternative pathways of respiration following energy deprivation in *Arabidopsis thaliana*. Plant Physiology 175: 62–76.

Birnberg, P.R. and M.L. Brenner. 1987. Effect of gibberellic acid on pod set in soybean. Plant Growth Regulation, Dordrecht 5: 195–206.

Borges, L.P., H.D. Torres, T.G. Neves, C.K.L. Cruvinel, P.G.F. Santos and F.S. Matas. 2014. Does Benzyladenine application increase soybean productivity? African Journal of Agricultural Research 9: 2799–2804.

Burn, W.A. and K.J. Betts. 1984. Source/Sink relations of abscising and nonabscising soybean flowers. Plant Physiology 75: 187–191.

Carlson, D.R., D.J. Dyer, C.D. Cotterman and R.C. Durley. 1987. The physiological basis for cytokinin induced increases in pod set in 1X93-100 soybean. Plant Physiology 84: 233–239.

Castro, P.R.C. and B.A. Appezzato-da-Glória. 1993. Efeitos de reguladores vegetais no desenvolvimento e na produtividade do amendoinzeiro (*Arachis hypogaea* L.). Scientia agricola 50: 176–184.

Cato, S.C. and P.R.C. Castro. 2006. Redução da altura de plantas de soja causada pelo ácido 2,3,5-triiodobenzóico. Ciência Rural 36: 981–984.

Chaudhuri, D., P. Basuchaudhuri and D.K. Das Gupta. 1980. Effect of growth substances on growth and yield of rice. Indian Agriculturist 24: 169–175.

Chudhary, N.Y. and A. Khan. 2000. Effect of growth hormones, i.e., GA3, IAA and kinetin on shoot of *Cicer arietinum* L. Pak. J. Biol. Sci. 3(8): 1263–1266.

Chen, M. and J.J. Thelen. 2010. The plastid isoform of triose phosphate isomerase is required for the postgerminative transition from heterotrophic to autotrophic growth in *Arabidopsis*. The Plant Cell 22: 77–90.

Chen, K. and R. Arora. 2011. Dynamics of the antioxidant system during seed osmopriming, postpriming germination, and seedling establishment in spinach (*Spinacia oleracea*). Plant Science 180: 212–220.

Chung, I.M. and K.J. Kim. 1989. Effect of plant growth regulator (TIBA, ABA, DGLP) treatment on growth and seed yield of soybean (*Glycine max* L.). Hangug jugmul haghoej (source: Korean Agricultural Science Digital Library), 34(1).

Dario, G.J.A., T.N. Matin, D.D. Neto, P.A. Manfrom, R.A.G. Bonnecarrere and P.E.N. Crespo. 2005. Influencia do us o defitor regulator no crescimento da soja. Revista da Faculdade da Zootecnia, Veterniary and Agronomy 12: 63–70.

Davies, P.J. 2010. The plant hormones: Their nature, occurrence and functions. pp. 1–15. *In*: Plant Hormones. Springer, Dordrecht, Netherland.

Deotale, R.D., V.G. Maske, N.V. Sorte, B.S. Chimurkar and A.Z. Yerne. 1998. Effect of GA3 and NAA on morpho-physiological parameters of soybean. Journal of Soils and Crops 8: 95–94.

Devi, K.N., A.K. Vyas, M.S. Singh and N.G. Singh. 2011. Effect of bioregulators on growth, yield and chemical constituents of soybean (*Glycine max*). Journal of Agricultural Science 3: 151–157.

Dhakne, A.S., I.A.B. Mirza, S.V. Pawar and V.B. Awasarmal. 2015. Yield and economics of soybean (*Glycine max* (L.) Merill) as influenced by different levels of sulphur and plant growth regulator. International Journal of Tropical Agriculture 33: 2645–2648.

Domingo, C.V. 1981. Gibberellic acid effect on the growth and yield components of soybean. TCA Research Journal (source: University of the Philippines Library).

Dybing, D.C., H. Ghiasi and C. Paech. 1986. Biochemical characterization of soybean ovary growth from anthesis to abscission of aborting ovaries. Plant Physiology 81: 1069–1074.

Dyer, D., D.R. Carlson, C.D. Cotterman and J.A. Sikorski. 1986. The role of cytokinin in soybean pod set regulation. In: Plant Growth Regulation, Proceedings. St. Petersburg Beach, p. 130.

Eastmond, P.J. 2006. SUGAR-DEPENDENT1 encodes a patatin domain triacylglycerol lipase that initiates storage oil breakdown in germinating *Arabidopsis* seeds. The Plant Cell 18: 665–675.

El-Araby, M.M., S.M.A. Moustafa, A.I. Ismail and A.Z.A. Hegazi. 2006. Hormone and phenol levels during germination and osmopriming of tomato seeds, and associated variations in protein patterns and anatomical seed features. Acta Agronomica Hungarica 54: 441–458.

Fincher, G.B. 1989. Molecular and cellular biology associated with endosperm mobilization in germinating cereal grains. Annual Review of Plant Physiology and Plant Molecular Biology 40: 305–346.

Fleming, M.B., C.M. Richards and C. Walters. 2017. Decline in RNA integrity of dry-stored soybean seeds correlates with loss of germination potential. Journal of Experimental Botany 68: 2219–2230.

Frey, A., C. Audran, E. Marin, B. Sotta and A. Marion-Poll. 1999. Engineering seed dormancy by the modification of zeaxanthin epoxidase gene expression. Plant Mol. Biol. 39: 1267–1274.

Frey, A., D. Effroy, V. Lefebvre, M. Seo, F. Perreau, A. Berger et al. 2012. Epoxycarotenoid cleavage by NCED5 fine-tunes ABA accumulation and affects seed dormancy and drought tolerance with other NCED family members. Plant J. 70: 501–512.

Gai, J., R.G. Palmer and W.R. Fehr. 1984. Bloom and pod set in determinate and indeterminate soybeans in China. Agronomy Journal 76: 979.

Geldner, N., J. Friml, Y.D. Stierhof, G. Jürgens and K. Palme. 2001. Auxin transport inhibitors block PIN1 cycling and vesicle trafficking. Nature 413: 425–428.

Gong Ping, G. 2001. Seed invigoration treatments for improved storability, field emergence and productivity of soybean. Crop Physiology Abst. 27(2): 1030.

Graham, I.A. 2008. Seed storage oil mobilization. Annual Review of Plant Biology 59: 115–142.

Grewal, H.S., J.S. Kdar, S.S. Cheema and S. Sing. 1993. Studies on the use of growth regulators in relation to nitrogen for enhancing sink capacity and yield of gobhisorson (*Brassica napus*). Indian Journal of Plant Physiology 36: 1–4.

Guan, C., X. Wang, J. Feng, S. Hong, Y. Liang, B. Ren et al. 2014. Cytokinin antagonizes abscisic acid-mediated inhibition of cotyledon greening by promoting the degradation of abscisic acid insensitive 5 protein in *Arabidopsis*. Plant Physiology 164: 1515–1526.

Gurdev, S and O.P. Saxena. 1991. Seed treatments with bioregulators in relation to wheat productivity. pp. 201–210. *In*: New Trends in Plant Physiology, Proceedings of National Symposium on Growth and Differentiation in Plants, New Delhi, India.

Gutierrez-Coronado, M.A., C. Trejo-Lopez and A.S. Karque-Saavedra. 1998. Effect of salicylic acid on the growth of roots and shoots in soybean. Plant Physiology and Biochemistry 36: 563.

Hayat, R., S. Ali, U. Amara, R. Khalid and I. Ahmed. 2010. Soil beneficial bacteria and their role in plant growth promotion: A review. Ann. Microbiol. 60: 579–598.

Hayati, R., D.B. Egli and S.J. Crafts–Brandner. 1995. Carbon and nitrogen supply during seed filling and leaf senescence in soybean. Crop Science 35: 1063–1069.

Heindl, J.C., D.R. Carlson, W.A. Brun and M.L. Brenner. 1982. Ontogenetic variation of four cytokinins in soybean root pressure exudates. Plant Physiology 70: 1619–1625.

Heitholt, J.J., D.B. Egli and J.E. Legget. 1986a. Characteristics of reproductive abortion in soybean. Crop Sci. 26: 589–595.

Heitholt, J.J., D.B. Egli, J.E. Legget and C.T. MacKown. 1986b. Role of assimilate and carbon-14 photosynthate partitioning in soybean reproductive abortion. Crop Sci. 26: 999–1004.

Hilhorst, H.W.M. 2007. Definition and hypotheses of seed dormancy. pp. 50–71. *In*: Bradford, K.J. and H. Nonogaki (eds.). Seed Development, Dormancy and Germination. Annual Plant Reviews. Vol. 27, Chap 4. Sheffield, UK: Blackwell Publishing.

Huff, A. and C.D. Dybing. 1980. Factors effecting shedding of flowers in soybean (*Glycine max* (L.) Merill). Journal of Experimental Botany 31: 751–762.

Ishibashi, Y., T. Tawaratsumida, K. Kondo, S. Kasa, M. Sakamoto, N. Aoki et al. 2012. Reactive oxygen species are involved in gibberellin/abscisic acid signaling in barley aleurone cells. Plant Physiol. 158: 1705–1714.

Jisha, K.C., K.J.T. Vijayakumari and J.T. Puthur. 2013. Seed priming for abiotic stress tolerance: An overview. Acta Physiologiae Plantarum 3: 1381–1396.

Kalarani, M.K., M. Thangaraj, R. Siva Kumar and V. Mallika. 2002. Effect of salicylic acid on tomato (*Lycopercicon esculentum*) productivity. Crop Research 23: 486–492.

Kalyankar, S.V., G.R. Kadam, S.B. Borgaonkar, D.D. Deshmukh and B.P. Kodam. 2007. Effect of foliar application of growth regulators on seed yield and yield components of soybean (*Glycine max* (L.) Merill). Asian Journal of Bioscience 3: 229–230.

Kamal, M., H. Takahashi, H. Mikoshiba and Y. Ota. 1998. Effect of S-abscisic acid application on soybean grain yield under different condition of water stress. Japan Journal of Tropical Agriculture 42: 1–6.

Kapgate, H.G., N.N. Pitkile, N.G. Zode and A.M. Dhopte. 1989. Persistance of physiological responses of upland cotton to growth regulators. Annals of Plant Physiology 3: 188–195.

Kaur, J. and G. Singh. 1991. Hormonal regulation of grain filling in relation to peduncle anatomy in rice cultivars. Indian Journal of Experimental Biology 25: 63–65.

Kelly, A.A., A.L. Quettier, E. Shaw and P.J. Eastmond. 2011. Seed storage oil mobilization is important but not essential for germination or seedling establishment in *Arabidopsis*. Plant Physiology 157: 866–875.

Khalil, S. and H.M. Mandurah. 1989. Growth and metabolic changes of cowpea plants as affected by water deficiency and indole-3-ylacetic acid. Journal of Agronomy and Crop Science 163: 160–166.

Khan, W., P. Balakrishnan and L.S. Donald. 2003. Photosynthetic responses of corn and soybean to foliar application of salicylates. Journal of Plant Physiology 160: 485–492.

Khaswa, S.L., R.K. Dubey, S. Singh and R.C. Tiwari. 2014. Growth, productivity and quality of soybean (*Glycine max* (L.) Merill) under different levels and sources of phosphorus and plant growth regulators in Subhumid Rajasthan. African Journal of Agricultural Research 9: 1045–1051.

Khatun, S., T.S. Roy, M.N. Haque and B. Alamgir. 2016. Effect of plant growth regulators and their time of application on yield attributes and quality of soybean. International Journal of Plant and Soil Science 11: 1–9.

Kokubun, M. and I. Honda. 2000. Intra-raceme variation in pod-set probability is associated with cytokinin content in soybeans. Plant Production Science 3: 354–359.

Kokubun, M. 2011. Physiological Mechanisms Regulating Flower Abortion in Soybean, Soybean - Biochemistry, Chemistry and Physiology, Prof. Tzi-Bun Ng (Ed.), ISBN: 978-953-307-219-7, InTech.

Koornneef, M., C.J. Hanhart, H.W. Hilhorst and C.M. Karssen. 1989. *In vivo* inhibition of seed development and reserve protein accumulation in recombinants of abscisic acid biosynthesis and responsiveness mutants in *Arabidopsis thaliana*. Plant Physiol. 90: 463–469.

Kubala, S., M. Garnczarska, L. Wojtyla, A. Clippe, A. Kosmala, A. Zmiénko et al. 2015. Deciphering priming-induced improvement of rape seed (*Brassica napus* L.) germination through an integrated transcriptomic and proteomic approach. Plant Science 231: 94–113.

Kumar, P., S.M. Hiremath, P.S. Deshmukh and S.R. Kushwaha. 2002. Effect of growth regulators on growth, yield and metabolism of soybean genotypes. Indian Journal of Agricultural Research 36: 254–258.

Lee, H.S. 1990. Effects of presowing seed treatments with GA3 and IAA on flowering and yield components in groundnuts. Korean Journal of Crop Science 35: 1–9.

Lee, S. et al. 2002. Gibberellin regulates Arabidopsis seed germination via RGL2, a GAI/RGA-like gene whose expression is up-regulated following imbibition. Genes Dev. 16: 646–658.

Lee, J., R. Welti, M. Roth, W.T. Schapaugh, J. Li and H.N. Trick. 2012. Enhanced seed viability and lipid compositional changes during natural ageing by suppressing phospholipase Dα in soybean seed. Plant Biotechnology Journal 10: 164–173.

Leite, V.M., C.A. Rosolem and J.D. Rodrigues. 2003. Gibberellin and cytokinin effects on soybean growth. Scientia Agricola 60537–541.

Li, Q.T., X. Lu, Q.X. Song et al. 2017. Selection for a zinc-finger protein contributes to seed oil increase during soybean domestication. Plant Physiology 173: 2208–2224.

Liu, X., S.J. Herbert, K. Baath and A.M. Hashemi. 2006. Soybean (*Glycine max*) seed growth characteristics in response to light enrichment and shading. Plant, Soil and Environment 52: 178–185.

Liu, W.-Z., D.-D. Kong, X.-X. Gu, H.B. Gao, J.-Z. Wang, M. Xia et al. 2013. Cytokinins can act as suppressors of nitric oxide in *Arabidopsis*. Proceedings of the National Academy of Sciences, USA 110: 1548–1553.

Liu, Y., N. Ye, R. Liu, M. Chen and J. Zhang. 2010. H_2O_2 mediates the regulation of ABA catabolism and GA biosynthesis in *Arabidopsis* seed dormancy and germination. Journal of Experimental Botany 61: 2979–2990.

Maske, V.G., R.D. Dotale, P.N. Sorte, B.D. Tale and C.N. Chore. 1997. Germination, root and shoot studies in soybean as influenced by GA3 and NAA. Journal of Soils and Crops 7: 147–149.

McGaw, B.A. and L.R. Burch. 1995. Cytokinin biosynthesis and metabolism. pp. 98–117. *In*: Davies, P.J. (ed.). Plant Hormones, Physiology, Biochemistry, and Molecular Biology (2nd ed.). Kluwer Academic, Dordrecht.

Melro, D., A. Soldati and E.R. Keller. 1987. Influence of growth regulators on abscission of flower and young pods of soybeans. Eurosaya 5: 31–38.

Meng, Y., F. Chen, H. Shuai, X. Luo, J. Ding, S. Tang et al. 2016. Karrikins delay soybean seed germination by mediating abscisic acid and gibberellin biogenesis under shaded conditions. Sci. Rep. 6: 2207.

Munz, E., H. Rolletschek, S. Oeltze-Jafra, J. Fuchs, A. Guendel, T. Neuberger, S. Ortleb, P.M. Jakob and L. Borisjuk. 2017. A functional imaging study of germinating oilseed rape seed. New Phytologist 216: 1181–1190.

Nagel, L., R. Brewster, W.E. Riedell and R.N. Reese. 2001. Cytokinin regulation of flower and pod set in soybeans (*Glycine max* (L.) Merr.). Annals of Botany 88: 27–31.

Nonokawa, K., M. Kokubun, T. Nakajim, T. Nakamura and R. Yoshida. 2007. Role of auxin and cytokinin in soybean pod setting. Plant Production Science 10: 199–206.

Nonokawa, K., T. Nakajima, T. Nakamura and M. Kokubun. 2012. Effect of synthetic cytokinin application on pod setting of individual florets with in raceme in soybean. Plant Production Science 15: 79–81.

Nooden, I.D., K.R.S. Santo and D.S. Letham. 1990. Correlation of xylem sap cytokinin levels with monocarpic senescence in soybean. Plant Physiology 93: 33–39.

Oberholster, S.D., C.M. Peterson and R.R. Dute. 1991. Pedicel abscission of soybean: Cytological and ultrastructural changes induced by auxine and ethephon. Canadian Journal of Botany 69: 2177–2186.

Ohashi, Y., N. Nakayama, H. Saneoka and K. Fujita. 2006. Effect of drought stress on photosynthetic gas exchange, chlorophyll fluorescence and stem diameter of soybean plants. Biologia Plantarum 50: 138–141.

Oracz, K., H. El-Maarouf-Bouteau, I. Kranner, R. Bogatek, F. Corbineau and C. Bailly. 2009. The mechanisms involved in seed dormancy alleviation by hydrogen cyanide unravel the role of reactive oxygen species as key actors of cellular signalling during germination. Plant Physiology 150: 494–505.

Oracz, K. and S. Karpinski. 2016. Phytohormones signaling pathways and ROS involvement in seed germination. Front Plant Sci. 7: 864.

Pankaj Kumar, S.M. Hiremath, P.S. Deshmukh, S.R. Kushwaha, T.M.M.A. Aftab Khan et al. 2001. Effect of growth regulators on growth yield and metabolism in soybean genotypes. J. Crop Sci. and Biotech. 13: 183–188.

Paparella, S., S.S. Araújo, G. Rossi, M. Wijayasinghe, D. Carbonera and A. Balestrazzi. 2015. Seed priming: State of the art and new perspectives. Plant Cell Reports 34: 1281–1293.

Park, J. et al. 2011. Integration of auxin and salt signals by the NAC transcription factor NTM2 during seed germination in *Arabidopsis*. Plant Physiol. 156: 537–549.

Passos, A.M.A.D., P.M.D. Rezende, A.A.D. Alvarenga, D.P. Baliza, E.R. Carvalho and H.P.D. Alcantara. 2011. Yield per plant and other characteristics of soybean plants treated with Kinetin and potassium nitrate. Ciencia Agrotechnologia 35: 965–972.

Phanophat, N., R. Phromsattha and S. Chokworawatthanakon. 1986. Effects of naphthalene acetic acid (NAA) on yield and quality of soybean. Available: http://ring/cird.net/node/10758.

Prinsen, E., M. Kaminek and H.A. van Onckelen. 1997. Cytokinin biosynthesis: A black box? Journal of Plant Growth Regulation 23: 3–15.

Piskurewicz, U., Y. Jikumaru, N. Kinoshita, E. Nambara, Y. Kamiya and L. Lopez-Molina. 2008. The gibberellic acid signaling repressor RGL2 inhibits *Arabidopsis* seed germination by stimulating abscisic acid synthesis and ABI5 activity. Plant Cell 20: 2729–2745.

Posmyk, M.M., F. Corbineau, D. Vinel, C. Bailly and D. Côme. 2001. Osmoconditioning reduces physiological and biochemical damage induced by chilling in soybean seeds. Physiologia Plantarum 111: 473–482.

Procko, C., C.M. Crenshaw, K.L. Jung, J.P. Noel and J. Chory. 2014. Cotyledon generated auxin is required for shade-induced hypocotyl growth in *Brassica rapa*. Plant Physiology 165: 1285–1301.

Puntarulo, S., R.A. Sánchez and A. Boveris. 1988. Hydrogen peroxide metabolism in soybean embryonic axes at the onset of germination. Plant Physiology 86: 626–630.

Puntarulo, S., M. Galleano, R.A. Sanchez and A. Boveris. 1991. Superoxide anion and hydrogen peroxide metabolism in soybean embryonic axes during germination. Biochimica et Biophysica Acta 1074: 277–283.

Quettier, A.L., E. Shaw and P.J. Eastmond. 2008. *SUGAR-DEPENDENT6* encodes a mitochondrial flavin adenine dinucleotide-dependent glycerol-3-P dehydrogenase, which is required for glycerol catabolism and post germinative seedling growth in *Arabidopsis*. Plant Physiology 148: 519–528.

Rademacher, W. 2000. Growth retardants: Effects on gibberellin biosynthesis and other metabolic pathways. Annual Review of Plant Physiology and Plant Molecular Biology 51: 501–531.

Ramaih, S., M. Guedira and G.M. Paulsen. 2003. Relationship of indoleacetic acid and tryptophan to dormancy and preharvest sprouting of wheat. Functional Plant Biology 30: 939–945.

Ravikumar, G.H. and G.N. Kulkarni. 1998. Effect of growth regulators on seed quality in soybean genotypes (*Glycine max* L. Merill). Seeds and Farms 14(2): 25–28.

Reese, R.N., C.D. Dybing, C.A. While, S.M. Page and J.F. Larson. 1995. Expression of vegetative storage protein (VSP-β) in soybean raceme tissues in response to flower set. Journal of Experimental Botany 46: 957–964.

Ren, Z. and X. Wang. 2016. SlTIR1 is involved in crosstalk of phytohormones, regulates auxin-induced root growth and stimulates stenospermocarpic fruit formation in tomato. Plant Sci. 253: 13–20.

Riedell, W.E., U. Khoo and G.E. Inglett. 1985. Effects of bioregulators on soybean leaf structure and chlorophyll retention. pp. 204–212. *In*: Plant Growth Regulation, Lake Alfred, Florida, 1985. Proceedings. Lake Alfred.

Sarkar, P.K., M.S. Haque and M.A. Karim. 2002. Effects of GA3 and IAA and their frequency of application on morphology, yield contributing characters and yield of soybean. Journal of Agronomy 1: 119–122.

Schmidt, M.A., W.B. Barbazuk, M. Sandford, G. May, Z. Song, W. Zhou, B.J. Nikolau and E.M. Herman. 2011. Silencing of soybean seed storage proteins results in a rebalanced protein composition preserving seed protein content without major collateral changes in the metabolome and transcriptome. Plant Physiology 156: 330–345.

Schussler, J.R., M.L. Brenner and W.A. Brun. 1984. Abscisic acid and its relationship to seed filling in soybeans. Plant Physiology 76: 301–306.

Senthil, A., G. Pathmanaban and P.S. Srinivasan. 2003. Effect of bioregulators on some physiological and biochemical parameters of soybean. Legume Research 26: 54–56.

Seo, M., A. Hanada, A. Kuwahara, A. Endo, M. Okamoto, Y. Yamauchi et al. 2006. Regulation of hormone metabolism in Arabidopsis seeds: Phytochrome regulation of abscisic acid metabolism and abscisic acid regulation of gibberellin metabolism. Plant J. 48: 354–366.

Shakirova, F.M., A.R. Skhabutdinova, M.V. Bezrukova, A. Fathutdinova and D.R. Fathutdinova. 2003. Changes in the hormonal status of wheat seedlings induced by salicylic acid and salinity. Plant Science 164: 317–323.

Shimizu-Sato, S., M. Tanaka and H. Mori. 2009. Auxin-cytokinin interactions in the control of shoot branching. Plant Molecular Biology 30: 271–280.

Shou, H., D. Zhu, C. Chen, W. Zhu and S. Zhu. 1991. The initial study of responses and physiological indexes for drought resistance in eight soybean varieties under drought condition. Acta Agriculture Zhejiangensis 2: 78–81.

Shu, K., H. Zhang, S. Wang, M. Chen, Y. Wu, S. Tang et al. 2013. ABI4 regulates primary seed dormancy by regulating the biogenesis of abscisic acid and gibberellins in *Arabidopsis*. PLoS Genet. 9: e1003577.

Shu, K., X.D. Liu, Q. Xie and Z.H. He. 2016. Two faces of one seed: hormonal regulation of dormancy and germination. Mol. Plant 9: 34–45.

Shu, K., X.D. Liu, Q. Xie and Z.H. He. 2016b. Two faces of one seed: Hormonal regulation of dormancy and germination. Mol. Plant 9: 34–45.

Singh, S.R., K.O. Rachio and K.E. Dashiell. 1987. Soybean for the Tropics: Research, Production and Utilization. John Wiley & Sons Ltd., Chichester, New York, USA.

Soeda, Y., M.C.J.M. Konings, O. Vorst, A.M.M.L. van Houwelingen, G.M. Stoopen, C.A. Maliepaard, J. Koddle, R.J. Bino, S.P.C. Groot and A.H.M. van der Geest. 2005. Gene expression programs during *Brassica oleracea* seed maturation, osmopriming, and germination are indicators of progression of the germination process and the stress tolerance level. Plant Physiology 137: 354–368.

Solomani, A., C. Sivakumar, S. Anbumani, T. Suresh and K. Arumugam. 2001. Role of plant growth regulators on rice production: A review. Agriculture Review 23: 33–40.

Strader, L.C. and Y. Zhao. 2016. Auxin perception and downstream events. Curr. Opin. Plant Biol. 33: 8–14.

Subramaniam, G. et al. 2016. Type B Heterotrimeric G protein gamma-subunit regulates auxin and ABA signaling in tomato. Plant Physiol. 170: 1117–1134.

Sudadi, S. 2012. Exogenous application of tryptophan and indole acetic acid (IAA) to reduce root nodule formation and increase yield of soybean. Agricultural Science Research Journal 2: 134–139.

Sun, J., D.A. Hutchins, Y. Feng, E.L. Seubert, D.A. Caron and F.-X. Fu. 2011. Effects of changing pCO_2 and phosphate availability on domoic acid production and physiology of the marine harmful bloom diatom *Pseudo-nitzschia multiseries*. Limnology and Oceanography 56: 829–840.

Taiz, L. and R. Zeiger. 2010. Plant Physiology (5th edition). Sinauer Associates Inc. Sunderland, MA, USA.

Takahashi, H., M. Kamal, H. Mikoshiba and Y. Ota. 1996. Effect of abscisic acid application on pod formation and yield of soybean plants. Japanese Journal of Tropical Agriculture 40: 1–6.

Teng, W., W. Li, Q. Zhang, D. Wu, X. Zhao, H. Li, Y. Han and W. Li. 2017. Identification of quantitative trait loci underlying seed protein content of soybean including main, epistatic, and QTL × environment effects in different regions of Northeast China. Genome 60: 649–655.

Tian, X., S. Song and Y. Lei. 2008. Cell death and reactive oxygen species metabolism during accelerated ageing of soybean axes. Russian Journal of Plant Physiology 55: 33–40.

Travaglia, C., H. Reinoso and R. Boltini. 2009. Application of abscisic acid promotes yield in field-cultured soybean by enhancing production of carbohydrates and their allocation in seed. Crop and Pasture Science 60: 1131–1135.

Upadhyay, R.G. and R. Ranjan. 2015. Effect of growth hormones on morphological parameters, yield and quality of soybean (*Glycine max* L.) during changing scenario of climate under midhill conditions of Uttarakhand. International Journal of Tropical Agriculture 33: 1899–1904.

Vikhi, S.B., D.B. Bangal, S.N. Deshmuk and V.A. Patil. 1983. Effect of growth regulators and urea on pod number of pigeonpea cv. 148. International Pigeonpea Newsletter 2: 39–40.

Vu, J.C.V., R.W. Gesch, A.H. Pennanen and L. Allenttartwell. 2001. Soybean photosynthesis, Rubisco and carbohydrate enzymes function at super optimal temperatures in elevated CO_2. Journal of Plant Physiology 158: 295–307.

Wang, R. and M. Estelle. 2014. Diversity and specificity: Auxin perception and signaling through the TIR1/AFB pathway. Curr. Opin. Plant Biol. 21: 51–58.

Wareing, P.F., M.M. Khalifa and K.J. Treharne. 1968. Rate-limiting processes in photosynthesis at saturating light intensities. Nature 220: 453–457.

Wiebold, W.J. 1990. Rescue of soybean flower destined to abscise. Agronomy Journal 82: 85–88.

Wiebold, W.J. and M.T. Panciera. 1999. Vasculature of soybean racemes with altered intraraceme competition. Crop Science 30: 1089–1093.

Yamaguchi, S. and Y. Kamiya. 2000. Gibberellin biosynthesis: Its regulation by endogenous and environmental signals. Plant and Cell Physiology 41: 251–257.

Yamauchi, Y., N. Takeda-Kamiya, A. Hanada, M. Ogawa, A. Kuwahara, M. Seo et al. 2007. Contribution of gibberellin deactivation by AtGA2ox2 to the suppression of germination of dark-imbibed *Arabidopsis thaliana* seeds. Plant Cell Physiol. 48: 555–561.

Yashima, Y., A. Kaihatsu, T. Nakajima and M. Kokubun. 2005. Source/sink ratio and cytokinin application on pod set in soybean. Plant Production Science 8: 139–144.

Yuan, L. and D.Q. Xu. 2001. Stimulation effect of gibberellic acid short-term treatment on leaf photosynthesis related to the increase in Rubisco content in broad bean and soybean. Photosynthesis Research 68: 39–47.

Zhang, M., L. Duan, Z. Zuho, J. Li, X. Yian, B. Wang, Z. He and Z. Li. 2004. Effects of plant growth regulators on water deficit induced yield loss in soybean. Proceedings of 4th International Crop Science Congress, pp. 571–575.

Zhang, J., D.L. Smith, W. Liu, X. Chen and W. Yang. 2011. Effects of shade and drought stress on soybean hormones and yield of main stem and branch. African Journal of Biotechnology 10: 14393–14398.

Zhu, Z. et al. 2016. Overexpression of AtEDT1/HDG11 in Chinese Kale (*Brassica oleracea* var. alboglabra) enhances drought and osmotic stress tolerance. Front. Plant Sci. 7: 1285.

CHAPTER 12
Abiotic Stress

Climate change has far-reaching implications for global food security and has already substantially impacted agricultural production worldwide through effects on soil fertility and carbon sequestration, microbial activity and diversity, as well as on plant growth and productivity.

Negative environmental impacts are exacerbated in current cropping systems by low diversity and the high intensity of inputs, climate associated yield instabilities being higher in grain legumes such as soybean and broad leaved crops than in autumn sown cereals (Reckling et al., 2018). The predicted increased frequency of drought and intense precipitation events, elevated temperatures, as well as increased salt and heavy metals contamination of soils, will often be accompanied by increased infestation by pests, and pathogens are also expected to take a major toll on crop yields, leading to enhanced risks of famine (Long et al., 2015). For example, the frequency and intensity of extreme temperature events in the tropics are increasing rapidly as a result of climate change. Tropical biomes are currently experiencing temperatures that may already exceed physiological thresholds. The ability of tropical species to withstand such "heat peaks" is poorly understood, particularly with regard to how plants prevent precocious senescence and retain photosynthesis in the leaves during these high temperature (HT) conditions. Such environmental stresses are among the main causes for declining crop productivity worldwide, leading to billions of dollars of annual losses.

Throughout history, farmers have adopted new crop varieties and adjusted their practices in accordance with changes in the environment. However, with the rising global temperatures, the pace of environmental change will likely be unprecedented. Furthermore, with the expansion of crop cultivation to non-optimal environments and non-arable lands, development of climate-specific resilient crops is becoming increasingly important for ensuring food security (Kathuria et al., 2007).

Global food demand is predicted to grow by 70–85% as the population increases to over 9 billion people by 2050 (FAO, 2017). Future agriculture requires tailored solutions that not only incorporate fundamental step changes in current knowledge and enabling technologies but also take into account of the need to protect the earth and respect societal demands.

12.1 Water Stress

Agriculture requires ~ 70% of the total fresh water resources (Cosgrove, 2009). Agricultural drought refers to the shortage of precipitation that causes a deficit in soil water and reduction of ground water or reservoir levels, which will hamper farming and crop production (Motha, 1992). Tremendous importance has been placed on the enhancement of drought tolerance of soybean, with a primary goal of enhancing yield under drought. Traditional breeding is a widely accepted strategy which will combine desirable agronomic traits from soybean germplasms via repeated crossing and selection processes. The recent advances in genomics, genetics, and molecular biology facilitate the identification of molecular markers and functional genes that are related to drought tolerance in soybean. Therefore, the ideas of enhancing drought tolerance by marker-assisted breeding and genetic modification have gained growing attention.

In addition to the genetic improvement programs for soybean, agronomic practices aimed at minimizing water input, reducing water loss, and increasing plant water usage efficiency have also been developed in a bid to cope with the problem of water scarcity. Some of these can be applied for soybean cultivation.

Numerous efforts have been put to examine the effects of drought and irrigation at various vegetative stages on soybean production. A 2-year field experiment by Brown et al. (1985) on 4 determinate cultivars, Davis, Lee 74, Sohoma and Centennial, demonstrated that moisture stress initiated at R2 or R4 reduced yield significantly (Brown et al., 1985).

An in-depth analysis of the effects of drought at various growth stages on seed yield of soybean cultivar Douglas was reported by Eck et al. (1987). In their study, yield loss was the most severe when drought stress was applied throughout the seed development period (R5–R7), resulting in a reduction of 45% and 88%, respectively, in two consecutive years (Eck et al., 1987). Besides, Desclaux et al. (2000) conducted a comprehensive analysis of yield components when drought stress was applied to soybean cultivar Weber at different developmental stages. In this experiment, the stress condition was attained by temporally withholding irrigation for 4 to 5 days until the plant available water reduced to 50% or 30% of the normal conditions. The most severe effect of this treatment was observed during the seed filling period.

Korte et al. (1983) conducted a 3-year study on 8 soybean cultivars to assess yield enhancement by irrigation, using non-irrigated soybean plants as the control group. The experimental groups were irrigated at different developmental stages (one stage or different stages in combination), including the flowering stage (R1–R2), the pod elongation stage (R3–R5), and the seed enlargement stage (R5–R6). Results of factorial analysis indicated that the yield was sensitive to the enhancement by irrigation, at pod elongation stage (R3–R4) and the seed enlargement stage (R5–R6). For 5 cultivars, the enhancement effect by irrigation followed the order: Seed enlargement stage (R5–R6) > pod elongation stage (R3–R4) > flowering (R1–R2). A separate experiment by Kadhem et al. (1985) supported the sensitivity toward irrigation at the pod elongation stage in determinate cultivars (R3.7 and R4.7).

Despite the general decreasing trend of seed weight under drought, the seed weight may not decrease uniformly as a function of drought intensity. Therefore, seed weight distribution has become another parameter employed to evaluate the effect of drought on seed weight, through the assessment of weight of seeds of different sizes. Dornbos and Mullen (1991) reported that, under severe drought, the proportion of seeds of diameter larger than 4.8 mm was reduced by 30%–40% while the proportion of seeds of diameter smaller than 3.2 mm was increased by 3%–15%. Under drought, soybean plants continued to produce heavy seeds. However, a greater portion of seeds were of low weight.

Drought stress reduced the number of nodes, which is a result of the reduced main stem height and the decreased node emergence rate. Length of internode is also a parameter for evaluating drought stress. However, the change in internode length is dependent on the timing of drought. In the experiment reported by Desclaux et al. (2000), only the internodes which initiated during drought stress showed reduction in length.

Various parameters have been adopted to assess drought tolerance. Due to different environmental and temporal factors, the results of assessment may be varied. Therefore, experiments have to be conducted consecutively for a few years in the same regions with large sampling size using various assessment parameters in order to achieve a more reliable classification.

Soybean is an important seed legume due to its high protein (35%), oil and carbohydrate contents (21% and 34%, respectively). Nitrogen fixing ability (17–127 kg/ha/year) is another advantage of soybean plants (Messina, 1997). Growth, development and yield of soybean are the result of a genetic potential interacting with environment. Soybean seed production may be limited by environmental stresses and minimizing these stresses will optimize seed yield.

One of the important environmental stresses that affect crop production worldwide is water stress. When the full crop requirements are not met, water deficit in the plant can develop to a point where crop growth and yield are affected. The need for water in soybean increases with plant development, peaking during the flowering and seed filling phases (7–8 mm day^{-1}) and decreasing thereafter (Bertolli et al., 2012). So, various growth stages of soybean respond differently to water stress (Egli and Bruening, 2004). It has been observed that maximum reduction in yield due to drought stress occurs during the pod set and seed filling period (Desclaux et al., 2000).

Seed filling and yield of soybean under water and radiation deficits were investigated during 2011 and 2012. Treatments were irrigations (I1, I2, I3 and I4 for irrigation after 60, 90, 120 and 150 mm evaporation from class A pan, respectively) in main plots and light interceptions (L1: 100%, L2: 65% and L3: 25% sunlight) in sub-plots. Seeds per plant under I1 and I2 decreased, but under I3 and I4 increased as a result of radiation deficit. Maximum seed weight and seed filling duration of plants under 25% light interception (L3) were higher than those under full sunlight (L1) and 65% light interception (L2).

In contrast, plants under full sunlight had the highest seed filling rate, particularly under water stress. Seed filling duration under severe light deficit (L3) was about 9 days longer than that under full sunlight (L1), leading to 15.8% enhancement in maximum seed weight. Decreasing seed yield of soybean under well-watered and mild water stress conditions and improving it under moderate and severe water deficit due to low solar radiation conditions are directly related to changes in seed filling duration and, consequently, in seed weight and number of seeds per plant (Fig. 12.1).

Drought induces morphological changes in plants, enabling them to sense and rapidly adapt to the stress. Root-related traits are crucial in maintaining crop yield in soybean (Bengough et al., 2011). Drought alters the root system architecture (root depth, root angle and root branching density) (Fenta et al., 2014).

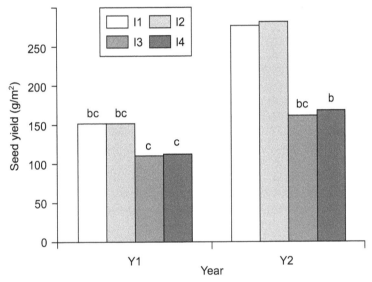

Fig. 12.1 Seed yield of soybean under different irrigation treatments in 2011 (Y1) and 2012 (Y2). I1, I2, I3, I4: Irrigation after 60, 90, 120 and 150 mm evaporation, respectively.

For instance, root architecture was characterized in the field under normal and water deficit conditions using three soybean cultivars (Jackson, Prima 2000 and A5409RG). As a result, Prima 2000 (drought-tolerant cultivar) has an intermediate root phenotype with a root angle of 40–60°, while a shallow root phenotype along with root angle of < 40° has been observed in drought-sensitive cultivar A5409RG (Fenta et al., 2014).

Depth of rooting system influenced by the elongation of taproot also plays an important role for plant survival under water deficit conditions (Fenta et al., 2014). An increase in the number of root tips, root length, root surface area and root volume was observed under water limited conditions. Several studies have proposed that roots having large xylem number, diameters, lateral root systems with more root hairs are indicators of drought tolerance (Prince et al., 2017). Jackson is considered as drought escaping cultivar, with long roots penetrating deep into the soil, permitting better water uptake compared with drought-sensitive cultivars (Fenta et al., 2014; Serraj et al., 1997). Under water-limited conditions, Plant Introduction (PI) 578477A and 088444 exhibited higher yield due to a higher lateral root number in clay soil (Prince et al., 2016). It was reported that deeper region of soil has high root density under seasonal drought as compared to dry surface of soil (Garay and Wilhelm, 1983). In addition, total root length/ plant weight, dry root weight/plant weight and root volume/plant weight were positively correlated with drought tolerance (Liu et al., 2005). Therefore, studying the relationship between root traits and drought is helpful to develop a drought-resistant cultivar.

Root-to-shoot ratio is also a good indicator to allocate the resources between different plant components. The water-limited environment increases the root-to-shoot ratio. For example, in soybean, root-to-shoot ratio increased by 13% indicates the flow of biomass towards roots (He et al., 2017).

The drought-tolerant soybean genotype (C12) showed a higher root-to-shoot ratio than the susceptible genotype (C08) under restricted soil water with application of exogenous ABA. To cope with drought stress, leaf morphology also plays an important role. Under water-limited conditions, plants reduce their leaf area by closing stomata. Due to water scarcity, reduction in soybean plant leaf area has been reported (Liu et al., 2003). In contrast, drought-tolerant soybean cultivar exhibited a greater leaf area than a less-tolerant cultivar under hydric stress conditions (Stolf-Moreira et al., 2010).

Aerenchyma formation is a major indicator that facilitates gas exchange between aerial and submerged plant parts (shoots and/or roots) to avoid flooding stress (Laan et al., 1989). Flooding stress induces two kinds of aerenchyma, i.e., primary (cortical) (Kawai et al., 1998) and secondary (white and spongy tissues) (Shimamura et al., 2003). A number of aquatic plants develop cortical aerenchymatous tissue by cell disintegration (lysigenous aerenchyma) and cell separation (schizogenous aerenchyma) (Seago et al., 2005). In rice, barley, maize and wheat, lysigenous aerenchyma is induced by flooding (Steffens et al., 2012). In some species, particularly soybean, secondary aerenchyma, having a spongy parenchyma cell layer, develops through cell division of phellogen (Bailey-Serres and Voesenek, 2008). Secondary aerenchyma is morphologically and anatomically different from cortical aerenchyma (lysigenous and schizogenous aerenchyma) (Shimamura et al., 2003). Waterlogging stimulated the formation of aerenchyma and adventitious roots in soybean plants facilitating transport of oxygen from shoot to root (Shimamura et al., 2010). Under waterlogging conditions, adventitious roots are formed in several flooded plants, including soybean (Steffens et al., 2012). However, adventitious roots are absent in soybean seedlings under complete submergence (Tamang et al., 2014). Under flooding conditions, secondary aerenchyma, consisting of white and spongy tissues, develops within a few weeks in stems, roots and root nodules of soybean (Shimamura et al., 2003). Aerenchyma formations initiated by ethylene, Ca^{2+}, and ROS signalling through a programmed cell death process are involved in aerenchyma development (Evans, 2004).

Rapid shoot elongation is another escape mechanism for adaptation in waterlogging stress (Phukan et al., 2014). It has been reported that lower stem of soybean having hypertrophic lenticels helps oxygen entry into the aerenchyma (Shimamura et al., 2010). Flooding also causes a significant reduction in leaf number, leaf area, canopy height and dry weight at maturity in soybean.

Water and high air temperature (AT) stresses that occur during soybean (*Glycine max* (L.) Merr.) seed fill greatly reduce seed yield, but their effects on seed germination and vigor are less clear. A study was made in order to determine the effect of water stress at optimum and high AT during seed fill on soybean seed yield and individual seed weight, and the subsequent germination and vigor of the seed. At daytime ATs of 21 and 35°C in 1985 and 29 and 33°C in 1986. Control, moderate and severe water-stress treatments were imposed by differential irrigation throughout seed fill on greenhouse-grown plants.

Water stress intensity, measured by accumulating stress degree days (SDD) during seed fill, increased linearly as the volume of irrigation water declined. The weight and number of seed produced by each plant, and individual seed weight, declined linearly as SDDs accumulated at each AT. Water stress at optimum ATs reduced seed number more than individual seed weight, but water stress at high ATs reduced individual seed weight more than seed number. Water or high AT stress caused fewer larger seed and more small seed to be produced. The germination percentage and vigor of the harvested seeds was reduced by water stress and high ATs, but by a smaller proportion than yield or seed number.

Individual seed weight, germination, and seedling growth rate were strongly correlated when reduced by water and high AT stress. Severe stress during seed fill caused soybean plants to exceed their capacity to buffer seed number, shifting seed weight distributions towards a larger proportion of small seed, resulting in poor seed lot germination and vigor.

Pod number per plant at maturity is a main yield determinant in soybean (Dybing et al., 1986). Drought stress occurring during flowering and early pod development significantly increases the rate of pod abortion, thus decreasing final seed yield (Liu et al., 2003b). Studies have shown that early pod expansion (3–5 days after anthesis, DAA) is the critical stage in soybean reproductive development under drought conditions (Liu et al., 2003b). This stage is characterized by an active cell division in the young ovules, which is particularly sensitive to soil water deficits (Peterson et al., 1992). It has been well recognized that the external size and weight of soybean pods are positively correlated with the internal features, including ovule and embryo cell numbers (Peterson et al., 1992), and that the growth rate of young soybean pods is often positively correlated with pod set (Vega et al., 2001). Therefore, a better understanding of the effects of soil water deficits on pod growth during the critical stage of pod development is required.

Accumulated evidence has shown that root-originated xylem sap ABA can move to crop reproductive structures and accumulate there to a high level under drought conditions (Liu et al., 2003b). This elevated ABA content in crop reproductive structures had been thought to be involved in controlling kernel/pod abortion, presumably via inhibition of cell division in the young ovaries (Liu et al., 2003b). In addition, exogenous application of ABA to develop maize ovaries inhibits cell division in the embryo and endosperm; this effect is probably due to a depression of cellcycle gene expression by high levels of ABA (Setter and Flannigan, 2001). Taken together, these studies suggest that drought-induced increase in xylem sap [ABA] might affect expansion growth of crop reproductive structures, resulting in a weak sink intensity, which fails to attract assimilate from source organs and eventually leads to abortion.

Root-originated chemical signals have been shown to regulate the response of vegetative shoot to drought in soybeans (*Glycine max* L. Merr.). However, their roles in the growth of soybean reproductive structures under drought stress have not yet been investigated. To explore this, a glasshouse experiment, in which potted soybeans were either well-watered (WW) or subjected to six levels of drought stress, was conducted. Irrigation was withheld in pots at six different dates before anthesis in order to induce drought of different severity (D1–D6) at sampling, viz., 4 days after anthesis (DAA). Root water potential, leaf water potential, pod water potential, xylem sap [ABA], pod fresh weight (FW), and pod set percentage were determined. Soil water status in the pot was expressed as the fraction of transpirable soil water (FTSW). Pod FW started to decrease at FTSW ¼ 0:43±0:02, when pod water potential was similar to that in the WW plants, while root water potential had decreased to –0.15 MPa and xylem sap [ABA] had increased 9-fold as compared with the WW plants. Pod set started to decrease at FTSW ¼ 0:30±0:01, and coincided with the decrease in pod water potential. Pod set started to decrease only when pod FW had decreased ca. 30%. Based on the results, a potential role of drought-induced increase in xylem sap ABA in affecting pod growth was suggested. It was proposed that a low pod water potential, which might have led to disruptions in metabolic activities in the pods, is important in determining pod abortion.

Early pod development of soybean is characterised by active cell division in the young ovules and is marked by rapid pod expansion; both processes are very sensitive to drought stress. Drought-induced carbohydrate deprivation and change in the concentration of endogenous abscisic acid (ABA) of the plants could have significant effects on pod growth and development, and may, therefore, be involved in inducing pod abortion. To test these hypotheses, four pot experiments with soybean (*cv.* Holladay) were conducted in an environmentally-controlled glasshouse during 2002 and 2003. The fraction of transpirable soil water (FTSW) was used as a measure of the soil water status in the pots. The first two experiments were designed to investigate the critical stage for pod abortion and the associated changes of biophysical and biochemical factors, viz., plant water relation characteristics, photosynthesis, endogenous ABA and carbohydrate concentrations under drought stress. The third experiment was set up to investigate the effect of drought stress on pod ABA concentration and to further study the relationships of pod set to the aforementioned factors at the critical, abortion-sensitive stage. Finally, in the last experiment, manipulation studies were carried out in order to verify the postulations formulated during the former experiments. The results showed that severe drought stress significantly decreased pod set up to 40% and the critical stage for pod abortion was 3–5 days after anthesis (DAA), when cell division was active in the ovaries. Drought at later stages, when pod filling had begun, reduced seed size but had no significant effect on pod set. Pod water potential was reduced by drought, however pod turgor was maintained at similar levels to the well-watered controls.

ABA concentration increased significantly in the xylem sap, leaves, and pods of drought stressed plants. Xylem-borne ABA and leaf ABA were seemingly the source of ABA accumulated in the drought-stressed pods.

Carbohydrate metabolism was disrupted by drought stress in both leaves and floral organs. In leaves, drought stress decreased photosynthetic rate, starch and sucrose concentrations but increased hexoses (glucose + fructose) concentrations, indicating a source limitation. In flowers and pods, drought stress increased sucrose and hexoses concentrations but decreased starch concentration, soluble invertase activity, and hexoses to sucrose ratio, indicating that the capacity of the pods to utilise the incoming sucrose was impaired by drought stress. As a consequence of both source and sink restrictions, non-structural carbohydrate (sucrose + hexoses + starch) accumulated in the pods was significantly reduced under drought stress.

The soil water thresholds for reduction in pod growth and pod set were 0.43 and 0.30 of FTSW, respectively. Pod growth was reduced before a significant decrease of pod water potential was detected, and the decrease of pod fresh weight was closely correlated with increasing xylem sap ABA concentration, implying that root signal and not pod water potential was detected, and the decrease of pod fresh weight was closely correlated with increasing xylem sap ABA concentration, implying that root signal and not pod water potential controlled pod growth during soil drying. Pod set began to decrease only when pod water potential had decreased and photosynthetic rate and pod fresh weight had decreased by 40% and 30% respectively; pod ABA concentration had increased 1.5-fold compared to the well-watered controls. Below the threshold water potential, pod set decreased further and correlated positively with photosynthetic rate and pod fresh weight, but negatively with pod ABA concentration.

Manipulation studies showed that application of 0.1 mM ABA on the canopy decreased gas exchange rates and pod set in well-watered soybeans. In drought-stressed plants, ABA treatment induced stomatal closure during early stage of soil drying, leading to higher leaf water potential. This maintained greater gas exchange rates, resulting in an increased pod set compared to the plants without ABA application. Application of 1 mM 6-benzylaminopurine (BA, an artificial cytokinin) on the canopy increased gas exchange rates and pod set in well-watered plants, but decreased leaf water potential, gas exchange rates and slightly decreased pod set in drought-stressed plants. In ABA- and BA-treated plants, pod set was linearly correlated with the leaf photosynthetic rate, implying that the two hormones exert their roles in altering pod set partly by modifying photosynthate availability.

In higher plants, the drought stimuli are presumably perceived by osmosensors (that are yet to be identified) and then transduced down the signaling pathways, which activate downstream drought responsive genes to display tolerance effects (Ahuja et al., 2010). The tolerance involves not only the activities of protein receptors, kinases, transcription factors, and effectors but also the production of metabolites as messengers for transducing the signals. Drought tolerance is of multigenic nature, involving complex molecular mechanisms and genetic networks.

The signaling pathway of drought stress largely overlaps with the signaling pathway of osmotic stresses; this has been extensively reviewed (Ahuja et al., 2010).

The perception of drought stimulus is presumably via unknown osmosensors. It is speculated that these sensors are associated with alterations in membrane porosity, integrity (Mahajan and Tuteja, 2005), and turgor pressure (Reiser et al., 2003). From the spatial perspective, membrane proteins, cell wall receptors, and cytosolic enzymes are all potential sensors for osmotic stress (Kader and Lindberg, 2010). For example, the families of THESEUS 1 and FERONIA receptor-like kinases (RLKs) in *A. thaliana* are putative stress sensors in cell wall to perceive changes in cell wall integrity and turgor pressure (Cheung and Wu, 2011). On the other hand, from the functional point of view, calcium ion (Ca^{2+}) channels, Ca^{2+} binding proteins, two-component histidine kinases, receptor-like protein kinases, Gprotein coupled receptors are also potential candidates for being osmosensors (Grene et al., 2011; Huang et al., 2011). For instance, AHK1 has been postulated as a cell surface sensor that activates the high-osmolarity glycerol response 1 (HOG1) mitogen-activated kinase (MAPK) cascade in transgenic yeast (Urao et al., 1999).

In soybean, two-component histidine kinases (GmHK07, GmHK08, GmHK09, GmHK14, GmHK15, GmHK16 and GmHK17) and receptor-like protein kinases (GmCLV1A, GmCLV1B, GmRLK1, GmRLK2, GmRLK3 and GmRLK4) have been identified as candidates of osmosensors (Le et al., 2011). However, direct evidence for their functions to perceive stress signals in soybean is still missing.

Abscisic acid (ABA) regulates the physiology (e.g., closure of stomata) and metabolism of plants (e.g., expression of enzymes) in order to rapidly cope with environmental challenges (Zhang et al., 2006). Biosynthesis, accumulation, and catabolism of ABA are all crucial for the transduction of ABA mediated signals. The accumulation of ABA in response to drought is associated with the changes in the levels of Ca^{2+} and ROS (Xiong and Zhu, 2002; Zhao et al., 2001). *In planta*, ABA is synthesized in various cell types, including root cells, parenchyma cells, and mesophyll cells. Under drought stress, ABA is transported to guard cells to control stomatal aperture (Wilkinson and Davies, 2010). ABA reaching the target tissues and cells will be recognized and the signals will be transduced down the ABA signalosome (Umezawa et al., 2010), including ABA receptors (PYR/PYL/RCAR), negative regulators (e.g., group A protein phosphatases 2C), and positive regulators (e.g., SnRK-type kinases).

Components of this system have been discovered in soybean. For example, GsAPK is a SnRK-type kinase from wild soybean that is up-regulated by drought stress in both leaves and roots, but down-regulated by ABA treatment in roots. *In vivo* assay revealed that the phosphorylation activities of GsAPK is activated by ABA in a Ca^{2+}-independent manner, suggesting that GsAPK may play a role in the ABA-mediated signal transduction. Activated SnRK-type kinases in rice and *A. thaliana* will phosphorylate target proteins, including bZIP transcription factors and membrane ion channels (Yang et al., 2012).

Perceived stress signals may trigger transient changes in the cytosolic Ca^{2+} level, which acts as a second messenger (McAinsh and Pittman, 2009). Ca^{2+} sensors, in turn, transmit and activate the signaling pathways for downstream stress responses (Xiong and Zhu, 2002). Ca^{2+} sensors include various types of Ca^{2+}-binding proteins: CaMs (calmodulins), CMLs (CaM-like proteins), CDPKs (Ca^{2+}-dependent protein kinases), and CBLs (calcineurin B-like proteins) (DeFalco et al., 2010). Among these Ca^{2+} sensors, all are plant and protist-specific with the exception of CaM. Expression of the soybean CaM (GmCaM4) in transgenic *A. thaliana* activated a R2R3 type MYB transcription factor which, in turn, up-regulated several drought-responsive genes, including *P5CS* (encoding a proline anabolic enzyme) (Yoo et al., 2005). While the application of Ca^{2+} affects the nodulation of soybean (Bell et al., 1989), the gene encoding a soybean CaM binding protein was found to be differentially expressed in soybean nodules under drought stress (Clement et al., 2008).

The drought tolerance-related CDPK family is well-studied in rice and *A. thaliana* (Zhu et al., 2007). In isolated soybean symbiosome membrane, a CDPK was demonstrated to phosphorylate an aquaporin called nodulin 26 and, hence, enhance the water permeability of the membrane. It was hypothesized that this is an integral part of the drought tolerance mechanism (Guenther et al., 2003). Besides Ca^{2+}, phosphatidic acid (PA) and the intermediates of inositol metabolism are also second messengers for signal transduction (Xue et al., 2009). However, there is only very limited evidence supporting the involvement of phospholipid signaling in drought stress response of soybean. The soybean nodulin gene *G93* encoding a ZR1 homologue was down-regulated under drought stress (Clement et al., 2006). Plant ZR1 homologue such as RARF-1 in *A. thaliana* may involve in lipid signaling via interaction with phosphatidylinositol 3-phosphate (Drobak, 2002).

When plants are subjected to drought stress, accumulation of cellular ROS will trigger the generation of hydrogen peroxide, a signaling molecule that will activate ROS scavenging mechanisms (Cruz, 2008). In soybean, exogenous application of hydrogen sulphide alleviates symptoms of drought stress, probably via triggering an antioxidant signaling mechanism (Zhang et al., 2010).

Many studies support the roles of protein kinases in stress signaling (Bartels et al., 2010). In plants, the drought responsive signal transduction of the MAPK family (MAPK, MAPKK/MEKK, MAPKKK/MKK) as well as the MAPK phosphatases (MKP) family have been relatively well-studied in *A. thaliana* and rice (Boudsocq and Lauriere, 2005), but remained under-explored in soybean, although a PA-responsive MAPK has been identified in soybean (Lee et al., 2001).

On the other hand, some non-MAPK type protein kinases found in soybean may be related to drought responses. The soybean gene encoding a serine/threonine ABA-activated protein kinase was found to be up-regulated by ABA, Ca^{2+}, and polyethylene glycol treatments (Luo et al., 2006). The With No Lysine protein kinase 1 of soybean is another serine/threonine protein kinase that is a putative osmoregulator (Wang et al., 2010).

The ubiquitin-mediated protein degradation pathway is also an integral part of the signal transduction network (Zhou et al., 2010). This pathway directs the degradation of target proteins by the 26S proteasome and is responsive to drought stress. Two ubiquitin genes and one gene encoding ubiquitin conjugating enzyme were identified as differentially expressed genes in nodulated soybean under drought stress (Clement et al., 2008). Overexpression of the ubiquitin ligase gene *GmUBC2* enhances drought tolerance in *A. thaliana* by up-regulating the expression of genes encoding ion transporters (AtNHX1 and AtCLCa), a proline biosynthetic enzyme (AtP5CS), and a copper chaperone (AtCCS) (Zhou et al., 2010).

Transcription regulation plays an important role in drought stress response. For instance, using oligo microarray analysis, transcriptions of 4,433 and 5,098 soybean genes were found to be significantly up-regulated and down-regulated, respectively, when subjected to a no-irrigation period for 4 days (Maruyama et al., 2012). The signal transduction pathways can ultimately regulate the expression of drought-responsive genes through diverse transcription factors. Transcription factors often target the corresponding *cis*-acting promoter elements, such as the drought stress-related elements DRE, ABRE, Gbox, and T/Gbox (Maruyama et al., 2012; Mochida et al., 2009).

In the soybean genome, ~ 500 transcription factors were *in silico* annotated. Increasing efforts have been made to characterize their importance and functions in relation to drought (Le et al., 2011; Mizoi et al., 2011; Nakashima et al., 2011; Perira et al., 2011; Phang et al., 2011).

Flooding ranks second after drought, causing yield reduction in soybean (Mittler, 2006; Valliyodon et al., 2014). Flooding stress can be categorized as waterlogging or submergence. In waterlogging stress, the roots are under water while shoots remain above ground, whereas during submergence, the plant is completely immersed in water-saturated soil. As plants are aerobic, hypoxia (insufficient oxygen) or anoxia (complete absence of oxygen) causes losses in crop production. Soybean is more sensitive to flooding stress, resulting in yield decline by reducing photosynthesis nitrogen fixation and biomass accumulation. Flooding stress can happen during any growing stage, but especially in the seed germination and vegetative stages, and leads to substantial decrease in soybean grain yield (Kokubun, 2013). In addition, flooding stress hampers yield production during vegetative (17–43%) and reproductive stage (50–56%) (Osterhuis et al., 1990).

To mitigate the negative impact of flooding stress, plants use a number of strategies for their survival, mainly escape and quiescence strategies (van Veen et al., 2013). In escape strategy, morphological (aerenchyma development, shoot elongation and adventitious root formation) and anatomical alterations allow the plant to exchange gas between cells and atmosphere. The Quiescence strategy suppresses morphological changes in order to save energy and resources and retard plant growth. This strategy depends on anaerobic energy production (Bailey-Serres and Voesenek, 2008).

Aerenchyma formation is a major indicator that facilitates gas exchange between aerial and submerged plant parts (shoots and/or roots) in order to avoid flooding stress (Laan et al., 1989). Flooding stress induces two kinds of aerenchyma, i.e., primary (cortical) (Kawai et al., 1998) and secondary (white and spongy tissues) (Shimamura et al., 2010). A number of aquatic plants develop cortical aerenchymatous tissue by cell disintegration (lysigenous aerenchyma) and cell separation (schizogenous aerenchyma) (Seago et al., 2005). In rice, barley, maize and wheat, lysigenous aerenchyma is induced by flooding (Steffens et al., 2012). In some species, particularly soybean, secondary aerenchyma, having a spongy parenchyma cell layer, develops through cell division of phellogen (Bailey-Serres and Voesenek, 2008). Secondary aerenchyma is morphologically and anatomically different from cortical aerenchyma (lysigenous and schizogenous aerenchyma) (Shimamura et al., 2010). Waterlogging stimulated the formation of aerenchyma and adventitious roots in soybean plants facilitating transport of oxygen from shoot to root (Shimamura et al., 2010). Under waterlogging conditions, adventitious roots are formed in several flooded plants, including soybean (Steffens et al., 2012). However, adventitious roots are absent in soybean seedlings under complete submergence (Tamang et al., 2014). Under flooding conditions, secondary aerenchyma consisting of white and spongy tissues develops within a few weeks in stems, roots and root nodules of soybean (Shimamura et al., 2003). Aerenchyma formations initiated by ethylene, Ca^{2+}, and ROS signalling through a programmed cell death process are involved in aerenchyma development (Evans, 2004).

Rapid shoot elongation is another escape mechanism adaptation to waterlogging stress (Phukan et al., 2014). It has been reported that lower stem of soybean having hypertrophic lenticels helps oxygen entry into the aerenchyma (Shimamura et al., 2010). Flooding also causes a significant reduction in leaf number, leaf area, canopy height and dry weight at maturity in soybean crops.

Under flooding stress, plants undergo different physiological and biochemical adaptations. For instance, in soybean, a significant reduction in photosynthetic activity and stomatal conductance was observed in Essex and Forrest within 48 h of flooding at vegetative and reproductive growth stages. Waterlogging also decreases biological nitrogen fixation, as nodules need adequate oxygen in order to maintain nitrogenase activity for aerobic respiration and contributing adenosine triphosphate (Osterhuis et al., 1990). As a consequence of flooding stress, a reduction in root hydraulic conductivity has also been reported (Else et al., 2001). Several studies have provided the correlation between stomatal conductance and carbon fixation. In flooded plants, photosynthetic activities were reduced by restricting CO_2 due to stomatal closure (Malik et al., 2001). Furthermore, due to the higher concentration of CO_2, assimilation in flooded soil, biomass and soybean root elongation eventually repressed (Boru et al., 2003).

Tamang et al. (2014) reported that submergence stimulates starch degradation, soluble carbohydrates and ATP in cotyledons and hypocotyls of soybean seedlings. Extensive submergence degrades the chlorophyll contents in aerial parts of several terrestrial plants (Fukao et al., 2006). However, under submergence, abundance of chlorophyll *a* and *b* remained nearly constant in soybean (Tamang et al., 2014). The decrease in photosynthetic activity with long-term flooding may be triggered by the reduction in chlorophyll, transpiration and ribulose-1,5-biphosphate (RuBP) carboxylase activity. These combined effects against flooding declined the crop growth, net assimilation and leaf expansion of plants. Blocking of hypertrophic lenticels at the base of stem restricted O_2 transport into the roots, resulting in a reduction of plant growth under hypoxic conditions (Jackson and Attwood, 1996).

Waterlogging blocks the oxygen supply to the root system, which inhibits respiration and greatly reduces the energy status of cells that affect important metabolic processes. This study evaluated fermentative metabolism and carbohydrate contents in the root system of two soybean (*Glycine max* L. Merril) genotypes under hypoxic and post-hypoxic conditions. Nodulated plants (genotypes Fundacep 53 RR and BRS Macota) were grown in vermiculite and transferred to a hydroponic system at the reproductive stage. The root system was submitted to hypoxia by flowing N_2 (nitrogen) gas in a solution for 24 and 72 h. For recovery, plants returned to normoxia condition by transfer to vermiculite for 24 and 72 h. Fermentative enzyme activity, levels of anaerobic metabolites and carbohydrate content were all quantified in roots and nodules. The activity of alcohol dehydrogenase, pyruvate decarboxylase and lactate dehydrogenase enzymes, as well as the content of ethanol and lactate, increased with hypoxia in roots and nodules, and subsequently returned to pre-hypoxic levels in the recovery phase in both genotypes. Pyruvate content increased in nodules and decreased in roots. Sugar and sucrose levels increased in roots and decreased in nodules under hypoxia in both genotypes. Fundacep RR 53 was more responsive to the metabolic effects caused by hypoxia and post-hypoxia than BRS Macota, and it is likely that these characteristics contribute positively to improving adaptations to oxygen deficiency (Fig. 12.2).

Flooding stress causes higher production of ROS, resulting in oxidative damage to proteins related to photosynthetic apparatus (Yordanova et al., 2004). As a result, the scavenging activity is overpassed under flooding stress (Yordanova et al., 2004).

Numerous studies reported that negative correlation exists between germination percentage and flooding stress (Maryam and Nasreen, 2012). Seeds are usually germinated under optimum conditions within 1 or 2 days. However, seed germination is delayed due to the quick absorption of water, collapse of seed structure, and outflow of internal seed contents under flooding stress. When seeds were flooded for 3 days after imbibition, germination percentage drastically dropped and seed injury was observed (Wuebker et al., 2001). Flooding causes mechanical damage on the soybean seeds and prohibits germination. Seed coat and seed weight are fundamental factors to evaluate a positive effect on seed flooding tolerance.

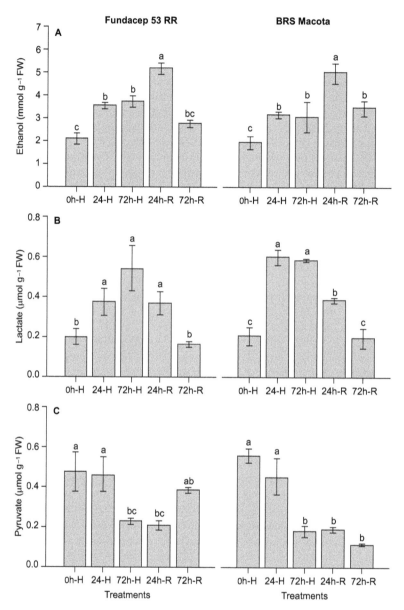

Fig. 12.2 Metabolites from roots of soybean genotypes (Fundacep 53 RR and BRS Macota) as induced by hypoxia and post-hypoxia conditions. Ethanol (A), lactate (B) and pyruvate (C). Hypoxia (0h-H, 24-H and 72-H) and recovery (24h-R and 72h-R). The *bars* mean value ± SD as determined from four biological replicates. Values that differ according to a one-way ANOVA ($p \leq 0.05$) and Tukey posthoc test are marked with different small letters.

For example, germination rate (GR) and normal seedling rate (NS) was higher in pigmented varieties as compared to yellow varieties of soybean. These parameters (GR and NS) were negatively correlated with seed weight (SW) in the combined population (Sayama et al., 2009). Therefore, pigmented seed coat and small seed weight could be key parameters in response to seed-flooding tolerance.

Essential traits, root length and shoot length are also important indicators in response to flooding stress. The insufficient allocation of water, minerals, nutrients, and hormones led to root and shoot damage (Jackson and Ricard, 2003). The first symptom that usually appears in soybean is wilting of leaves in response to flooding. Soybean shoot growth under flooded conditions is significantly reduced due to inability of the root system regarding water transport, hormones, nutrients and assimilates. Flooding tolerance in soybean is strongly correlated to root surface area, root length and dry weight (Sallam and Scott, 1987). It has been reported that root tips are extremely sensitive to flooding in soybean and pea seedlings (Nanjo et al., 2013). Under complete submergence, soybean root growth is absolutely repressed due to the death of root tips.

One of the major traits conferring tolerance to waterlogging is yield and production of good quality seeds (Van Toai et al., 1994). A significant decline in pod number, pods per node, branch number, and seed size was observed following 7 days of flooding at different vegetative and regenerative development stages (Linkemer et al., 1998). Sullivan et al. (2001) confirmed reduction in pod number and plant height at early vegetative growth stages. Soybean crops flooded with excessive water at early flowering stage showed severe chlorosis and stunted growth (Griffin and Saxton, 1988). Schöffel et al. (2001) showed a decreased number of pods per plant at the reproductive stage (R4) in pot trails. A field experiment was conducted in flooded soil and obtained yield reduction from 20–39% in the different soybean cultivars when subjected during the R5 stage. During flooding, a significant reduction in soybean yield was observed at R5 stage as compared to the R2 stage (Rhine et al., 2010).

Crop growth rate has been usually affected only when the waterlogging stress was applied for more than 2 d (Griffin and Saxton, 1988). Greater sensitivity to the stress was shown during the early reproductive (R1–R5) vs. vegetative periods (emergence-R1). However, waterlogging sensitivities at specific developmental stages throughout soybean's life cycle have not been identified. Soybean may not be as tolerant of waterlogging at late compared with optimal planting dates, because of the shorter vegetative growth period (emergence-R5) at late plantings (Board and Settimi, 1986).

Greenhouse and field studies were conducted during 1993 and 1994 near Baton Rouge, LA, (30°N Lat) on a Commerce silt loam. Waterlogging tolerance was assessed in cultivar Centennial (Maturity Group VI) at three vegetative and five reproductive growth stages by maintaining the water level at the soil surface in a greenhouse study. Using the same cultivar, the effect of drainage in the field for late-planted soybean was evaluated. Rain episodes determined the timing of waterlogging; redox potential and oxygen concentration of the soil were used to quantify the intensity of waterlogging stress.

Results of the greenhouse study indicated that the early vegetative period (V2) and the early reproductive stages (R1, R3, and R5) were most sensitive to waterlogging. Three to 5 cm of rain per day falling on poorly drained soil was sufficient to reduce crop growth rate, resulting in a yield decline from 2453 to 1550 kg ha-t. Yield loss in both field and greenhouse studies was induced primarily by decreased pod production, resulting from fewer pods per reproductive node. In conclusion, waterlogging was determined to be an important stress for late-planted soybean in high rainfall areas (Tables 12.1 and 12.2).

The analysis of quantitative trait loci (QTLs) for water-logging tolerance in soybean is usually challenging. However, several studies have been done on QTLs associated to flooding tolerance, focused on injury score and tolerance index in soybean (Sayama et al., 2009). For instance, a single QTL located on Chr. 18 (Sat_064) was identified using 208 lines of two recombinant inbred (RI) populations, for soybean growth and grain yields under water-logging conditions (Van Togai et al., 2001).

Table 12.1 Plot yield, sample yield, seed size, seed number, pod Dumber, pods per reproductive node, and total dry matter (R7) for Centennial soybean planted on drained and undrained sites near Baton Rouge, LA, 1993 and 1994. Data are averaged across row and years.

Parameter	Drained	Undrained	LSD0.05
Plot yield (kg/ha)	2452	1550	255
Sample yield (g/m)	295	219	38
Seed size (g/100 seeds)	13.43	11.88	0.24
Seed no. per m^2	2249	1842	263
Pod no. per m	1129	912	121
Pods per rep. node	2.43	1.90	0.14
Total dry matter (R7) (g/m)	493	362	63

Table 12.2 Yield of soybean cultivar Hartz 5000 flooded for one week at one vegetative and three reproductive stages averaged over two years (1999 and 2000).

Waterlogging treatment	Yield (kg/ha)
Control (no waterlogging)	4153a
1 wk waterlogging at V4	3454b
1 wk waterlogging at R1	3618b
1 wk waterlogging at R3	3485b
1 wk waterlogging at R5	3380b

Means not followed by the same letter are significantly different at 0.05 probability level.

RNA-seq based transcriptomic analysis resulted in detection of 729 and 255 genes in the flooding-tolerant line and ABA-treated soybean, respectively, which were significantly changed under stress condition. Transcript profiles also revealed that a total of 31 genes, included 12 genes involved in the regulation of RNA and protein metabolism, were commonly altered between the flooding-tolerant line and ABA-treated soybean under flooding stress (Yin et al., 2017). On the basis of the above findings, it can be concluded that transcript profiles can be helpful as an adaptive mechanism for soybean survival under water stress.

12.2 Temperature Stress

Variability, along with occurrences of short episodes of high temperature (HT), will be one of the important characteristics of future climates. Climate models predict that global surface temperatures will continue to rise (IPCC, 2007). Lobell and Asner (2003) reported that each 1°C increase in the average growing season temperature leads to a 17% decrease in soybean yield. Photosynthesis was the main physiological process that was affected by HT, often before other cell functions were impaired. The photosynthetic rate in C3 plants is generally maximum at about 30°C, with significant reductions for each additional degree increase in temperature (Wise et al., 2004). Leaf photosynthesis was usually inhibited when leaf temperatures exceeded approximately 38°C (Berry and Bjorkman, 1980). In general, chlorophyll fluorescence is also a good measure of the photosynthetic activity in leaves (Govindjee, 1995). Photosystem II (PS II) is temperature sensitive and down-regulates photosynthesis under HT conditions. High-temperature stress can also cause loss of grana stacking and denaturation of the PS-II complexes and light-harvesting complex II (Shutilova et al., 1995). However, the exact mechanisms causing lower photosynthesis under HT stress in soybeans are still not clearly understood and need attention.

Earlier research in soybean showed that, under HT stress, the stomatal conductance and photosynthetic rate decreased (Djanaguiraman and Prasad, 2010). The lower photosynthetic rate can be caused by decreased stomatal conductance or by nonstomatal factors. The carbon isotope ratio can provide information on stomatal and nonstomatal factors influencing photosynthetic rate. The ratio of ^{13}C to ^{12}C in plant tissue is less than the ratio of ^{13}C to ^{12}C in the atmosphere, indicating that plants discriminate against ^{13}C during photosynthesis. The ratio of ^{13}C to ^{12}C in C3 plants ($\delta\ ^{13}C$) varies mainly because of the discrimination that occurs during diffusion and enzymatic processes.

The rate of diffusion of $^{13}CO_2$ across the stomatal pore is lower than that of $^{12}CO_2$ by a factor of 4.4‰ (Farquhar et al., 1989). Hence, measuring the carbon isotope ratio in leaves can help determine if the decreased photosynthetic rate is an effect of stomatal conductance or of the activation state and/or content of the enzyme Rubisco.

In addition, HT stress also decreased the activity of the antioxidant defense system. Premature leaf senescence under HT stress was caused by the increased production of reactive oxygen species (ROS) (Djanaguiraman and Prasad, 2010). There is limited anatomical evidence (ultrastructural changes in the chloroplast, thylakoid membranes, mitochondria, and plasma membrane) to explain the mechanisms of decreased activity of PS-II photochemistry (*Fv/Fm*, the ratio of variable fluorescence to maximum fluorescence) and decreased photosynthetic rate under HT stress. Anatomical changes under HT stress showed smaller cell size, closure of stomata, and greater stomatal density (Anon et al., 2004) in other species. Zhang et al. (2005) showed that HT stress severely damaged the mesophyll cells and increased the permeability of the plasma membrane in grapes (*Vitis vinifera*; Linn.). At the subcellular level, major structural modifications occurred in the chloroplasts and mitochondria, leading to significant changes in photosynthesis (Zhang et al., 2005). Such studies on crop species, and particularly on soybean, are limited in number. Earlier studies showed that photosynthetic rates in plants were related to the ratio Fo/Fm (Ristic et al., 2008; Djanaguiraman and Prasad, 2010). An increased Fo/Fm ratio (which indicates thylakoid membrane damage) is associated with a lower chlorophyll content in wheat (Ristic et al., 2008) and soybean (Djanaguiraman and Prasad, 2010).

Soybean (*Glycine max* L. Merr.) genotype K03-2897 was grown in a controlled environment in order to study the effects of high-temperature (HT) stress on ultrastructural changes in leaves and its relationship with photosynthetic rate. The objectives of this study were to (i) quantify the effects of HT stress during flowering stage on photosynthetic rate and (ii) observe the anatomical and ultrastructural changes in leaves of soybean grown under HT. Plants were exposed to HT (38/28°C) or optimum temperature (OT; 28/18°C) for 14 d at flowering stage R2. High-temperature stress significantly decreased the leaf photosynthetic rate and stomatal conductance by 20.2 and 12.8%, respectively, compared with those at OT. However, HT stress significantly increased the thicknesses of the palisade and spongy layers and the lower epidermis above those at OT. In addition, HT stress damaged the plasma membrane, chloroplast membrane, and thylakoid membranes. The mitochondrial membranes, cristae, and matrix were distorted under HT stress. High-temperature stress increased the content of leaf reducing sugars by 82.6% above that at OT. Leaves under HT stress had a higher carbon isotope ratio compared with that at OT. Decreases in photosynthesis at HT stress were mediated through anatomical and structural changes in the cell and cell organelles, particularly the chloroplast and mitochondria (Table 12.3 and Fig. 12.3).

Under HT stress, there is a general tendency towards reduced cell size, closure of stomata and curtailed water loss, increased stomatal density and trichomatous densities, and larger xylem vessels in both roots and shoots. Several lines of study indicate that exposure of plants to HT results in the disintegration of ultrastructural characteristics, mainly attributed to a lower stomatal density, larger stomatal chamber with a larger stomatal opening area, thinner leaves, loose arrangement of mesophyll cells, a partially developed vascular bundle and unstable organelle structure.

It is difficult to obtain a high yield in soybean in successive years due to its compensatory growth and alterations in plant development, though many attempts have been made to obtain a high yield (Nakaseko et al., 1984). It was previously reported that the yield of soybean was very high in Xinjiang, China in 2002, 2003 and 2004 seasons (Isoda et al., 2006). In 2005, however, the seed yield was very low. Such a fluctuation in seed yield among years is influenced largely by climatic conditions, especially by air temperature (Kitano et al., 2006). On the other hand, pod number is considered to be one of the main factors for determining seed yield, and the variation in seed yield was often explained by pod number (Saitoh et al., 1998). A high night air temperature has been reported to increase pod number (Zheng et al., 2002), but reduce seed yield (Seddigh and Jolliff, 1984).

Table 12.3 Effect of optimum temperature (OT) and high-temperature (HT) stress on leaf anatomical traits. Each datum is the mean ± SEM of five images ($n = 5$). Means with different letters in each row are significantly different at $P \leq 0.05$.

Trait	OT (28/18°C)	HT (38/28°C)
Upper epidermis thickness (μm)	5.2a ± 1.1	7.2b ± 1.3
Palisade layer I thickness (μm)	81.4a ± 1.8	99.3b ± 2.2
Palisade layer II thickness (μm)	45.3a ± 2.1	52.5b ± 1.8
Spongy cell thickness (μm)	71.4a ± 1.6	83.6b ± 1.9
Lower epidermis thickness (μm)	4.2a ± 2.3	8.4b ± 1.5
Total leaf thickness (μm)	207.5a ± 10.2	251b ± 12.3
Stomatal density (number of stomata per mm^2)	107a ± 10.2	101a ± 12.4
Stomata diameter (μm)	64a ± 2.4	51b ± 2.1

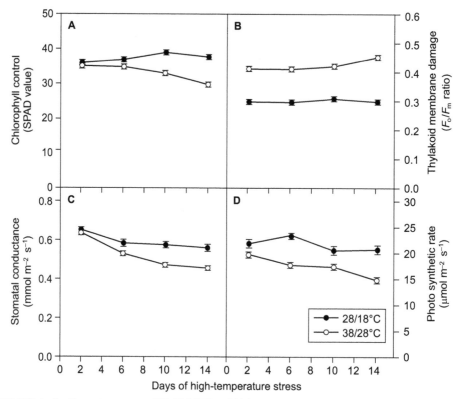

Fig. 12.3 Effect of optimum temperature (OT, 28/18°C) and high temperature (HT, 38/28°C) on (a) chlorophyll content (SPAD value), (b) thylakoid membrane damage (F_o/F_m ratio), (c) stomatal conductance (mol m^{-2} s^{-1}), and (d) photosynthetic rate (μmol m^{-2} s^{-1}) in soybean. Each point is mean of two independent experiments. Vertical bars denote ± SE of means ($n = 6$).

The growth and seed yields of 2 Japanese and 3 Chinese cultivars of soybeans cultivated in 2002–2005 using a drip irrigation system in the arid area of Xinjiang, China, were analyzed with respect to growth parameters and air temperature. Seed yield was very high in 2002, 2003 and 2004, but relatively low in 2005. The variation among years in seed yield clearly depended on pod number. The mean leaf area index (LAI) and crop growth rate (CGR) in 2005 was lower than those in the other 3 years. CGR showed significant positive correlations with mean LAI at the early growing stages, and with net assimilation rate (NAR) at the later growing stages. The increasing rate of pod number (IRP) was positively correlated with the mean LAI and CGR at the pod setting period, suggesting that an adequate supply of photosynthates would be required for pod setting. It was noted that excellent growth in the years with high yields was supported by the large LAI before the pod setting periods and by high NAR and vigorous pod growth at the latter half of the growing season.

Night temperature for example is one of the major important factors for the growth and development of soybeans (*Glycine max* (L.) Merr.). It is well known that differences between day and night temperatures enhances the seed filling rate and finally increases seed size. Actually increase in night temperature from 20°C to 25°C after flowering decreased seed size and seed yield (Sato and Ikeda, 1979).

Other experiments showed that a high night temperature brought the flowering time forward and decreased seed size and seed yield (Thomas and Raper, 1978).

Egli and Wardlaw (1980) reported that the reduction in seed size with a high night temperature was associated with a lower rate of seed growth. On the other hand, the reduction of seed yield has been reported to be more affected by a high day temperature than night temperature (Gibson and Mullen, 1996). However, the effects of night temperature on soybean reproductive growth are still difficult to precisely determine.

Soybean (*Glycine max* (L.) Merr. cv. Enrei) plants were grown in pots (15-L volume) placed in a greenhouse with ventilation. At the time when the first flower opened, pots were transferred to growth chambers with natural lighting under day temperature of 30°C and night temperatures of 20, 25 or 30°C. The numbers of flowers opened and pods set each day were recorded and the seed yield and yield components were investigated after harvest.

The increase in night temperature decreased the seed size and increased the number of flowers and pods. As a result, the seed weight per plant was unaffected by night temperature. However, high night temperatures increased the number of flowers on the secondary and tertiary racemes. These flowers opened after the 18th day of the flowering period and showed a high rate of pod setting. These results suggest that a high night temperature stimulated flower opening and pod setting in the secondary and tertiary racemes. The increases in the numbers of flowers and pods could serve to moderate the reduction of seed yield caused by a high night temperature (Table 12.4).

Low positive temperature has an inhibiting effect on the growth, development and other physiological processes of cold-sensitive plants, including soybean. Huang and Yang (1995) report that soybean seeds generally germinate at a temperature between 10 and 30°C. However, the rate of germination increases with increasing temperature and reaches its maximum at 30°C. Other authors also report that the optimal temperature for germination and hypocotyl elongation in soybean is around 30°C (Liao Fang Lei et al., 2011). Bharati et al. (1983) also found that earlier hydration of seeds at higher temperatures accelerated the rate of germination, while at lower temperatures (10°C) it slowed the rate of germination.

Soybean plants are also sensitive to chilling in the juvenile period; temperatures above zero but below 10°C cause damage to soybean plants resulting from temperature induced physiological and biochemical changes. A distinct increase in proline content belongs to the most frequently encountered changes and it has been observed in soybean seedlings by Yadegari et al. (2007), and in other species by Koc et al. (2010).

Chilling stress also causes the leakage of intracellular electrolytes from tissues as a result of the loss of cytoplasmic membrane integrity (Borowski and Blamowski, 2009). The destructive effect of chilling on the membranes is even greater in light than in the dark (Szalai et al., 1996). This is undoubtedly associated with the negative effect of stress also on the process of photosynthesis, which has been found both in soybean (Heerden et al., 2003a, b) and in other species (Starck et al., 2000).

Similarly to other types of environmental stress, chilling induces the production of H_2O_2 and other reactive oxygen species (ROS) in plants. Under such conditions, plants activate the enzymatic system that prevents ROS accumulation. One of the elements of this system is catalase (CAT EC 1.11.1.6). Increased catalase activity under chilling conditions has been observed in germinating soybean seeds by Posmyk

Table 12.4 The numbers of flowers opened and pods set, and the rate of pod set at different periods during flowering period in 1998.

Temperature (Day/night)	1–10 DAFFO	11–20 DAFFO	21–30 DAFFO	31–40 DAFFO	41–50 DAFFO	Total
30/20°C						
Flowers/plant	78.8 ± 16.6	29.0 ± 8.6	3.3 ± 2.8	3.5 ± 2.4	0.3 ± 0.5	114.8 ± 25.8
Pods/plant	35.0 ± 3.6	18.3 ± 4.0	1.3 ± 1.9	1.3 ± 1.0	0 ± 0	55.8 ± 3.9
% pod set	44.4	62.9	38.5	35.7	0	48.6
30/25°C						
Flowers/plant	71.5 ± 7.1	26.5 ± 9.0	17.0 ± 5.0	11.0 ± 0.8	0 ± 0	126 ± 13.3
Pods/plant	33.5 ± 2.6	15.3 ± 2.6	9.5 ± 4.4	5.5 ± 3.5	0 ± 0	63.8 ± 2.8
% pod set	46.9	57.5	55.9	50.0	0	50.6
30/30°C						
Flowers/plant	80.0 ± 5.2	21.8 ±6.8	41.5 ± 9.9	26.0 ± 6.2	2.5 ± 3.3	171.8 ± 14.5
Pods/plant	42.8 ± 6.8	14.0 ± 5.2	17.3 ± 3.9	9.5 ± 5.5	0 ± 0	84.3 ± 5.7
% pod set	53.4	64.4	41.6	36.5	0	49.1

DAFFO: Days after first flower opened.

et al. (2001), and in seedlings of soybean and other species by Posmyk et al. (2005). Increased synthesis of proline and other substances serving as osmoprotectants and antioxidants (ROS) under the influence of chilling requires large energy inputs from plants, which causes the inhibition of growth and development under stress conditions (Borowski and Blamowski, 2009).

Superoxide radical (O_2 •–) is formed in many photooxidation reactions (flavoprotein, redox cycling), Mehler reaction in chloroplasts, mitochondrial ETCs reactions, glyoxisomal photorespiration, NADPH oxidase in plasma membranes and xanthine oxidase and membrane polypeptides. Hydroxyl radical (OH•) is formed due to the reaction of H_2O_2 with O_2 •– (Haber-Weiss reaction), reactions of H_2O_2 with Fe^{2+} (Fenton reaction) and decomposition of O_3 in apoplastic space (Halliwell, 2006; Moller et al., 2007). Hydroxyl radicals (OH•) can potentially react with all biomolecules, like, pigments, proteins, lipids and DNA, and almost with all constituent of cells. Hydroxyl radical is not considered to have signaling function, although the products of its reactions can elicit signaling responses, and cells sequester the catalytic metals to metallochaperones efficiently avoiding OH• (Halliwell, 2006; Moller et al., 2007). Singlet oxygen ($^1O_2$) is formed during photoinhibition, and PS II electron transfer reactions in chloroplasts. This radical directly oxidizes protein, polyunsaturated fatty acids, and DNA (Karuppanapandian et al., 2011a, b).

During the early germination process of soybean, low temperatures can significantly reduce the rate of imbibition, the ability of embryo tissue to expand, and mitochondrial respiration (Vertucci and Leopold, 1983; Duke et al., 1977).

Furthermore, susceptibility to chilling injury increases with decreasing initial moisture content in the embryo (Vertucci and Leopold, 1983). The rate of hypocotyl elongation significantly decreases with decreasing temperature below 30°C (Hatfield and Egli, 1974). Interestingly, after effects of low temperatures during the seedling stage can substantially extend the vegetative growth rate, and increase number of axillary branches, the rate of dry weight per plant and pod setting (Skrudlik and Koscielniak, 1996). Whereas, the effects of high temperatures are mostly studied and considered in terms of the reproductive growth and yield potential in soybean, especially under the CSPS system (Hatfield and Prueger, 2015). Many argue that the success of the ESPS system was due to continuously increasing global air temperatures over the years (Thuzar et al., 2010) and they emphasize the importance of determining heat/cold tolerance among the available soybean cultivars during early-growth stages. A study was conducted in a controlled-environment facility to quantify 64 soybean cultivars from Maturity Groups III to V, to low (LT; 20/12°C), optimum (OT; 30/22°C), and high (HT; 40/32°C) temperature treatments during the seedling growth stage. Several shoot, root, and physiological parameters were assessed at 20 days after sowing. The study found a significant decline in the measured root, shoot, and physiological parameters at both low and high temperatures, except for root average diameter (RAD) and lateral root numbers under LT effects. Under HT, shoot growth was significantly increased, however, root growth showed a significant reduction. Maturity group (MG) III had significantly lower values for the measured root, shoot, and physiological traits across temperature treatments when compared with MG IV and V. Cultivar variability existed and reflected considerably through positive or negative responses in growth to LT and HT. Cumulative stress response indices and principal component analysis were used to identify cultivar-specific tolerance to temperatures. Based on the analysis, cultivars CZ 5225 LL and GS47R216 were identified as most sensitive and tolerant to LT, while, cultivars 45A-46 and 5115LL identified as most tolerant and sensitive to HT, respectively. The information on cultivar-specific tolerance to low or high temperatures obtained would help in cultivar selection to minimize stand loss in present production areas (Fig. 12.4).

Water and high air temperature (AT) stresses that occur during soybean (*Glycine max* (L.) Merr.) seed fill greatly reduce seed yield, but their effects on seed germination and vigor are less clear. Thus, this study was to determine the effect of water stress at optimum and high AT during seed fill on soybean seed yield and individual seed weight, and the subsequent germination and vigor of the seed. At daytime ATs of 21°C and 35°C in 1985 and 29 and 33°C in 1986, control, moderate and severe water-stress treatments were imposed by differential irrigation throughout seed fill on greenhouse-grown plants.

Water stress intensity, measured by accumulating stress degree days (SDD) during seed fill, increased linearly as the volume of irrigation water declined. The weight and number of seed produced

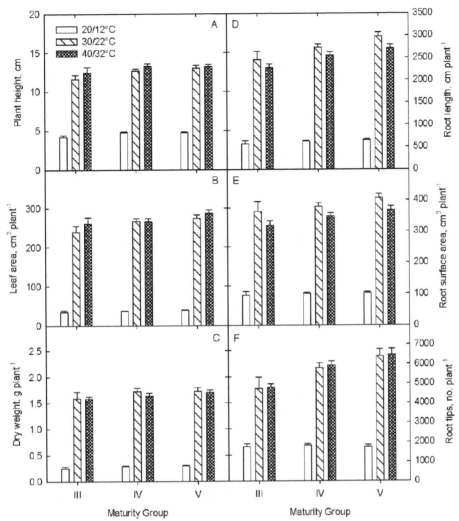

Fig. 12.4 Temperature and maturity group interaction for (A) plant height, (B) leaf area, (C) dry weight, (D) root length, (E) root surface area, and (F) root tips for soybean 64 cultivars harvested 20 days after sowing. Data shows mean + SE.

by each plant, and individual seed weight, declined linearly as SDDs accumulated at each AT. Water stress at optimum ATs reduced seed number more than individual seed weight, but water stress at high ATs reduced individual seed weight more than seed number. Water or high AT stress caused fewer larger seed and more small seed to be produced. The germination percentage and vigor of the harvested seeds was reduced by water stress and high ATs, but by a smaller proportion than yield or seed number.

Individual seed weight, germination, and seedling growth rate were strongly correlated when reduced by water and high AT stress. Severe stress during seed fill caused soybean plants to exceed their capacity to buffer seed number, shifting seed weight distributions towards a larger proportion of small seed, resulting in poor seed lot germination and vigor. Similar numbers of SDDs accumulated at the high ATs in both experiments. Neither seed number nor seed weight were reduced by the high AT in exp. II to the extent of exp. I, suggesting that the plants in exp. II were exposed to less high AT stress. On average, the AI was 2°C lower in exp. II than exp. I. The 33°C average temperature of exp. II may have been moderate enough to avoid the severity of stress effects noted in exp. I (Table 12.5).

The cellular changes induced by either HT or LT include responses those lead to the excess accumulation of toxic compounds, especially reactive oxygen species (ROS). The end result of ROS

Table 12.5 Distribution of soybean seed number and seed mass after exposure to high temperature stress throughout seed fill.

Year	Air temperature (C)	Seed fill					
		Sieve size (diameter, mm)					
		4.8	4.4	4.0	3.6	3.2	< 3.2
1985		*Seed number per plant*					
	29	91	6	2	1	1	2
	35	42	10	6	5	4	9
	LSD$_{0.05}$	28	4	4	1	3	3
		Seed mass (mg seed^{-1})					
	29	195	119	96	77	71	31
	35	172	132	112	94	81	50
	LSD$_{0.05}$	11	19	8	6	13	31
1986		*Seed number per plant*					
	27	83	9	3	1	1	0
	33	53	16	8	6	3	4
	LSD$_{0.05}$	19	7	3	2	1	4
		Seed mass (mg seed^{-1})					
	27	185	124	99	79	61	46
	33	166	127	107	90	73	50
	LSD$_{0.05}$	37	17	21	28	12	13

accumulation is oxidative stress (Yin et al., 2008). In response to HT, the reaction catalyzed by ribulose-1,5-bisphosphate carboxylase oxygenase (RuBisCO) can lead to the production of H_2O_2 as a consequence of increases in its oxygenase reactions (Kim and Portis, 2004). On the other hand, LT conditions can create an imbalance between light absorption and light use by inhibiting the activity of the Calvin–Benson cycle. Enhanced photosynthetic electron flux to O_2 and over-reduction of the respiratory electron transport chain (ETC) can also result in ROS accumulation during chilling, which causes oxidative stress (Hu et al., 2008). Plants have evolved a variety of responses to extreme temperatures that minimize damages and ensure the maintenance of cellular homeostasis (Kotak et al., 2007). A considerable amount of works have explored the idea that there is a direct link between ROS scavenging and plant stress tolerance under temperature extremes (Suzuki and Mittler, 2006). Thus, the improvement of temperature stress tolerance is often related to enhanced activities of enzymes involved in antioxidant systems of plants. Plants exposed to extreme temperatures use several non-enzymatic and enzymatic antioxidants to cope with the harmful effects of oxidative stress; higher activities of antioxidant defense enzymes are correlated with higher stress tolerance. Different plant studies have revealed that enhancing antioxidant defense confers stress tolerance to either HT or LT stress (Babu and Devraj, 2008).

12.3 Salt Stress

During the onset and development of salt stress within a plant, all the major processes, such as photosynthesis, protein synthesis and energy and lipid metabolism, are affected (Parida and Das, 2005). During initial exposure to salinity, plants experience water stress, which, in turn, reduces leaf expansion. The osmotic effects of salinity stress can be observed immediately after salt application and are believed to continue for the duration of exposure, resulting in inhibited cell expansion and cell division, as well as stomatal closure (Flowers, 2004; Munns, 2002). During long-term exposure to salinity, plants experience ionic stress, which can lead to premature senescence of adult leaves, and, thus, a reduction in the photosynthetic area available to support continued growth (Cramer and Nowak, 1992). In fact, excess sodium and, more importantly, chloride has the potential to affect plant enzymes and cause cell swelling, resulting in reduced energy production and other physiological changes (Larcher, 1980). Ionic stress results in premature senescence of older leaves and in toxicity symptoms (chlorosis, necrosis) in mature leaves due to high Na^+ which affects plants by disrupting protein synthesis and interfering with enzyme activity (Munns, 2002). Many plants have evolved several mechanisms either to exclude salt from their cells or to tolerate its presence within the cells. A fundamental biological understanding and knowledge of the effects of salt stress on plants is necessary to provide additional information for the dissection of the plant response to salinity and try to find natural genetic variations and generation of transgenic plants with novel genes or altered expression levels of the existing genes.

In the first, osmotic phase, which starts immediately after the salt concentration around the roots increases to a threshold level, making it harder for the roots to extract water, the rate of shoot growth falls significantly. An immediate response to this effect, which also mitigates ion flux to the shoot, is stomatal closure. However, because of the water potential difference between the atmosphere and leaf cells and the need for carbon fixation, this is an untenable long-term strategy of tolerance (Hasegawa et al., 2000). Shoot growth is more sensitive than root growth to salt-induced osmotic stress, probably because a reduction in the leaf area development relative to root growth would decrease the water use by the plant, thus allowing it to conserve soil moisture and prevent salt concentration in the soil. Reduction in shoot growth due to salinity is commonly expressed by a reduced leaf area and stunted shoots (Läuchli and Epstein, 1990). The growth inhibition of leaves sensitive to salt stress appears to be also a consequence of inhibition by salt of symplastic xylem loading of Ca^{2+} in the root (Läuchli and Grattan, 2007). Final leaf size depends on both cell division and cell elongation. Leaf initiation, which is governed by cell division, was shown to be unaffected by salt stress in sugar beet, but leaf extension was found to be a salt-sensitive process (Papp et al., 1983), depending on Ca^{2+} status.

Moreover the salt-induced inhibition of the uptake of important mineral nutrients, such as K^+ and Ca^{2+}, further reduces root cell growth (Larcher, 1980) and, in particular, compromises root tips expansion. Apical region of roots grown under salinity show extensive vacuolization and lack of typical organization of apical tissue. A slight plasmolysis due to a lack of continuity and adherence between cells is present, with a tendency to the arrest of growth and differentiation. Otherwise, control plants root tips are characterized by densely packed tissues with only small intercellular spaces.

The second phase, ion specific, corresponds to the accumulation of ions, in particular Na^+, in the leaf blade, where Na^+ accumulates after being deposited in the transpiration stream, rather than in the roots (Munns, 2002). Na^+ accumulation turns out to be toxic especially in old leaves, which are no longer expanding and so no longer diluting the salt arriving in them as young growing leaves do. If the rate at which they die is greater than the rate at which new leaves are produced, the photosynthetic capacity of the plant will no longer be able to supply the carbohydrate requirement of the young leaves, which further reduces their growth rate. In photosynthetic tissues, in fact, Na^+ accumulation affects photosynthetic components, such as enzymes, chlorophylls, and carotenoids (Davenport et al., 2005). The derived reduction in photosynthetic rate in the salt sensitive plants can increase also the production of reactive oxygen species (ROS). Normally, ROS are rapidly removed by antioxidative mechanisms, but this removal can be impaired by salt stress (Foyer and Noctor, 2003). ROS signalling has been shown to be an integral part of acclimation response to salinity. ROS play, in fact, a dual role in the response of plants to abiotic stresses, functioning as toxic by-products of stress metabolism, as well as important signal transduction molecules integrated in the networks of stress response pathway mediated by calcium, hormone and protein phosphorylation (Miller et al., 2010).

ABA plays an important role in the response of plants to salinity and ABA-deficient mutants perform poorly under salinity stress (Xiong et al., 2001). Salt stress signalling through Ca^{2+} and ABA mediate the expression of the late embryogenesis–abundant (LEA)-type genes including the dehydration-responsive element (DRE)/C-repeat (CRT) class of stress-responsive genes Cor. The activation of LEA-type genes may actually represent damage repair pathways (Xiong et al., 2002).

Salt and osmotic stress regulation of Lea gene expression is mediated by both ABA dependent and independent signalling pathways. Both the pathways use Ca^{2+} signalling to induce Lea gene expression during salinity. It has been shown that ABA-dependent and -independent transcription factors may also cross talk to each other in a synergistic way in order to amplify the response and improve stress tolerance (Shinozaki and Yamaguchi-Shinozaki, 2000).

Results have indicated that salinity affects growth and development of plants through osmotic and ionic stresses. Because of accumulated salts in soil under salt stress condition plant wilts apparently while soil salts such as Na^+ and Cl^- disrupt normal growth and development of plant (Farhoudi et al., 2007). The difference in a plant response to a given level of salinity is dependent on the concentration and composition of ions in solution as well as the genotype that is exposed to the salinity (Cramer, 1992). Seed germination is usually the most critical stage in seedling establishment, determining successful crop production. Factors adversely affecting seed germination may include sensitivity to drought stress (Wilson et al., 1985) and salt tolerance (Sadeghian and Yavari, 2004).

Numerous studies have been conducted on the management and identification of salinity tolerant crops, such as cotton or cereals (Leidi and Saiz, 1997). Soybean is an important agricultural crop and has, among its genotypes, a relatively wide variation of salt tolerance. As measured by vegetative growth and yield, however, the emergence or failure of a high emergence ratio and seedling establishment on saline soils can have significant economic implications in areas where soil salinity is a potential problem for soybean.

Fifteen soybean genotypes were tested in sand culture experiment. The seeds were irrigated with saline waters of different EC levels (0, 3, 6, 7.2, 10, 12, 14 dSm^{-1}). Length and dry weight of root and shoot as well as PR were evaluated under salinity at 7 DAS. Salinity significantly reduced dry matter accumulation in both roots and shoots in all the cultivars, though declension was more pronounced in PS 1347 and PS 1024. Shoot growth was affected more adversely than root growth. Cultivars showed a wide range of variation in their salinity tolerance as mediated by PR (percent reduction in seedling dry weight over control) and SSI (salinity susceptibility index). PK 1029 and PK 416 exhibited higher levels of tolerance to salinity compared to the other cultivars.

Netondo et al. (2004) reported that photosynthetic activity decreases when plants are grown under saline conditions, leading to reduced growth and productivity. The reduction in photosynthesis under salinity can be attributed to a decrease in chlorophyll content (Jamil et al., 2007) and activity of photo-system II (Ganivea et al., 1998). Salinity can affect chlorophyll content through inhibition of chlorophyll synthesis or an acceleration of its degradation (Reddy and Vora, 1986). Fluorescence of chlorophyll reflected the photochemical activities of photo-system II (Ganivea et al., 1998). Photochemical efficiency of photo-system II (fv/fm) could be reduced by salinity stress (Jamil et al., 2007; Netondo et al., 2004) (Table 12.7).

Plants have evolved complex mechanisms that contribute to the adaptation to osmotic stress caused by high salinity (Meloni et al., 2004). Osmotic adjustment has undoubtedly gained considerable recognition as a significant and effective mechanism of salinity tolerance in crop plants (Pakniyat and Armion, 2007). In salt stressed plants, osmotic potential of vacuole decreased by proline accumulation (Yoshiba et al., 1997). Several possible roles have been attributed to supra-optimal level of proline including osmoregulation under salinity, stabilization of proteins and prevention of heat denaturation of enzymes and conservation of nitrogen and energy for a post-stress period (Aloni and Rosenshtein, 1984) (Table 12.8).

A factorial experiment based on completely randomized design (CRD) with three replications was conducted in 2007, to evaluate grain filling (four harvests), yield and yield components of three soybean cultivars under a non-saline (control) and three saline (3, 6 and 9 dS/m NaCl) conditions in the greenhouse. Six seeds were sown in each pot filled with 900 g perlite, using 144 pots. After emergence, seedlings were thinned and 4 plants were kept in each pot. Grain filling rate and duration and, consequently, maximum grain weight decreased with increasing salinity. Grain filling rate also varied among cultivars and was positively associated with maximum grain weight. However, variation in grain filling duration had little effect on final grain weight of soybean cultivars. Mean number of pods and grains per plant and mean grain weight and grain, oil and protein yields per plant under non-saline conditions were considerably higher than those under saline conditions. Although grain yield of soybean cultivars was statistically similar, grain yield of Zan was about 11% less than that of L17 and Williams. L17 produced more, but

Table 12.6 Effect of salinity on germination and seedling growth in some cultivars of soybean at 7 days after germination.

| Salinity (dSm⁻¹) | % Germ. | Length (cm) | | Dry wt. (mg) | | Seedling | | SSI |
		Shoot	Root	Shoot	Root	Ht. (cm)	Dry wt. (mg)	
PK327								
Control	92	13.60	6.00	18.00	8.00	19.60	23.00	
3.0	80	9.00	5.20	14.00	4.46	14.20	18.46	0.511
6.0	68	3.30	2.20	6.00	1.40	5.50	7.40	1.050
7.2	56	2.50	1.20	4.20	0.80	3.70	5.00	1.030
10.0	32	1.50	0.00	3.20	0.00	1.50	3.20	1.010
12.0	24	0.80	0.00	2.80	0.00	0.80	2.80	0.987
14.0	10	0.60	0.00	2.40	0.00	0.60	2.40	0.965
CD (0.05)	1.98	0.45	0.75	0.77	0.69	0.54	0.74	
PK472								
Control	84	19.80	4.30	26.00	4.60	24.10	30.60	
3.0	72	11.60	3.60	18.00	3.60	15.20	21.60	0.761
6.0	60	5.00	3.00	10.00	3.30	8.00	13.00	0.890
7.2	36	2.60	1.80	5.60	1.70	4.40	7.30	0.997
10.0	28	1.80	0.00	4.20	0.00	1.80	4.20	1.020
12.0	16	0.80	0.00	3.60	0.00	0.80	3.60	0.992
14.0	8	0.60	0.00	1.80	0.00	0.60	1.80	1.010
CD (0.05)	1.15	0.68	0.74	1.53	0.16	0.30	1.15	
PK1029								
Control	100	15.80	3.60	22.0	3.80	19.40	25.80	
3.0	100	12.80	3.40	18.00	3.60	16.20	21.60	0.421
6.0	92	6.50	3.20	10.80	2.80	9.70	13.60	0.732
7.2	84	3.80	1.50	7.60	1.40	5.50	9.40	0.833
10.0	72	2.30	0.00	5.40	0.00	2.30	5.40	0.931
12.0	56	2.00	0.00	4.40	0.00	2.00	4.40	0.933
14.0	48	1.30	0.00	3.60	0.00	1.30	3.60	0.927
CD (0.05)	1.62	0.22	0.15	1.48	1.48	0.35	0.67	

Table 12.7 Comparison of means chlorophyll content index (CCI) and fluorescence of chlorophyll (fv/fm) of three cultivars of soybean under salinity stress.

Treatment		CCI	Fv/fm
Year	1	13.66	0.779
	2	10.50	0.728
	0	14.06	0.792
Salinity	3	12.81	0.768
(dS/m)	6	11.50	0.742
	9	9.97	0.713
	L17	12.63	0.764
Cultivar	Zan	11.19	0.739
	Williams	12.43	0.758

Table 12.8 Comparison of means of proline content and grain yield per plant of three cultivars of soybean under salinity.

Treatment		Proline content (Mm/g)	Grain yield (g/plant)
	0	19.40	1.250
Salinity	3	26.12	0.892
(dS/m)	6	39.28	0.516
	9	45.89	0.274
	L17	31.71	0.782
Cultivar	Zan	35.36	0.651
	Williams	30.96	0.766

smaller grains per plant, compared with other cultivars. However, Williams produced the largest grains, particularly under salinity stress. Decreasing oil and protein yields per plant with increasing salinity were mainly attributed to the large reductions in grain yield per plant under saline conditions. It was concluded that the soybean plant is sensitive to salinity stress and both environment and genotype are responsible for the variation in yield and yield components of this crop (Tables 12.9 and 12.10).

Accumulation of different compatible solutes has been reported during salt stress in different plants. The compatible solutes protect the plants from stress through different courses, including contribution to cellular osmotic adjustment, detoxification of reactive oxygen species, protection of membrane integrity and stabilization of enzymes/proteins. Some solutes perform an extra function of protection of cellular components from dehydration injury and are called "osmoprotectants". These solutes include proline, sucrose, polyols, trehalose and quaternary ammonium compounds (QACs), such as glycine betaine, alanine betaine, and proline.

Apart from osmotic stress, salt is responsible for the generation of ROS in cells that leads to oxidative stress. The generation of these ROSs is due to imbalance between the production and scavenging machinery of ROS. The unquenched ROS react spontaneously with the organic molecules and cause membrane lipid peroxidation, protein oxidation, enzyme inhibition and DNA and RNA damage.

Under severe stress condition, this ROS ultimately leads to cell death. Plants have evolved mechanisms that allow them to adapt and survive under abiotic stress. The production of ROS is, however, kept under tight control by a versatile and cooperative antioxidant system that modulates intracellular ROS concentration and sets the redox status of the cell. Plants overexpressing antioxidant enzymes have been engineered with the aim of increasing stress tolerance by directly modifying the expression of these ROS-scavenging enzymes. The positive effects of SOD, CAT, APX, GR, MDHAR, AsA, Glutathione, etc., in combating oxidative damage to the cell has been reported. There can be no doubt that transgenic plants will be invaluable in assessing the precise role that main antioxidants and ROS play in the functional network that controls stress tolerance.

Table 12.9 Comparison of means of grain filling period, grain filling rate and maximum grain weight of three soybean cultivars under salinity stress.

Treatment		Grain filling rate (mg/day)	Grain filling duration (day)	Maximum grain weight (mg)
Salinity (dS/m)	0	2.087	66.6	135.90
	3	2.197	58.5	121.97
	6	2.119	50.5	105.35
	9	1.846	49.6	89.04
Cultivar	L17	2.099	49.4	102.19
	Zan	1.750	66.0	113.52
	Williams	2.337	53.6	123.47

Table 12.10 Yield and yield components of three soybean cultivars under salinity stress.

Treatment	Pods per plant	Grains per plant	Grain weight (mg)	Grain yield per plant (g)	Oil yield per plant (mg)	Protein yield per plant (mg)
Salinity						
0	4.203	10.041	138	1.388	241	533
3	3.878	7.444	125	0.925	163	352
6	3.359	5.359	109	0.578	104	216
9	2.783	3.751	90	0.330	60	121
Cultivar						
L17	3.920	7.567	105	0.836	150	311
Zan	3.290	6.069	115	0.746	132	282
Williams	3.458	6.309	127	0.834	143	342

The knowledge on how Na[+] is sensed is still very limited in most cellular systems. Theoretically, Na[+] can be sensed either before or after entering the cell, or both. Extracellular Na[+] may be sensed by a membrane receptor, whereas intracellular Na[+] may be sensed either by membrane proteins or by any of the many Na[+]-sensitive enzymes in the cytoplasm. In spite of the molecular identity of Na[+] sensor(s) remaining elusive, the plasma-membrane Na[+]/H[+] antiporter SALT OVERLY SENSITIVE1 (SOS1) is a possible candidate (Silva and Gerós, 2009). In fact, in *Arabidopsis*, ion homeostasis is mediated mainly by the SOS signal pathway (Yang et al., 2009). SOS proteins are sensor for calcium signal that turn on the machinery for Na[+] export and K[+]/Na[+] discrimination (Zhu, 2007).

In particular, SOS1, encoding a plasma membrane Na[+]/H[+] antiporter, plays a critical role in Na[+] extrusion and in controlling long-distance Na[+] transport from the root to shoot (Shi et al., 2000; 2002). This antiporter forms one component in a mechanism based on sensing of the salt stress that involves an increase of cytosolic [Ca^{2+}], protein interactions and reversible phosphorylation with SOS1 acting in concert with other two proteins, known as SOS2 and SOS3 (Oh et al., 2010) (Fig. 12.5).

Excess Na[+] and high osmolarity are separately sensed by unknown sensors at the plasma membrane level, which then induce an increase in cytosolic [Ca^{2+}]. This increase is sensed by SOS3 which activates SOS2. The activated SOS3-SOS2 protein complex phosphorylates SOS1, the plasma membrane Na[+]/H[+] antiporter, resulting in the efflux of Na[+] ions. SOS2 can regulate NHX1 antiport activity and V-H+-ATPase activity independently of SOS3, possibly by SOS3-like Ca^{2+}-binding proteins (SCaBP) that target it to the tonoplast. Salt stress can also induce the accumulation of ABA, which, by means of ABI1 and ABI2, can negatively regulate SOS2 or SOS1 and NHX1. Adapted from Silva and Gerós (2009).

Enhanced salt tolerance in crop plants may be achieved via traditional and molecular breeding and transgenic approaches. However, genetic and physiological complexity of salinity tolerance does not lend itself easily to traditional breeding that may, for example, use a pedigree approach, which consists of screening germplasm for donors of salt tolerance, crossing a donor with an elite line and advancing the F1 hybrid to about the F8 generation while selecting for an elite trait. While the efficiency of this type of breeding is not sufficient (Ashraf and Akram, 2009), the use of wild relatives for breeding gives useful physiological information about the salt tolerance traits (Colmer et al., 2006). In addition, cell and tissue

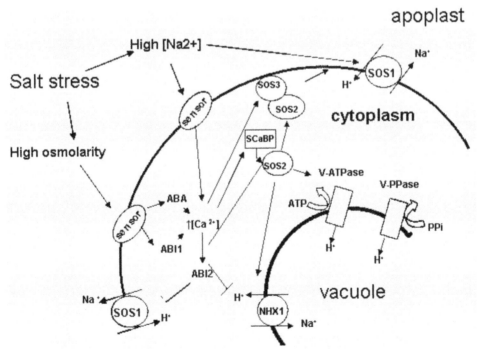

Fig. 12.5 Signalling pathways responsible for Na[+] extrusion in *Arabidopsis* under salt stress.

culture techniques are used to identify somaclonal variants (Zhu et al., 2000) and screen germplasm for salt tolerance *in vitro* (Arzani, 2008).

Molecular and transgenic breeding is more expensive than conventional, but represents an efficient way to produce salt-tolerant lines. Advances in genomics are underpinning an alternative approach in which a pre-breeding phase is used to pyramid several known genes and finely-mapped major quantitative trait loci (QTLs) for complementary aspects of salt tolerance. DNA-based selection protocols that are used to pyramid these genes are again employed during the breeding phase to transfer the entire set of genes for salt tolerance into any elite line by backcrossing (Benneth and Khush, 2003; Flowers, 2004).

Given the complexity of salt tolerance, only a few QTLs are identified within any given genome (Yeo et al., 2000). The fact that a QTL represents many genes complicates the task of finding key loci within the QTL. Quesada et al. (2002) found five QTLs associated with salinity in *Arabidopsis* and identified the location of two of them. The other genetic approach currently used to enhance salt tolerance includes generation of transgenic plants to introduce new genes or to alter expression levels of existing genes to affect the degree of tolerance (avoid or reduce deleterious effects, re-establish the homeostatic conditions and maintain active growth in saline environments; Zhu, 2001).

Important deleterious effects under saline conditions might be due to reactive oxygen species (ROS), which cause oxidative stress as one of the most general stress types (Lee et al., 2001). Plants under salt stress produce stress protein and specific osmolytes for scavenging ROS (Xiong and Zhu, 2002; Zhu et al., 1997). Oxidative stress tolerance is genetically controlled; improvements can be provided by conventional breeding and transgenic techniques (Asraf and Akram, 2009) or by adopting physiological approaches (Afzal et al., 2006). Most transgenic improvements in plant salt tolerance are focused on overexpressing enzymes involved in oxidative protection (Allen et al., 1997). In addition, engineering with the regulatory protein NPK1, a nitrogen-activated protein kinase that mediates oxidative stress responses, is also an efficient way to support antioxidant defence (Kovtun et al., 2000). Improvements provided via other proteins, like barley late embriogenesis abundant proteins and C-repeat-binding/dehydration-responsive element-binding proteins (CBF/DREBs) in transgenic plants may have ROS detoxifying effects, but are not specific for salt tolerance (Liu et al., 1998; Stockinger et al., 1997).

There are many proteins that appear to be related to salt stress response in plants. These proteins can directly regulate the levels of osmolytes and control ion homeostasis. Osmolytes, such as mannitol, fructans, proline and glycinebetaine, are also active in scavenging ROS. Genetic engineering of these osmolytes resulted in increasing salt tolerance (Zhu, 2001).

For improved tolerance to salinity, it is important to re-establish homeostasis by controlling Na^+, K^+, Cl-transporters mediating influx and efflux to fine-tune ion concentrations in the cytoplasm (Zhu, 2001). Nonselective cation channels mediate Na^+ entry into the cell, and their molecular identity is becoming clear. Anion and cation transporters are a frequent target of genetic engineering to improve crop salt tolerance (Yamaguchi and Blumwald, 2005). Transcriptome analysis can pinpoint genes associated with regulation of RNA and protein metabolism that have a significant role in regulating salt stress tolerance (Sahi et al., 2006). Mian et al. (2011) identified a large number of important genes using forward and reverse genetics, yeast complementation and transcriptomics.

Microarray analysis has clearly shown that transcripts encoding RNA-binding proteins, helicases, cyclophilins, F-box proteins, dynamin-like proteins, and ribosomal proteins are linked to the salt-stress response in *Arabidopsis* (Sottosanto et al., 2004).

Benefitting from the advances of next-generation sequencing (NGS), genome-wide association studies (GWAS) were broadly used to detect the associated single nucleotide polymorphism (SNP) markers for many agronomy and resistance traits in soybean (Zhang et al., 2015). Recently, GWAS was successfully performed to identify the SNP markers associated with salt tolerance in soybean. Kan et al. (2016) detected eight SNP-trait associations and 13 potential SNP-trait associations with salt tolerance during the seed germination stage by GWAS using a mixed linear model and TASSEL 4.0 software. Eight SNPs or potential SNPs were co-associated with salt tolerance in soybean. Based on the soybean genome database, nine candidate genes were located in or near the salt tolerance associated SNP marker region. Subsequently, five candidate genes Glyma08g12400.1, Glyma08g09730.1, Glyma18g47140.1, Glyma09g00460.1, and Glyma09g00490.3 were validated in response to salt stress during the soybean germination stage. Based on phenotype and the SoySNP50K BeadChip database of 106 soybean lines, Patil et al. (2016) performed GWAS for salt tolerance using an expedited single-locus mixed model. The SNPs identified by GWAS pin-pointed a single and highly significant association for salt tolerance on chr. 03. In this associated region, GWAS found that the soybean salt tolerance gene Glyma03g32900 overlapped the associated locus (Patil et al., 2016). The whole-genome resequencing (WGRS) data of the 106 soybean lines were also used to detect the SNPs associated with salt tolerance during the seedling stage. Finally, a significant SNP within the salt tolerance gene Glyma03g32900 was identified and determined to explain up to 63% of the phenotypic variation. Genome-wide analysis showed that natural variation associated with the Glyma03g32900 gene has a major impact on salt tolerance in soybean. A total of 283 worldwide soybean germplasm lines and the SoySNP50K BeadChip database were used to perform the GWAS for salt tolerance (Zeng et al., 2017). A total of 45 SNPs representing nine genomic regions on nine chromosomes were significantly associated with salt tolerance based on GWAS analysis. Most of the SNPs significantly associated with salt tolerance were located within or near the major salt tolerance gene Glyma03g32900 on chr. 03. Moreover, seven putative novel QTLs represented by significant SNPs were identified for salt tolerance in soybean. GWAS has been a powerful tool for identifying all candidate loci conferring target traits at the whole genome level. For soybean salinity traits, more related QTLs/QTNs (quantitative trait nucleotide) contributing to salt tolerance will be identified and can be used to improve salinity through genomic selection strategies.

References

Afzal, I., S.M.A. Basra, A. Hameed and M. Farooq. 2006. Physiological enhancement for alleviation of salt stress in wheat. Pak. J. Bot. 38: 1649–1659.

Ahuja, I., R.C.H. de Vos, A.M. Bones and R.D. Hall. 2010. Plant molecular stress responses face climate change. Trends in Plant Science 15(12): 664–674.

Aloni, B. and G. Rosenshtein. 1984. Proline accumulation: A parameter for evaluation of sensitivity of tomato varieties to drought stress. Physiol. Plant. 61: 231–235.

Anon, S., J.A. Fernandez, J.A. Franco, A. Torrecillas, J.J. Alarcon and M.J. Sanchez-Blanco. 2004. Effects of water stress and night temperature preconditioning on water relations and morphological and anatomical changes of *Lotus creticus* plants. Sci. Hortic. (Amsterdam, Neth.) 101: 333–342.

Arzani, A. 2008. Improving salinity tolerance in crop plants: A biotechnological view. *In Vitro* Cellular and Developmental Biology. Plant. 44: 373–383.

Asraf, M. and N.A. Akram. 2009. Improving salinity tolerance of plants through conventional breeding and genetic engineering: An analytical comparison. Advance in Biotechnology 27: 744–752.

Babu, N.R. and V.R. Devraj. 2008. High temperature and salt stress response in French bean (*Phaseolus vulgaris*). Australian Journal of Crop Science 2: 40–48.

Bailey-Serres, J. and L. Voesenek. 2008. Flooding stress: Acclimations and genetic diversity. Annual Review of Plant Biology 59: 313–339.

Bartels, S., M.A.G. Besteiro, D. Lang and R. Ulm. 2010. Emerging functions for plant MAP kinase phosphatases. Trends in Plant Science 15(6): 322–329.

Bell, R.W., D.G. Edwards and C.J. Asher. 1980. External calcium requirements for growth and nodulation of six tropical food legumes grown in flowing solution culture [peanut; pigeon pea; guar; soybean; cowpea cv Vita 4 and CPI 28215]. Australian Journal of Agricultural Research, 40.

Bengough, A.G., B. McKenzie, P. Hallett and T. Valentine. 2011. Root elongation, water stress, and mechanical impedance: A review of limiting stresses and beneficial root tip traits. Journal of Experimental Botany 62: 59–68.

Benneth, J. and G.S. Khush. 2003. Enhancing salt tolerance in crops through molecular breeding: A new strategy. Journal of Crop Production 7: 11–65.

Berry, J.A. and O. Bjorkman. 1980. Photosynthetic response and adaptation to temperature in higher plants. Annu. Rev. Plant Physiol. 31: 491–543.

Bertolli, S.C., G.L. Rapchan and G.M. Souza. 2012. Photosynthetic limitations caused by different rates of water-deficit induction in *Glycine max* and *Vigna unguiculata*. Photosynthetica 50: 329–336.

Bharati, M.P., R.J. Lawn and D.E. Byth. 1983. Effects of seed hydration–dehydration pre-treatment on germination of soybean lines at sub-optimal temperatures. Aust. J. Exp. Agric. Anim. Husb. 23: 309–317.

Borowski, E. and Z.K. Blamowski. 2009. The effects of triacontanol 'TRIA' and Asahi SL on te development and metabolic activity of sweet basil (*Ocimum basilicum* L.) plants treated with chilling. Folia Hort. 21/1: 39–48.

Board, J.E. and J.R. Settini. 1996. Photoperiod effect before and after flowering on branch development in determinate soybean. Agron. J. 78: 995–1002.

Boru, G., T. Vantoai, J. Alves, D. Hua and M. Knee. 2003. Responses of soybean to oxygen deficiency and elevated root-zone carbon dioxide concentration. Annals of Botany 91: 447–453.

Boudsocq, M. and C. Laurière. 2005. Osmotic signaling in plants. Multiple pathways mediated by emerging kinase families. Plant Physiology 138(3): 1185–1194.

Brown, E.A., C.E. Cavines and D.A. Brown. 1985. Response of selected soybean cultivars to soil moisture deficit. Agron J. 77: 274–278.

Cheung, A.Y. and H.M. Wu. 2011. THESEUS 1, FERONIA and relatives: A family of cell wall sensing receptor kinases? Current Opinion in Plant Biology 14(6): 632–641.

Clement, M., E. Boncompagni, J. de Almeida-Engler and D. Herouart. 2006. Isolation of a novel nodulin: A molecular marker of osmotic stress in *Glycine max/Bradyrhizobium japonicum* nodule. Plant, Cell and Environment 29(9): 1841–1852.

Clement, M., A. Lambert, D. Heroulart and E. Boncompagni. 2008. Identification of new upregulated genes under drought stress in soybean nodules. Gene 426: 15–22.

Colmer, T.D., T.J. Flowers and R. Munns. 2006. Use of wild relatives to improve salt tolerance in wheat. Journal of Experimental Botany 57: 1059–1078.

Cornelious, B., P. Chen, Y. Chen, N. De Leon, J. Shannon and D. Wang. 2005. Identification of QTLs underlying water-logging tolerance in soybean. Molecular Breeding 16: 103–112.

Cosgrove, W. 2009. Water in a changing world, in The United Nations World Water Development Report, Programme W.W.A., Editor 2009.

Cramer, G.R. 1992. Response of a Na-excluding cultivar and a Na-including cultivar to varying Na/Ca. J. Exp. Bot. 43: 857–864.

Cramer, G.R. and R.S. Nowak. 1992. Supplemental manganese improves the relative growth, net assimilation and photosynthetic rates of salt-stressed barley. Physiol. Plant. 84: 600–605.

Cruz, C.M.H. 2008. Drought stress and reactive oxygen species: Production, scavenging and signaling. Plant Signaling and Behavior 3(3): 156.

Davenport, R., R. James, A. Zakrisson-Plogander, M. Tester and R. Munns. 2005. Control of sodium transport in durum wheat. Plant Physiology 137: 807–818.

DeFalco, T., K. Bender and W. Snedden. 2010. Breaking the code: Ca^{2+} sensors in plant signalling. Biochemical Journal 425: 27–40.

Desclaux, D., T.T. Huynh and P. Roumet. 2000. Identification of soybean plant characteristics that indicate the timing of drought stress. Crop Science 40(3): 716–722.

Djanaguiraman, M. and P.V.V. Prasad. 2010. Ethylene production under high temperature stress causes premature leaf senescence in soybean. Funct. Plant Biol. 37: 1071–1084.

Dornbos, D.L. and R.E. Mullen. 1991. Influence of stress during soybean seed fill on seed weight, germination, and seedling growth rate. Canadian Journal of Plant Science 71(2): 373–383.

Drøbak, B.K. 2002. PARF1: An *Arabidopsis thaliana* FYVE-domain protein displaying a novel eukaryotic domain structure and phosphoinositide affinity. Journal of Experimental Botany 53(368): 565–567.

Duke, S.H., L.E. Schrader and M.G. Miller. 1977. Low temperature effects on soybean (*Glycine max* L. Merr. cv. Wells) mitochondrial respiration and several dehydrogenases during imbibition and germination. Plant Physiol. 60: 716–722.

Dybing, C.D., H. Ghiasi and C. Paech. 1986. Biochemical characterization of soybean ovary growth from anthesis to abscission of aborting ovaries. Plant Physiol. 81: 1069–1074.

Eck, H.V., A.C. Mathers and J.T. Musick. 1987. Plant water stress at various growth stages and growth and yield of soybeans. Field Crops Research 17(1): 1–16.

Egli, D.B. and I.F. Wardlaw. 1980. Temperature response of seed growth characteristics of soybean. Agron. J. 72: 560–564.

Else, M.A., D. Coupland, L. Dutton and M.B. Jackson. 2001. Decreased root hydraulic conductivity reduces leaf water potential, initiates stomatal closure and slows leaf expansion in flooded plants of castor oil (*Ricinus communis*) despite diminished delivery of ABA from the roots to shoots in xylem sap. Physiologia Plantarum 111: 46–54.

Evans, D.E. 2004. Aerenchyma formation. New Phytologist 161: 35–49.

Farhoudi, R., F. Sharifzadeh, M. Makkizadehand and M. Kochakpour. 2007. The effects of NaCl priming on salt tolerance in canola (*Brassica napus*) seedlings grown under saline conditions. Seed Science and Technology 35: 754–759.

Farquhar, G.D., J.R. Elheringer and K.T. Hubick. 1989. Carbon isotope discrimination and photosynthesis. Ann. Rev. Plant Physiol. Plant Mol. Biol. 40: 503–537.

Fenta, B.A., S.E. Beebe, K.J. Kunert, J.D. Burridge, K.M. Barlow, J.P. Lynch et al. 2014. Field phenotyping of soybean roots for drought stress tolerance. Agronomy 4: 418–435.

Flowers, T.J. 2004. Improving crop salt tolerance. Journal of Experimental Botany 55(396): 307–319.

Food and Agriculture Organization of the United Nations. 2017. The Future of Food and Agriculture—Trends and Challenges. Rome: Food and Agriculture Organization of the United Nations.

Foyer, C. and G. Noctor. 2003. Redox sensing and signalling associated with reactive oxygen in chloroplasts, peroxisomes and mitochondria. Physiologia Plantarum 119: 355–364.

Fukao, T., K. Xu, P.C. Ronald and J. Bailey-Serres. 2006. A variable cluster of ethylene response factor–like genes regulates metabolic and developmental acclimation responses to submergence in rice. The Plant Cell 18: 2021–2034.

Ganivea, R.A., S.R. Allahverdiyev, N.B. Guseinova, H.I. Kavakli and S. Nafisi. 1998. Effect of salt stress and synthetic hormone polystimuline K on the photosynthetic activity of cotton (*Gossypium hirsutum*). Tr. J. Botany 22: 217–221.

Garay, A. and W. Wilhelm. 1983. Root system characteristics of two soybean isolines undergoing water stress conditions. Agronomy Journal 75: 973–977.

Gibson, L.R. and R.E. Mullen. 1996. Influence of day and night temperature on soybean seed yield. Crop Sci. 36: 98–104.

Govindjee, G. 1995. Sixty-three years since Kautsky: Chlorophyll a fluorescence. Australian J. Plant Physiol. 22: 131–160.

Grene, R., C. Vasquez-Robinet and H.J. Bohnert. 2011. Molecular biology and physiological genomics of dehydration stress. Plant Desiccation Tolerance 255–287.

Griffin, J.L. and A.M. Saxton. 1988. Response of solid-seeded soybean to flood irrigation. II. Flood duration. Agronomy Journal 80: 885–888.

Guenther, J.F., N. Chanmanivone, M.P. Galetovic, I.S. Wallace, J.A. Cobb and D.M. Roberts. 2003. Phosphorylation of soybean nodulin 26 on serine 262 enhances water permeability and is regulated developmentally and by osmotic signals. The Plant Cell 15(4): 981–991.

Halliwell, B. 2006. Reactive species and antioxidants: Redox biology is a fundamental theme of aerobic life. Plant Physiology 141: 312–322.

Hasegawa, P.M., R.A. Bressan, J.K. Zhu and H.J. Bohnert. 2000. Plant cellular and molecular responses to high salinity. Annual Review of Plant Physiology and Plant Molecular Biology 51: 463–499.

Hatfield, J.L. and D.B. Egli. 1974. Effect of temperature on the rate of soybean hypocotyl elongation and field emergence. Crop Sci. 14: 423–426.

Hatfield, J.L. and J.H. Prueger. 2015. Temperature extremes: Effect on plant growth and development. Weather Clim. Extrem. 10: 4–10.

He, J., Y. Jin, Y.-L. Du, T. Wang, N.C. Turner, R.-P. Yang et al. 2017 Genotypic variation in yield, yield components, root morphology and architecture, in soybean in relation to water and phosphorus supply. Frontiers in Plant Science 8: 1499.

Heerden, P.D.R., M.M. Tsimilli, G.H.J. Krüger and R.J. Strasser. 2003a. Dark chilling effects on soybean genotypes during vegetative development; parallel studies of CO_2 assimilation, chlorophyll a kinetics O-J-I-P and nitrogen fixation. Physiol. Plant. 117(4): 476–491.

Heerden, P.D.R., G.H.J. Krüger, J.E. Loveland, M.A.J. Parry and C.H. Foyer. 2003b. Dark chilling imposes metabolic restrictions on photosynthesis in soybean. Plant Cell Environ. 26: 323–337.

Hu, W.H., X.S. Song, K. Shi, X.J. Xia, Y.H. Zhou and J.Q. Yu. 2008. Changes in electron transport, superoxide ismutase and ascorbate peroxidase isoenzymes in chloroplasts and mitochondria of cucumber leaves as influenced by chilling. Photosynthetica 46: 581–588.

Huang, C.-H. and C.-M. Yang. 1995. Use of Weibull function to quantify temperature effect on soybean germination. Chinese Agron. J. 5: 25–34.

Huang, C., S. Ding, H. Zhang, H. Du and L. An. 2011. CIPK7 is involved in cold response by interacting with CBL1 in *Arabidopsis thaliana*. Plant Sci. 181: 57–64.

IPCC. 2007. Intergovernmental Panel on Climate Change Fourth Assessment Report. Climate Change 2007. Cambridge University Press, Cambridge, UK.

Isoda, A., M. Mori, S. Matsumoto, Z. Li and P. Wang. 2006. High yielding performance of soybean in northern Xinjiang, China. Plant Prod. Sci. 9: 401–407.

Jackson, M. and B. Ricard. 2003. Physiology, biochemistry and molecular biology of plant root systems subjected to flooding of the soil. pp. 193–213. *In*: Visser, E.J.W. and H. de Kroon (eds.). Root Ecology. Heidelberg: Springer-Verlag.

Jackson, M.B. and P.A. Attwood. 1996. Roots of willow [*Salix viminalis* L.] show marked tolerance to oxygen shortage in flooded soils and in solution culture. Plant and Soil 187: 37–45.

Jamil, M., S. Rehman, K.J. Lee, J.M. Kim, H.S. Kim and E.S. Rha. 2007. Salinity reduced growth PSII photochemistry and chlorophyll content in radish. Sci. Agric. 64: 1–10.

Kader, M.A. and S. Lindberg. 2010. Cytosolic calcium and pH signaling in plants under salinity stress. Plant Signaling and Behavior 5(3): 233.

Karuppanapandian, T., H.W. Wang, N. Prabakaran, K. Jeyalakshmi, M. Kwon, K. Manoharan and W. Kim. 2011. Dichlorophenoxyacetic acid-induced leaf senescence in mung bean (*Vigna radiata* L. Wilczek) and senescence inhibition by co-treatment with silver nanoparticles. Plant Physiology and Biochemistry 49: 168–177.

Karuppanapandian, T., J.-C. Moon, C. Kim, K. Manoharan and W. Kim. 2011. Reactive oxygen species in plants: Their generation, signal transduction, and scavenging mechanisms. Australian Journal of Crop Science 5: 709–725.

Kathuria, H., J. Giri, H. Tyagi and A.K. Tyagi. 2007. Advances in transgenic rice biotechnology. Critical Reviews in Plant Sciences 26: 65–103.

Kawai, M., P. Samarajeewa, R. Barrero, M. Nishiguchi and H. Uchimiya. 1998. Cellular dissection of the degradation pattern of cortical cell death during aerenchyma formation of rice roots. Planta 204: 277–287.

Kim, K. and J. Portis. 2004. Oxygen-dependent, H_2O_2 production by Rubisco. FEBS Letters 571: 124–128.

Kitano, M., K. Saitoh and T. Kuroda. 2006. Effects of high temperature on flowering and pod set in soybean. Sci. Rep. Fac. Agric. Okayama Univ. 95: 49–55.

Koc, E., C. Islek and A.S. Üstün. 2010. Effect of cold on protein, proline, phenolic compounds and chlorophyll content of two pepper (*Capsicum annum* L.) G. U. J. Sci. 23(1): 1–6.

Korte, L.L., J.H. Williams, J.E. Specht and R.C. Sorensen. 1983. Irrigation of soybean genotypes during reproductive ontogeny. I. Agronomic responses. Crop Science 23(3): 521–527.

Kotak, S., J. Larkindale, U. Lee, P. Von Koskull-döring, E. Vierling and K.D. Scharf. 2007. Complexity of the heat stress response in plants. Current Opinion in Plant Biology 10: 310–316.

Kovtun, Y., W.L. Chiu, G. Tena and J. Sheen. 2000. Functional analysis of oxidative stress activated mitogen-activated protein kinase cascade in plants. Proc. Natl. Acad. Sci. U.S.A. 97: 2940–2945.

Kokubun, M. 2013. Genetic and cultural improvement of soybean for waterlogged conditions in Asia. Field Crops Research 152: 3–7.

Laan, P., M. Berrevoets, S. Lythe, W. Armstrong and C. Blom. 1989. Root morphology and aerenchyma formation as indicators of the flood-tolerance of Rumex species. The Journal of Ecology 77: 693–703.

Larcher, W. 1980. Physiological plant ecology. In 2nd Totally Rev. Edition ed. (pp. 303). Berlin and New York: Springer-Verlag.

Läuchli, A. and E. Epstein. 1990. Plant responses to saline and sodic conditions. pp. 113–137. *In*: Tanji, K.K. (ed.). Agricultural Salinity Assessment and Management. New York: American Society of Civil Engineers.

Läuchli, A. and S.R. Grattan. 2007. Plant growth and development under salinity stress. pp. 1–32. *In*: Jenks, M.A., P.M. Hasegawa and S.M. Jain (eds.). Advances in Molecular Breeding Toward Drought and Salt Tolerant Crops. Springer Netherlands.

Le, D.T., R. Nishiyama, Y. Watanabe, K. Mochida, K. Yamaguchi-Shinozaki, K. Shinozaki and L.S.P. Tran. 2011. Genome-wide survey and expression analysis of the plant-specific NAC transcription factor family in soybean during development and dehydration stress. DNA Research 18(4): 263–276.

Le, D.T., R. Nishiyama, Y. Watanabe, K. Mochida, K. Yamaguchi-Shinozaki, K. Shinozaki and L.S.P. Tran. 2011. Genome-wide expression profiling of soybean two-component system genes in soybean root and shoot tissues under dehydration stress. DNA Research 18(1): 17–29.

Lee, D.H., Y.S. Kim and C.B. Lee. 2001. The inductive responses of the antioxidant enzymes by salt stress in the rice (*Oryza sativa* L.). Journal of Plant Physiology 158: 737–745.

Leidi, E.O. and J.F. Saiz. 1997. Is salinity tolerance related to Na accumulation in upland cotton (*Gossypium hirsutum*) seedlings? Plant Soil 190: 67–75.

Liao Fang Lei, Jiang Wu, Zheng Yue Ping, Xu Hang Lin, Li Li Qing and Lu Hong Fei. 2011. Influences of temperature regime on germination of seed of wild soybean (*Glycine soja*). Agric. Sci. Technol. 12: 480–483.

Linkemer, G., J.E. Board and M.E. Musgrave. 1998. Waterlogging effects on growth and yield components in late-planted soybean. Crop Science 38: 1576–1584.

Liu, F., M.N. Andersen and C.R. Jensen. 2003. Loss of pod set caused by drought stress is associated with water status and ABA content of reproductive structures in soybean. Functional Plant Biology 30: 271–280.

Liu, Y., J.-H. Gai, H. Lu, Y.-J. Wang and S.-Y. Chen. 2005. Identification of drought tolerant germplasm and inheritance and QTL mapping of related root traits in soybean (*Glycine max* [L.] Merr). Acta Genetica Sinica 32: 855–863.

Liu, Q., M. Kasuga, Y. Sakuma, H. Abe, S. Miura, K. Yamaguchi-Shinozaki and K. Shinozaki. 1998. Two transcription factors, DREB1 and DREB2, with an REBP/AP2DNA binding domain separate two cellular signal transduction pathways in drought and low temperature-responsive gene expression, respectively, in *Arabidopsis*. Plant Cell 10: 1391–1406.

Lobell, D.B. and G.P. Asner. 2003. Climate and management contributions to recent trends in U.S. agricultural yields. Science 299: 1032.

Long, S.P., A. Marshall-Colon and X.-G. Zhu. 2015. Meeting the global food demand of the future by engineering crop photosynthesis and yield potential. Cell 161: 56–66.

Lee, S., H. Hirt and Y. Lee. 2001. Phosphatidic acid activates a wound-activated MAPK in *Glycine max*. The Plant Journal 26(5): 479–486.

Luo, G.Z., Y.J. Wang, Z.M. Xie, J.Y. Gai, J.S. Zhang and S.Y. Chen. 2006. The putative Ser/Thr protein kinase gene GmAAPK from soybean is regulated by abiotic stress. Journal of Integrative Plant Biology 48(3): 327–333.

Messina, M. 1997. Soyfoods: Their role in disease prevention and treatment. pp. 442–466. *In*: Liu, K. (ed.). Soybeans: Chemistry, Technology and Utilization, Chapman and Hall, New York, USA.

Mahajan, S. and N. Tuteja. 2005. Cold, salinity and drought stresses: An overview. Archives of Biochemistry and Biophysics 444(2): 139–158.

Malik, A.I., T.D. Colmer, H. Lambers and M. Schortemeyer. 2001. Changes in physiological and morphological traits of roots and shoots of wheat in response to different depths of waterlogging. Functional Plant Biology 28: 1121–1131.

Maryam, A. and S.A. Nasreen. 2012. Review: Water logging effects on morphological, anatomical, physiological and biochemical attributes of food and cash crops. International Journal of Water Resources and Environmental Sciences 1: 113–120.

Maruyama, K., D. Todaka, J. Mizoi, T. Yoshida, S. Kidokoro, S. Matsukura, H. Takasaki, T. Sakurai, Y.Y. Yamamoto and K. Yoshiwara. 2012. Identification of cis-acting promoter elements in cold-and dehydration-induced transcriptional pathways in Arabidopsis, rice, and soybean. DNA Research 19(1): 37–49.

McAinsh, M.R. and J.K. Pittman. 2009. Shaping the calcium signature. New Phytologist 181(2): 275–294.

Meloni, D.A., M.R. Gulotta, C.A. Martinez and M.A. Oliva. 2004. The effects of salt stress on growth, nitrate reduction and proline and glycine-betaine accumulation in Prosopis alba. Braz. J. Plant Physiol. 16: 39–46.

Mian, A.A., P. Senadheera and F.J.M. Maathuis. 2011. Improving crop salt tolerance: anion and cation transporters as genetic engineering targets. Global Science Books, Plant Stress 5(Special Issue 1): 64–72.

Miller, G.A.D., N. Suzuki, S. Ciftci-Yilmaz and R.O.N. Mittler. 2010. Reactive oxygen species homeostasis and signalling during drought and salinity stresses. Plant, Cell and Environment 33(4): 453–467.

Mittler, R. 2006. Abiotic stress, the field environment and stress combination. Trends in Plant Science 11: 15–19.

Mizoi, J., K. Shinozaki and K. Yamaguchi-Shinozaki. 2011. AP2/ERF family transcription factors in plant abiotic stress responses. Biochimica et Biophysica Acta (BBA)-Gene Regulatory Mechanisms 2011.

Mochida, K., T. Yoshida, T. Sakurai, K. Yamaguchi-Shinozaki, K. Shinozaki and L.S.P. Tran. 2009. *In silico* analysis of transcription factor repertoire and prediction of stress responsive transcription factors in soybean. DNA Research 16(6): 353–369.

Motha, R. 1992. Monitoring, assessment and combat of drought and desertification. *In*: Commission for Agricultural Meteorology Reports 1992, World Meteorological Organization: Geneva.

Munns, R. 2002. Comparative physiology of salt and water stress. Plant, Cell and Environment 25(2): 239–250.

Nakashima, K., H. Takasaki, J. Mizoi, K. Shinozaki and K. Yamaguchi-Shinozaki. 2011. NAC transcription factors in plant abiotic stress responses. Biochimica et Biophysica Acta (BBA)-Gene Regulatory Mechanisms 2011.

Nakaseko, K., N. Nomura, K. Gotoh, T. Ohnuma, Y. Abe and S. Konno. 1984. Dry matter accumulation and plant type of the high yielding soybean grown under converted rice paddy fields. Jpn. J. Crop Sci. 53: 510–518.

Nanjo, Y., T. Nakamura and S. Komatsu. 2013. Identification of indicator proteins associated with flooding injury in soybean seedlings using label-free quantitative proteomics. Journal of Proteome Research 12: 4785–4798.

Netondo, G.W., J.C. Onyango and E. Beck. 2004. Sorghum and salinity: II. Gas exchange and chlorophyll fluorescence of sorghum under salt stress. Crop Sci. 44: 806.

Oh, D.-H., S.Y. Lee, R.A. Bressan, D.-J. Yun and H.J. Bohnert. 2010. Intracellular consequences of SOS1 deficiency during salt stress. Journal of Experimental Botany 61: 1205–1213.

Oosterhuis, D., H. Scott, R. Hampton and S. Wullschleger. 1990. Physiological responses of two soybean (*Glycine max* [L.] Merr) cultivars to short-term flooding. Environmental and Experimental Botany 30: 85–92.

Pakniyat, H. and M. Armion. 2007. Sodium and proline accumulation as osmoregulators in tolerance of sugar beet genotypes to salinity. Pakistan J. Biological Sci. 10: 4081–4086.

Papp, J.C., M.C. Ball and N. Terry. 1983. A comparative study of the effects of NaCl salinity on respiration, photosynthesis, and leaf extension growth in *Beta vulgaris* L. (sugar beet). Plant, Cell and Environment 6(8): 675–677.

Parida, A.K. and A.B. Das. 2005. Salt tolerance and salinity effects on plants: A review. Ecotoxicology and Environmental Safety 60(3): 324–349.

Patil, G., T. Do, T.D. Vuong, B. Valliyodan, J.D. Lee, J. Chaudhary, J.G. Shannon and H.T. Nguyen. 2016. Genomic-assisted haplotype analysis and the development of high-throughput SNP markers for salinity tolerance in soybean. Scientific Reports 6: 19199.

Pereira, S.S., F.C.M. Guimarães, J.F.C. Carvalho, R. Stolf-Moreira, M.C.N. Oliveira, A.A.P. Rolla, J.R.B. Farias, N. Neumaier and A.L. Nepomuceno. 2011. Transcription factors expressed in soybean roots under drought stress. Genetics and Molecular Research 10(4): 3689–3701.

Peterson, C.M., C.O.H. Mosjidis, R.R. Dute and M.E. Westgate. 1992. A flower and pod staging system for soybean. Ann. Bot. 69: 59–67.

Phang, T.H., M.W. Li, C.C. Cheng, F.L. Wong, C. Chan and H.M. Lam. 2011. Molecular responses to osmotic stresses in soybean. pp. 215–240. *In*: Sudaric, A. (ed.). Soybean—Molecular Aspects of Breeding. InTech: Rijeka, Croatia.

Prince, S.J., M. Murphy, R.N. Mutava, Z. Zhang, N. Nguyen, Y.H. Kim et al. 2016. Evaluation of high yielding soybean germplasm under water limitation. Journal of Integrative Plant Biology 58: 475–491.

Prince, S.J., M. Murphy, R.N. Mutava, L.A. Durnell, B. Valliyodan, J.G. Shannon et al. 2017. Root xylem plasticity to improve water use and yield in water-stressed soybean. Journal of Experimental Botany 68: 2027–2036.

Phukan, U.J., S. Mishra, K. Timbre, S. Luqman and R.K. Shukla. 2014. *Mentha arvensis* exhibit better adaptive characters in contrast to *Mentha piperita* when subjugated to sustained waterlogging stress. Protoplasma 251: 603–614.

Posmyk, M.M., F. Corbineau, D. Vinel, C. Bailly and D. Come. 2001. Osmoconditioning reduces physiological and biochemical damage induced by chilling in soybean seeds. Physiol. Plant. 111(4): 473–482.

Posmyk, M.M., C. Bailly, K. SzafraËska, K.M. Janas and F. Corbineau. 2005. Antioxidant enzymes and isoflavonoids in chilled soybean (*Glycine max* (L.) Merr) seedlings. J. Plant Physiol. 162: 403–412.

Quesada, V., S. Garcia-Martinez, P. Piqueras, M.R. Ponce and J.L. Micol. 2002. Genetic architecture of NaCl tolerance in *Arabidopsis*. Plant Physiology 130: 951–957.

Reckling, M., T.F. Doring, G. Berkvist, F.-M. Chmielewski, F.L. Stoddard, C.A. Watson and J. Bachinger. 2018. Grain legume yield instability has increased over 60 years in long-term field experiments as measured by a scale-adjusted coefficient of variation. Aspects of Applied Biology 138: 15–20.

Reddy, M.P. and A.B. Vora. 1986. Changes in pigment composition, hill reaction activity and saccharides metabolism in bajra (*Pennisetum typhoides* S & H) leaves under NaCl salinity. Photosynthetica 20: 50–55.

Rhine, M.D., G. Stevens, G. Shannon, A. Wrather and D. Sleper. 2010. Yield and nutritional responses to waterlogging of soybean cultivars. Irrigation Science 28: 135–142.

Ristic, Z., U. Bukovnik, I. Momcilovic, J. Fu and P.V.V. Prasad. 2008. Heat-induced accumulation of chloroplast protein synthesis elongation factor, EF-Tu, in winter wheat. J. Plant Physiol. 165: 192–202.

Sadeghian, S.Y. and N. Yavari. 2004. Effects of water deficit stress on germination and seedling growth in sugar beet. J. Agron. Crop Sci. 190(2): 138–144.

Sahi, C., A. Singh, E. Blumwald and A. Grover. 2006. Beyond osmolytes and transporters:novel plant salt-stress tolerance-related genes from transcriptional profiling data. Physiologia Plantarum 127: 1–9.

Sallam, A. and H. Scott. 1987. Effects of prolonged flooding on soybeans during early vegetative growth. Soil Science 144: 61–66.

Sato, K. and T. Ikeda. 1979. The growth responses of soybean plant to photoperiod and temperature IV. The effect of temperature during the ripening period on the yield and characters of seeds. Jpn. J. Crop Sci. 48: 283–290.

Sayama, T., T. Nakazaki, G. Ishikawa, K. Yagasaki, N. Yamada, N. Hirota et al. 2009. QTL analysis of seed-flooding tolerance in soybean (*Glycine max* [L.] Merr.). Plant Science 176: 514–521.

Schöffel, E.R., A.V. Saccol, P.A. Manfron and S.L.P. Medeiros. 2001. Excesso hídrico sobre os components do rendimento da cultura da soja. Ciência Rural 31: 7–12.

Seago, J.R.J.L., L.C. Marsh, K.J. Stevens, A. Soukup, O. Votrubova and D.E. Enstone. 2005. A re-examination of the root cortex in wetland flowering plants with respect to aerenchyma. Annals of Botany 96: 565–579.

Seddigh, M. and G.D. Jolliff. 1984. Night temperature effects on morphology, phenology, yield components of indeterminate field grown soybean. Agron. J. 76: 824–828.

Serraj, R., S. Bona, L.C. Purcell and T.R. Sinclair. 1997. Nitrogen accumulation and nodule activity of field-grown 'Jackson'soybean in response to water deficits. Field Crops Research 52: 109–116.

Setter, T.L. and B.A. Flannigan. 2001. Water deficit inhibits cell division and expression of transcripts involved in cell proliferation and endoreduplication in maize endosperm. J. Exp. Bot. 52: 1401–1408.

Shi, H., M. Ishitani, C. Kim and J.-K. Zhu. 2000. The *Arabidopsis thaliana* salt tolerance gene SOS1 encodes a putative Na^+/H^+ antiporter. Proceedings of the National Academy of Sciences 97(12): 6896–6901.

Shi, H., F.J. Quintero, J.M. Pardo and J.-K. Zhu. 2002. The putative plasma membrane Na^+/H^+ antiporter SOS1 controls long-distance Na^+ transport in plants. The Plant Cell Online 14(2): 465–477.

Shimamura, S., T. Mochizuki, Y. Nada and M. Fukuyama. 2003. Formation and function of secondary aerenchyma in hypocotyl, roots and nodules of soybean (*Glycine max*) under flooded conditions. Plant and Soil 251: 351–359.

Shimamura, S., R. Yamamoto, T. Nakamura, S. Shimada and S. Komatsu. 2010. Stem hypertrophic lenticels and secondary aerenchyma enable oxygen transport to roots of soybean in flooded soil. Annals of Botany 106: 277–284.

Shinozaki, K. and K. Yamaguchi-Shinozaki. 2000. Molecular responses to dehydration and low temperature: Differences and cross-talk between two stress signaling pathways. Current Opinion in Plant Biology 3(3): 217–223.

Shutilova, N., G. Semenova, V. Klimov and V. Shnyrov. 1995. Temperature-induced functional and structural transformations of the photosystem II oxygen-evolving complex in spinach subchloroplast preparations. Biochem. Mol. Biol. Int. 35: 1233–1243.

Silva, P. and H. Gerós. 2009. Regulation by salt of vacuolar H^+-ATPase and H^+-pyrophosphatase activities and Na^+/H^+ exchange. Plant Signaling Behavior 4(8): 718–726.

Skrudlik, G. and J. Koscielniak. 1996. Effects of low temperature treatment at seedling stage on soybean growth, development and final yield. J. Agron. Crop Sci. 176: 111–117.

Sottosanto, J.B., A. Gelli and E. Blumwald. 2004. DNA array analyses of *Arabidopsis thaliana* lacking a vacuolar Na/H antiporter: Impact AtNHX1 on gene expression. Plant Journal 40: 752–771.

Steffens, B., A. Kovalev, S.N. Gorb and M. Sauter. 2012. Emerging roots alter epidermal cell fate through mechanical and reactive oxygen species signaling. The Plant Cell 24: 3296–3306.

Stockinger, E.J., S.J. Gilmour and M.F. Thomashow. 1997. *Arabidopsis thaliana* CBF1 encodes an AP2 domain-containing transcriptional activator that binds to the C repeat/DRE, a *cis*-acting DNA regulatory element that stimulates transcription in response to low temperature and water deficit. Proc. Natl. Acad. Sci. U.S.A. 94: 1035–1040.

Stolf-Moreira, R., M. Medri, N. Neumaier, N. Lemos, J. Pimenta, S. Tobita et al. 2010. Soybean physiology and gene expression during drought. Genetics and Molecular Research 9: 1946–1956.

Sullivan, M., T. Van Toai, N. Fausey, J. Beuerlein, R. Parkinson and A. Soboyejo. 2001. Evaluating on-farm flooding impacts on soybean. Crop Science 41: 93–100.

Starck, Z., B. Niemyska, J. Bogdan and R.N. Akour Tawalbeh. 2000. Response of tomato plant to chilling stress in associated with nutrient or phosphorus starvation. Plant Soil 226: 99–106.

Suzuki, N. and R. Mittler. 2006. Reactive oxygen species and temperature stresses: A delicate balance between signaling and destruction. Physiologia Plantarum 126: 45–51.

Szalai, G., T. Janda, E. Paldi and Z. Szigeti. 1996. Role of light in the development of post-chilling symptoms in maize. J. Plant Physiol. 148: 378–383.

Tamang, B.G., J.O. Magliozzi, M.S. Maroof and T. Fukao. 2014. Physiological and transcriptomic characterization of submergence and reoxygenation responses in soybean seedlings. Plant, Cell and Environment 37: 2350–2365.

Thomas, J.F. and C.D. Raper. 1978. Effect of day and night temperatures during floral induction on morphology of soybeans. Agron. J. 70: 893–898.

Thuzar, M., A.B. Puteh, N.A.P. Abdullah, M.B.M. Lassim and K. Jusoff. 2010. The effects of temperature stress on the quality and yield of soya bean (*Glycine max* L.) Merrill. J. Agric. Sci. 2: 172–179.

Umezawa, T., K. Nakashima, T. Miyakawa, T. Kuromori, M. Tanokura, K. Shinozaki and K. Yamaguchi-Shinozaki. 2010. Molecular basis of the core regulatory network in ABA responses:sensing, signaling and transport. Plant and Cell Physiology 51(11): 1821–1839.

Urao, T., B. Yakubov, R. Satoh, K. Yamaguchi-Shinozaki, M. Seki, T. Hirayama and K. Shinozaki. 1999. A transmembrane hybrid-type histidine kinase in Arabidopsis functions as an osmosensor. The Plant Cell 11(9): 1743–1754.

Valliyodan, B., T.T. Van Toai, J.D. Alves, P. de Fátima, P. Goulart, J.D. Lee, F.B. Fritschi et al. 2014. Expression of root-related transcription factors associated with flooding tolerance of soybean (*Glycine max*). International Journal of Molecular Sciences 15: 17622–17643.

van Veen, H., A. Mustroph, G.A. Barding, M. Vergeer-van Eijk, R.A. Welschen-Evertman, O. Pedersen et al. 2013. Two Rumex species from contrasting hydrological niches regulate flooding tolerance through distinct mechanisms. The Plant Cell 25: 4691–6707.

Vega, C.R.C., F.H. Andrade and V.O. Sadras. 2001. Reproductive partitioning and seed set efficiency in soybean sunflower and maize. Field Crops Research 72: 163–175.

Vertucci, C.W. and A.C. Leopold. 1983. Dynamics of imbibition by soybean embryos. Plant Physiol. 72: 190–193.

Wilkinson, S. and W.J. Davies. 2010. Drought, ozone, ABA and ethylene: New insights from cell to plant to community. Plant, Cell and Environment 33(4): 510–525.

Wilson, D.R., P.D. Jamieson, W.A. Jermyn and R. Hanson. 1985. Models of growth and water use of field pea (*Pisum sativum* L.). *In*: Hebblethwaite, P.D., M.C. Heath and T.C.K. Dawkins (eds.). The Pea Crop. Butterworths, London.

Wise, R.R., A.J. Olson, S.M. Schrader and T.D. Sharkey. 2004. Electron transport is the functional limitation of photosynthesis in field-grown Pima cotton plants at high temperature. Plant Cell Environ. 27: 717–724.

Wuebker, E.F., R.E. Mullen and K. Koehler. 2001. Flooding and temperature effects on soybean germination. Crop Science 41: 1857–1861.

Xiong, L. and J.K. Zhu. 2002. Molecular and genetic aspects of plant responses to osmotic stress. Plant, Cell and Environment 25(2): 131–139.

Xiong, L., K.S. Schumaker and J.-K. Zhu. 2002. Cell signaling during cold, drought, and salt stress. The Plant Cell Online 14(suppl 1): S165–S183.

Xue, H.W., X. Chen and Y. Mei. 2009. Function and regulation of phospholipid signalling in plants. Biochemical Journal 421(Pt 2): 145.

Yadegari, L.Z., R. Heidari and J. Carapetian. 2007. The influence of cold acclimation on proline, malondialdehyde (MDA), total protein and pigments contents in soybean (*Glycine max*) seedlings. J. Biol. Sci. 7(8): 1436–1441.

Yang, Q., Z.-Z. Chen, X.-F. Zhou, H.-B. Yin, X. Li, X.-F. Xin, X.-H. Hong, J.-K. Zhu and Z. Gong. 2009. Overexpression of SOS (Salt Overly Sensitive) genes increases salt tolerance in transgenic *Arabidopsis*. Molecular Plant 2(1): 22–31.

Yamaguchi, T. and E. Blumwald. 2005. Developing salt-tolerant crop plants: Challenges and opportunities. Trends in Plant Science 10: 615–620.

Yang, L., W. Ji, P. Gao, Y. Li, H. Cai, X. Bai, Q. Chen and Y. Zhu. 2012. GsAPK, an ABA-activated and calcium-independent SnRK2-type kinase from G. soja, mediates the regulation of plant tolerance to salinity and ABA stress. PLoS ONE 7(3): e33838.

Yeo, A.R., M.L. Koyama, S. Chinta and T. Flowers. 2000. Salt tolerance at the whole plant level. pp. 107–123. *In*: Cherry, J.H. (ed.). Plant Tolerance to Abiotic Stresses in Agriculture: Role of Genetic Engineering. Vol. 3. Netherlands: Kluwer.

Yin, H., Q.M. Chen and M.F. Yi. 2008. Effects of short-term heat stress on oxidative damage and responses of antioxidant system in *Lilium longiflorum*. Plant Growth Regulation 54: 45–54.

Yin, X., S. Hiraga, M. Hajika, M. Nishimura and S. Komatsu. 2017. Transcriptomic analysis reveals the flooding tolerant mechanism in flooding tolerant line and abscisic acid treated soybean. Plant Molecular Biology 93: 479–496.

Yoo, J.H., C.Y. Park, J.C. Kim, W. Do Heo, M.S. Cheong, H.C. Park, M.C. Kim, B.C. Moon, M.S. Choi and Y.H. Kang. 2005. Direct interaction of a divergent CaM isoform and the transcription factor, MYB2, enhances salt tolerance in Arabidopsis. Journal of Biological Chemistry 280(5): 3697–3706.

Yordanova, R.Y., K.N. Christov and L.P. Popova. 2004. Antioxidative enzymes in barley plants subjected to soil flooding. Environmental and Experimental Botany 51: 93–101.

Yoshiba, Y., T. Kiyosue, K. Nakashima, K.Y. Yamaguchi and K. Shinozaki. 1997. Regulation of leaves of proline as an osmolyte in plants under water stress. Plant Cell Physiol. 38: 1095–1102.

Zeng, A., P. Chen, K. Korth, F. Hancock, A. Pereira, K. Brye, C. Wu and A. Shi. 2017. Genome-wide association study (GWAS) of salt tolerance in worldwide soybean germplasm lines. Molecular Breeding 37: 30.

Zhang, H., H. Jiao, C.X. Jiang, S.H. Wang, Z.J. Wei, J.P. Luo and R.L. Jones. 2010. Hydrogen sulphide protects soybean seedlings against drought-induced oxidative stress. Acta Physiologiae Plantarum 32(5): 849–857.

Zhang, J., W. Jia, J. Yang and A.M. Ismail. 2006. Role of ABA in integrating plant responses to drought and salt stresses. Field Crops Research 97(1): 111–119.

Zhang, J.H., W.D. Huang, Y.P. Liu and Q.H. Pan. 2005. Effects of temperature acclimation pretreatment on the ultrastructure of mesophyll cells in young grape plants (*Vitis vinifera* L. cv. Jingxiu) under cross-temperature stresses. J. Integr. Plant Biol. 47: 959–970.

Zhang, J., Q. Song, P.B. Cregan, R.L. Nelson, X. Wang, J. Wu and G. Jiang. 2015. Genome-wide association study for flowering time, maturity dates and plant height in early maturing soybean (*Glycine max*) germplasm. BMC Genomics 16: 217.

Zhao, Z., G. Chen and C. Zhang. 2001. Interaction between reactive oxygen species and nitric oxide in drought-induced abscisic acid synthesis in root tips of wheat seedlings. Functional Plant Biology 28(10): 1055–1061.

Zheng, S., H. Nakamoto, K. Yoshikawa, T. Furuya and M. Fukuyama. 2002. Influence of high night temperature on flowering and pod setting in soybean. Plant Prod. Sci. 5: 215–218.

Zhou, G.A., R.Z. Chang and L.J. Qiu. 2010. Overexpression of soybean ubiquitin-conjugating enzyme gene GmUBC2 confers enhanced drought and salt tolerance through modulating abiotic stress-responsive gene expression in Arabidopsis. Plant Molecular Biology 72(4-5): 357–367.

Zhu, S.Y., X.C. Yu, X.J. Wang, R. Zhao, Y. Li, R.C. Fan, Y. Shang, S.Y. Du, X.F. Wang and F.Q. Wu. 2007. Two calcium-dependent protein kinases, CPK4 and CPK11, regulate abscisic acid signal transduction in Arabidopsis. The Plant Cell 19(10): 3019–3036.

Zhu, G.Y., J.M. Kinet, P. Bertin, J. Bouharmont and S. Lutts. 2000. Crosses between cultivars and tissue culture-selected plants for salt resistance improvement in rice, *Oryza sativa*. Plant Breeding 119: 497–504.

Zhu, J.K., P.M. Hasegawa and R.A. Bressana. 1997. Molecular aspects of osmotic stress in plants. Crit. Rev. Plant Sci. 16: 253–277.

Zhu, J.K. 2001. Plant salt tolerance. Trends in Plant Science 6: 66–71.

CHAPTER 13
Source-Sink Relationships

Inspite of the adequate researches on soybean and the available management practices, the productivity of the soybean is not high. During the reproductive phase, though vegetative growth is high enough, the light interception by leaves are comparatively less. Hence, photosynthesis in reproductive phase took place in a diffused light condition. It is noted that the productivity of soybean is related to the number of seeds per square meter and the weight of the seeds. Hence, high number of seeds and the high seed weight are important for high productivity. Medium duration and medium height plants are most favored. Average production is about 2 t/ha. As it is a leguminous plant, much nitrogen fertilization is not necessary. Early stage weed management is one important concept. Soybean is usually grown under rain-fed condition but, if necessary, irrigation can be arranged. Microbial inoculation no doubt enhances the health and productivity of the soybean crop. It is a typical C3 plant and photosynthetic rate of leaves changes widely depending on the environmental changes. Soybean seed oil is good in terms of quality. Protein content in the seed is much higher in seeds than cereals and many pulses. This is a good quality food raw material and the finished products serve as food for human consumption.

In the plant system, source-sink relationship is a complex phenomenon. In many plants source is limited and in other plants sink may be limited; in some, both source and sink may be limited. With the emergence of high yielding plant types in crops, the changes were arranged from those of traditional varieties so that the source and sink are balanced. It is interesting to understand the underlying basis of source and sink in soybean plant as both source and sink strength under different agro-climatic conditions and under abiotic stress may affect the yield of soybean. In this crop, it is necessary to optimize the source in order to maximize the yield through manifestation of yield contributing components or enhance the sink strength to an extent so that source may be limited, which can be further manipulated by the efficient cultural management practices. Thus, an attempt has been made to elaborate the source-sink relationships in soybean.

13.1 Source Strength

In a plant system, to understand the source-sink relationship it is necessary to know the source strength. In soybean plant, sources are usually the leaf area, which is better represented by leaf area index, i.e., leaf area per unit of land area. However, the source strength is the product of leaf area index and the efficiency of photosynthetic capacity of the leaves. Sometimes net assimilation rate is also considered as the photosynthetic capacity. Thus, source strength may be given as: Source strength = Leaf area index × photosynthetic capacity or net assimilation rate. Egli and Zhen-Wen (1991) suggested that seeds per unit area were related to canopy photosynthesis during flowering and pod set and canopy photosynthesis rate is determined through Leaf area index (LAI) and Crop growth rate (CGR). A plant with optimum LAI and Net assimilation rate (NAR) may have higher biological yield as well as seed yield (Mondal et al., 2007).

13.2 Leaf Area Index (Source Size)

Soybean genotypes differ significantly in leaf area index at different growth stages. However, leaf area index of soybean within the range of 4.5–5.5 is better correlated with high seed yield. Leaf area index followed a typical sigmoid pattern with respect to time and increased with age till 80 days after sowing in most of the genotypes, followed by a decline in green leaves because of abscission of old leaves and, at physiological maturity, suddenly all leaves dry and heavy leaf shading occurs. Another aspect is that the changes of leaf area index in soybean have a differential pattern at different growth stages. However, the leaf area index changes along up the height is of sigmoid type, at pod setting stage maximum leaf area being at three-fourth of plant height decreasing sharply down the canopy (Basuchaudhuri, 1987). This causes a mutual shading of sunlight in the canopy and reduced photosynthesis by the full grown leaves. Hence, leaf area index and the plant architecture are important in consideration of source of soybean plant.

13.3 Photosynthetic Capacity (Source Activity)

In soybean, biological yield depends on source-strength by photosynthetic capacity (Egli and Crafts-Brandner, 1996). Increasing canopy photosynthesis by increasing levels of atmospheric CO_2 (Hardman and Burn, 1971) or irradiance (Schou et al., 1978) increased pod and seed number, while reducing photosynthesis by shading (Egli and Zhen-Wen, 1991; Andrade and Ferreiro, 1996) or defoliation (Board and Tan, 1995) decreased pod and seed number. In Soybean, a significant negative correlation exists between leaf photosynthetic carbohydrate (sources of starch) content and photosynthetic rate (Sawada et al., 1986; Kasai, 2008). There have also been findings of photosynthetic carbohydrate-mediated decrease in the activity or the amount of Rubisco, the CO_2 fixing enzyme in leaves (Paul and Pellny, 2003), although the detailed mechanism is still unclear. On the other hand, the reproductive stage of growth may be attributed to excessive mutual shading as the leaf area was maximum during this period and increased number of old leaves could have lowered the photosynthetic efficiency (Salam et al., 1987). In grain legume, excess LA was reported to have lower relative growth rate and resulted in a decrease of dry matter production and net assimilation rate, which probably resulted from excessive mutual shading (Pandey et al., 1978).

It has gained renewed focus in efforts to increase yields (Stitt et al., 2010; Foyer et al., 2017). Photosynthesis is well described and efforts to improve efficiency tend to focus on weak links with yield production or transposing different mechanisms into the pathway (for example, Long et al., 2015). Recently, support for the link between increasing photosynthesis and yield has been driven by studies performed under elevated CO_2 conditions, which have suggested a need to increase source strength in order to improve yields (Ainsworth and Bush, 2011). Such increases in photosynthetic rates are attributable to increased substrate availability rather than photosynthetic performance with limited interpretation outside of systems supplemented with water and nutrient supply. More broadly, there is a lack of clear evidence to support the relationship between net-photosynthesis and yield beyond the concept of yield potential. Equations for yield potential, outlined by Monteith (1977), describe the efficiency with which a plant intercepts light, converts intercepted radiation to biomass and partitions this biomass into the harvested product only when the given crop is grown in ideal conditions where ample nutrients, water and all biological stresses are controlled (Evans and Fischer, 1999; Long et al., 2006; Amthor, 2007).

Increases in yield potential over the past 50 years have essentially been achieved through increases to harvest index (i.e., increased partitioning of biomass into the harvest product), greater responses to additional nitrogen fertilizers and increased canopy development, allowing for increased light interception (Long et al., 2006). Several authors have, therefore, suggested that, if two out of the three components of the theoretical yield potential equation are approaching their upper limits (Lobell et al., 2009; Zhu et al., 2010), the efficiency with which light energy is converted into biomass, i.e., photosynthesis, is the next target in efforts to increase yield potential (Zhu et al., 2010; Foyer et al., 2017). Little evidence supports this notion. Whilst improvements in light acquisition and utilization to drive photosynthetic performance is likely important to improving yield, there is no evidence suggesting an exhaustion in the capacity of harvest index to achieve yield gains. Beyond calculations for yield potential, correlations between crop yield and photosynthesis are weak (see Evans, 1997) and yield is typically limited by sink capacity rather than source strength (i.e., photosynthesis) in the major crops of wheat, maize and soybean (see Borrás et al., 2004). There is, however, some coordination between photosynthesis and yield in ideal environments where other resources are not limited (see Long et al., 2006 and references therein).

13.4 Sink Strength

Sink is the plant part where the assimilates moves to be utilized without synthesizing *in situ*. Growing and developing leaves and stems are also the sink. Seeds are the major sink, but developing buds are also sink. Roots are potential sink throughout the life span of the plant. So, the concept of sink at different stages of growth is different and complex. For simplicity, seeds will be considered as the sink, especially at reproductive stage.

13.5 Sink Size

Number of seeds is the sink size. There are variations among genotypes in average number of seeds per plant. However, management practices, such as plant density, nutrient management, sowing date also have some influences on seeds per square meter of land area. Though, pod and seed size varies occasionally, Mehta et al. (2000) observed that seed yield of soybean had no positive relationship with pod and seed size. In a study, among the genotypes, BAU-70 produced the highest seed yield per plant (9.95 g) and per ha (3.31 t) due to production of higher number of pods per plant (31.23) and greater dry matter partitioning to seeds (harvest index 37.73%), though it produced slightly smaller seed size than the others (Malek et al., 2012). So, it is necessary to optimize the management practices to obtain maximum seeds per unit land area and achieve the highest yield in soybean (Basuchaudhuri et al., 1986).

13.6 Sink Activity

The greater plasticity of soybean seeds for establishing final seed sink potential may be the most important trait behind its higher seed dry weight response to increased assimilate availability during seed filling. The efficiency of remobilization of assimilates temporarily stored in plant for seed production may be an important aspect determining seed dry weight response when assimilate availability is reduced. Seed growth rate (SGR) is considered as the sink activity. Experiments suggest that soybean SGR is generally sink limited if photosynthesis increases during seed filling, but source limited if photosynthesis is reduced. Usually, under normal conditions of growth, photosynthesis at seed filling stage is slightly reduced. Absolute growth rate of seeds is, therefore, important, which creates potential sink and SGR is positively associated with seed yield. Thus, seed number as well as SGR combination is the sink strength.

Ambiguity surrounding sink strength occurs due to an inability to directly measure (quantify) it and lack of understanding of the processes that drive sink activity. Given the number of factors that relate to sink activity (growth, metabolism), it is incongruous that sink size has a proportional influence on sink activity over sink strength. Whilst it is recognized that sink size has some influence over total metabolic activity, rates of metabolism vary according to ontological and tissue development. Understanding the processes and conditions governing changes to sink strength, along with improvements in technology that allow for the direct measurement of sink demand, will lead to greater accuracy in the way sink strength is described.

Improving the definition of sink strength will also allow for greater consistency in the literature and ensure that researchers are implementing a useful, though inevitably complex definition. Measuring sink strength is difficult. For provision of photoassimilates, source activity can be well characterized by measurement of net photosynthetic rate. However, due to the complexity of sinks, measurement is typically confined to a quantification of sink size, typically via the removal of sink tissue, along with some measure or estimate of sink activity. Issues with this technique of mass balance measurement and others, such as isotope labeling, have been raised regarding the definition and measurement of sink strength (see Farrar, 1993). In conclusion to the above, Farrar (1993) suggests that measurement of sinks should incorporate the transport system and sources. In essence, the ability to explore the complexity of sinks in the context of the whole plant without altering the system requires non-invasive technologies. Despite advances in phenotyping technologies (Rascher et al., 2011; Dhondt et al., 2013; Furbank et al., 2015), until recently, sink strength viewed in terms of carbon demand by individual sinks had still not been measured. This has prevented the exploration of questions that have interested researchers since the early discussions surrounding sink strength, including the abortion of sinks and a full exploration of fluxes between sources and sinks under varying environmental conditions (see Farrar, 1993 and references therein).

Rates of phloem unloading are an important component in this framework and offer a further point of potential regulation of photoassimilate movement. Regulation and mechanisms of phloem unloading vary between species, developmental stage and sink function (Werner et al., 2011; Braun et al., 2014). While there are no direct measures of phloem loading and unloading, Patrick (2013) cites "considerable indirect evidence" that phloem unloading capacity is exceeded by photosynthesis and phloem loading, particularly under optimal conditions (Voitsekhovskaja et al., 2009). Thus, using the high pressure manifold model, as proposed by Fisher (2000), "resource partitioning between sinks is finely regulated by their relative hydraulic conductance of plasmodesmata linking sieve element/companion cell complexes with the surrounding phloem parenchyma cells" (Patrick, 2013).

13.7 Harvest Index

Within this model, transporters play a role in moving and partitioning photoassimilates from sources to sinks. There are numerous transporters that have been found to aid in this process. As recently demonstrated by Wang et al. (2015) in transgenic rice, enhancing transporters is another mechanism to potentially increase yields. Expanding our research into the capacity for transport and a sinks ability to accept photoassimilates is vital to ensure the resilience of yield. The production of yield in cropping systems is a consequence of many biological processes, culminating in biomass and/or seed production. Harvest index, the ratio of harvested grain to total shoot dry matter (Donald and Hamblin, 1976), is a trait that is the cumulative result of allocation of acquired resources and is used in efforts to improve yields in seed producing crops (Reynolds and Langridge, 2016). Improvements to harvest index have historically increased yield potentials in major staple food crops (Hay, 1995) leading to broad economic gains for farmers. More specifically, harvest index represents the result of plant efficiency, including a range of processes governing the packaging, transport and deposition of photoassimilates and nutrients into the seed. Whilst much is known regarding the processes behind how this is achieved (Pritchard, 2007), few studies have sought to exploit these properties to improve yield production. Improvements in harvest index are credited with inducing large increases in yield potential in important food crops (Long et al., 2015), yet the specific mechanisms behind how this occurred are not well understood (Amthor, 2007). The fundamental basis of harvest index in seed producing crops is carbon centric and dictates that total shoot dry matter determines aboveground "sources" of photoassimilate and harvested grain represents the "sinks." Harvest index is the proportion of biomass invested into grain (Donald and Hamblin, 1976; Gifford and Evans, 1981) and reflects the balance between source and sink (Luo et al., 2015). Measurement of harvest index does not capture the efficiency of resource investment and confounds the processes and pathways that regulate the transfer of these resources from the total shoot biomass into grain. It is, therefore, not surprising that harvest index correlates with various yield-related traits in important crop species though, generally, these are interrelated (Luo et al., 2015), further confounding the underlying mechanisms that drive increases in this important trait.

Harvest index has high heritability under both ideal and stressed environments (Hay, 1995). Conservation of the trait across multiple environments and genotypes in different crops has led harvest index to be one of the most highly studied traits in plant breeding (Unkovich et al., 2010). Much of the variation observed in harvest index values results from the diverse range of climates and soils, which are a feature of the cropping region.

13.8 Source-Sink Relationships

The partitioning of recently fixed ^{14}C to setting and abscising flowers within the axilary raceme of 'Clark' isoline Elt soybeans (*Glycine max* L. Men.) was examined as a function of time after anthesis of individual flowers. In such racemes, the first four flowers showed a 17% abscission while the next four flowers showed 47% abscission.

Source/sink relations of flowers I–IV (normally setting) were compared to those of flowers V–VIII (normally abscising) by pulse labeling source leaves with $^{14}CO_2$ and determining the radioactivity of individual flowers after a 4-hour chase period. The relative specific activity (RSA; % disintegrations per minute per % dry weight), sink strength (% disintegrations per minute) and its components, sink

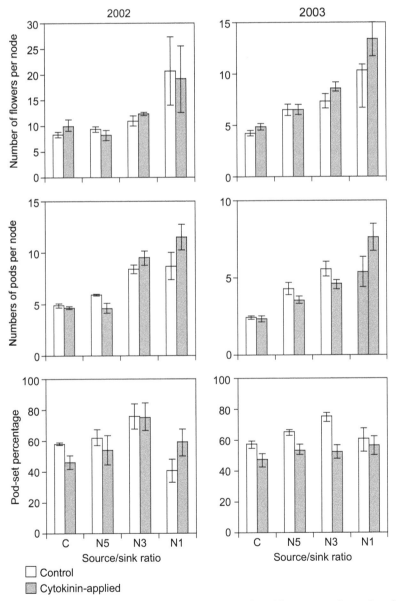

Fig. 13.1 Effect of source/sink ratio and cytokinin application on the number of flowers per node, number of pods per node and pod-set percentage in 2002 and 2003. Values represent the mean SE of six plants (5, 3 and 1 was the number of racemes in the main stem).

size (milligrams dry weight) and sink intensity (% disintegrations per milligram dry weight), were then calculated as a function of days after anthesis.

Sink intensity (i.e., the competitive ability to accumulate photoassimilate per unit mass) was very high prior to anthesis of both setting and abscising flowers. Sink intensity then became very low for the first 3 days following anthesis, after which it recovered in normally setting flowers, but failed to recover in normally abscising flowers. It is concluded that soybean reproductive abscission is determined at or very near the day of anthesis.

Experiments were conducted to investigate the effect of TIBA (2,3, S-Triiodobenzoic acid) foliar spray on soybean cv Co. 1 in winter and summer seasons. The source and sink components were increased by TIBA at 50 ppm in both the seasons.

The field experiments were conducted at Tamil Nadu Agricultural University, Coimbatore, during winter and summer season of 1984–85 on well drained sandy loam soil under irrigated condition. The cultivar Co. 1 of soybean (*Glycine max* L. Merril), which is determinate, photo insensitive and erect in growing habit, was used in both the seasons. The experiments were laid in split plot design with three replications. Different plant density levels (6.66, 8.10 and 3.3 lakh plants per hectare) were assigned to main plots and 4 concentrations of TIBA, viz., 25 ppm (T3) 50 ppm (T4) 75 ppm (T5) 100 ppm (T6) water spray (T2) and no spray (TI) were used as sub plot treatments. Only the sub plot treatments are presented in this paper. The TIBA foliar application was given at pre-flowering stage. The mean maximum and minimum temperatures were 30°C and 21.2°C and 34.6°C and 21.5°C during the experimental periods of winter and summer seasons, respectively.

The dry matter accumulation (DMA) during summer was 13.3 per cent more than in winter season. The foliar application of 50 ppm TIBA had a curvilinear response to the DMA in soybean. The source size in terms of LAI showed an increase of 2% during summer over winter. The application of TIBA reduced the LAI irrespective of season and greater reduction was noticed at 100 ppm. The reduction of

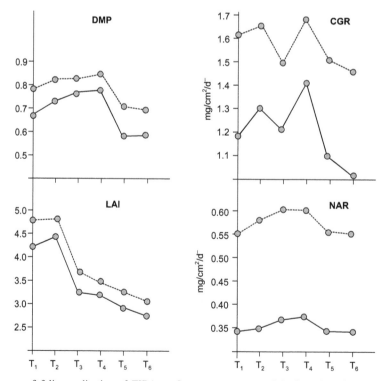

Fig. 13.2 Influence of foliar application of TIBA on Source components of Soybean in Winter (0-0) and Summer (0 ••. 0) season. DMP(kg/m2), LAI no unit, CGR(mg/d), NAR(mg/cm2/d) Tr-No spray. Tr-Water spray. Ta-2S ppm, *T,-SO* ppm, T.-7S ppm and To-IOO ppm.

Table 13.1 Influence of TIBA on yield and yield components in soybean.

Treatment	Pods per plant		Seeds per pod		Test weight (g)		Seed yield (kg/ha)	
	Winter	Summer	Winter	Summer	Winter	Summer	Winter	Summer
Control	39.1	42.8	2.33	2.35	99.2	109.1	1213	1341
Water spray	41.7	43.6	2.46	2.41	99.1	101.2	1290	1362
TIBA 25 ppm	51.4	55.1	2.13	2.12	96.6	101.8	1548	1116
TIBA 50 ppm	53.5	59.3	2.82	2.98	96.1	100.3	1589	1161
TIBA 75 ppm	50.3	54.9	2.55	2.61	92.3	98.6	1500	1698
TIBA 100 pm	48.5	51.6	2.48	2.52	90.1	91.1	1417	1601
CD (0.05)	2.2	2.2	0.24	0.22	0.16	0.10	21.1	21.8

source size did not influence the DMA and yield up to 50 ppm. This may be due to the improved crop architecture brought through the chemical manipulation suggested by Bauer et al. (1969). The increased NAR, to the magnitude of 60%, was observed during summer as compared to winter. Seed yield was positively and significantly correlated with NAR and CGR, as reported by Kenneth and Hell (1980). The results indicated that 50 ppm TIBA application increased all the source components in summer season.

Among various sink components the number of pods per plant was altered more as a result of seasonal and chemical treatment (Table 13.1). The highest number of pods per plant was achieved by foliar application of 50 ppm TIBA during summer season. Though the number of seeds per pod increased in both the seasons at 50 ppm, it was comparatively more in summer season. The application of TIBA reduced test weight irrespective of season, which presumably is due to the inverse relationship between pod number and weight. The foliar application of 50 ppm TIBA increased seed yield considerably by increasing the number of pods per plant during summer.

Source-to-sink transport of sugar is one of the major determinants of plant growth and relies on the efficient and controlled distribution of sucrose (and some other sugars, such as raffinose and polyols) across plant organs through the phloem. However, sugar transport through the phloem can be affected by many environmental factors that alter source/sink relationships. Here, current knowledge about the phloem transport mechanisms can be summarized and the effects of several abiotic (water and salt stress, mineral deficiency, CO_2, light, temperature, air, and soil pollutants) and biotic (mutualistic and pathogenic microbes, viruses, aphids, and parasitic plants) factors are reviewed. Concerning abiotic constraints, alteration of the distribution of sugar among sinks is often reported, with some sinks as roots favoured in case of mineral deficiency. Many of these constraints impair the transport function of the phloem but the exact mechanisms are far from being completely known. Phloem integrity can be disrupted (e.g., by callose deposition) and, under certain conditions, phloem transport is affected, earlier than photosynthesis. Photosynthesis inhibition could result from the increase in sugar concentration due to phloem transport decrease. Biotic interactions (aphids, fungi, viruses...) also affect crop plant productivity. Recent breakthroughs have identified some of the sugar transporters involved in these interactions on the host and pathogen sides.

For the past several decades, effect of altered source-sink relationship on soybean yield has been extensively studied (Board and Harville, 1998; Liu et al., 2006). Former studies indicated that when equilibrium is broken between sources and sinks in individual soybean plant through pods or leaf removal, the direction of assimilate transport is changed (Board and Harville, 1998). A positive correlation between leaf area and seed weight across the main axis in soybean was reported, and was defined as source-sink parallelism (Dong et al., 1993).

Assimilates of pods and seeds in a certain node were mainly supplied from the leaf developed from the same node, namely, local supplying characteristic (Pate et al., 1977). Wang et al. (1983) stated that pods gained assimilates not only from the attached leaf but also from the leaves at the adjacent nodes (above or below). They found a strong adjacent compensation in assimilate distribution process, because pod weight at the nodes with leaf removal was 60% of the same nodes with leaf attached. Fu et al. (1999)

found that a mature trifoliate leaf from a certain node supplied most assimilates to the pods attached in the axis of the same leaf petiole, and only a small part of the assimilates transported to the adjacent node, particularly to the adjacent pods at same side of the stem using 14C technique. They then proposed that assimilates in soybean plant had the characteristic of same side transport.

Effect of source-sink distance on soybean yield was studied. Hypothesis was that soybean can redistribute assimilates from lower leaves to young pods formed at youngest part of stems across the main stem to maintain yield. Two, four, six, eight, and ten-nodes of source-sink distances in soybean plant were artificially created. A reduction in the seed yield of remained sink occurred with the increase in distance from the source. This was mostly due to a reduction of pod number caused by the increase of the distance from the source. No pod reduction was found at the 2-node distance compared with control plant. Reduction in seed size occurred for the 2- and 10-node distance from the source. When source-sink distance was four, six, eight nodes, seed size was similar to control. Thus, findings suggest soybean plants are able to transport and use assimilates as far as 8 nodes distance. The successful translocation of assimilate from lower nodes is mainly used for pod formation and seed filling over a short distance, but can be translocated over a long distance and used for growth of remaining seed. This suggests the reproductive sink of soybean plants has an internal mechanism to off-set yield loss and ensure seed survival.

Limitations to crop yields are frequently sought in either photosynthesis, the source of assimilates or in the sink the site of assimilate utilization. This division recognizes the two major processes involved in the accumulation of grain yield, the production of assimilate in the leaves and utilization of this assimilate by the developing seed. Focusing on sources and sinks provides what appears to be a simple two-component system; unfortunately, analysis of this system does not always clearly identify the yield-limiting processes (Evans, 1993).

Modification of source activity of soybean during flowering and pod set usually results in a corresponding change in pod and seed number, indicating a source limitation. Increasing canopy photosynthesis by increasing levels of atmospheric CO_2 (Hardman and Brun, 1971) or irradiance (Schou et al., 1978) increased pod and seed number, while reducing photosynthesis by shading (Egli and Zenwen, 1991; Andrade and Ferreiro, 1996) or defoliation (Board and Tan, 1995) decreased pod and seed number. The effects of source-sink alterations during seed filling, after pod and seed number are fixed, are more complex. De-podding to increase assimilate supply to the remaining seed usually increases seed size (weight per seed) (Munier-Jolain et al., 1998), but does not always change individual seed growth rates (SGR) (Egli et al., 1985; Munier-Jolian et al., 1998). Reducing assimilate supply by defoliation (Board and Harville, 1998) reduced seed size, but the effect of shade was more variable (Egli et al., 1985; Egli, 1999).

Source limitations during seed filling seem to be relatively common, based on changes in seed size, but SGR is not as responsive to changes in source activity and can be sink limited. Recent analysis of changes in starch levels in soybean leaves during seed filling supports this contention by suggesting that intermittent sink limitations can occur (Egli, 1999).

The response of the sink (the seed) to source-sink alterations during seed filling depends upon the effect on the assimilate level in the seed and the ability of the seed to respond to a change in assimilate supply (Jenner et al., 1991). The response could be manifested in changes in SGR or seed fill duration, with their interaction determining final seed size. The response of seed fill duration to assimilate availability is not well defined (Egli, 1998). However, studies with *in vitro* culture systems demonstrated that SGR exhibits a classic saturation response to increasing sucrose concentrations (Thompson et al., 1977; Egli et al., 1989).

Unfortunately, the effect of gross manipulations of source-sink ratios *in planta* on seed sucrose levels are rarely reported, greatly limiting the use of the *in vitro* response in the interpretation of the effects of these manipulations (Jenner, 1980).

Field experiments using two soybean (*Glycine max* L. Merrill) cultivars ('Elgin 87' and 'Essex') were conducted for 2 years near Lexington, KY, USA in order to evaluate the effect of source-sink alterations on seed carbohydrate status and growth. Sucrose concentrations in developing cotyledons of control plants were consistently low (550 mM) early on in seed development, but they increased to 100 ± 150 mM by physiological maturity. The concentrations increased in both years by 47 to 59% when 90% of the pods were removed from 'Elgin 87', but the increase had no effect on individual seed growth rate (SGR). Shading (80%) reduced cotyledon sucrose levels and SGR in both years. The critical cotyledon sucrose concentration (the concentration providing 80% of the maximum cotyledon growth rate) was estimated from *in vitro* cotyledon growth at sucrose concentrations of 0 ± 200 mM. These critical concentrations varied from 72 ± 124 mM; *in planta* control cotyledon sucrose concentrations were below this critical level during the first half of seed growth but exceeded it in the later stages of growth in all experiments. The estimated critical concentration was consistent with the failure of *in planta* SGR to respond to an increase in assimilate supply and with the reduction in SGR associated with a decrease in assimilate supply. The results suggest that soybean SGR is generally sink limited if photosynthesis increases during seed filling, but source limited if photosynthesis is reduced (Fig. 13.3).

Shading of all side leaflets of a determinate soybean cultivar during pod filling significantly increased rates of photosynthesis in the unshaded centre leaflets, compared to centre leaflets of controls. Higher rates were associated with both higher stomatal and mesophyll conductances, and were reversible within 2 days of shades being removed. These higher rates of photosynthesis were not associated with decreased percentage enhancement by low oxygen, indicating that treatment effects were probably not associated with changes in photorespiration relative to photosynthesis. Percentage enhancement did, however, increase as the plants approached physiological maturity, chiefly because of a decrease in photosynthesis.

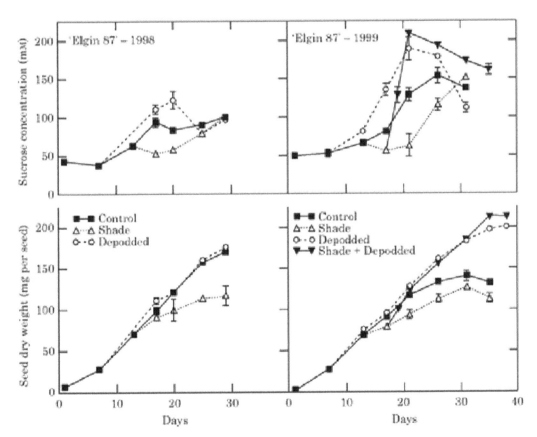

Fig. 13.3 The effect of shade (80%) and de-podding (90%) on seed growth and cotyledon sucrose concentrations in *Glycine max* L. Merr. 'Elgin 87', in 1998 and 1999. Plants were depodded on day 6 in 1998 and day 7 in 1999 and the shade treatment began on day 13 in both years. Plants under the shade were depodded on day 17 in 1999. Sampling stopped before physiological maturity in 1998 and the weight per seed at maturity was 169 mg per seed for the control, 200 mg for the depodded treatment and 120 mg in the shade treatment. Bars represent +S.E. of the mean. Some bars were smaller than the symbols and others were omitted to avoid excessive clutter.

In spite of these increases in rates of photosynthesis, seed weight per plant was decreased by 37% in plants with side leaflets shaded for the entire pod-filling period and by 28% in plants shaded for only the second half of the period. In plants where shades were removed during the second half of pod filling, seed yield was reduced by only 19% because shade removal delayed leaf senescence. The four treatments reduced yield by different mechanisms. Plants shaded continuously during pod filling produced fewer seeds than controls, but the weight per seed was similar. When shading was applied during the second half of pod fillings, seed number was unchanged but weight per seed was significantly reduced. In contrast, when shades were removed for the second half of pod filling, seed number remained similar to that of continuously shaded plants, but seed weight increased.

Potential yield of soybean was estimated to be around 7.0–8.0 t ha^{-1} (Specht et al., 1999; Sinclair, 1999). From a long term research, 6000–7000 kgha^{-1} were found from individual lines in a favorable environment at Wooster, US (Cooper, 2003). The highest yield of soybean at research level (favorable conditions) reached > 8 t ha^{-1} in Australia (Cooper, 2003).

It is of great importance to direct soybean breeding to a much higher potential which seems unlikely in Indonesia even with an effective growing duration of 80 days and a day length of 12 hours, which are often considered responsible for a low productivity of soybean in Indonesia compared with others in the subtropics. Varieties with a net CO_2 exchange rate (CER) of 20 μmol CO_2.m^{-2}.s^{-1} at light saturation (Pmax) and a quantum efficiency of ϕ = 0.05 μmol CO_2.μmol quanta^{-1} would produce $\geq$ 4 t ha^{-1} of seed yields with a daily average light of about 800 μmol quanta.m^{-2}.s^{-1}, based on a simulation modeling (Sinclair, 1991).

The breeding of soybean in Indonesia for high yield, based on seed yield as a selection criterion, seems difficult to further increase the potential yield. This method of selection limits the opportunity of best lines or genotypes to be selected in terms of high pod number and photosynthetic rate. This is based on the low heritability of seed yield, in general, at the early filial generations in combination with a high rate of interaction between genetic and environmental factors (Toledo et al., 2000). The approach of genetic engineering at molecular level is a powerful way of generating genetic diversity that would complement classical breeding. It is, however, hard to expect in a short-term due to the complexity of genes and gene interaction involved in yield formation (Sinclair et al., 2004). Other problems relate to uncertainty on many aspects of gene transfer, such as the site of insertion in the genome and the expression of trans genes (Gepts, 2002).

The integration of physiological parameters into the breeding program is an alternative that has received considerable attention (Gillbert et al., 2011). VanToai and Specht (2004) approached physiological traits through the model of Sinclair (1998), showing harvest index and biomass or RUE (radiation use efficiency) as alternative selection criteria. Liu et al. (2005) concluded that biomass production and leaf area are the key to future yield increase. The direct and indirect involvement of biomass as a selection criterion is impractical for huge numbers of lines. Other approaches to reach a high potential yield (i.e., Q0 = 4 t ha^{-1}) as proposed in the present, are called pod approach and photosynthesis approach. The pod approach, based on pod number (N), is supported by evidence showing a close relationship between pod number and seed yield in soybean studies involving genotypes, locations, and light treatments. The use of pod number as a parameter of indirect selection for seed yield was suggested. The photosynthesis approach is based on findings suggesting that the supply of photosynthate is a major factor directly influencing the number of pods and the seed yield (Schou et al., 1978; Egli, 1993; Board and Tan, 1995; Jiang and Egli, 1995).

The essence of both physiological parameters as selection criteria is the evidence of pod number, as the primary determinant of seed yield, directly dependent upon the rate of photosynthesis (Egli, 2005). The ratio of pod number per plant and Pmax varied between lines and were used to group lines resulting in close relationships between Pmax and pod number. It is concluded that the use of pod number and CER (Pmax) as selection criteria offers an alternative approach in soybean breeding for high yield (Fig. 13.4).

Environmental conditions prevailing during the reproductive period are important determinants of soybean yield and yield components (Board and Harvill et al., 1996; Liu et al., 2010; Liu et al., 2013). Board and Harville (1996) suggested that intensity and quality of solar radiation intercepted by the canopy influenced yield by changing survival rate of flowers in soybean plant. Liu et al. (2010) stated that light

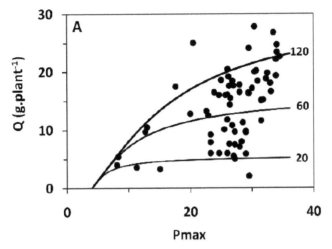

Fig. 13.4 Observed (round symbols) and estimated (lines) seed weight (Q) in relation to Pmax (A). The lines on Figure A were based on eq (4) for 30, 60 and 120 pods plant^{-1} as indicated in the figure.

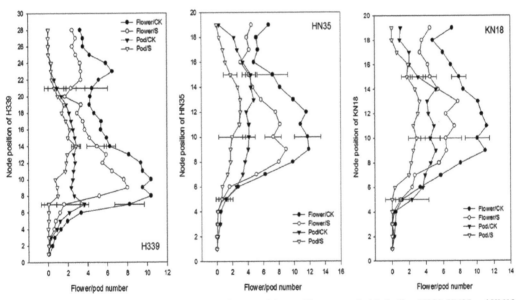

Fig. 13.5 Distribution of flower/pod number in the main stem of three cultivars treated with shading. H339, HN35 and KN18 are Hai339, Heinong35 and Kennong18, respectively. CK and S are natural light and shade treatment. Bar indicates standard error of the mean.

enrichment and shading significantly decreased and increased abortive rate of flower, resulting in change of pod number per plant (Fig. 13.5).

Much of the soybean yield variation is related to changes in flower number that survived to mature. From 32 to 81% of the flowers of field-grown soybean do not develop into mature pods (Wiebold et al., 1981).

Flowering in soybean is a dynamic system in which flower survival may depend on where a flower is located and when it is initiated. Egli and Bruening (2006b) stated that the temporal distribution of flower and pod production plays an important role in determining pod or seed number at maturity. Both flowering and pod production periods at individual node continue for 30 days or more, and they are nearly the same length (Egli and Bruening, 2006b). The timing of a flower initiate during the bloom or seed-filling period was important and late developing flowers may abort because large, rapidly-growing pods and seeds from early flowers consume most of the assimilate (Bruening and Egli, 2000).

Although effects of environmental factors on flowering and pod reproduction patterns have been reported (Liu et al., 2010), little information is available for the distribution of flower abscission across main axis under shading condition. More detailed research is still needed in order to analyze the effects of environmental conditions on flower abscission (Table 13.2).

The study was conducted at the BenHur Research Farm near Baton Rouge, LA, (30°N lat), on a soil (fine-silt, mixed, non-acid, thermic, Acric, Flavaquent). Planting dates were 3 June 1997 and 10 June 1998. The experimental treatments consisted of a factorial design of 10 soybean cultivars and three defoliation treatments: A non defoliated control, partial defoliation at the temporal midpoint of seed filling, and total defoliation at the temporal three-fourths point of seed filling. Leaf area index (LAI) and canopy light interception (LI) were measured after defoliation treatments. Grain yield was determined at maturity by machine harvest. Partial defoliation at mid-seed filling significantly (P < 0.05) reduced LAI, LI, and yield in 8 out of 10 cultivars in at least 1 study. Total defoliation at the three-fourths point of seed filling also reduced yields for almost every cultivar-year treatment combination. Results tended to support the original criteria related to LI and leaf area criteria for maintenance of optimal yield (Tables 13.3 and 13.4).

Split-split plots experimental design based on randomized complete block design was conducted at the Baiekola Agricultural Research Center in 2010 in order to study the effects of changes in source–sink on the yield and yield components of soybean cultivars planted at different dates. Factors studied in this experiment included planting date, as the main plot, at two levels (June 6, June 27), cultivars (Line 032, 033, Sari or JK, and Telar or BP), as the sub plot, and five levels, including the removal of the top, the middle, and the bottom one thirds of the leaves, removal of one third of the flowers, and the control treatment as the sub-sub plot. Results of analysis of variation showed that the various cultivars were significantly different in all the studied traits at (P < 0.01) probability level. Seed yield at the first planting date (184.03 g. m^{-2}) was 11.18% lower than that at the second planting date (163.45 g.m^{-2}). The highest seed yield was obtained in Line 033 with 219.96 g.m^{-2}, which was statistically different from those of all the other cultivars except that of Line 032, which was 186.19 g.m^{-2}. The higher seed yields

Table 13.2 Effect of shading on flower, pod number, flower abscission and yield in three soybeans.

Cultivar	Treatment	Flowers/plant	Pods/plant	Abscission (%)	Yield (g/plant)
H339	CK	147.3	41.8	72	24.8
	Shade	89.4	23.3	74	16.0
HN35	CK	114.9	48.4	58	17.2
	Shade	78.9	26.4	67	9.9
KN18	CK	147.5	52.7	64	17.0
	Shade	97.9	26.0	73	8.2

H339, HN35 and KN18 are Hai339, Heinong35 and Kennong18, respectively.

Table 13.3 Maximal pod number, final pod number, and percentage of aborted pods of control and defoliation treatments (DF) for soybean averaged for 1991 and 1992.

Defoliation treatment	Pods per reproductive node Maximal	Final	Pod abortion (%)	Reduced pod due to Decreased pod initiation	Increased pod abortion (%)
Control	3.41	2.45	28		
DF-R3	2.86	2.13	26	100	0
DF-R4	3.03	1.99	34	59	41
DF-R5	3.02	1.87	38	48	52
DF-R6.5	3.00	1.87	38	50	50

Table 13.4 Leaf area index (LAI), light interception (LI), yield, seed number per area, and seed size for 10 soybean cultivars receiving no defoliation (no def.), partial defoliation (part. def.) at mid-seed filling (R6.3), and total defoliation (tot. def.) at late seed filling (R6.6), grown near Baton Rouge, LA, 1997.

MG	Cultivar	Treatment	LAI	LI (%)	Yield (kg ha^{-1})	Seeds m^{-2}	Seed size (g/100 seed)
IV	DP3478	No def.	3.45	95	4060	2326	15.20
	HBK49	Part. def. (R6.3)	2.89	82	4074	2431	14.58
	RVS499	Tot. def. (R6.6)	-	-	3071	2129	12.56
		No def.	4.16	97	3869	2306	14.59
		Part. def. (R6.3)	3.74	91	3859	2383	14.10
		Tot. def. (R6.6)	-	-	3510	2293	13.42
		No def.	4.21	98	4189	2537	14.38
		Part. def. (R6.3)	3.08	91	3750	2295	14.24
		Tot. def. (R6.6)	-	-	3001	2298	11.34
V	Hutch	No def.	3.70	92	3789	2184	15.13
	HYP574	Part. def. (R6.3)	2.48	77	3869	2390	14.10
		Tot. def. (R6.6)	-	-	3506	2232	13.67
		No def.	3.56	92	4746	2703	15.32
		Part. def. (R6.3)	2.16	76	4150	2582	13.98
		Tot. def. (R6.6)	-	-	3606	2497	12.62
VI	A6961	No def.	4.79	97	4265	2913	12.72
	P9641	Part. def. (R6.3)	3.10	85	3982	2753	12.57
		Tot. def. (R6.6)	-	-	3433	2800	10.69
		No def.	4.48	96	4286	2467	15.12
		Part. def. (R6.3)	2.89	73	3701	2273	14.18
		Tot. def. (R6.6)	-	-	3815	2467	13.46
VII	DG3682	No def.	5.32	99	4566	2502	15.87
	H7190	Part. def. (R6.3)	3.45	92	3926	2196	15.55
	Stwall	Tot. def. (R6.6)	-	-	3332	2168	13.40
		No def.	4.51	96	4359	2529	14.99
		Part. def. (R6.3)	3.02	91	4168	2525	14.37
		Tot. def. (R6.6)	-	-	3859	2575	13.07
		No def.	4.76	98	4293	2460	15.18
		Part. def. (R6.3)	3.08	84	3912	2318	14.70
		Tot. def. (R6.6)	-	-	3446	2229	13.45

in Line 033, as compared with other treatments, were accompanied by the highest 1000 seed weight (242.93 g), the highest number of pods per plant on the main stem (43.31) and on the auxiliary stems (33.37), and a relatively high number of seeds per pod (2.37). Results of applying limitations of the sink–source treatments on seed yield showed that the highest seed yield was obtained in the control treatment (212.17 g.m^{-2}).

With the removal of the top one third of the leaves, the yield decreased severely so that the least seed yield in this treatment (138.08 g.m^{-2}) was 35% less than that of the control. In the treatments of removing the middle one third of the leaves and one third of the flowers, the seed yield was reduced by 28.5 and 21.8%, respectively, as compared with the control. The least effect on seed yield was observed in the treatment of removing the bottom one third of the leaves, in which the yield was 201 g.m^{-2}, or only 5.25% less than that of the control. The high yields in the control treatment of applying source-sink limitations were accompanied by high 1000 seed weight (220.217 g), the maximum number of seeds per pod (2.53), the maximum number of pods on the main stem (46.40), and, finally, the maximum number of pods per stem (74). The least seed yield in applying the treatment of the removal of the top one third of the leaves was also accompanied by the least 1000 seed weight (199.59 g), the fewest number of seeds per pod (2.27), the fewest number of pods per main stem (30.62), and the fewest number of pods per plant (53).

The results obtained in the experiments with soybean grown in field conditions support the idea that ABA enhances yield by a combination of factors. Therefore, foliar application of ABA may be an alternative tool for enhancing yield of short-cycle soybean, since it gives relief to temporary situations of water stress, such as the stress that happens in the hours of maximum irradiance, where an imbalance between water transpiration and absorption is frequently produced. ABA seems to improve a combination of factors that contribute to increasing the number of lateral roots and the density of the radical system, protecting the photosynthetic apparatus, keeping the stomata conductance more stable over the time, and enhancing carbon allocation and partitioning to the seeds. The results presented here are also related to those obtained for wheat and other species, and open the possibility for the future use of this hormone in commercial products. Although nowadays its relative cost is high, it has decreased remarkably in the last years and some commercial products are now registered around the world; besides, its application will not represent an environmental threat since ABA is a natural compound produced by plants, fungus and bacteria.

The increase in the potential capacity of photoassimilate supply (source) is an important matter for high crop yield; however, high yield results not only from the increase in source alone but also from the potential capacity of photoassimilate accepter (sink). The sink accepts and consumes the photoassimilate for its own growth and maintenance. Soybean major sink is the economically important harvest components (seed); soybean seed yield can be expressed as the function of yield components as follows:

Yield = Pod number × Seed number per pod × Seed size

Seed size (g per seed) is also expressed as 100 seed weight. Pod number (no. m^{-2}) can be separated into plant density (hill number, plant m^{-2}) and pod number per plant (no. $plant^{-1}$); plant density is relatively easy to control among yield components through sowing. Meanwhile, pod number can be also separated into pod per reproductive node (no.), reproductive node number per area (no. m^{-2}), percent reproductive nodes (%) and node number per area (no. m^{-2}). Many researchers examine the determinant period of seed yield in relation to the manipulation of light interception by shading, defoliation and wide row spacing at various growth periods and indicate that light interception during the period between R1 and R6 affects seed yield strongly through the response of pod and seed number (Board et al., 2010) (Table 13.5).

Sink capacity is like a process of photoassimilate accumulation in a container; high volume for accumulation can accept a large amount of photoassimilate, and this results in high seed yield. Similar to other crop plants, seed number plays a determinant role in sink capacity, and this increase is often associated with high seed yield (Fig. 13.6A) (Board et al., 2010). The period between R1 and R6

Table 13.5 Comparison of the means of yield and yield components as affected by planting dates, varieties, and limitations on source-sink.

Treatment	Seed yield (gm⁻²)	Test weight (g)	Seeds per pod	Pods on auxiliary stems	Pods on main stem
Planting date					
A1: May 27	184.031	221.540	2.563	27.115	42.476
A2: July 7	163.458	214.767	2.266	21.219	38.430
Varieties					
B1: Line 032	186.193	185.797	2.598	29.025	41.521
B2: Line 033	219.969	242.933	2.371	33.372	43.318
B3: VarietyJK	162.425	226.613	2.197	18.710	37.251
B4: VarietyBP	126.411	217.271	2.492	15.560	39.723
Source-sink limitations					
C1: Removing Top 1/3 leaves	138.082	199.596	2.270	22.988	30.620
C2: Removing Middle1/3 leaves	151.615	214.013	2.360	19.624	40.507
C3: Removing Bottom 1/3 leaves	201.026	222.275	2.470	27.090	42.116
C4: Removing 1/3 of flowers	165.853	234.668	2.435	23.263	42.615
C5: Control	212.173	220.217	2.537	27.868	46.408

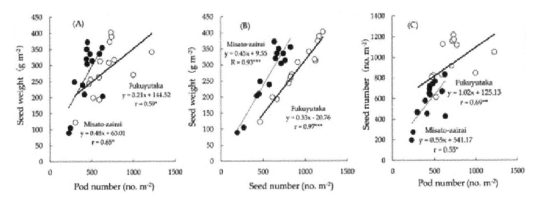

Fig. 13.6 Variation in seed weight as a function of pod number (A), seed number (B) and the relationship between seed number and pod number (C) at maturity (R8) in Fukuyutaka (open) and Misato-zairai (close) cultivated at Mie Prefecture in Japan in 7 years (unpublished data). Experimental field, plant density, fertilization and the measurement of yield components in these experiments were the same as the previous studies. Main data are those of normal sowing (early-middle July sowing at Mie) from 2008 to 2014 (excluding 2012), and the same include early sowing (middle May and middle June in 2009) and irrigation treatment (from blooming in 2009 and from 1 month after sowing in 2013 and 2014). r is correlation coefficient.
*, **, ***: significant at 0.5, 0.1 and 0.01%, respectively.

determines two numbers; pod number is determined critically by light interception during the period between V5 and R3 (Mathew et al., 2001) and seed number is during R3–R6 (Egli, 2010). Although pod number does not control seed yield as strongly as seed number (Fig. 13.6A, B), the occurrence of seed number depends on pod number and reproductive development (Fig. 13.6C).

High canopy photosynthesis during the period between R1 and R6 would affect soybean seed yield with the increase in pod number and seed number. Sink capacity is also associated with source in a volumatic flow of photoassimilate, phloem loading and unloading. For example, the removal of wheat ear at grain filling period reduced about 50% flag leaf photosynthesis within 3–15 h, and the outflow of 14C-labeled assimilates from the flag leaves (indicator of loading) also decreased remarkably (King et al., 1967). Sink activity, such as formation of flower, pod and seed, is sensitive to environmental stress, and this decline decreases leaf photosynthesis through the restriction of phloem loading. Maintenance of sink activity contributes to high yield not only by the increases in pod number and seed number but also by activating leaf photosynthesis. However, another major sink, *Rhizobium japonicum* in the root nodules of soybean plants, is more sensitive to environmental condition than the host plants. For example, respiration of root nodules decreases below −0.4 MPa of root nodule water potential (Pankhust and Sprent, 1975) even though leaf photosynthesis is kept until −1.1 MPa of leaf water potential. Root nodules consume much photoassimilates to fix atmospheric nitrogen into ammonium and, subsequently, ureides for long-range transport (Collier and Tegeder, 2012). The major source of nitrogen accumulation is atmospheric nitrogen fixation by root nodule (Jin et al., 2011); 100% (Brazil) (Dobereiner, 1997) or 40–50% (Midwestern United States) (Ham, 1978) of nitrogen need depends on biological nitrogen fixation. The decline in root nodule activity would decrease not only canopy photosynthesis and seed filling directly by nitrogen deficiency but also leaf photosynthesis indirectly through the restriction of phloem loading and unloading, and this may be associated with the complex of soybean seed production.

Key issues emerging from this are:

1. Seeds of wheat, maize and soybean are usually growing within different assimilate availability ranges, independently of the specific genotype or growing condition (Fig. 13.7).

   Because of this, crops differ in their relative source/sink yield limitation during the seed filling period. Growth of wheat seeds is apparently more sink- than source-limited in most conditions. Soybean seeds seem to experience a large degree of co-limitation, with a large degree of variation due to genotypes and environments (Fig. 13.8).

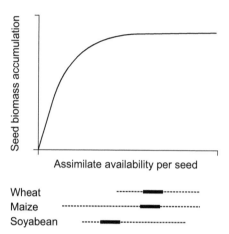

Fig. 13.7 Schematic diagram showing the hyperbolic relationship describing the dependence of seed storage accumulation on the level of assimilates available (Jenner et al., 1991) together with the expected average limits (plain line) and possible extreme situations (dotted line) that experience wheat, maize and soybean seeds during seed filling.

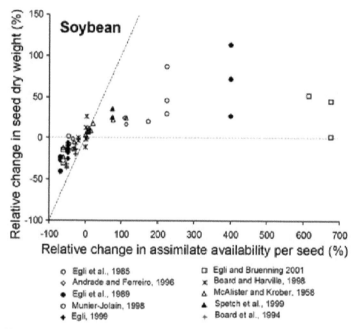

Fig. 13.8 Relationship between soybean relative change in seed dry weight and the relative change in potential assimilate availability per seed produced during seed filling in a number of experiments. Dashed lines stand for the theoretical slopes of 1 (full source limitation) and 0 (full sink limitation).

Maize shows a consistent trend towards a dramatic reduction in seed dry weight if the post-anthesis availability of assimilates is reduced, accompanied by a virtual lack of responsiveness to improvements in availability of assimilates per seed.

2. The greater plasticity of soybean seeds for establishing final seed sink potential may be the most important trait behind its higher seed dry weight response to increased assimilate availability during seed filling when compared to wheat and maize. These differences may be associated with differences among species in their timing of achieving maximum seed volume, as well as with differences in the

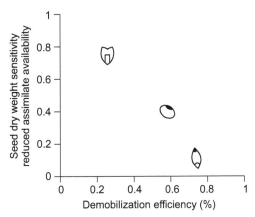

Fig. 13.9 Wheat, maize and soybean seed dry weight sensitivity to reductions in assimilate availability per seed produced during the seed filling period plotted against the mean values reported in the literature of remobilization efficiency for each crop.

environmental (radiation and temperature) conditions each crop generally experiences during seed filling.

3. Differences in the efficiency of remobilization of assimilates temporarily stored in the stem for seed production may be an important aspect determining seed dry weight response when assimilate availability is reduced. Wheat, maize and soybean differ markedly in their capacity to utilize assimilates stored before seed filling for seed biomass deposition, and this may be the cause of their divergent seed dry weight sensitivity to a shortage in current assimilate production during this phase (Fig. 13.9).

Limitations to crop yields are frequently sought in either photosynthesis, the source of assimilates or in the sink, the site of assimilate utilization. This division recognizes the two major processes involved in the accumulation of grain yield, the production of assimilates in the leaves and utilization of this assimilates by the developing seed. Focusing on sources and sinks provides what appears to be a simple two-component system; unfortunately, analysis of this system does not always clearly identify the yield limiting processes (Evans, 1993). Modification of source activity of soybean during flowering and pod set usually results in a corresponding change in pod and seed number, indicating a source limitation. Increasing canopy photosynthesis by increasing levels of atmospheric CO_2 (Hardman and Brun, 1971) or irradiance (Schou et al., 1978) increased pod and seed number, while reducing photosynthesis by shading (Egli and Zhen-wen, 1991) or defoliation (Board and Tan, 1995) decreased pod and seed number. The effects of source-sink alterations during seed filling, after pod seed numbers are fixed, are more complex. De-podding to increase assimilate supply to the remaining seed usually increases seed size (weight per seed) (Munier-Jolain et al., 1998), but does not always change individual seed growth rates (SGR) (Egli et al., 1985; Munier-Jolain et al., 1998). Reducing assimilate supply by defoliation (Board and Harville, 1998) reduced seed size, but the effect of shade was more variable (Egli et al., 1985; Egli, 1999). Source limitations during seed filling seem to be relatively common, based on changes in seed size, but SGR is not as responsive to changes in source activity and can be sink limited. Recent analysis of changes in starch levels in soybean leaves during seed filling supports this contention by suggesting that intermittent sink limitation can occur (Egli, 1999). The response of the sink (the seed) to source-sink alterations during seed filling depends upon the effect on the assimilate level in the seed and the ability of the seed to respond to a change in assimilate supply (Jenner et al., 1991). The response could be manifested in change in SGR or seed fill duration, with their interaction determining final seed size. The response of seed fill duration to assimilate availability is not well understood. In a study, in Iran, to know the effects of changes in source-sink on the yield and yield components of soybean cultivars planted at different dates as the main plot (June 6, June 27), cultivars (Line 032, 033, Sare or JK and Telar or BP) as the sub-plot, and five levels, including the removal of the top, the middle and bottom one thirds of the leaves, removal of one third of the flowers, and the control treatment as the sub-sub plot. Seed yield at the first planting date (184.03 gm^{-2}) was 11/18% lower than that of second planting date (163.45 gm^{-2}). The highest seed yield was obtained in Line 033 with 219.96 gm^{-2}, which was statically different from those of all other cultivars except Line 032, which produced 186.19 gm^{-2}. The higher seed yield in Line 033 as compared with other treatments were accompanied by the highest 1000 seed weight (242.93 g), the highest number of pods per plant on the main stem (43.31) and on the auxillary stems (33.37) and the relatively high number of seeds per pod (2.37).

The newly developed soybean genotypes (*Glycine max* L. Merril) grown in summer season, which have different yielding ability and duration, were tested for their better source-sink relationship in terms of leaf dry weight, total dry weight, leaf area, leaf area index, to overcome problem of quality seed production. The source-sink relationship among genotypes is different. The genotype MAUS 61-2 had better source strength because of maximum number of leaves, high dry matter of leaves and high leaf area and leaf area index. Furthermore, MAUS-61-2 had better sink, i.e., high number of pods, to accumulate more photosynthates as compared to other genotypes. JS-335 was the second best genotype, having best source and sink strength (Tables 13.6 and 13.7).

In a series of experiments to assess the balance between growth and yield of soybean, the application of growth substances, viz., GA$_3$ (500 and 1000 ppm) and CCC (200 and 400 ppm), as foliar spray 15 days before flowering as well as defoliation (50 per cent), deflowering (removal of first flash) and decapitation (cent per cent) influenced source-sink relationships, yield and yield components (Table 13.8). Optimum leaf area index for maximum seed yield was calculated to be 5.5 (Basuchaudhuri et al., 1986).

Results of applying limitations of sink-source treatments on seed yield showed that the highest seed yield was obtained in the control plants (212.17 gm^{-2}). With the removal of top one third of the leaves, the yield decreased severely (138.08 gm^{-2}) to about 35% less than that of control. In removing one third of leaves in the middle and one third of flowers, the seed yield was reduced by 28.5 and 1.8%, respectively, in relation to the control. The least effect on seed yield was observed in the treatment of removing the bottom one third of the leaves, in which the yield was 201 gm^{-2}, only 5.25% less than that of control. The least seed yield associated with the removal of top one third of leaves was accompanied by the least 1000 seed weight (199.59 g), the fewest number of seeds per pod (2.27), the fewest number of pods per main

Table 13.6 Different source strength of soybean genotypes.

Genotypes	Leaf dry weight (g/plant)			Leaf area (cm²/plant)			LAI		
	30DAS	60DAS	Harvest	30DAS	60DAS	Harvest	30DAS	60DAS	Harvest
MAUS81	1.10	4.72	5.15	3.85	14.37	18.27	1.69	6.12	8.01
MAUS61-2	1.11	4.95	5.02	4.14	15.28	19.76	1.84	6.78	8.78
MAUS71	1.13	5.07	5.15	3.47	13.66	16.93	1.53	6.06	7.52
MAUS32	1.19	5.10	5.29	3.39	14.32	17.32	1.50	6.36	7.69
MAUS47	1.11	4.82	5.18	3.88	13.87	16.85	1.72	6.16	7.48
JS-335	1.15	5.35	5.35	4.26	14.83	19.37	1.88	6.89	8.61
CD (0.05)	0.058	0.316	0.215	NS	0.686	1.70	NS	0.38	0.82

Table 13.7 Different sink strength of soybean genotypes.

Genotypes	Pods/plant	Seeds/pod	Seeds/plant	Seed wt. (g/plant)	Harvest index (%)
MAUS81	23.66	2.27	75.22	5.20	49.35
MAUS61-2	51.41	2.43	95.74	9.10	51.09
MAUS71	39.99	2.29	85.80	8.25	50.34
MAUS32	47.49	2.31	84.15	8.75	51.06
MAUS47	28.48	2.26	81.62	7.52	50.24
JS-335	37.65	2.33	88.22	8.85	50.32
CD(0.05)	3.81	NS	6.38	1.53	NS

Table 13.8 Source-sink relationships in soybean.

Treatment	Leaf area (m²perm²)	No. of pods per plant	Seeds per plant	1000 seed weight (g)	Seed yield (g per m²)
Control	6.12	45	80	140.0	235.5
Defoliation	3.20	25	44	148.8	165.0
Deflowering	6.15	46	80	141.0	232.5
Decapitation	5.00	61	87	145.1	257.4
GA₃ 1000 ppm	6.78	33	58	163.7	212.8
GA₃ 500 ppm	6.60	37	61	152.4	214.8
CCC 400 ppm	5.32	55	91	139.5	252.3
CCC 200 ppm	5.60	39	76	151.4	244.8
Mean	5.72	42.63	72.13	147.14	226.8
CD (0.05)	1.12	8.51	13.95	28.21	29.75

stem (30.62) and the fewest number of pods per plant (53) (Yasari et al., 2011). Shading of all side leaflets of a determinate soybean cultivar during pod filling significantly increased rates of photosynthesis in the un-shaded centre leaflets, compared to centre leaflets of controls. Higher rates were associated with both higher stomatal and mesophyll conductances, and were reversible within 2 days when shades were removed. These higher rates of photosynthesis were not associated with decreased percentage enhancement by low oxygen, indicating that treatment effects were probably not associated with changes in photorespiration relative to photosynthesis. Percentage enhancement did, however, increase as the plants approached physiological maturity, chiefly because of a decrease in photosynthesis. Despite these increases in rates of photosynthesis, seed weight per plant decreased by 37% in plants with side leaflets shaded for the entire pod filling period and by 28% in plants shaded for only the second half of the period. In plants where shades were removed during the second half of pod filling, seed yield was reduced by only 19% because shade removal delayed leaf senescence. The four treatments reduced yield by different mechanisms. Plants shaded continuously during pod filling produced fewer seeds than controls, but the weight per seed was similar. When shading was applied by 37% in plants with side leaflets shaded for the entire pod filling period and by 28% in plants shaded for only the second half of the period. In plants where shades were removed during the second half of pod filling, seed yield was reduced by only 19% because shade removal delayed leaf senescence. The four treatments reduced yield by different mechanisms. Plants shaded continuously during pod filling produced fewer seeds than controls, but the weight per seed was similar. When shading was applied during second half of pod fillings, seed number was unchanged but weight per seed was significantly reduced. In contrast, when shades were removed for second half of pod filling, seed number remained similar to that of continuously shaded plants, but seed weight increased. Although all shading treatments reduced yield, the reduction was not proportional to the 63% reduction in leaf area available for photosynthesis (Peet and Kramer, 1980). Defoliation treatments decreased yield in soybean in following order as 100% defoliation, 50% defoliation of top leaves, defoliation of top 4 leaves, defoliation of top 3 leaves, 50% defoliation of based leaves and defoliation of top 1 or 2 leaves did not affect yield. With defoliation of either 1 or 2 top leaves, yield production remained same as compared to control. This indicates that reduced source size and strength might have proportional changes on sink activities (Islam et al., 2014). An experiment on the effect of shading, defoliation and seeding rates on source and sink reactions, in two determinate and two indeterminate cultivars and consisted of three different seeding rates, shade at two different periods during reproduction, and manipulation of the canopy during the reproductive period. Five plants from each treatment were taken at intervals from R5 to maturity. Biomass of entire plants and individual pods were measured along with pods per plant. As grow AG5605 was the least source limited at all locations. Treatments that were source limited varied some by locations, however, both the defoliated and non defoliated low seeding rate, the high seeding rate, and the medium seeding rate shaded late all exhibited source limitations at each location. In another experiment, conducted to investigate the effect of defoliation on some morphological characters and seed yield in two soybean varieties, viz., BARI Soybean-5 and

BARI Soybean-6, four defoliation treatments were imposed at flower initiation stage, viz., (i) control (no leaf removal), (ii) 25% leaf removal, (iii) 50% leaf removal and (iv) 75% leaf removal. All morphological and yield attributes had shown significant differences among the treatments in two soybean varieties. Morphological characteristics, such as plant height, root length, leaf area, total dry matter and yield attributes, such as pod number, pod length, seeds per pod and seed yield, decreased with increasing defoliation levels, whereas the reverse trend was observed in number of branches per plant, nitrate reductase activity, photosynthesis and harvest index. The higher morphological and yield contributing characters were recorded in control and 25% defoliation which resulted in higher seed yield, thereafter further increment of defoliation decreased significantly yield attributes and seed yield in soybean (Ali, 2011). In a determination of starch, sucrose and triose phosphate in a whole homogenate sample and non-aqueously isolated chloroplast sample from plants of different source-sink ratios, the amounts of starch in chloroplasts, the number of starch granules per chloroplast and their size in the sink limited treatment were significantly higher than in the source limited and normal treatments. Thus, it could be concluded that starch acts as a messenger from sink to source and thereby acts as a regulator of source-sink relationships. Although there are some indications of regulation by triose phosphate and sucrose, their role is less clear (Bandara, 1992). Seed filling and yield of soybean under water and radiation deficits were investigated. Treatments were irrigations (I, I2, I3 and I4 for irrigation after 60, 90, 120 and 150 mm evaporation from class A pan, respectively) in main plots and light interceptions (L1-100%, L2-65% and L3-25% of sunlight) in sub-plots. Seeds per plant under I1 and I2 decreased, but increased under I3 and I4 as a result of irrigation deficit. Maximum seed weight and seed filling duration of plants under 25% light interception were higher than those under full sunlight (L1) and 65% light interception (L2). In contrast, plants under full sunlight had the highest seed filling rate, particularly under water stress. Seed filling duration under severe light deficit (L3), was about 9 days longer than that under full sunlight (L1), leading to 15.8% enhancement in maximum seed weight. Decreasing seed yield of soybean under well-watered and mild water stress and improving it under moderate and severe water deficit due to low solar radiation are directly related with changes in seed filling duration and, consequently, in seed weight and number of seeds per plant under these conditions. It is noted that early pod development of soybean is characterized by active cell division in the young ovules and is marked by rapid pod expansion; both processes are very sensitive to drought stress. Drought-induced carbohydrate deprivation and change in the concentration of endogenous abscisic acid of the plants could have significant effects on pod growth and development, and may, therefore, be involved in inducing pod abortion (Liu, 2004). It had also been noted that fruiting soybean plants at 24°C had close to double the CO_2 uptake rates of barren plants; whereas at 27°C, the rates were the same. Removing the pods reduced the CO_2 uptake at 24°C to close to that of barren plants. Sink activity appeared to limit CO_2 uptake over a very narrow temperature range below optimum temperatures for growth, whereas at optimum temperatures, CO_2 uptake appeared to limit growth (Hofstra, 1984). Kobraee et al. (2011) showed that, in soybean, grain yield is limited by the sink, while fewer changes in 100 seed weight at the seed filling period indicated that seed growth and seed yield were not sink limited. High change in seed weight may not always be indicative of source limitation. Regulatory mechanism of plant photosynthesis under sink limitation occurring under various environmental conditions is still insufficient. In a study, two different degrees of sink limitation were imposed on soybean plants by removing either half or all developing pods, then photosynthetic rate and various other characteristics were investigated in fully expanded source leaves of the de-podded and control plants. It was found that a larger degree of pod removal (sink limitation) resulted in lower leaf photosynthetic rate, stomatal conductance and activation ratio (a percentage of initial activity to total activity) of Rubisco and higher leaf sucrose and starch contents without decreasing leaf intercellular CO_2 concentration and affecting leaf water, chlorophyll, total protein and Rubisco protein contents. In addition, there were agreements between activation ratios of Rubisco and photosynthetic rates of the de-podded plants, relative to controls. These results showed that a down regulation mechanism of leaf photosynthetic rate in soybean, which depends on a decrease in activation ratio of Rubisco rather than those leaf Rubisco content or leaf intercellular CO_2 concentration under sink limitation (Kasai, 2008).

    The regulation of carbon partitioning between source and sink tissues in higher plants is not only important for plant growth and development, but insight into the underlying regulatory mechanism is also prerequisite to modulating assimilates partitioning in transgenic plants. Hexoses, as well as sucrose, have been recognized as important signal molecules in source-sink regulation. Components of the underlying signal transduction pathways have been identified and parallels, as well as distinct differences, to know pathways in yeast and animals have become apparent. There is accumulating evidence for crosstalk, modulation and integration between signaling pathways responding to phytohormones, phosphate, light, sugars and biotic and abiotic stress-related stimuli. These complex interactions at the signal transduction levels and coordinated regulation of gene expression seem to play a central role in source-sink regulation (Roitsch, 1999).

Source-sink relationship is a very complex regulation process in considering assimilate supply, translocation and remobilization systems operating side by side. In soybean, both source and sink may be limited and affect the seed yield. Thus, soybean is known as a co-limited crop plant. But, in common, soybean crop shows source limited until environment is of disadvantage.

References

Ainsworth, E.A. and D.R. Bush. 2011. Carbohydrate export from the leaf: A highly regulated process and target to enhance photosynthesis and productivity. Plant Physiol. 155: 64–69.

Ali, M.R. 2011. Effect of Defoliation on Some Morphological Features and Yield of Soybean. M.S. thesis in Crop Botany, Bangladesh Agricultural University, Mymensingh.

Amthor, J.S. 2007. Improvement of Crop Plants for Industrial End Uses. Ranalli, P. (ed.). (Dordrecht: Springer), pp. 27–58.

Andrade, F.H. and M.A. Ferreiro. 1996. Reproductive growth of maize, sunflower and soybean at different source levels during grain filling. Field Crops Research 48: 155–165.

Arshad, M., N. Ali and A. Ghafoor. 2006. Character correlation and path coefficient in soybean *Glycine max* (L.) Merrill. Pak. J. Bot. 38(1): 121–130.

Bandara, D.C. 1992. The role of starch, sucrose and triose phosphates in the source-sink relationship of soybean (*Glycine max* L. Merrill). Ceylon Journal of Science (Biological Sciences) 22: 22–28.

Basuchaudhuri, P., G.C. Munda and C.S. Patel. 1986. Some aspects of source-sink relations in soybean. Annals of Agricultural Research 7: 271–274.

Basuchaudhuri, P. 1987. Above ground characteristics of soybean crop. Annals of Agricultural Research 8: 135–140.

Bauer, M.B., T.O. Sherback and A.L. Ohlrogge. 1969. Effects of rate, time and methods of application of TIBA on soybean production. Agron. J. 61: 604–606.

Board, J.E. and Q. Tan. 1995. Assimilatory capacity effects on soybean yield components and pod number. Crop Science 35: 846–851.

Board, J.E. and B.G. Harville. 1998. Late-planted soybean yield response to reproductive source-sink stress. Crop Science 38: 763–771.

Board, J.E., S. Kumudini, J. Omielan, E. Prior and C.S. Kahlon. 2010. Yield response of soybean to partial and total defoliation during the seed-filling period. Crop Science 50: 703–712.

Borrás, L., G.A. Slafer and M.E. Otegui. 2004. Seed dry weight response to source–sink manipulations in wheat, maize and soybean: a quantitative reappraisal. Field Crops Res. 86: 131–146.

Braun, D.M., L. Wang and Y.L. Ruan. 2014. Understanding and manipulating sucrose phloem loading, unloading, metabolism, and signalling to enhance crop yield and food security. J. Exp. Bot. 65: 1713–1735.

Bruening, W.P. and D.B. Egli. 2000. Leaf starch accumulation and seed set at phloem-isolated nodes in soybean. Field Crops Res. 68: 113–120.

Collier, R. and M. Tegeder. 2012. Soybean ureide transporters play a critical role in nodule, development, function and nitrogen export. The Plant Journal 72: 355–367.

Cooper, R.L. 2003. A delayed flowering barrier to higher soybean yields. Field Crop Res. 82(1): 27–35.

Dhondt, S., N. Wuyts and D. Inze. 2013. Cell to whole-plant phenotyping: The best is yet to come. Trends Plant Sci. 18: 433–444.

Dobereiner, J. 1997. Biological nitrogen fixation in the tropics–Social and economic contributions. Soil Biology and Biochemistry 29: 771–774.

Donald, C.M. and J. Hamblin. 1976. The biological yield and harvest index of cereals as agronomic and plant breeding criteria. Adv. Agron. 28: 361–405.

Dong, Z., G.Q. Na, R.X. Wang and F.T. Xie. 1993. Correlative performance between leaf and seed in soybeans. Soybean Sci. 12: 1–7 (In Chinese).

Egli, D.B., E.L. Ramseur, Y. Zhen-Wen and C.H. Sullivan. 1989. Source–sink alterations affect the number of cells in soybean cotyledons. Crop Sci. 29: 732–735.

Egli, D.B. and Y. Zhen-Wen. 1991. Crop growth rate and seeds per unit area in soybean. Crop Science 31: 439–442.

Egli, D.B. and S.J. Crafts-Brandner. 1996. Soybean. pp. 595–623. *In*: Zamaski, E. and A.A. Schaffer (eds.). Photoassimilate Distribution in Plants and Crops: Source-Sink Relationship. Marcel Dekker Inc., New York, USA.

Egli, D.B. 1998. Seed Biology and the Yield of Grain Crops. CAB International, Wallingford, UK.

Egli, D.B. 1999. Variation in leaf starch and sink limitations during seed filling in soybean. Crop Science 31: 439–442.

Egli, D.B. 2005. Flowering, pod set and reproductive success in soya bean. J. Agron. Crop. Sci. 191(4): 283–291.

Egli, D.B. and W.P. Bruening. 2006b. Temporal profiles of pod production and reproductive success in soybean. Eur. J. Agron. 24: 11–18.

Egli, D.B. 2010. Soybean reproductive sink size and short-term reductions in photosynthesis during flowering and pod set. Crop Science 50: 1971–1977.

Evans, L.T. 1997. Adapting and improving crops: The endless task. Philos. Trans. R. Soc. Lond. B. Biol. Sci. 352: 901–906.

Evans, L.T. and R.A. Fischer. 1999. Yield potential: Its definition, measurement, and significance. Crop Sci. 39: 1544–1551.

Farrar, J.F. 1993. Sink strength—what is it and how do we measure it—a summary. Plant Cell Environ. 16: 1045–1046.

Fisher, D.B. 2000. Biochemistry and Molecular Biology of Plants. Buchanan, B. (ed.). Rockville, MD: American Society of Plant Physiologists, 730–78.

Foyer, C.H., A.V. Ruban and P.J. Nixon. 2017. Photosynthesis solutions to enhance productivity. Philos. Trans. R. Soc. Lond. B. Biol. Sci. 372: 20160374.

Fu, J.M., G.L. Zhang, F. Su, Z.L. Wang, Y. Dong and C.Y. Shi. 1999. Partitioning of 14C-Assimilates and effects of source-sink manipulation at seed-filling in soybean. Acta Agronomica Sinica 25: 170–173 (In Chinese).

Furbank, R.T., W.P. Quick and X.R.R. Sirault. 2015. Improving photosynthesis and yield potential in cereal crops by targeted genetic manipulation: Prospects, progress and challenges. Field Crops Res. 182: 19–29.

Gepts, P. 2002. A comparison between crop domestication, classical plant breeding and genetic engineering. Crop Sci. 42: 1780–1790.

Gifford, R.M. and L.T. Evans. 1981. Photosynthesis, carbon partitioning, and yield. Annu. Rev. Plant Physiol. Plant Mol. Biol. 32: 485–509.

Gilbert, M.E., N.M. Holbrook, M.A. Zwieniecki, W. Sadok and T.R. Sinclair. 2011. Field confirmation of genetic variation in soybean transpiration response to vapor pressure deficit and photosynthetic compensation. Field Crop Res. 124: 85–92.

Ham, G.E. 1978. Interactions of *Glycine max* and *Rhizobium japonicum*. pp. 289–296. *In*: Advances in Legume Science. Kew: Royal Botanic Gardens.

Hardman, L.L. and W.A. Burn. 1971. Effects of atmospheric carbon dioxide enrichment at different development stages on growth and yield components of soybean. Crop Science 11: 886–888.

Hay, R.K.M. 1995. Harvest index—a review of its use in plant-breeding and crop physiology. Ann. Appl. Biol. 126: 197–216.

Hofstra, G. 1984. Response of source-sink relationship in soybean to temperature. Canadian Journal of Botany 62: 166–169.

Islam, M.T. 2014. Effects of defoliation on photosynthesis, dry matter production and yield in soybean. Bangladesh Journal of Botany 43: 261–265.

Jenner, C.F. 1980. Effects of shading or removing spikelets in wheat: Testing assumptions. Austr. J. Plant Physiol. 7: 113–121.

Jenner, C.F., T.D. Ugalde and D. Aspinall. 1991. The physiology of starch and protein deposition in the endosperm of wheat. Australian Journal of Plant Physiology 18: 211–226.

Jiang, H. and D.B. Egli. 1995. Soybean seed number and crop growth rate during flowering. Agron. J. 87: 264–267.

Jin, J., X. Liu, G. Wang, J. Liu, L. Mi, X. Chen et al. 2011. Leaf nitrogen status as a main contributor to yield improvement of soybean cultivars. Agronomy Journal 103: 441–448.

Kasai, M. 2008. Regulatory mechanism of photosynthesis that depends on the activation state of Rubisco under sink limitation. International Journal of Agriculture and Biology 10: 283–287.

Kenneth, L.L. and A.B. Hell. 1980. Drought adoption of cowpea. Influence of drought on plant growth and relation with seed yield. Agron. J. 72: 428–433.

King, R.W., I.F. Wardlaw and L.T. Evans. 1967. Effect of assimilate utilization on photosynthetic rate in wheat. Planta 77: 261–276.

Kobraee, S., G. Noormohamadi, H. Heidarisharifabad, F. Darvishkajori and B. Delkhosh. 2011. Influence of micronutrient fertilizer on soybean nutrient composition. Indian Journal of Science and Technology 4: 763–769.

Liu, F. 2004. Physiological regulation of pod set in soybean (*Glycine max* L. Merr.) during drought at early reproductive stages. Ph.D. Dissertation, Department of Agricultural Sciences, The Royal Veterinary and Agricultural University, Copenhagen.

Liu, X.B., J. Jin, S.J. Herbert, Q.Y. Zhang and G.H. Wang. 2005. Yield components, dry matter, LAI and LAD of soybeans in Northeast China. Field Crop Res. 93: 85–93.

Liu, X.B., S.J. Herbert, M. Hashemi, G.V. Litchfield, Q.Y. Zhang and A.R. Barzegar. 2006. Yield and yield components responses of old and new soybean cultivars to source-sink manipulation under light enrichment. Pant Soil Environ. 52: 150–158.

Liu, B., X.B. Liu, C. Wang, Y.S. Li, J. Jin and S.J. Herbert. 2010. Soybean yield and yield component distribution across the main axis in response to light enrichment and shading under different densities. Plant Soil Environ. 56: 384–392.

Liu, B., X.B. Liu, Y.S. Li and S.J. Herbert. 2013. Effect of enhanced UV-B radiation on seed growth characteristics and yield components in soybean. Field Crops Res. 154: 158–163.

Lobell, D.B., K.G. Cassman and C.B. Field. 2009. Crop yield gaps: Their importance, magnitudes, and causes. Annu. Rev. Environ. Resour. 34: 179–204.

Long, S.P., X.-G. Zhu, S.L. Naidu and D.R. Ort. 2006. Can improvement in photosynthesis increase crop yields? Plant Cell Environ. 29: 315–330.

Long, S.P., A. Marshall-Colon and X.G. Zhu. 2015. Meeting the global food demand of the future by engineering crop photosynthesis and yield potential. Cell 161: 56–66.

Luo, X., C.Z. Ma, Y. Yue, K.N. Hu, Y.Y. Li, Z.Q. Duan et al. 2015. Unravelling the complex trait of harvest index in rapeseed (*Brassica napus* L.) with association mapping. BMC Genomics 16: 379.

Malek, M.A., M.M.A. Mondal, M.R. Ismail, M.Y. Rafii and Z. Berahim. 2012. Physiology of seed yield in soybean: Growth and dry matter production. African Journal of Biotechnology 11: 7643–7649.

Mathew, J.P., S.J. Herbert, S. Zhang, A.F. Rautenkranz and G.V. Litchfield. 2001. Differential response of soybean yield components to the timing of light enrichment. Agronomy Journal 92: 1156–1161.

Mehta, N., A.B.L. Bohar, G.S. Raneat and Y. Mishra. 2000. Variability and character association in soybean. Bangladesh Journal of Agricultural Research 25: 1–7.

Mondal, M.M.A., M.H.K. Howlader, B. Akter and R.K. Dutta. 2007. Evaluation of five advanced lentil mutants in relation to morphophysiological characters and yield. Bangladesh Journal of Crop Science 18: 367–372.

Monteith, J.L. 1977. Climate and efficiency of crop production in Britain. Philos. Trans. R. Soc. Lond. B. Biol. Sci. 281: 277–294.

Munier-Jolain, N.G., N.M. Munier-Jolain, B. Ney, R. Roche and C. Duthion. 1998. Seed growth rate in grain legumes I. Effect of photoassimilate availability on seed growth rate. Journal of Experimental Botany 49: 1963–1969.

Pandey, R.K., M.C. Saxena and V.B. Singh. 1978. Growth analysis of black gram genotypes. Indian Journal of Agricultural Science 48: 466–473.

Pankhurst, C.E. and J.I. Sprent. 1975. Effects of water stress on the respiratory and nitrogen-fixing activity of soybean root nodules. Journal of Experimental Botany 26: 287–304.

Pate, J.S., P.J. Sharkey and C.A. Atkins. 1977. Nutrition of a developing legume fruit. Plant Physiol. 59: 506–510.

Patrick, J.W. 2013. Does don fisher's high-pressure manifold model account for phloem transport and resource partitioning? Front. Plant Sci. 4: 184.

Paul, M.T. and T.K. Pellny. 2003. Carbon metabolite feedback regulation of leaf photosynthesis and development. Journal of Experimental Botany 54: 539–547.

Peet, M.M. and P.J. Kramer. 1980. Effects of decreasing source/sink ratio in soybeans on photosynthesis, photorespiration, transpiration and yield. Plant Cell and Environment 3: 201–206.

Pritchard, J. 2007. Plant Solute Transport. Yeo, A.R. and T.J. Flowers (eds.). Oxford: Blackwell Publishing.

Rascher, U., S. Blossfeld, F. Fiorani, S. Jahnke, M. Jansen, A.J. Kuhn et al. 2011. Non-invasive approaches for phenotyping of enhanced performance traits in bean. Funct. Plant Biol. 38: 968–983.

Reynolds, M. and P. Langridge 2016. Physiological breeding. Curr. Opin. Plant Biol. 31: 162–171.

Roitsch, T. 1999. Source-sink regulation by sugar and stress. Current Opinion in Plant Biology 2: 198–206.

Salam, M.A., A.F.M. Moniruzzaman and S.I. Chowdhury. 1987. Growth analysis in mung bean. Bangladesh Journal of Nuclear Agriculture 3: 58–64.

Sawada, S., T. Hayakawa, K. Fukushi and M. Kasai. 1986. Influence of carbohydrates on photosynthesis in singles rooted soybean leaves used as source-sink model. Plant Cell Physiology 27: 591–600.

Schou, J.B., D.L. Jeffers and J.G. Streeter. 1978. Effects of reflectors, black boards or shades applied at different stages of plant development on yield of soybean. Crop Science 18: 29–34.

Sinclair, T.R. 1991. Canopy carbon assimilation and crop radiation-use efficiency dependence on leaf nitrogen content. pp. 95–107. *In*: Boote, K.J. and R.S. Loomis (eds.). Modeling Crop Photosynthesis-from Biochemistry to Canopy. Crop Sci. Society of America Special Publ. No. 19. Madison, USA.

Sinclair, T.R. 1998. Historical changes in harvest index and crop nitrogen accumulation. Crop Sci. 38: 638–643.

Sinclair, T.R. 1999. Limits of crop yield. *In*: Plants and population: Is there time? Proceedings of the National Academy of Science Colloquium. Dec. 5–6 1998. National Academy of Sci. Washington DC.

Sinclair, T.R., L.C. Purcell and Clay H. Sneller. 2004. Crop transformation and the challenge to increase yield potential. Trends in Plant Sci. 9(2): 70–75.

Specht, J.E., D.J. Hume and S.V. Kumudini. 1999. Soybean yield potential—a genetic and physiological perspective. Crop Sci. 39(6): 1560–1570.

Stitt, M., J. Lunn and B. Usadel. 2010. Arabidopsis and primary photosynthetic metabolism—more than the icing on the cake. Plant J. 61: 1067–1091.

Thompson, J.F., J.T. Madison and A.E. Muenster. 1977. *In vitro* culture of immature cotyledons of soyabean (*Glycine max* L. Merrill). Annals of Botany 41: 29–39.

Toledo, J.F.F. de, C.A.A. Arias, M.F. de Oliveira, C. Triller and Z.D.F.S. Miranda. 2000. Genetical and environmental analyses of yield in six biparental soybean crosses. Pesq. Agropec. Bras. 35(9): 1783–1796.

Unkovich, M., J. Baldock and M. Forbes. 2010. Variability in harvest index of grain crops and potential significance for carbon accounting. Adv. Agron. 105: 173–219.

VanToai, T.T. and J.E. Specht. 2004. The physiological basis of soybean yield potential and environmental adaptation. *In*: Nguyen, H.T. and A. Blum (eds.). Physiology and Biotechnology Integration for Plant Breeding. CRC Press. New York.

Voitsekhovskaja, O.V., E.L. Rudashevskaya, K.N. Demchenko, M.V. Pakhomova, D.R. Batashev, Y.V. Gamalei et al. 2009. Evidence for functional heterogeneity of sieve element–companion cell complexes in minor vein phloem of Alonsoa meridionalis. J. Exp. Bot. 60: 1873–1883.

Wang, T., S.Y. Sun and C.L. Chen. 1983. Primary report of the relationship between leaf-pods and yield investigation. Soybean Sci. 2: 63–73 (In Chinese).

Werner, D., N. Gerlitz and R. Stadler. 2011. A dual switch in phloem unloading during ovule development in Arabidopsis. Protoplasma 248: 225–235.

Wiebold, W.J., D.A. Ashley and H.R. Boerma. 1981. Reproductive abscission levels and patterns for eleven determinate soybean cultivars. Agron. J. 73: 43–47.

Zhu, X.G., S.P. Long and D.R. Ort. 2010. Improving photosynthetic efficiency for greater yield. Annu. Rev. Plant Biol. 61: 235–261.

Index

Milton Keynes UK
Ingram Content Group UK Ltd.
UKHW050449071024
449327UK00014B/301

9 780367 544003